Advances in Intelligent Systems and Computing

Volume 708

Series editor

Janusz Kacprzyk, Systems Research Institute, Polish Academy of Sciences, Warsaw, Poland
e-mail: kacprzyk@ibspan.waw.pl

The series "Advances in Intelligent Systems and Computing" contains publications on theory, applications, and design methods of Intelligent Systems and Intelligent Computing. Virtually all disciplines such as engineering, natural sciences, computer and information science, ICT, economics, business, e-commerce, environment, healthcare, life science are covered. The list of topics spans all the areas of modern intelligent systems and computing such as: computational intelligence, soft computing including neural networks, fuzzy systems, evolutionary computing and the fusion of these paradigms, social intelligence, ambient intelligence, computational neuroscience, artificial life, virtual worlds and society, cognitive science and systems, Perception and Vision, DNA and immune based systems, self-organizing and adaptive systems, e-Learning and teaching, human-centered and human-centric computing, recommender systems, intelligent control, robotics and mechatronics including human-machine teaming, knowledge-based paradigms, learning paradigms, machine ethics, intelligent data analysis, knowledge management, intelligent agents, intelligent decision making and support, intelligent network security, trust management, interactive entertainment, Web intelligence and multimedia.

The publications within "Advances in Intelligent Systems and Computing" are primarily proceedings of important conferences, symposia and congresses. They cover significant recent developments in the field, both of a foundational and applicable character. An important characteristic feature of the series is the short publication time and world-wide distribution. This permits a rapid and broad dissemination of research results.

More information about this series at http://www.springer.com/series/11156

Pankaj Kumar Sa · Sambit Bakshi
Ioannis K. Hatzilygeroudis
Manmath Narayan Sahoo
Editors

Recent Findings in Intelligent Computing Techniques

Proceedings of the 5th ICACNI 2017, Volume 2

Editors
Pankaj Kumar Sa
Department of Computer Science
and Engineering
National Institute of Technology, Rourkela
Rourkela, Odisha
India

Sambit Bakshi
Department of Computer Science
and Engineering
National Institute of Technology, Rourkela
Rourkela, Odisha
India

Ioannis K. Hatzilygeroudis
Department of Computer Engineering
and Informatics
University of Patras
Patras, Greece

Manmath Narayan Sahoo
Department of Computer Science
and Engineering
National Institute of Technology, Rourkela
Rourkela, Odisha
India

ISSN 2194-5357 ISSN 2194-5365 (electronic)
Advances in Intelligent Systems and Computing
ISBN 978-981-10-8635-9 ISBN 978-981-10-8636-6 (eBook)
https://doi.org/10.1007/978-981-10-8636-6

Library of Congress Control Number: 2018934925

Printed on acid-free paper

This Springer imprint is published by the registered company Springer Nature Singapore Pte Ltd.
The registered company address is: 152 Beach Road, #21-01/04 Gateway East, Singapore 189721, Singapore

Foreword

Message from the General Chairs Dr. Modi Chirag Navinchandra and Dr. Pankaj Kumar Sa

Welcome to the 5th International Conference on Advanced Computing, Networking, and Informatics. The conference is hosted by the Department of Computer Science and Engineering at National Institute of Technology Goa, India, and co-organized with Centre for Computer Vision & Pattern Recognition, National Institute of Technology Rourkela, India. For this fifth event, held on June 1–3, 2017, the theme is security and privacy, which is a highly focused research area in different domains.

Having selected 185 articles from more than 500 submissions, we are glad to have the proceedings of the conference published in the *Advances in Intelligent Systems and Computing* series of Springer. We would like to acknowledge the special contribution of Prof. Udaykumar R. Yaragatti, Former Director of NIT Goa, as the chief patron for this conference.

We would like to acknowledge the support from our esteemed keynote speakers, delivering keynotes titled "*On Secret Sharing*" by Prof. Bimal Kumar Roy, Indian Statistical Institute, Kolkata, India; "*Security Issues of Software Defined Networks*" by Prof. Manoj Singh Gaur, Malaviya National Institute of Technology, Jaipur; "*Trust aware Cloud (Computing) Services*" by Prof. K. Chandrasekaran, National Institute of Technology Karnataka, Surathkal, India; and "*Self Driving Cars*" by Prof. Dhiren R. Patel, Director, VJTI, Mumbai, India. They are all highly accomplished researchers and practitioners, and we are very grateful for their time and participation.

We are grateful to advisory board members Prof. Audun Josang from Oslo University, Norway; Prof. Greg Gogolin from Ferris State University, USA; Prof. Ljiljana Brankovic from The University of Newcastle, Australia; Prof. Maode Ma, FIET, SMIEEE from Nanyang Technological University, Singapore; Prof. Rajarajan Muttukrishnan from City, University of London, UK; and Prof. Sanjeevikumar Padmanaban, SMIEEE from University of Johannesburg, South

Africa. We are thankful to technical program committee members from various countries, who have helped us to make a smooth decision of selecting best quality papers. The diversity of countries involved indicates the broad support that ICACNI 2017 has received. A number of important awards will be distributed at this year's event, including Best Paper Awards, Best Student Paper Award, Student Travel Award, and a Distinguished Women Researcher Award.

We would like to thank all of the authors and contributors for their hard work. We would especially like to thank the faculty and staff of National Institute of Technology Goa and National Institute of Technology Rourkela for giving us their constant support. We extend our heartiest thanks to Dr. Sambit Bakshi (Organizing Co-Chair) and Dr. Manmath N. Sahoo (Program Co-Chair) for the smooth conduction of this conference. We would like to specially thank Dr. Pravati Swain (Organizing Co-Chair) from NIT Goa who has supported us to smoothly conduct this conference at NIT Goa.

But the success of this event is truly down to the local organizers, volunteers, local supporters, and various chairs who have done so much work to make this a great event.

We hope you will gain much from ICACNI 2017 and will plan to submit to and participate in the 6th ICACNI 2018.

Best wishes,

Goa, India
Rourkela, India

Dr. Modi Chirag Navinchandra
Dr. Pankaj Kumar Sa
General Chairs, 5th ICACNI 2017

Preface

It is indeed a pleasure to receive an overwhelming response from academicians and researchers of premier institutes and organizations of the country and abroad for participating in the 5th International Conference on Advanced Computing, Networking, and Informatics (ICACNI 2017), which makes us feel that our endeavor is successful. The conference organized by the Department of Computer Science and Engineering, National Institute of Technology Goa, and Centre for Computer Vision & Pattern Recognition, National Institute of Technology Rourkela, during June 1–3, 2017, certainly marks a success toward bringing researchers, academicians, and practitioners in the same platform. We have received more than 600 articles and very stringently have selected through peer review 185 best articles for presentation and publication. We could not accommodate many promising works as we tried to ensure the highest quality. We are thankful to have the advice of dedicated academicians and experts from industry and the eminent academicians involved in providing technical comments and quality evaluation for organizing the conference in good shape. We thank all people participating and submitting their works and having continued interest in our conference for the fifth year. The articles presented in the three volumes of the proceedings discuss the cutting-edge technologies and recent advances in the domain of the conference.

We conclude with our heartiest thanks to everyone associated with the conference and seeking their support to organize the 6th ICACNI 2018 at National Institute of Technology Silchar, India, during June 4–6, 2018.

Rourkela, India	Pankaj Kumar Sa
Rourkela, India	Sambit Bakshi
Patras, Greece	Ioannis K. Hatzilygeroudis
Rourkela, India	Manmath Narayan Sahoo

In Memoriam: Prof. S. K. Jena (1954–2017)

A man is defined by the deeds he has done and the lives he has touched; he is defined by the people who have been inspired by his actions and the hurdles he has crossed. With his deeds and service, Late Prof. Sanjay Kumar Jena, Department of Computer Science and Engineering, has always remained an epitome of inspiration for many. Born in 1954, he breathed his last on May 17, 2017, due to cardiac arrest. He left for his heavenly abode with peace while on duty. He is survived by his loving wife, beloved son, and cherished daughter.

He is known for his ardent ways of problem-solving right from his early years. Even at 62 years of age, his enthusiasm and dedication took NIT Rourkela community by surprise. From being the Superintendent of S. S. Bhatnagar Hall of Residence to Dean of SRICCE to Head of the Computer Science Department to a second term as the Head of Training and Placement Cell, he not only has contributed to the growth of the institute, but has been a wonderful teacher and researcher guiding a generation of students and scholars. Despite this stature, he was an audience when it came to hearing out problems of students, colleagues, and subordinates, which took them by surprise being unbiased in judgments. His kind and compassionate behavior added splendidly to the beloved teacher who could be approached by all. His ideas and research standards shall continue to inspire generations of students to come. He will also be remembered by the teaching community for the approach and dedication he has gifted to the NIT community.

Committee: ICACNI 2017

Advisory Board Members

Audun Josang, Oslo University, Norway
Greg Gogolin, Ferris State University, USA
Ljiljana Brankovic, The University of Newcastle, Australia
Maode Ma, FIET, SMIEEE, Nanyang Technological University, Singapore
Rajarajan Muttukrishnan, City, University of London, UK
Sanjeevikumar Padmanaban, SMIEEE, University of Johannesburg, South Africa

Chief Patron

Udaykumar Yaragatti, Director, National Institute of Technology Goa, India

Patron

C. Vyjayanthi, National Institute of Technology Goa, India

General Chairs

Chirag N. Modi, National Institute of Technology Goa, India
Pankaj K. Sa, National Institute of Technology Rourkela, India

Organizing Co-chairs

Pravati Swain, National Institute of Technology Goa, India
Sambit Bakshi, National Institute of Technology Rourkela, India

Program Co-chairs

Manmath N. Sahoo, National Institute of Technology Rourkela, India
Shashi Shekhar Jha, SMU Lab, Singapore
Lamia Atma Djoudi, Synchrone Technologies, France
B. N. Keshavamurthy, National Institute of Technology Goa, India
Badri Narayan Subudhi, National Institute of Technology Goa, India

Technical Program Committee

Adam Schmidt, Poznan University of Technology, Poland
Akbar Sheikh Akbari, Leeds Beckett University, UK
Al-Sakib Khan Pathan, SMIEEE, UAP and SEU, Bangladesh/Islamic University in Madinah, KSA
Andrey V. Savchenko, National Research University Higher School of Economics, Russia
B. Annappa, SMIEEE, National Institute of Technology Karnataka, Surathkal, India
Biju Issac, SMIEEE, FHEA, Teesside University, UK
Ediz Saykol, Beykent University, Turkey
Haoxiang Wang, GoPerception Laboratory, USA
Igor Grebennik, Kharkiv National University of Radio Electronics, Ukraine
Jagadeesh Kakarla, Central University of Rajasthan, India
Jerzy Pejas, Technical University of Szczecin, Poland
Laszlo T. Koczy, Szechenyi Istvan University, Hungary
Mithileysh Sathiyanarayanan, City, University of London, UK
Palaniappan Ramaswamy, SMIEEE, University of Kent, UK
Patrick Siarry, SMIEEE, Université de Paris, France
Prasanta K. Jana, SMIEEE, Indian Institute of Technology (ISM), Dhanbad, India
Saman K. Halgamuge, SMIEEE, The University of Melbourne, Australia
Sohail S. Chaudhry, Villanova University, USA
Sotiris Kotsiantis, University of Patras, Greece
Tienfuan Kerh, National Pingtung University of Science and Technology, Taiwan
Valentina E. Balas, SMIEEE, Aurel Vlaicu University of Arad, Romania
Xiaolong Wu, California State University, USA

Organizing Committee

Chirag N. Modi , National Institute of Technology Goa, India
Pravati Swain, National Institute of Technology Goa, India
B. N. Keshavamurthy, National Institute of Technology Goa, India
Damodar Reddy Edla, National Institute of Technology Goa, India

B. R. Purushothama, National Institute of Technology Goa, India
T. Veena, National Institute of Technology Goa, India
S. Mini, National Institute of Technology Goa, India
Venkatanareshbabu Kuppili, National Institute of Technology Goa, India

Contents

About the Editors

Pankaj Kumar Sa received his Ph.D. degree in Computer Science in 2010. He is currently serving as an assistant professor in the Department of Computer Science and Engineering, National Institute of Technology Rourkela, India. His research interests include computer vision, biometrics, visual surveillance, and robotic perception. He has co-authored a number of research articles in various journals, conferences, and chapters. He has co-investigated some research and development projects that are funded by SERB, DRDOPXE, DeitY, and ISRO. He has received several prestigious awards and honors for his excellence in academics and research. Apart from research and teaching, he conceptualizes and engineers the process of institutional automation.

Sambit Bakshi is currently with Centre for Computer Vision & Pattern Recognition of National Institute of Technology Rourkela, India. He also serves as an assistant professor in the Department of Computer Science and Engineering of the institute. He earned his Ph.D. degree in Computer Science and Engineering. He serves as an associate editor of *International Journal of Biometrics* (2013–), *IEEE Access* (2016–), *Innovations in Systems and Software Engineering* (2016–), *Plos One* (2017–), and *Expert Systems* (2018–). He is a technical committee member of IEEE Computer Society Technical Committee on Pattern Analysis and Machine Intelligence. He received the prestigious Innovative Student Projects Award 2011 from the Indian National Academy of Engineering (INAE) for his master's thesis. He has more than 50 publications in journals, reports, and conferences.

Ioannis K. Hatzilygeroudis is an associate professor in the Department of Computer Engineering and Informatics, University of Patras, Greece. His research interests include knowledge representation (KR) with an emphasis on integrated KR languages/systems; knowledge-based systems, expert systems; theorem proving with an emphasis on classical methods; intelligent tutoring systems; intelligent e-learning; natural language generation; and Semantic Web. He has several papers published in journals, contributed books, and conference proceedings. He has over

25 years of teaching experience. He is an associate editor of *International Journal on AI Tools* (IJAIT), published by World Scientific Publishing Company, and also serving as an editorial board member to *International Journal of Hybrid Intelligent Systems* (IJHIS), IOS Press, and *International Journal of Web-Based Communities* (IJWBC), Inderscience Enterprises Ltd.

Manmath Narayan Sahoo is an assistant professor in the Department of Computer Science and Engineering, National Institute of Technology Rourkela, Rourkela, India. His research interest areas are fault-tolerant systems, operating systems, distributed computing, and networking. He is the member of IEEE, Computer Society of India, and The Institution of Engineers, India. He has published several papers in national and international journals.

Part I
Research on Optical Networks, Wireless Sensor Networks, VANETs, and MANETs

Multichannel Assignment Algorithm for Minimizing Imbalanced Channel Utilization in Wireless Sensor Networks

Abhinav Jain, Shivam Singh and Sanghita Bhattacharjee

Abstract Interference management is extremely important in wireless sensor networks (WSNs). Due to the sharing of spectrum, interference resulting from neighboring transmissions may degrade significantly the network performance. Use of multiple non-overlapping channels improves the network capacity or mitigates the interference by allowing multiple transmissions simultaneously on different channels. In this paper, we propose a centralized channel assignment method for single radio-equipped WSNs, which take advantage of multiple channels. The main objective is to minimize the imbalanced channel utilization among nodes in the network. Key components of the channel assignment procedure are root selection, node ordering, and channel selection. We divide the channel set of each node into three categories and assign channel to the node based on channel capacity, load of the node, and load of nodes in 2-hop local neighborhood. The simulation results demonstrate the effect of node ordering and root selection metrics on the network performance, in terms of channel utilization balancing and number of nodes accessing channel ratio.

Keywords Multichannel · Channel utilization balancing · Free channel Primary channel · Secondary channel · Minimum assignable channel

---

A. Jain (✉) · S. Singh · S. Bhattacharjee
Department of Computer Science and Engineering, National Institute of Technology, Durgapur, India
e-mail: abhinavjain0809@gmail.com

S. Singh
e-mail: shivam023.singh@gmail.com

S. Bhattacharjee
e-mail: sanghita.b@gmail.com

P. K. Sa et al. (eds.), *Recent Findings in Intelligent Computing Techniques*, Advances in Intelligent Systems and Computing 708,
https://doi.org/10.1007/978-981-10-8636-6_1

1 Introduction

In the past two decades, wireless sensor networks (WSNs) are popularly used as wireless technology for monitoring and controlling various real-time applications. The performance of WSNs is generally impaired by interference. High interference minimizes the overall network throughput and increases unnecessary energy consumption at the node. As a result, the power of a node depletes quickly and the network becomes partitioned. Therefore, interference reduction is an important issue in sensor network design. Transmission power control can minimize interference at the node. However, the power of a node affects the network connectivity as some links become inaccessible [1]. Multichannel technologies eliminate interference at node effectively and improve the overall capacity of the network by allowing concurrent transmissions on different channels. Although such schemes effectively improve the network performance, the number of channels is limited in wireless networks [2].

Researchers have proposed many multiple channel allocation algorithms [3–7] for wireless networks. However, these protocols are not suitable for WSNs because of the typical characteristics of the sensor. Generally, each sensor node is equipped with a single half-duplex radio [2], which cannot transmit and receive simultaneously, but can switch to different channels. MMSN [8] is a first single radio-based multichannel protocol for WSNs. In frequency assignment [8], frequency is assigned to each node for receiving the data, while in media access, the slot allocation is done for scheduling the nodes. In [9], the authors proposed routing-aware channels allocation algorithms for WSNs with an aim to minimize the maximum interference in the network. Hybrid-MAC (Hy-MAC) proposed in [10] utilizes multiple channels to get high throughput. An interference-aware MAC protocol (IM-MAC) for WSNs was introduced in [11]. IM-MAC improves the network throughput by minimizing the hidden terminal problem. In IM-MAC, channel assignment is done based on node ID, i.e., the node with the highest ID executes the channel allocation process. A game theory-based channel assignment protocol, known as ACBR, was proposed in [12] for prolonging the network lifetime. ACBR uses multiparameter-based utility function to choose a suitable channel for each node.

In this paper, we propose a multichannel assignment technique for WSNs, where each node is equipped with a single radio, which is capable of transmitting and receiving on a different channel. The goal of the work is to minimize imbalanced channel utilization among nodes, so that balanced channel utilization is achieved in the network. Our channel assignment scheme has three major components: (i) *root selection*, (ii) *node ordering,* and finally, (iii) *channel selection.* At first, root is selected and assigned the smallest numbered channel. Node ordering shows the order of assignment of nodes (except root), while in channel selection, each node selects its channel based on the channel capacity, traffic load of the node, and load of the nodes in 2-hop local neighborhood. The channel set of each node is divided into three categories: primary channel, secondary channel, and free channel. Free channels are interference free and are not chosen by the others in 2-hop local neighborhood. Primary channels can be used by 1-hop neighbors and might be assigned to the node if no free channel is available. Lastly, the secondary channels can be used by the interfering nodes and can be allocated only if no valid primary channel is available. Simulations are performed to evaluate the results that our scheme produces.

The rest of the paper is organized as follows: Sect. 2 describes the model and problem definition. The proposed multichannel assignment algorithm is described in Sect. 3, and results are discussed in Sect. 4. Finally, Sect. 5 concludes the paper.

2 Model and Problem Definition

2.1 Network Model and Assumptions

In this paper, wireless sensor network is represented as an undirected communication graph $G = (V, E)$ where V is set of static nodes and $E \subset V \times V$ is the set of edges. Each node is characterized by transmission range R_{max} and interference range I_{max} where $R_{max} \leq I_{max}$. Two nodes u and v are connected by an edge if they are 1-hop neighbors. Let, $N_1(u)$ be the set of 1-hop neighbors of node u. Any node $v \in N_1(u)$ if $d_{uv} \leq R_{max}$ where d_{uv} is Euclidian distance between nodes u and v. To find the interfering nodes of a node, we follow Protocol Interference Model [13]. Two nodes u and v will be interfering if they are operating in the same channel and $u \in N_2(v) \wedge v \in N_2(u)$ where $N_2(u)$ and $N_2(v)$ are the set of 2-hop neighbors of u and v, respectively. In this paper, we have made following assumptions:

- K: The set of different and non-overlapping channels in the network. The capacity of each channel is same, and it is denoted by CT_c where $c \in K$.
- C: A $1 - D$ channel assignment array and is initialized to 0.
- $L(u)$: The traffic load of node u and it is known a priori.

2.2 Problem Definition

The objective of the channel assignment algorithm presented in this paper is to assign one of K channels to each node in such a way that minimizes imbalanced channel utilization among nodes in the network.

To maximize the channel utilization of node u, we define a binary channel assignment variable $C^c_{u,v}$ for each node $v \in N_2(u)$. Binary variable $C^c_{u,v}$ will be 1 if v is within 2-hop away and both u and v are operating on the same channel c.

$$C^c_{u,v} = \begin{cases} 1, & \textit{if } v \in N_2(u) \wedge C(u) = C(v) = c \\ 0, & \textit{otherwise} \end{cases} \tag{1}$$

Now, we define the load on the channel of node u by

$$L_u(c) = \sum_{\substack{C(u)=C(v)=c \\ c \in K}}^{v \in N_2(u)} C^c_{u,v} \times L(v) + L(u) \tag{2}$$

Using Eq. (2), we calculate the channel utilization of node u and it is given as

$$Z_u(c) = \frac{CT_c}{L_u(c)} \tag{3}$$

We use standard deviation $(Z(G))$ given in Eq. (5) to calculate channel utilization balancing in G. Standard deviation is a metric that finds the deviation of channel utilization in G. Lower value implies better balancing. $\bar{Z}(V)$ is defined in Eq. (4).

$$\bar{Z}(V) = \sum_{c \in K}^{V} u = 1 Z_u(c) \tag{4}$$

$$Z(G) = \sqrt{\frac{1}{|V|} \sum_{u=1}^{V} (Z_u(c) - \bar{Z}(V))^2} \tag{5}$$

Therefore, our channel assignment problem becomes

$$\textit{minimize}(Z(G)) \tag{6}$$

Subject to: 1. $C(u) \in K, \forall u \in V$

2. $CT_c - C^c_{u,v} \times L(v) - L(u) > 0, \forall u \in V \& v \in N_2(u)$

3 Proposed Multichannel Assignment Algorithm

In this section, we present a centralized channel assignment algorithm which assigns a different channel to each node in WSNs. The key aim is to minimize imbalanced channel utilization among nodes. The proposed method has three phases, namely root selection, node ordering, and channel selection which are described subsequently.

3.1 Phase I: Root Selection

At first, we select the root and assign the smallest numbered channel to it. One way to select the root is lowest ID, i.e., the node with the smallest ID becomes the root. The next alternative to choose the root is maximum degree which includes both 1-hop and 2-hop neighbors. Another criterion is minimum degree, i.e., the node with the minimum degree is selected as the root for the channel assignment process.

3.2 Phase II: Node Ordering

The second important issue is node ordering which basically reflects the order in which the nodes are considered for channel assignment. Here, we discuss three ordering alternatives. First one is random ordering (RO) which uses randomization to organize the nodes (except the root) for channel assignment. The next alternative uses some fixed order (FO) to arrange the nodes. The channel assignment with random and fixed ordering of nodes is given in Algorithm 1 and Algorithm 2, respectively. Another alternative that we consider is minimum assignable channel (MAC)-based node ordering. Previous two techniques (i.e., RO and FO) use the original graph G to order the nodes, while MAC-based technique utilizes the interference graph, denoted by G^2, for organizing the nodes for the channel allocation. Interference graph G^2 consists of V nodes and E^2 edges. Two nodes u and v are connected by an edge, i.e., $(u, v) \in E^2$ if $u \in N_2(v) \wedge v \in N_2(u)$. In MAC-based ordering, we use *BFS*-based traversal to traverse all nodes in G^2 and then assign labels to them. Here, the root is initialized to level 0. MAC generally works in top to bottom fashion, i.e., nodes in lower level get their channels earlier than upper layer. In each layer, we calculate the number of free channels (FCs) for each node and select the node with the minimum value for channel allocation. If a tie happens, the node ID is used to break it. The channel assignment when MAC-based ordering is used is briefly shown in Algorithm 3.

Algorithm 1: RO-based Channel Assignment

```
1.  C(v) := 0, ∀v ∈ V
2.  root := Root Selection(V)
3.  C(root) := Channel Selection(root, K)
4.  Permutate(V\{root})
5.  while (V! = Empty)
6.          v := Removehead(V)
7.          C(v) :=Channel Selection(v, K)
8.          V := V-1
9.  end while
10. return(C)
```

Algorithm 2: FO-based Channel Assignment

```
1.  C(v) := 0, ∀v ∈ V
2.  root := Root Selection(V)
3.  C(root) := Channel Selection(root, K)
4.  U := {}
5.  for i:= 0 to |V| − 2 do
6.          U := U ∪ { (root + i)mod V + 1}
7.  end for
8.  while (U! = Empty)
11.          v := Removehead(U)
12.          C(v) := Channel Selection(v, K)
9.       U := U-1
10. end while
11. return(C)
```

3.3 Phase III: Channel Selection

The goal of this phase is to assign a different channel to each node in the network. Similar to [11], the entire channel set of a node is divided into three categories: Free channel set (*FCS*), primary channel set (*PCS*), and secondary channel set (*SCS*).

FCS of node u, $FCS(u)$, contains set of channels which are not assigned in u's neighborhood. If $FCS(u)$ is available, u selects the smallest numbered channel. *PCS* of node u, $PCS(u)$, is a set of channels which are used by 1-hop neighbors of u. If $FCS(u)$ is empty, then u can choose its channel from $PCS(u)$ only if it has valid channel. A channel is said to be valid if the current capacity of the channel is greater than the load of the node where the current capacity is complementary of total capacity and load of neighbor nodes using that channel. If more than one channel is available, the channel with the least load is selected. Node u can select the secondary channel only if *PCS* is empty. The secondary channel set contains the channels which are used by 2-hop neighbors of the node. If valid secondary channels are available, node u chooses the least loaded channel. Since $SCS(u)$ is *PCS* of $N_1(u)$, node u discards those primary channels which are in $SCS(u)$. After selection, node u broadcasts the assigned channel and ID within its 2-hop local neighborhood. All nodes in the neighborhood update their primary, secondary, and free channel set accordingly. Algorithm 4 describes our channel assign method in brief.

Algorithm 3: MAC-based Channel Assignment

1. $C(v) := 0,\ \forall v \in V$
2. $root :=$ Root Selection(V)
3. Build Interference Graph $G^2(V,E^2)$ from $G(V,E)$
4. $level(root) := 0$
5. $LS :=$LevelStructure($G^2, root$)
6. $C(root) :=$ Channel Selection($root, K$)
7. **for** each $l \in LS$ **do**
8. $\quad nodeset := \{v \mid level(v) == l\ \&\ C(v) = 0\}$
9. **while** ($nodeset\ != Empty$)
10. $\qquad v := arg\ min_{v \in V}\ |FCS(v)|$
11. $\qquad C(v) :=$ Channel Selection(v, K)
12. $\quad nodeset := nodeset - 1$
12. $\quad$ **end while**
13. **end for**
14. return(C)

Algorithm 4: Channel Selection

```
1. Get FCS,PCS and SCS for node v
2. if (FCS(v) != Empty)
3.     c :=Findsmallestnumberchannel(FCS)
4.     C(v) := c
5. end if
6. if (FCS == Empty & PCS != Empty)
7.     for each channel c ∈ PCS(v) do
8.         if(channel c ∈ SCS(v)) | (CT_c − Σ_{u∈N1(v), C(u)=c} L(u) < L(v))
9.             remove c from PCS(v)
10.                continue
11.        end if
12.    end for
13. c :=Findleastloadedchannel(PCS)
14. C(v) := c
15. end if
16. if(PCS == Empty & SCS != Empty)
17.    for each channel c ∈ PCS(v) do
18.        if(CT_c − Σ_{u∈N2(v), C(u)=c} L(u) < L(v))
19.            remove c from SCS(v)
20.            continue
21.        end if
22.    end for
23. c := Findleastloadedchannel(SCS)
24. C(v) := c
25. end if
26. Node v broadcasts <c,ID> in its local neighborhood
27. N1(v) updates FCS and PCS; N2(v) updates FCS and SCS
28. return(c)
```

4 Results and Discussion

In this section, we conduct simulations to evaluate RO-, FO-, and MAC-based channel assignment approaches and then compare their performances with respect to channel utilization balancing and number of nodes accessing channel ratio. All algorithms are simulated through MATLAB and C programming language. We consider $1000 \times 1000\,m^2$ square field where sensor nodes are randomly distributed. Number of nodes, $|V|$, varies from 100 to 400. Transmission range of each node is 50 m, and interference range is 100 m. The number of available channels is 4 and 12. The channel is assumed to be error-free, and capacity of each channel is set to 10 Mz. Load at each node is assigned randomly in the range of [0.1–1.0].

Figure 1 shows the performance of channel utilization balancing for RO-, FO-, and MAC-based channel assignment approaches for 100–400 nodes but for different number of channels. It is clear that the channel utilization decreases as number of nodes increases. The increasing network size brings more interference at node and degrades the performance of the network. We have also seen that balancing value is much higher using 12 channels. Increased number of channels not only minimizes number of conflicting nodes, but also reduces load on channels. As a result, channel utilization balancing is increased.

Figure 1a depicts the performance of channel utilization balancing when the root is the smallest ID node. MAC-based channel assignment achieves better channel utilization than RO and FO. In MAC, the node is selected based on its free channels and thereby minimizes imbalanced channel utilization. From Fig. 1a, it is clear that

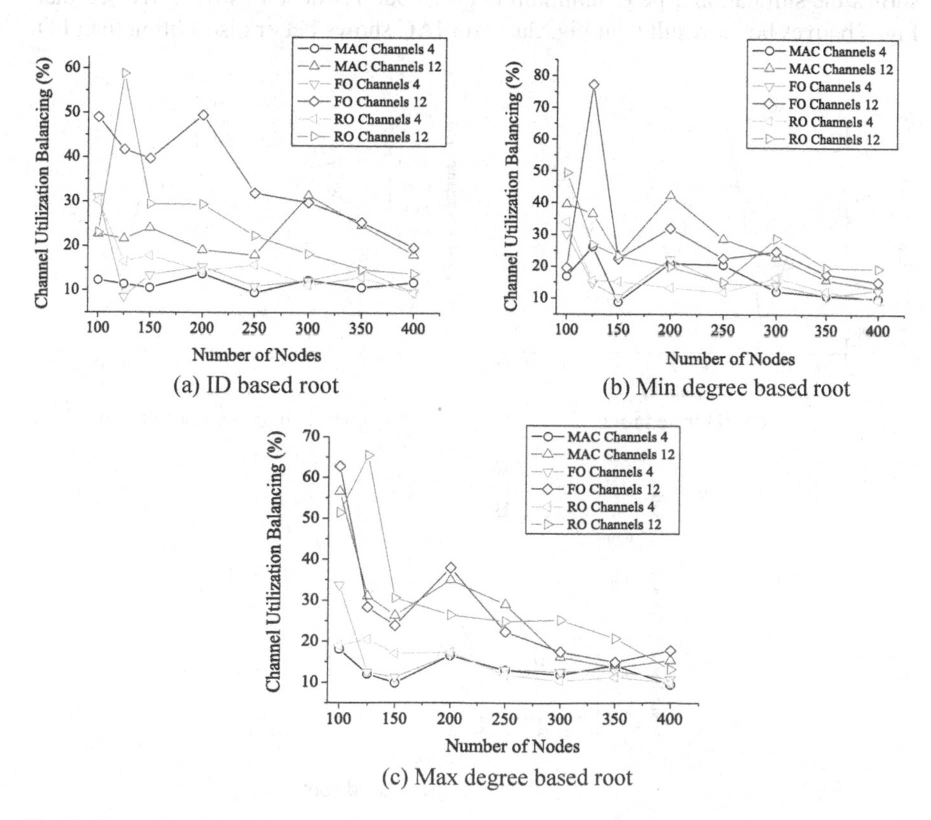

(a) ID based root

(b) Min degree based root

(c) Max degree based root

Fig. 1 Comparing MAC, FO, and RO in terms of channel utilization balancing

MAC using 4 channels achieves better balancing. However, the performance gap between the algorithms become thinner with increasing network size and available channels, while it was bigger in smaller network size.

In Fig. 1b, the minimum node degree is used to select the root. Similar performance improvement is also exhibited in Fig. 1b. In sparse network, RO achieves better balancing. However, it degrades as number of nodes increases. Initially, the performance of MAC-based ordering was lower than FO and RO. But MAC gets better improvement with the deployment of more nodes. Moreover, in larger networks, MAC is more effective than the result shown in Fig. 1a. In contrast to Fig. 1a–c shows that the performance gap between MAC and other algorithms increases with network size and available channels. The result is close to RO and FO using 4 channels. In maximum degree-based root, we get better the channel utilization by allocating free channels to the node having the maximum conflict.

Figure 2 displays number of nodes in various channels. In this experiment, number of available channels is 12 and number of nodes is 200. In Fig. 2a, we select the smallest indexed node as root and compare the performance of RO, FO, and MAC. Number of nodes in various channels using FO differ much than RO and MAC. However, the performance of MAC is slightly lower than RO. Figure 2b shows the simulation where minimum degree node is chosen as root. We see that Fig. 2b gives better result than Fig. 2a, and MAC shows better distribution than FO

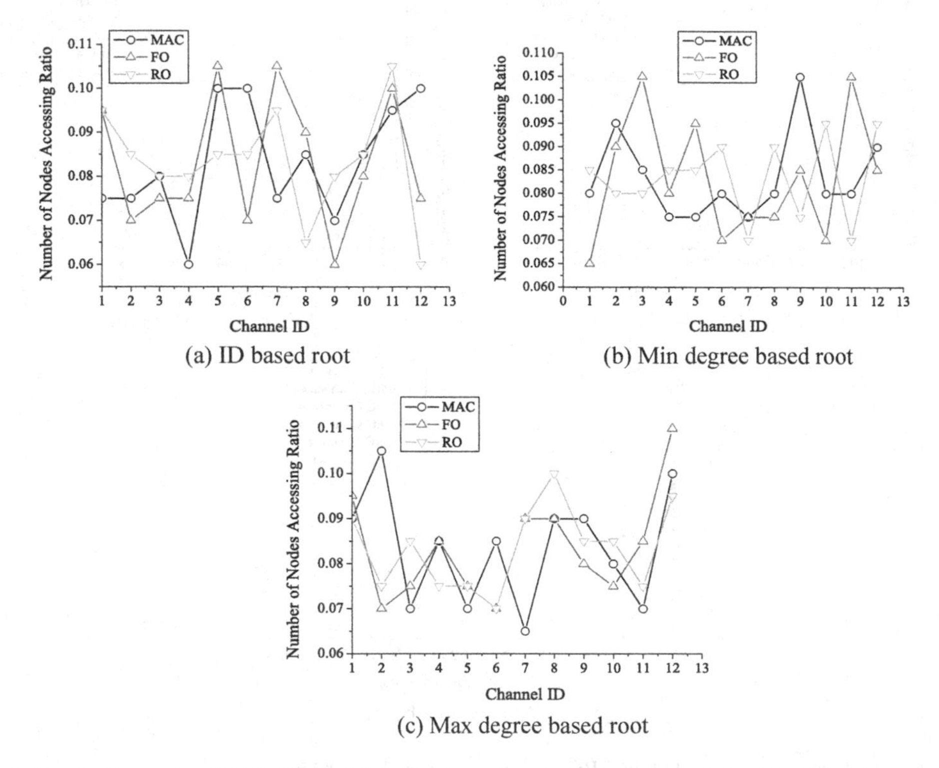

(a) ID based root

(b) Min degree based root

(c) Max degree based root

Fig. 2 Comparing MAC, FO, and RO in terms of number of nodes accessing channel ratio

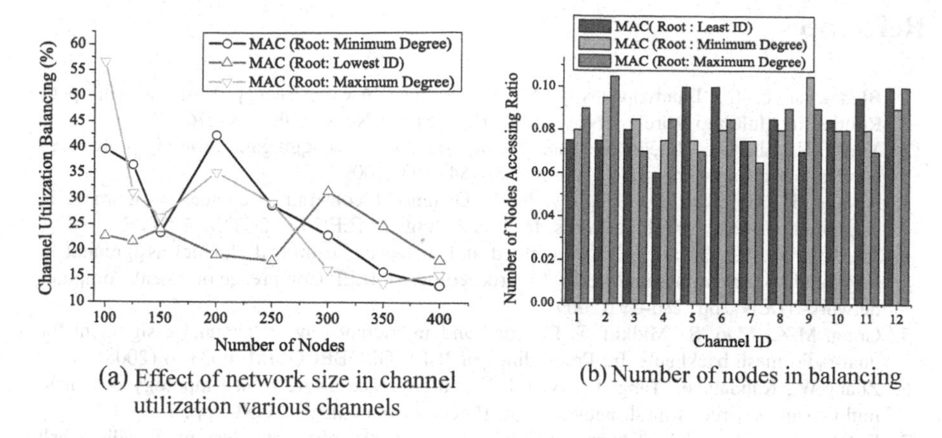

(a) Effect of network size in channel utilization various channels

(b) Number of nodes in balancing

Fig. 3 Performance of MAC using various root selection techniques

and RO. Figure 2c illustrates the number of nodes in various channels for different algorithms when a maximum degree node is selected as the root. Figure 2c shows that the distribution slightly falls for some channels, while it does not differ much in Fig. 2b. Note that, channel utilization balancing is better using maximum degree-based root selection than others.

Figure 3a, b displays the performance of MAC in terms of channel utilization balancing and number of nodes in various channels considering different root selection, respectively. As shown in Fig. 3a, MAC with ID-based root achieves better balancing in small network sizes, while its performance degrades as number of nodes increases. On the other hand, MAC with maximum degree-based root selection improves the channel balancing among nodes than the other two techniques significantly in large network sizes. From Fig. 3b, we find that MAC with minimum degree-based root gives better distribution of channels among nodes, while it differs much in ID-based selection. Note that, the performance curve of MAC with maximum degree-based root is very close to MAC with minimum degree-based root.

5 Conclusions

In this paper, we have presented a centralized channel allocation algorithm for WSNs for minimizing imbalanced channel utilization among nodes. We have used three root selection metrics and node ordering techniques to organize the nodes for channel assignment. The proposed channel selection procedure is basically a load-based channel selection which assigns the channel to the node based on traffic load in and around it. The experimental results show the impact of root selection and node ordering on channel utilization balancing and number of nodes accessing channel ratio.

References

1. Bhattacharjee, S., Bandyopadhyay, S.: Interference Aware Energy Efficient Multipath Routing in Multihop Wireless Networks. J High Speed Netws. **20**, 263–276 (2014)
2. Yen, H.H., Lin, C.L.: Integrated channel assignment and data aggregation routing problem in wireless sensor networks. IET Commun. **3**, 784–793 (2009)
3. Aryafar, E., Gurewitz, O., Knightly, E.W.: Distance-1 constrained channel assignment in single-radio wireless mesh networks. In: Proceedings of IEEE INFOCOMM (2008)
4. Jin, J., Zhao, B., Zhou, H.: Link-weighted and distance-constrained channel assignment in single-radio wireless mesh networks. In: Proceedings of IEEE Conference on Local Computer networks (LCN), pp. 786–791(2009)
5. Gong, M.X., Mao, S., Midkiff, S. F.: Load and interference aware channel assignment for dual-radio mesh backhauls. In: Proceedings of IEEE GLOBECOMM, pp. 1–6 (2008)
6. Zhang,W., Kandah, F., Tang, J., Nygard, K.: Interference aware robust topology design in multi-channel wireless mesh networks. In: Proceedings of IEEE CCNC, pp. 1–5 (2010)
7. Delakis, M., Siris, V.A.: Channel assignment in a metropolitan wireless multi-radio mesh networks: In: Proceedings of IEEE BROADNETS, pp. 610–617 (2008)
8. Zhou, G., et al.: MMSN: multi frequency media access control for wireless sensor networks. In: Proceedings of IEEE INFOCOMM, pp. 1–13 (2006)
9. Saifullah, A., Xu, Y., Lu, C., Chen, Y.: Distributed channel allocation algorithm for wireless sensor networks. IEEE Trans Parallel Distrib Syst **25**, 2264–2274 (2014)
10. Salajeghed, M.,Soroush, H., Kalis, A.: HYMAC: Hybrid TDMA/FDMA medium access control protocol for wireless sensor networks. In: Proceedings of IEEE PIMRC, pp. 1–5 (2007)
11. Yuanyuan, Z., Xiong, N., Park, J.H., Yang, L.T.: An interference-aware multichannel media access control protocol for wireless sensor networks. J. Supercomput. **60**, 437–460 (2012)
12. Hao, X.C., Yao, N., Li, X.D., Liu, W.J.: Multi-channel allocation algorithm for anti-interference and extending lifetime in wireless sensor network. Wirel Pers. Commun. **87**, 1299–1318 (2016)
13. Shi, Y., Hou, Y. T., Y., Liu, J., Kompella, S.: How to correctly use the protocol interference model for multihop wireless networks. In: Proceedings of ACM Symposium on Mobile Ad hoc Networking and Computing, pp. 239–248 (2009)

MWPR: Minimal Weighted Path Routing Algorithm for Network on Chip

E. Lakshmi Prasad, A. R. Reddy and M. N. Giri Prasad

Abstract Due to the cumbersome routing links in Network on Chip (NoC), routing congestion and latency issues have taken place. To minimize the traffic congestion and latency issues, minimal weighted path routing (MWPR) algorithm and virtual channel (VC) router are introduced. The introduced methods are applied and experimented with respect to various bit lengths for $4 \times 4 \times 8$ Star over Mesh. Simulation and synthesis results have carried out by using Xilinx 14.3 and executed in vertex-4 FPGA. The empirical result analysis exhibits that the time period for 8 bit is 0.884 ns, for 16 bit is 0.965 ns, and for 32 bit is 1.171 ns and the amount of power consumption for 8 bit is 240.96 mW, for 16 bit is 286.72 mW, and for 32 bit is 419.58 mW.

Keywords Routing algorithm · System on chip · Network on chip
Multiprocessors

1 Introduction

NoC is a paradigm of advanced System on Chips (SoCs). In this section, begin with the necessary building blocks of Network on Chip. NoC design consists of router interface with local processing element (PE) and connected to adjacent routers. An example design of 3 X 3 mesh NoC structure and their interdependences are illustrated in Fig. 1.

E. L. Prasad (✉) · A. R. Reddy · M. N. G. Prasad
Department of ECE, JNTUA, Anantapuramu 515002, India
e-mail: lakshmi_prasad2@yahoo.com

A. R. Reddy
e-mail: dean-rrc@mits.ac.in

M. N. G. Prasad
e-mail: mahendragiri1960@gmail.com
URL: http://www.jntua.ac.in

P. K. Sa et al. (eds.), *Recent Findings in Intelligent Computing Techniques*,
Advances in Intelligent Systems and Computing 708,
https://doi.org/10.1007/978-981-10-8636-6_2

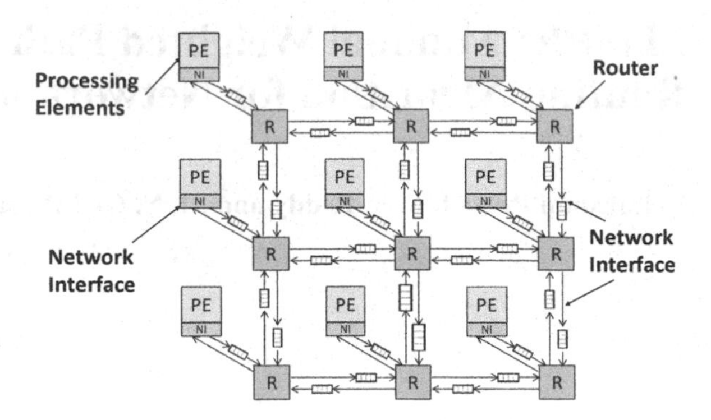

Fig. 1 3 × 3 2D NoC MESH

The basic challenges of NoC are discussed below:

1. To determine an application as a bottom-line step in NoC. Homogeneous or heterogeneous multiprocessor can be opted based upon the application needs.
2. To design a specific router design to route the data packets and to manage the traffic congestion.
3. To design a network interface to set up between the router to router and IP (intellectual property) core. Design a specific topology to establish the connection among all routers in NoC.
4. To design the shortest path routing algorithm to scale down the latency and the issue of congestion.

In NoC, difficulties may come at any point, and it may be in router architecture or in routing algorithm [1]. These problems are occurring because of cumbersome routing, and it leads to cause traffic congestion and latency. To resolve these issues, an efficient routing method is required.

The primary goal of this paper is to design an efficient router design and routing method to achieve lower latency, to handle the traffic congestion, and to meet the system performance. In this section, the difficulties of an NoC as discussed and the rest of the paper organized are as follows: related work described in section-2, in section-3 described the router architecture with their importance, in section-4 minimal weighted path routing algorithm, in section-5 discussion on implementation results and final discussion on conclusion with future scope.

2 Related Work

In this section, related work of NoC with different methods is presented. These different methods helpful to identify the problem in NoC. In every corner of NoC, most of the cases struggle with latency and handling the traffic congestion. Due to these two issues, the system performance does not meet up to the mark.

Liu et al. [2] suggested a square spiral routing method to resolve the redundant path. This routing algorithm helps to select the shorter distance route rather than the redundant path. This routing algorithm occupied less area overhead, which is 0.74%, and the power overhead is 0.52mW. There is another method called support vector regression (SVR) introduced by Zhi-Liang Qian [3]. SVR estimated the waiting time of queue instead of facing a traffic congestion in clumsy network. This technique serves to predict the latency and traffic flow in a mesh network.

Tang et al. [4] proposed a local congestion estimated in NoC. The local issues of traffic congestion occur in bunch of network region. In most of the cases, system performance decreases because of clumsy network. Divide and conquer technique can mitigate the issue of congstion in regional networks. In Table 1, performance comparison of related work is presented.

In the next section will be discussed with router architecture and their importance.

3 Virtual Channel(VC) Router Architecture

Router architecture is a key role in NoC. In order to control the hardware IP cores in multiprocessor System on Chip, thus, it requires an additional support called as NoC. NoC can handle the number of IP cores with the help of router architecture and its routing algorithm.

There are various existing router architectures available like a wormhole and look-ahead router architecture. In the wormhole router design, no buffering system is available due to that congestion problems are occurring. This issue can be resolved by replacing with virtual channels to each input port. We have proposed a Diagonal Virtual Channel router, that consists of six directional ports and out of which one dedicated to local port for processing element (PE) and remaining ports are East, West, North, South and Diagonal ports [9]. The virtual channel (VC) router design is illustrated in Fig. 2.

VC router can help to avoid the problem of traffic congestion and routing latency in the network. The proposed VC router design aimed to achieve the bandwidth efficiency and better throughput. This architecture has several features incorporated such as low latency, managing the traffic congestion, and buffer the packets based on volatile storage system.

The initial step in the virtual channel router writing a data into the buffer can be called as buffer queue (QW). In the second step, routing computation (RC) block assigns the destination address to the header flit. Then, based on the header flit information, the packets are routed to the destination. In reality, Switch allocator and crossbar network monitors the nearby router to check whether they are busy or free. With this, router will decide to transmit the packet to nearby routers packets [10]. suppose, if it is busy, then it will stall the packets in virtual channels. In the third step, VC allocates the packets, based upon the grants received by the switch allocator (SA). SA can update the grant status given to the VC and VIP allocators. Based on the grants, VC allocator allocates the packet into the buffer queue. In case

Table 1 Review comparison of related work in Network on Chip

Design	Authors et al.	Method of data transmission			Network Computations			Tools used	Year	References
		Channel type	Types of topological connection	Routing algorithm	Throughput	Timing delay	Power consumption			
Swift	Postman	packet switching	4 × 4 Mesh topology	LAH routing	**2Gb/s**	4ns	116.5 mW	Noxim	2013	[5]
PCAR For NoC	Enjuicheng	64-bit virtual packet switching	4 × 4 2D Mesh	PCAR	3.8Gb/s	1.6ns	58.26 mW	Cycle Accurate Simulator	2014	[6]
Star type NoC	Kaun-ju Cheng	Packet switching	12 × 12 star topology	XY-routing algorithm	–	4–5 ops	67.2 mW	ORION 2.0	2015	[7]
PDCR algorithm	Rabab EZZ-Eldin	Packet switching	4 × 4 Torus topology	PDCR routing	0.9Gb/s	42.56ns	–	OmNet++	2016	[8]

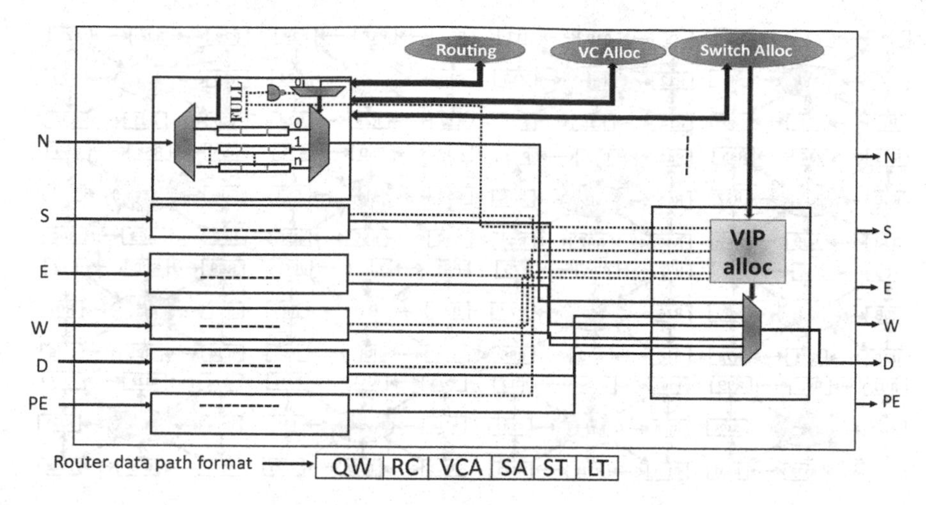

Fig. 2 Virtual channel router design

buffers are full, then it stalls the packets; otherwise, it passes the packets to the crossbar network. In the fourth step, switch traversal (ST) traverses the packets to the corresponding output port and link traversal (LT) is connected to the corresponding next router. Switch traversal (ST) and link traversal (LT) are part of VIP. This router will work based on the dimension distributed routing algorithm. A little variation is done in VC router design by adding an extra diagonal port to the router.

4 Minimal Weighted Path Routing Algorithm (MWPR):

MWPR is similar to Dijkstra's algorithm. This algorithm can able to find out the shortest distance based upon the weighted path. Dimension distributed routing algorithm followed the MWPR commands. We applied this shortest path algorithm for $4 \times 4 \times 8$ Star over Mesh. In this Star over Mesh, 4×4 topology is global Mesh as represented with M and each global node has eight local subnodes connected like star network as represented with R. Due to the additional direction of diagonal in the proposed router, the communication latency becomes reduced to reach the large distance node in $4 \times 4 \times 8$ star over mesh topology. $4 \times 4 \times 8$ Star over Mesh (SoM) topology is shown in Fig. 3.

Dimension distributed routing algorithm (DDRA) is incorporated within MWPR. To transfer the packets to the next router decided by the MWPR. It estimates the finest route and status given to the DDRA. Then, it follows the XYD-direction based on the status of MWPR. In detailed operation of MWPR and XYD, Dimension Distributed Routing Algorithm is given below:

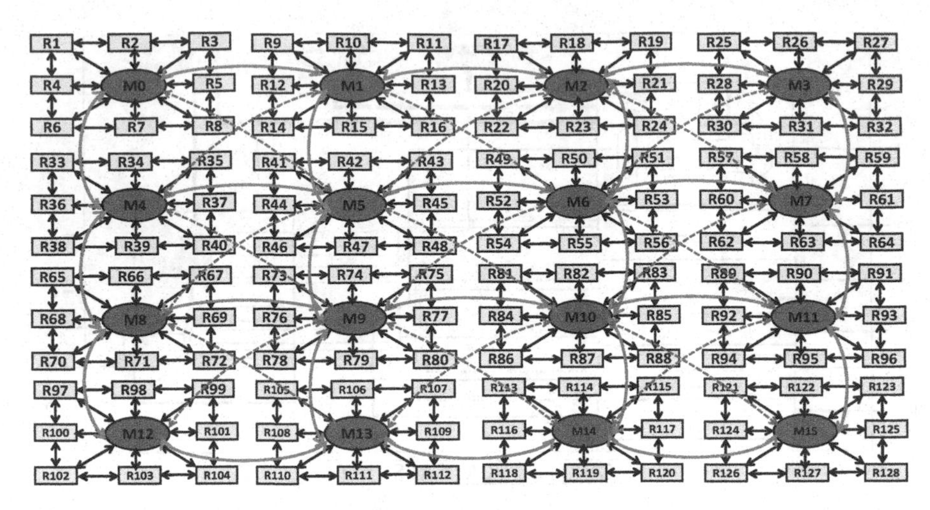

Fig. 3 4×4 × 8 2D Star over Mesh (SoM)

Algorithm-1: XYD Dimension Distributed routing algorithm:

X-plane, Y-plane, diagonal plane

1. If $x = 0, y = 0, d = 0$ then
   Dest= local;
2. If $x = 0, y > 0, d = 0$then
   Dest= east then
3. If $x = 0, y < 0, d = 0$ then
   Dest =west then;
4. If $x > 0, y = 0, d = 0$ then
   Dest =north then;
5. If $x < 0, y = 0, d = 0$ then
   Dest =south then;
6. If $x = 0, y > 0, d > 0$ then
   Dest =ES_d then ; // diagonal port b/w east and south
7. If $x = 0, y < 0, d < 0$ then
   Dest =ws_d then ; // diagonal port b/w west and south
8. If $x > 0, y = 0, d > 0$ then
   Dest =en_d then ; // diagonal port b/w east and north
9. If $x < 0, y = 0, d < 0$ then
   Dest =wn_d then ; // diagonal port b/w west and north

ALGORITHM-2: MWPR global & local routing for 4x4 star over mesh :

1. Function SoM(network, source):
2. *dist*[*source_g*, *Source_l*] − − > 0 // Initialization
3. Create *dest_g.info*, *dest_l.info* − − > header flit // destination information appended to each packet
4. For each vertex v_g, v_l in network: //global vertices (v_g) refers communication among mesh nodes, local vertices (v_l) refers communication among star over sub-nodes
5. determine the shortest path based on the distance weight (Similar to the Dijkstra's algorithm)
6. If$v_g = o$
7. dest=local sub node (*v_l*) //enable data transmission for local sub nodes of star
8. Apply dimension distributed routing algorithm
9. else $v_g = 1$
10. dest= node (*v_g*) //enable data transmission for nodes in the global mesh
11. Apply dimension distributed routing algorithm
12. Repeat the step-5 to step-11 until to reach the destination

5 Experimental Results

MWPR verified with various bit lengths for $4 \times 4 \times 8$ star over mesh. This algorithm is examined for large distance node in Star over Mesh. MWPR experiments with various bit lengths for mesh. This algorithm is examined for large distance node in a star over mesh. MWPR can serve to find out the least weighted path, and here, we distinguished three shortest paths such as best, average, and worst. The best shortest path is the most little path, the average shortest path is bit longer than the best path, and the final one is the critical path called as a worst-case shortest path. First, it gives an importance for the best shortest path; in case it fails, then it selects the average path; even average fails, then it selects the worst-case shortest path. Here, the simulation and synthesis results have carried out by using various bit lengths like 8, 16, 32 bits. The empirical result analysis exhibits that the time period for 8 bit is 0.884 ns, for 16 bit is 0.965 ns, and for 32 bit is 1.171 ns. The amount of power consumption for 8 bit is 240.96 mW, for 16 bit is 286.72 mW, and for 32 bit is 419.58 mW. The performance of $4 \times 4 \times 8$ Star over Mesh is shown in Fig. 4.

Therefore, an experimental work carried out by using Xilinx 14.3 software, and target to the vertex-4 board. The above result showed that area exceeds when bit length increases. Hence, the resultant output showed better performance in terms of less power with low latency even after increasing the bit length. Note that latency is measured in terms of minimum time period(ns) and power measured is measured

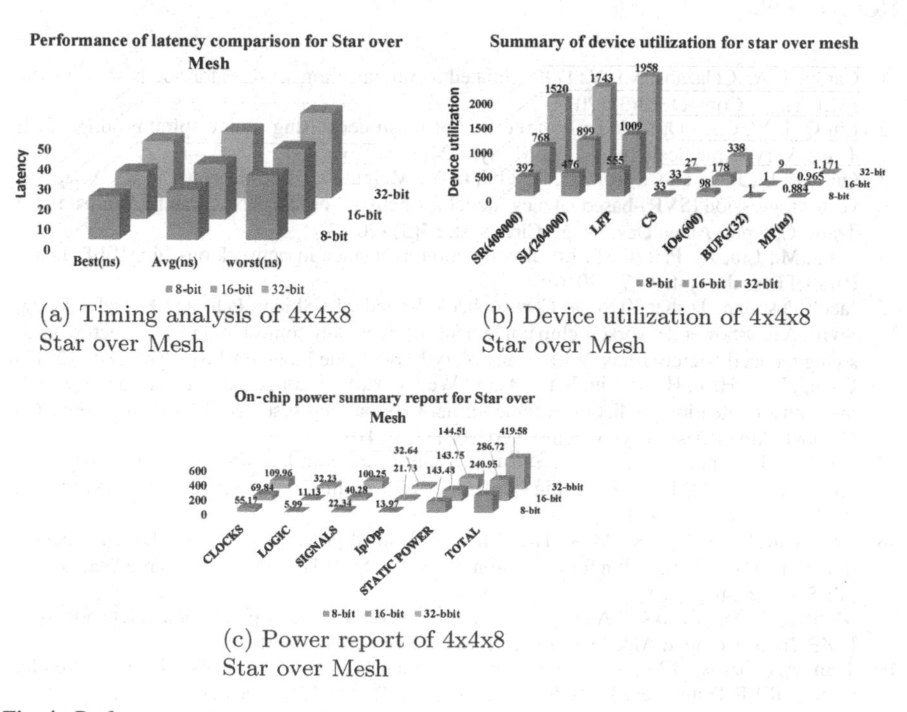

(a) Timing analysis of 4x4x8 Star over Mesh

(b) Device utilization of 4x4x8 Star over Mesh

(c) Power report of 4x4x8 Star over Mesh

Fig. 4 Performance of $4 \times 4 \times 8$ Star over Mesh

in terms of mW. When the size of bit length increases, then its time period also slightly increases. Power consumption is more in 32-bit length than that compared to 8, 16 bits. Due to diagonal directional port incorporated to the router, the latency got reduced. As a result, MWPR algorithm showed better efficiency in terms of reducing latency and power.

6 Conclusion

In this paper, MWPR presented a weighted oriented routing algorithm that controls the colossal amount of routing congestion and latency. This MWPR algorithm is tested with various bit lengths like 8, 16, and 32 bits, and their experimental results are clearly represented. As per the Xilinx reports, MWPR shows better improvement in reducing the latency and congestion for parallel enabled Star over Mesh. In future, we would like to extend this work by approaching with different shortest path routing algorithms for various topologies.

Acknowledgements The authors would like to thank the Principal and Management of Madanapalle Institute of Science & Technology for their kind support to carry out the research work, and also, they offered economical support on behalf of TEQIP-II World Bank Organization.

References

1. Carara, E.A., Calazans, N.L.V.: Differentiated communication services for noc-based mpsocs. IEEE Trans. Comput. **63**(3) (2014)
2. Liu Q, Ji W, Chen Q, Mak T.: IP protection of mesh nocs using square spiral routing. IEEE Trans. Very Large Scale Integr. (VLSI) Syst. **24**(4) (2016)
3. Qian, Z.-L., Juan, D.-C., Bogdan, P., Tsui, C.-Y., Marculescu, D., Marculescu, R.: A support vector regression (SVR)-based latency model for network-on-chip (NoC) architectures. IEEE Trans. Comput. Aided Des. Integr. Circ. Syst. **35**(3) (2016)
4. Tang, M., Lin, X., Palesi, M.: Local congestion avoidance in network-on-chip. IEEE Trans. Parallel Distrib. Syst. **27**(7) (2016)
5. Jacob Postman, Tushar Krishna, Christopher Edmonds, Li-Shiuan Peh, and Patrick Chiang, Swift: A low-power networkon-chip implementing the to- key control route ararchitecture swift swing reduced interconnects. IEEE Trans. Very Large Scale Integr. (VLSI) Syst. **21**(8) (2013)
6. Chang, E.-J., Hsin, H.-K., Lin, S.Y., (Andy) Wu, A.-Y.: Path congestion- aware adaptive routing with a contention prediction scheme for network-on-chip systems. IEEE Transactions On Comput.-Aided Des. Integr. Circuits Syst. **33**(1) (2014)
7. Chen, K.-J., Peng, C.-H., Lai, F.: Star-type architecture with low transmission latency for a 2d mesh noc. In: IEEE Asia Pacific Conference on Circuits and Systems (APCCAS) (2010). ISBN 978-1-4244-7456-1
8. Ezz-Eldin, R., El-Moursy, M.A., Hesham F.A.: Hamed process variation delay and congestion aware routing algorithm for asynchronous NoC. Des IEEE Trans Very Large Scale Integr (VLSI) Syst **24**(3) (2016)
9. Modarressi, M., Tavakkol, A., Sarbazi-Azad, H.: Virtual point-to-point connections for NoC. IEEE Trans. Comput. Aided Des Integr Circuits Syst. **29**(6) (2010)
10. Tran, A.T., Baas, B.M.: Achieving high-performance on-chip networks with shared-buffer routers. IEEE Trans. Very Large Scale Integr. (VLSI) Syst. **22**(6) (2014)

An IoT-VANET-Based Traffic Management System for Emergency Vehicles in a Smart City

Lucy Sumi and Virender Ranga

Abstract With the initiatives of Smart Cities and growing popularity of Internet of Things, traffic management system has become one of the most researched areas. Increased vehicle density on roads has resulted in plethora of challenges for traffic management in urban cities worldwide. Failure in transporting accident victims, critical patients and medical equipment on time has led to loss of human lives which are the results of road congestions. This paper introduces a framework with the fused concept of vehicular ad hoc networks and Internet of things that aims to prioritize emergency vehicles for smooth passage through the road traffic. The proposed system navigates ambulances in finding the nearest possible path to their destination based on real-time traffic information. A simulation of the designed framework has been demonstrated using CupCarbon simulator.

Keywords Intelligent traffic management system (ITMS) · Vehicular ad hoc network (VANET) · Internet of things (IoT) · Road side units (RSUs) Smart Cities

1 Introduction

High population rate and growth in business activities have led to greater demand of vehicles for transportation which has resulted in road traffics getting more and more congested [1]. With the escalation of vehicle density, monitoring traffic has become one of the crucial problems in metropolitan cities globally. Not only has traffic overcrowding resulted in wastage of time, money, property damage and environmental pollution but also in loss of lives. Several researches have been done on road

---

L. Sumi (✉) · V. Ranga
Department of Computer Engineering, National Institute of Technology, Kurukshetra, Harayana, India
e-mail: lucysumi866@gmail.com

V. Ranga
e-mail: virender.ranga@nitkkr.ac.in

P. K. Sa et al. (eds.), *Recent Findings in Intelligent Computing Techniques*, Advances in Intelligent Systems and Computing 708,
https://doi.org/10.1007/978-981-10-8636-6_3

traffic management, accident collision, congestion avoidance, etc., [1–12] but very few on emergency vehicles. To address these aspects of emergency services, we propose an IOT-VANET-based traffic management system that will ease traffic movements for ambulances. Building an intelligent traffic management system based on IOT can aid in collecting, integrating processes and analysing all types of traffic-related data on roads intelligently and automatically, thereby enhancing traffic conditions, reducing road jams and road safety. Hence, the system becomes extra intelligent and self-reliable. The objective of the paper is to provide a framework that gathers data using various advanced fundamental technologies such as wireless sensor network (WSN), vehicular ad hoc network (VANET), Internet of things (IoT), cloud computing.

The remainder of the paper is organized as follows: Sect. 2 presents related works while Sect. 3 introduces our proposed framework. Implementation and results of the framework are covered in Sect. 4. Finally, Sect. 5 concludes the paper and the future work.

2 State-of-the-Art Reviews

Smitha et al. [13] developed an efficient VANET-based navigation system for ambulances that addresses the problem of ascertaining the shortest path to the destination in order to get rid of unexpected congestions based on real-time traffic information updates and historical data. The system also includes a metro rail network with road transport system to guide ambulances in real-time scenarios. Similarly, Soufiene et al. [14] also presented an adaptive framework for an efficient traffic management of emergency vehicles that dynamically adjusts the traffic lights, recommend drivers the required behaviour changes, driving policy changes and exercise necessary security controls. Rajeshwari et al. [15] introduced an intelligent way of controlling traffic for clearing ambulances, detect stolen vehicles and control congestions. This was done by attaching RFID tags on vehicles which enables it to count the number of vehicles passing on a particular path, detect stolen vehicle and broadcast message to the police control room. The author in [16] introduced an intelligent traffic management system that prioritized emergency vehicles by using a different approach, i.e. by categorizing them on the basis of the type of incident occurred and the priority levels. They have also proposed a secure method to detect and respond hacking of traffic signals.

3 Proposed Framework Model

In this section, we present a block diagram of our model, as shown in Fig. 1, which consists of RSUs, traffic management server, sensors, vehicles of two types: normal and emergency cars and vehicles interconnected in a VANET system that

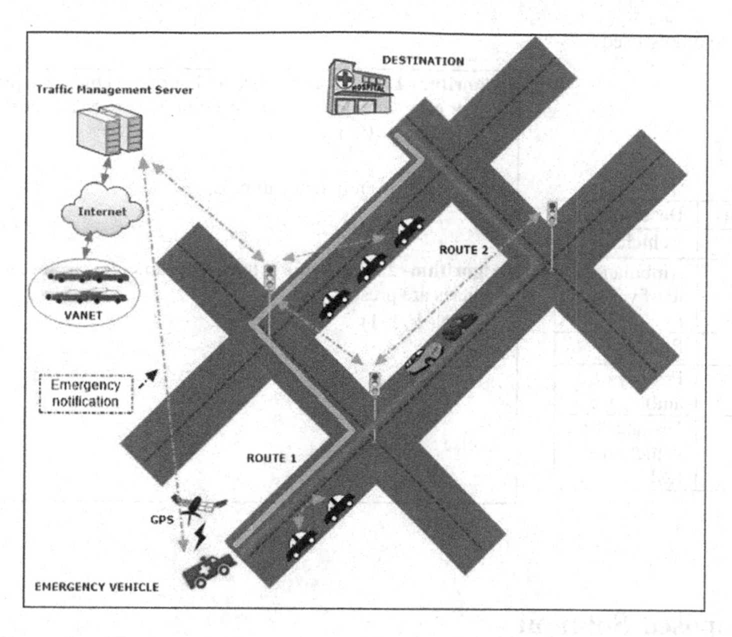

Fig. 1 Proposed traffic management system architecture for emergency vehicles

communicates with each other via vehicle-to-vehicle (V2V) and vehicle to infrastructure (V2I). Every ambulance is tagged with its unique ID that distinguishes it from the rest of the vehicles on the road. When an emergency car has to bypass a heavy traffic, it sends out emergency notifications to the nearest RSU. The TMS maintains database of information from sensors deployed in each lane at every intervals in order to achieve a global view of the entire road conditions in an area. This information is collected from heterogeneous sources such as wireless sensors and CCTV cameras. The RSUs deployed at every intersection manages and synchronizes information in such a manner to direct the vehicles. Now, when an emergency situation arises and an ambulance car is detected by the sensors of a particular junction, information will be sent to RSU of that particular lane which will automatically turn the traffic signals green to avoid traffic. The same RSU informs neighbouring RSUs about the approaching car and direct ambulance to its destination. Once the emergency car clears from the road, the system resumes to normal way of functioning. In our proposed system, every component has its distinct level of adaptability and dynamicity. For an instance, RSU controls the traffic signals depending on the timing constraints which may change dynamically by virtue of emergency levels and congestions. Likewise, for vehicles, the maximum and minimum speed control is adjusted by the RSU dynamically. Figure 1 depicts an ambulance that announces an emergency notification to TMS, which will broadcast the information further to all the nearby RSU based on the GPS location of the car. The RSU receiving the information authenticates ambulance via its ID and clears the route ahead while ensuring safe coordination of other vehicles as well.

Table 1 Symbols used

Symbols	Description
V_{id}	Vehicle id
V_{A_i}	Ambulance id of vehicle *i*
P_r	Priority
P_{A_i}	Priority of ambulance *i*
S	Distance of vehicle from RSU

```
Algorithm 1:When ambulance and normal vehicle are present
  For each vehicle i passing through RSU in S distance
    If ( V_id==V_Ai)
    {
      Allow ambulance to pass;
      P_Ai++;
    }
Algorithm 2:When more than one ambulance and normal vehicles are present
  If (N( V_A)>1)
    {
      If ( S_Ai<S_Aj)
        P_Ai++;
  else    P_Aj++;
    }
```

4 Proposed Solution

The algorithm proposed to help ambulances bypass traffic congestions has been considered under two cases: Algorithm 1 when there is a single emergency vehicle in a particular lane and Algorithm 2 when there is more than one emergency vehicle. At a fixed distance S, if a vehicle ID, V_{id} matches with ambulance ID, V_{A_i}, then the priority of ambulance will be increased and let to bypass traffic. When more than one ambulance car exists, say V_{A_i} and V_{A_j}, the car with shortest distance towards RSU will be given priority followed by the next. The descriptions of the used symbols in the algorithm are given in Table 1.

5 Implementation and Results

In this section, simulation work has been carried out in a simulated environment using CupCarbon U-One. It is a free and open source network simulator for Smart City and IoT. A total number of ten intersections, five RSUs between the starting point (i.e. the point where the ambulance starts moving) and hospital (i.e. its destination) have been taken for simulation. The values taken for various parameters in this work are summarized in Table 2.

As shown in the screenshot in Fig. 2, an ambulance is moving towards the hospital (destination). Its destination is pre-assigned and as it moves along the lane, the RSU in the current intersection detects ambulance in its vicinity as shown in Fig. 3. We assumed that there are two routes to reach the destined location: Route 1 and 2. On discovering Route 1 to be broken, ambulance chooses Route 2. When an

Table 2 Simulation parameters

Parameter	Value
Average speed of vehicles	50 km/h
Lane_id	L_1001, L_1002
Number of vehicles on existing lane	20
Simulation time	10 s
Simulation area	100 * 100 m^2 10 road intersections
MAC protocol	IEEE 802.11
Routing protocol	AODV
Transmission range	250 m

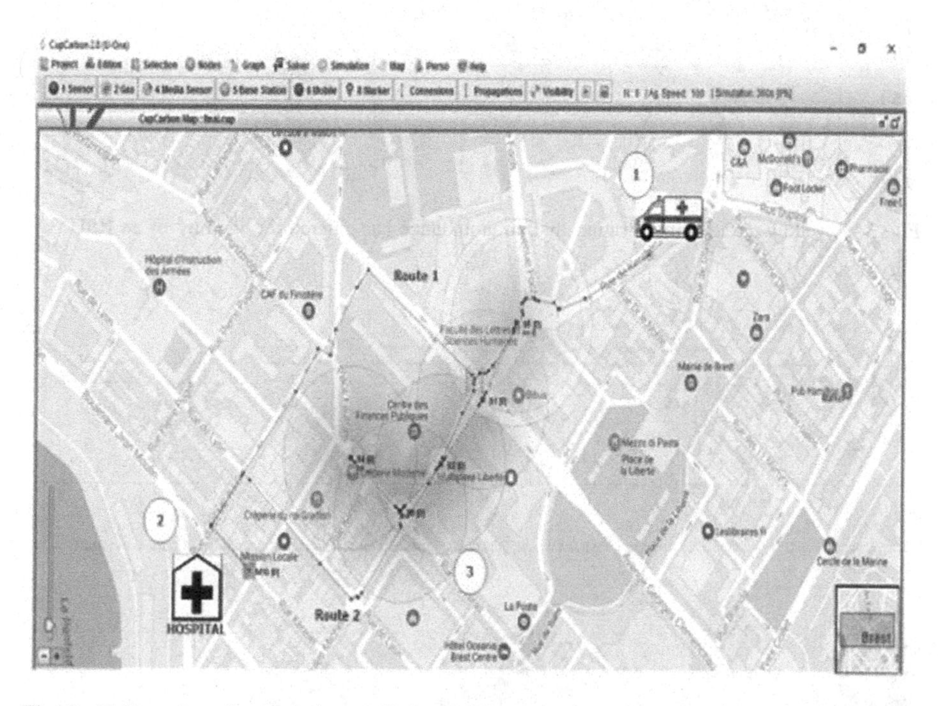

Fig. 2 (1) Location of an ambulance; (2) five RSUs deployed at five intersections; (3) location of hospital

approaching ambulance is detected, RSU of that particular lane informs other RSUs. Figure 4 displays a message “AMBULANCE” being sent by an RSU to another. Finally, Fig. 5 demonstrates that the ambulance has reached its destination.

We evaluate the performance of vehicles moving at an average speed delay. The graph in Fig. 6 shows that as the number of vehicles increases on road, average speed delay may increase initially but becomes constant at a certain time. Hence, even when the number of vehicles increments, it does not affect the average moving

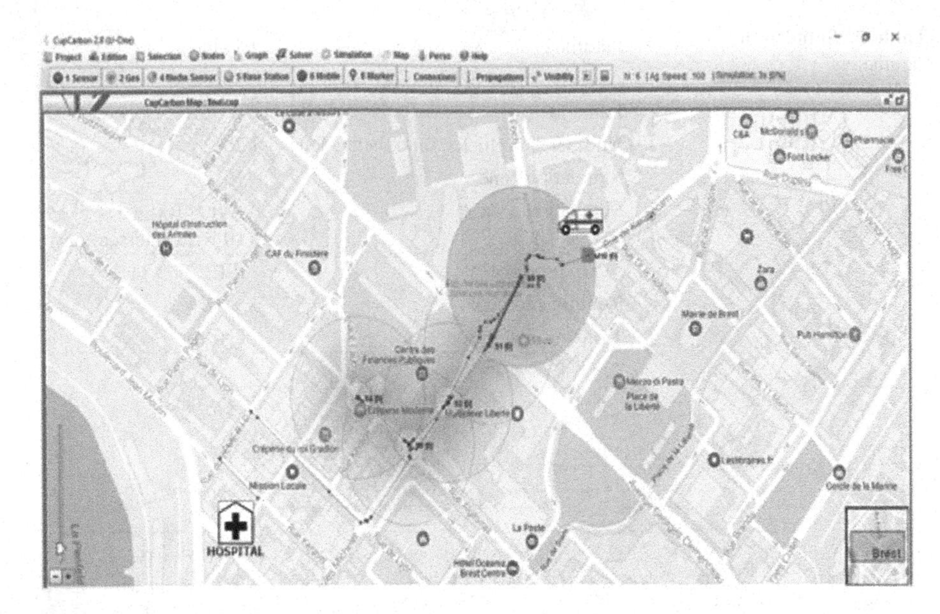

Fig. 3 A yellow circle demonstrating that an ambulance has entered the vicinity of an RSU

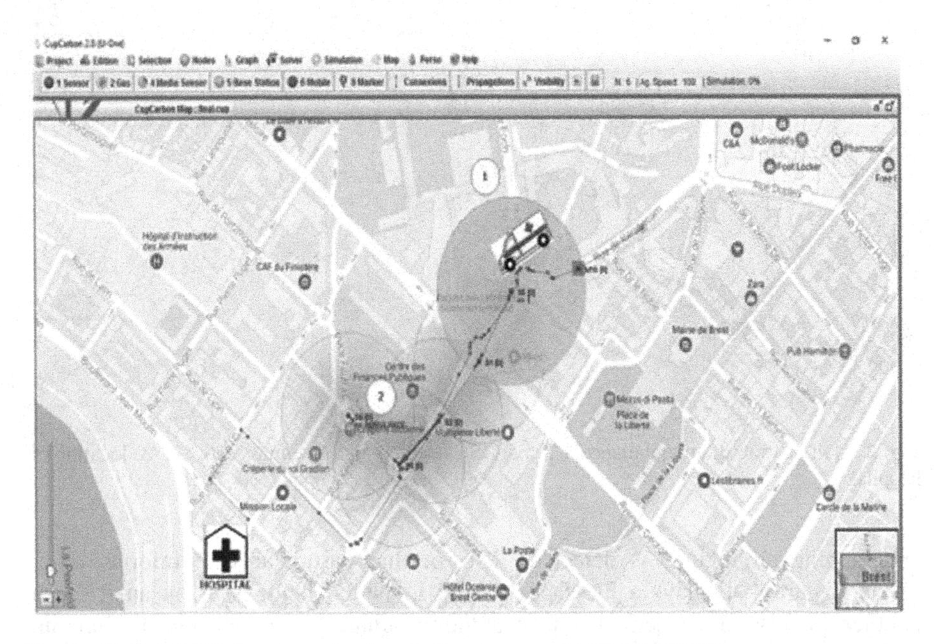

Fig. 4 (2) One RSU alarms another by sending message "AMBULANCE"

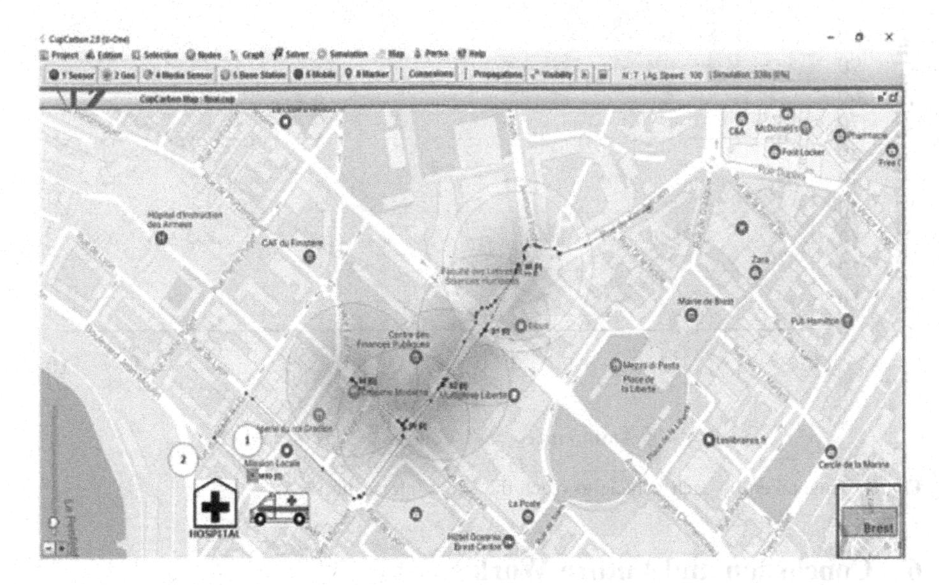

Fig. 5 Ambulance finally reaches its destination

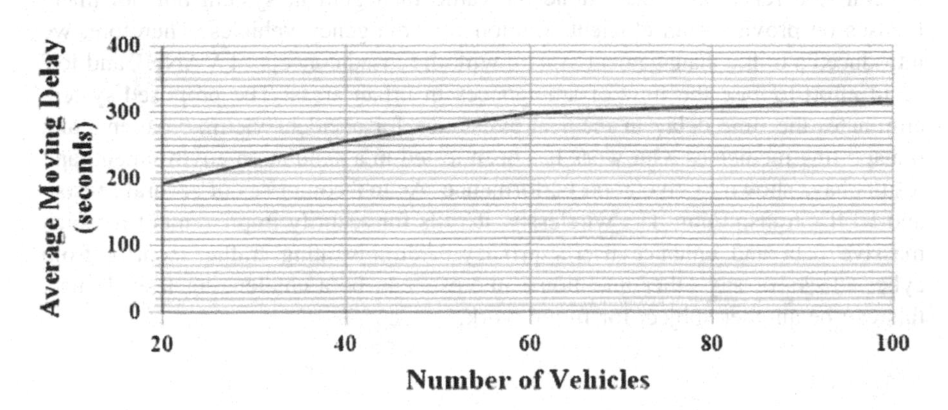

Fig. 6 Average moving delay measured by the number of vehicles

delay on road. In Fig. 7, we assumed three vehicles (one ambulance, two normal vehicles) to be travelling towards an intersection. Each vehicle starts at an equal distance of 55 m away from the intersection. It is evident that the ambulance received a priority as it reaches intersection in lesser time comparatively than the two normal vehicles.

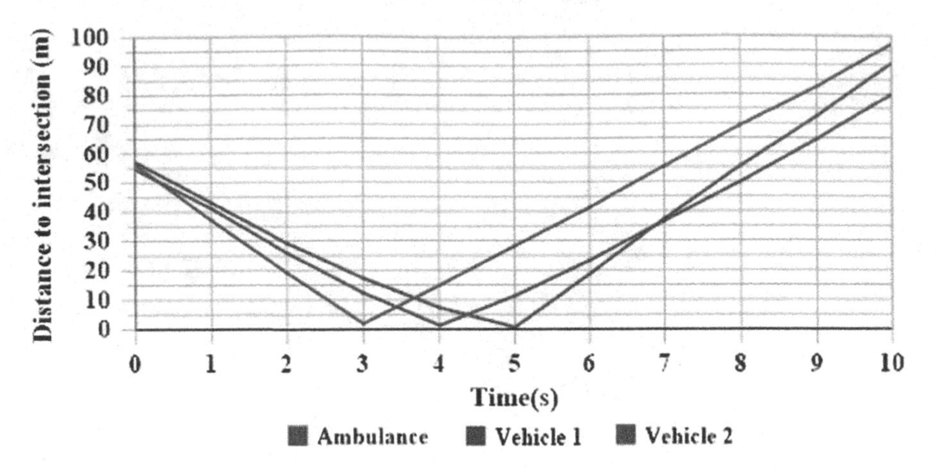

Fig. 7 Time taken by each vehicle to reach the same intersection

6 Conclusion and Future Work

Several researches have been done on traffic management system but not many focuses on providing an efficient solution for emergency vehicles. Therefore, we introduced a traffic management system with the fused concept of VANET and IoT in an effort to ease the flow of ambulances in urban areas. The proposed system eliminates the time delay in medical assistance for accident victims, patients and transporting medicines. Our work has been tested in a simulation environment, and results have shown to give good performance. As IoT comprises of several sensors and RFIDs transmitting data wirelessly, it calls for security improvement for such massive data and enhance user's privacy. Administrating traffic security from cyber-attacks or any other intentional interests can be a challenging task. Hence, this can be another subject for future work.

References

1. Francisco, J., Chai-Keong, Jaun-Carlos, Carlos, T., Pietro, M.: Emergency services in future intelligence transportation system based on vehicular communication networks. IEEE Intell. Transp. Syst. **2**, 07–20 (2010)
2. Minghe, Y., Dapeng, Z., Yurong, C., Mingshun, W.: An RFID electronic tag based automatic vehicle identification system for traffic IoT applications. In: Control and Decision Conference (CDC). IEEE, pp. 4192–4197, 01 Aug 2011
3. Luca, F., Tarik, T., Antonio, C., Dario, B.: M2M-based metropolitan plateform for IMS-enabled road traffic management in IoT. IEEE Commun. Mag. **49**, 50–56 (2011)
4. Vicente, M., Jorge, V., Jorge, G., Javier, S., Jishue, P., Enrique, O.: An intelligent V2I-based traffic management system. IEEE Trans. Intell. Transp. Syst. **13**(1), 49–58 (2012)

5. Carolina, T., Miguel, M., Pblo, R., Ahmad, M., Monica, A.: Smart city for VANET using warning messages, traffic statistics and intelligent traffic lights. In: 2012 Intelligent Vehicles Symposium, pp. 902–907, Spain, 5 July 2012
6. Naseer, K., Abdul, H.: A survey on intelligent transportation system. Middle-East J. Sci. Res. 629–642 (2013)
7. Chau, I.-C., Tai, H.-T., Yeh, F.-H., Hseih, D.-L., Chang, S.-H.: A VANET-based A* route planning algorithm for travelling time-and energy-efficient GPS navigation app. Int. J. Distrib. Sens. Netw. 1–14 (2013)
8. Miao, W., Hangguan, S., Rongxing, Ran Z., Xuemin, S., Bai, F.: Real-time path planning based on hybrid-VANET-enhanced transportation system. IEEE Trans. Veh. Technol. **64**(5), 1664–1678 (2014)
9. Hasan, O.: Intelligent traffic information system based on integration of internet of things and agent technology. Int. J. Adv. Comput. Sci. Appl. (IJACSA) **6**(7), 37–43 (2015)
10. Chatrapathi, C., Newlin, R.: VANET based integrated framework for smart accident management system. In: 2015 International Conference on Soft-Computing and Networks Security (ICSNS), Coimbatore, 08 Oct 2015
11. Sumi, L., Ranga, V.: Sensor enabled IoT for smart cities. In: Parallel Grid and Distributed Computing (PGDC-2016), Himachal Pradesh (2016) (to be appeared)
12. Cynthia, J., Sujith, R.: Road traffic congestion management using VANET. In: Advances in Human Machine Interaction (HMI-2016), Bangalore, 11 Apr 2016
13. Smitha, S., Narender, K., Usha, R., Divyashree, C., Gayatri, Aparajitha, M.: GPS based shortest path for ambulances using VANETs. In: International Conference on Wireless Networks (ICWN 2012), vol. 49, pp. 190–196, Singapore (2012)
14. Soufiene, D., Mazeriar, S., Irina, T., Pooyan, J.: Adaptive traffic management for secure and efficient emergency services in smart cities. In: IEEE Conference on Pervasive Computing and Communications Workshops, pp. 340–343, San Diego, 08 July 2013
15. Rajeshwari, S., Santhoshs, H., Varaprasad, G.: Implementating intelligent traffic control system for congestion control, ambulance clearance, and stolen vehicle detection. IEEE Sens. **15**(2), 1109–1113 (2015)
16. Abdullahi, C.: Priority based and secured traffic management system for emergency vehicle using IOT. In: International Conference on Engineering & MIS (ICEMIS). IEEE, 17 Nov 2016
17. CupCarbon U-one Simulator: http://www.cupcarbon.com

Analysis of Wormhole Attacks in Wireless Sensor Networks

Manish Patel, Akshai Aggarwal and Nirbhay Chaubey

Abstract Owing to their deployment characteristics, wireless sensor networks are vulnerable to many more attacks than wired sensor networks. The wormhole attack is a serious threat against wireless sensor networks. Launching this type of attack is easy for an attacker; it requires no breaking of a cryptographic mechanism. However, detecting the attack is very difficult. Moreover, after conducting the wormhole attack, an attacker can create more dangerous attacks. In this paper, we analyze the effects of wormhole attacks, discuss various detection features, and provide a comparison of various approaches. This paper presents various research challenges in the area of wormhole detection in wireless sensor networks.

Keywords Wormhole · Threat · Cryptographic · Vulnerable

1 Introduction

Security is crucial for wireless sensor networks. Because sensor nodes are resource-limited devices, traditional security algorithms, which are resource-hungry, are therefore not applicable to wireless sensor networks. A lightweight security framework is thus required for these types of networks.

The wormhole attack is one of many serious threats to the security of wireless sensor networks. An attacker installs two radio transceivers connected by a

M. Patel (✉)
Smt. S R Patel Engineering College, Gujarat Technological University, Gujarat, India
e-mail: it43manish@gmail.com

A. Aggarwal
School of Computer Science, University of Windsor, Ontario, Canada
e-mail: akshai.aggarwal@gmail.com

N. Chaubey
S. S. Agrawal Institute of Computer Science,
Gujarat Technological University, Gujarat, India
e-mail: nirbhay.chaubey@ieee.org

P. K. Sa et al. (eds.), *Recent Findings in Intelligent Computing Techniques*,
Advances in Intelligent Systems and Computing 708,
https://doi.org/10.1007/978-981-10-8636-6_4

high-capacity wired or wireless link, which is known as a tunnel. The existence of a tunnel creates a set of shortcut paths in the network. A large amount of traffic is drawn to this link, which fosters more dangerous attacks. By attracting traffic, a malicious node can collect, analyze, drop, and modify packets. It fundamentally changes the network topology and disturbs the routing process. Cryptographic techniques cannot prevent wormhole attacks. After overhearing the data packets, the malicious nodes obtain routing information and then move toward the specific path. Accordingly, a dynamic wormhole is created. Detecting dynamic wormholes is difficult compared to static wormholes. The main objective of this paper is comparison of various approaches and presents various research challenges for wormhole attack detection.

The remainder of this paper is organized as follows. Section 2 describes the effects of wormhole attacks in wireless sensor networks. In Sect. 3, various wormhole attack modes are compared. Section 4 describes various features for wormhole detection approaches. In Sect. 5, a comparison of various security approaches to preventing wormhole attacks is presented. Finally, our conclusions and future research are presented in Sect. 6.

2 Wormhole Attacks

After launching a wormhole attack, the attacker has unauthorized access to the given network. It enables the attacker to disrupt routing and launch various other attacks, such as blackhole, grayhole, and denial-of-service attacks. Furthermore, an attacker can launch a cryptanalysis attack, which enables cracking of communication keys. Symptoms that can be observed in the network on account of a wormhole are the following:

- A decrease in network utilization;
- An abrupt decrease in hop counts;
- One link receiving notably more traffic than others;
- Data receipt from a distant node;
- Return receipt of one's own message.

Some observations relating to wormhole attacks are outlined below.

Observation 1: If a node is under a wormhole attack, its neighbor nodes will approximately double.

As shown in Fig. 1, malicious nodes M1 and M2 form a wormhole tunnel. Node A is located in the area of M1. Thus, it believes that all the nodes in M2 are their neighbors.

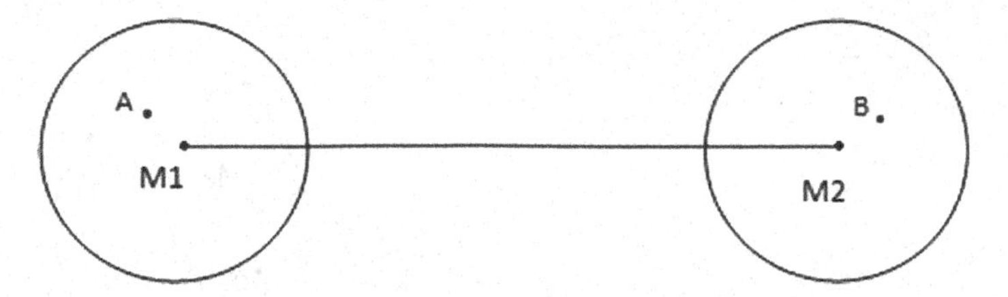

Fig. 1 Wormhole attack with two malicious nodes

$$N'_A = N_A \cup N_{M2}$$

where N'_A is the total number of neighbor nodes of node A, including remote nodes. In addition, N_A is the real neighbor set of node A, and N_{M2} is the neighbor set of malicious node M2. Similarly, node B is located in the area of M_2. Thus, it believes that all nodes in the area of M_2 are its neighbors.

$$N'_B = N_B \cup N_{M1}$$

where N'_B is the total number of neighbor nodes of node B, including remote nodes. In addition, N_B is the real neighbor set of node B, and N_{M1} is the neighbor set of malicious node M1.

$$|N'_A| = |N_A \cup N_{M2}| \approx |N_A \cup N_A| \approx |2N_A|$$

Observation 2: In a normal wireless network, for any pair of nodes (x, y), the set of common nodes between x and y are not more than two, which are also not one-hop neighbors.

As shown in Fig. 2, the distance between any two nodes in the area xyc is not more than R (radius); therefore, any two nodes in the area xyc are one-hop neighbors. The same scenario occurs for the area xyd.

In Fig. 2, node a exists in the area xyc, and node b exists in the area xyd. Nodes a and b are not one-hop neighbors. Suppose any node, p, exists in the area xcyd. Here, $p \neq a$, $p \neq b$, and p is a one-hop neighbor of neither a nor b. Because p is a

Fig. 2 Node neighborhood, where $d_{xy} = R$

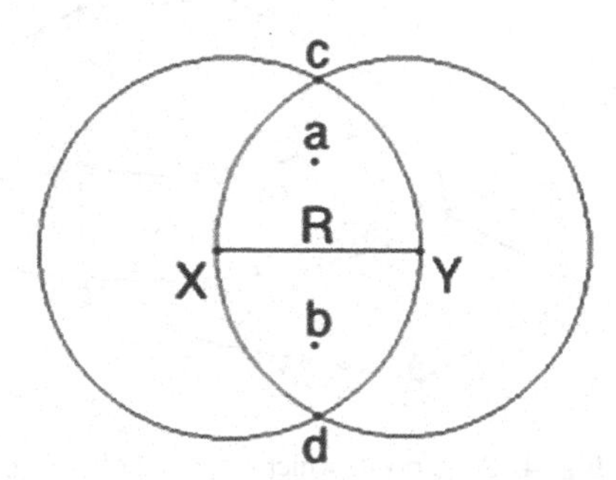

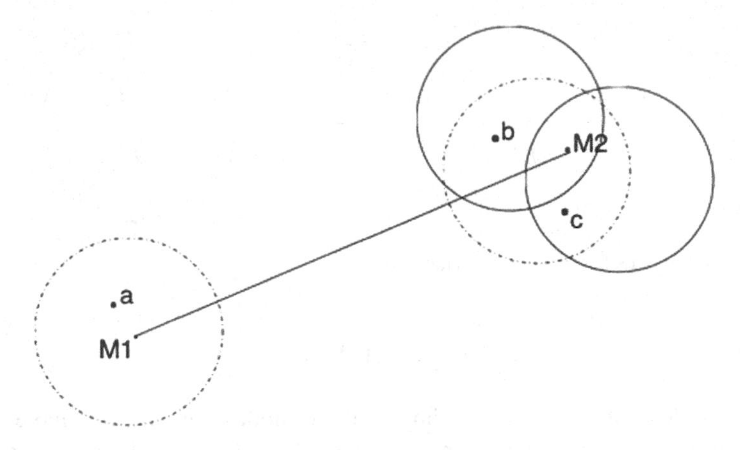

Fig. 3 Node neighborhood in the presence of a wormhole

one-hop neighbor of neither a nor b, it does not belong to the area xyc and xyd. This means that node p is not in the area xcyd. The hypothesis that node p exists in the area xcyd is incorrect.

Observation 3: Two nodes are either one- or two-hop neighbors if they exist in the range of the same wormhole.

As shown in Fig. 3, M1 and M2 are malicious nodes. Nodes a and b are one-hop neighbors connected through a wormhole. Node b is not a two-hop neighbor of node c. Nevertheless, owing to the wormhole, node b becomes a two-hop neighbor of node c.

Observation 4: Two nodes (neither of which are one-hop neighbors of each other) cannot have more than two common neighbors, which are also not one-hop neighbors.

As shown in Fig. 4, nodes x and y are not one-hop neighbors. Because of the wormhole, they are neighbors of nodes p, q, and z. These three nodes are located in another area than nodes x and y; moreover, these three nodes are not one-hop neighbors of each other.

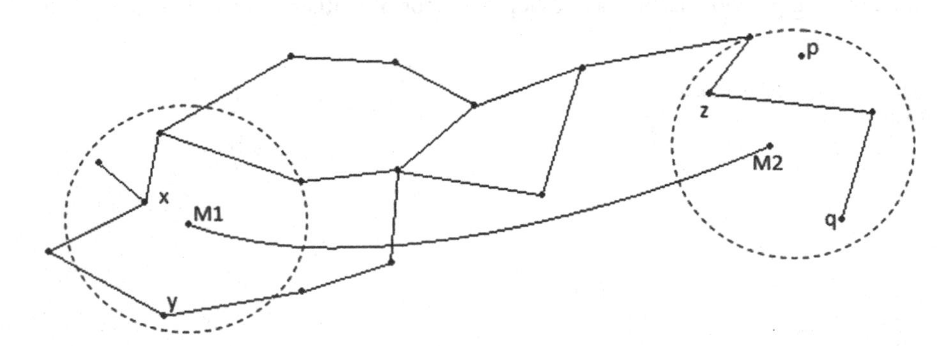

Fig. 4 Scenario in which a wormhole disturbs the routing process

Table 1 Comparison of various approaches for wormhole detection

Attack mode	Attack launching method	Challenges	Disadvantages
Encapsulation	One malicious node encapsulates the packets at one end; the other malicious node decapsulates them at the other end	Sending encapsulated packets to the other malicious node	Time and resource consumption in packet encapsulation and decapsulation
Out of band channel	A direct wireless or wired link is created between two malicious nodes	Requires special hardware and arrangements for tunneling	To launch this mode of attack, specialized hardware is needed
High power	Malicious node broadcasts with high power to increase the chance in the route establishment process	Requires power adjustments	Requires a high-power source
Packet relay	A malicious node relays packets between two distant nodes	Inserting malicious node at proper position by hiding its name	Resources are spent by relaying nodes for processing route request packets

3 Comparison of Wormhole Attack Modes

Wormhole attack modes are based on the techniques used for launching them. Table 1 shows a comparison of wormhole attack modes.

4 Detection Mechanism Analysis

4.1 Using Location Information

To identify the node location, each node is equipped with a Global Positioning System (GPS) device. The location information can be obtained using a directional antenna, which can provide the approximate direction based on the received signals. Approaches, respectively, based on GPS and a directional antenna requires additional hardware, which increases the network cost. The approaches presented in [1–8] are location-based approaches for detecting wormhole attacks in wireless sensor networks.

4.2 Using Neighborhood Information

The main characteristic of a wormhole attack is representing two distant nodes as neighbors. Detection techniques based on the neighborhood maintains immediate neighbor information. Some detection techniques maintain two-hop neighbor information. Additional storage and processing power are required for storing and analyzing two-hop neighbors. These methods have problems in dense networks because each node in dense networks has many neighbors. In a dynamic wireless sensor network, the neighbor list frequently changes. Thus, these techniques do not efficiently work. To address this issue, an approach based on neighborhood information in association with distance is presented in [4, 9, 10].

4.3 Using Hop Count/Distance

A path through a wormhole node is attractive because it contains a smaller number of hop counts compared to a normal path. When the message passes through the tunnel between malicious nodes, the actual hop count does not increase. A hop count technique in association with time is presented in [11]. Distance-based approaches are presented in [12–16].

4.4 Using Time Information

During a wormhole attack, the route in the attack has a greater than average time per hop compared to a normal route. To calculate the time between the source and destination, some approaches require a tightly synchronized clock [17, 18]. However, implementing a tightly synchronized clock is expensive for a wireless sensor network. The time difference between the source and destination is calculated in [19]. The source node sends a hello message to the destination. When the destination node receives the message, it returns a hello-reply message. The total round-trip time is calculated and divided by two. The average time for each hop is calculated. If the average time per hop count is greater than that for the normal route, it indicates the existence of a wormhole.

4.5 Using Connectivity Information

A wormhole attack is detected by examining the network connectivity [20–22], specifically by identifying fundamental topology deviations. A malicious node results in incorrect connectivity information.

5 Security Comparison of Various Approaches

In this section, we present a comparison of the wormhole detection approaches discussed in Sect. 4. Table 2 outlines the comparison. For wormhole detection, each node must perform extra processing, which causes an overuse of resources. Location- and time-based approaches require special hardware. Methods based on hop count cause routing delays, while methods based on neighborhoods create network congestion. The neighborhood-based approach does not properly operate when nodes are mobile. Moreover, methods, respectively, based on neighborhood, time and location are more expensive compared to the hop count-based approach.

Table 2 Comparison of various approaches for wormhole detection

Methods	Requirements	Comments
Geographical leashes [1, 2]	Loosely synchronized clocks	Location information may require more bits to represent and thereby increase the network overhead
Temporal leashes [1, 2]	Tightly synchronized clocks	Requires time synchronization level and may not detect physical layer wormholes
Graph-theoretic model [3]	Requires a combination of cryptography and location information	Efficient because it is based on symmetric cryptography; it does not require time synchronization; it requires location-aware guard nodes
Location-based keys [4]	Requires a group of mobile robots with GPS capabilities	Location-based keys can act as efficient countermeasures against wormhole attacks. It requires low memory
Secure localization and key distribution [5]	Two neighbor sensor nodes share a communication key	It is practical, has a low cost, and is scalable for large-scale wireless sensor network deployments
SeRLoc [6]	Directional antenna	It does not distinguish duplex and simplex wormhole attacks
HiRLoc [7]	Fewer locators are required to obtain the desired localization accuracy	To estimate the sensor location, range measurements are not required
Directional antenna [8]	Nodes use specific sectors of their antennas to communicate with each other	It efficiently uses bandwidth and energy
MDS-based detection [9]	Additional hardware is not required; it incurs low overhead	When both ends of two wormholes are very close, the approach fails to detect the attack
Neighbor Number and All Distance Test [10]	Additional hardware is not required	ADT performs better than NNT when the wormhole radius is small

(continued)

Table 2 (continued)

Methods	Requirements	Comments
SeRWA [4]	Additional hardware is not required	Symmetric key cryptography is used, which is suitable to wireless sensor networks
Delay per Hop Indicator [11]	Position information and clock synchronization are not required	Path suffers from a wormhole attack if the per-hop delay value is high
Distance consistency approach [12]	Locators, sensors, and attackers are deployed in the network	Even when the number of malicious locators is greater than normal, it has good performance
Ranging-based neighbor discovery [13]	Each node requires a microsecond precision clock	Negligible chance of an adversary creating a wormhole
Range-free anchor-free localization [14]	No reference nodes and distance measurement are needed	Diameter feature is used to detect a wormhole
Geographical wormhole detection [15]	A pair of public and private keys is required by each sensor node	A pairwise key pre-distribution protocol is used for detection
Using rank information [16]	Easy to implement. Does not need complex computing	Malicious nodes are detected when unreasonable rank values are found
Challenge-response delay measurement [17]	A symmetric key is shared by each pair of nodes	It uses distance bounding techniques and one-way hash chains
Wormhole resistant hybrid technique [23]	Does not require additional hardware and high computation complexity	It takes advantages of both Watchdog and Delphi methods and has good detection accuracy
Timing-based measurement [18]	Does not require synchronized clocks	When the node sends or receives packets, it is assumed to record the time
RTT-based approach [19]	To store RTT, additional memory is required	The approach is based on RTT and covers the multi-rate transmission problem
Local connectivity test [20]	The communication cost for the test is low	Network connectivity is examined
Visualization-based approach [21]	Does not require any hardware devices	More complex conditions need to be considered instead of flat plane
Passive and real-time detection scheme [22]	Minimal network overhead	Detection and localization are not performed at sensor nodes; they are conducted at the sink node

6 Conclusion

In this paper, we presented various methods for wormhole detection. Most of the methods presented in the literature for wormhole detection require additional hardware, which increases the sensor node cost. Detection methods based on distance and time must use cryptographic information because a malicious node can modify the distance and time information. Most of the existing methods consider static wireless sensor networks. However, detecting a wormhole attack in a dynamic wireless sensor network remains a challenging issue. The presence of multiple attackers sharply degrades the network throughput. Simultaneously detecting multiple wormholes is a challenging endeavor.

Acknowledgements The authors are highly thankful to the Gujarat Technological University (Smt. S.R. Patel Engineering College, Unjha) for providing the opportunity to conduct this research work.

References

1. Hu, Y.C., Perrig, A., Johnson, D.B.: Packet leashes: a defense against wormhole attacks in wireless networks. In: IEEE Computer and Communications Societies. IEEE, vol. 3, pp. 1976–1986 (2003)
2. Hu, Y.-C., Perrig, A., Johnson, D.B.: Wormhole attacks in wireless networks. IEEE J. Sel. Areas Commun. **24**(2), 370–380 (2006)
3. Poovendran, R., Lazos, L.: A graph theoretic framework for preventing the wormhole attack in wireless ad hoc networks. Wirel. Netw. **13**, 27–59 (2007)
4. Zhang, Y., Liu, W., Lou, W., Fang, Y.: Location-based compromise—tolerant security mechanisms for wireless sensor networks. IEEE J. Sel. Areas Commun. **24**(2) (2006)
5. Khalil, I., Bagchi, S., Shroff, N.B.: MOBIWORP: mitigation of the wormhole attack in mobile multihop wireless networks. J. Ad Hoc Netw. **6**, 344–362 (2008)
6. Lazos, L., Poovendran, R.: SeRLoc: robust localization for wireless sensor networks. ACM Trans. Sens. Netw. 73–100 (2005)
7. Lazos, L., Poovendran, R.: HiRLoc: high-resolution robust localization for wireless sensor networks. IEEE J. Sel. Areas Commun. **24**(2), 233–246 (2006)
8. Hu, L., Evans, D.: Using directional antennas to prevent wormhole attacks. In: Network and Distributed System Security Symposium (NDSS), pp. 131–141 (2004)
9. Lu, X., Dong, D., Liao, X.: MDS-based wormhole detection using local topology in wireless sensor networks. Int. J. Distrib. Sens. Netw. **2012**, Article ID 145702, 9 pp. (2012)
10. Buttyan, L., Dora, L., Vajda, I.: Statistical wormhole detection in sensor networks. In: SAS 2005, pp. 128–141. Springer
11. Chiu, H.S., Lui, K.S.: DelPHI: wormhole detection mechanism for ad hoc wireless networks. In: 1st IEEE International Symposium on Wireless Pervasive Computing (2006)
12. Chen, H., Lou, W., Sun, X., Wang, Z.: A secure localization approach against wormhole attacks using distance consistency. EURASIP J. Wirel. Commun. Netw. **2010**, 11 pp. (2010)
13. Shokri, R., Poturalski, M.: A practical secure neighbor verification protocol for wireless sensor networks. In: WiSec'09. ACM, Zurich, Switzerland, 16–18 Mar 2009
14. Xu, Y., Ouyang, Y., Le, Z., Ford, J., Makedon, F.: Analysis of range-free anchor-free localization in a WSN under wormhole attack. In: MSWiM'07. ACM, Chania, Greece, 22–26 Oct 2007
15. Sookhak, M., Akhundzada, A., Sookhak, A., Eslaminejad, M., Gani, A., Khan, M.K., Li, X., Wang, X.: Geographic wormhole detection in wireless sensor networks. J. PLOS ONE, 20 Jan 2015. https://doi.org/10.1371/journal.pone.0115324
16. Lai, G.-H.: Detection of wormhole attacks on IPv6 mobility-based wireless sensor network. EURASIP J. Wirel. Commun. Netw. (2016)

17. Capkun, S., Buttyan, L., Hubaux, J.P.: SECTOR: secure tracking of node encounters in multi-hop wireless networks. In: Proceedings of the 1st ACM workshop on Security of Ad-hoc and Sensor Networks (SASN 03), pp. 21–32, Oct 2003
18. Khabbazian, M., Mercier, H., Bhargava, V.K.: Severity analysis and countermeasure for the wormhole attack in wireless ad hoc networks. IEEE Trans. Wirel. Commun. **8**(2), 736–745 (2009)
19. Qazi, S., Raad, R., Mu, Y., Susilo, W.: Securing DSR against wormhole attacks in multirate ad hoc networks. J. Netw. Comput. Appl. 582–593 (2013)
20. Ban, X., Sarkar, R., Gao, J.: Local connectivity tests to identify wormholes in wireless networks. In: MobiHoc'11. ACM, Paris, France, 16–20 May 2011
21. Wang, W., Bhargava, B.: Visualization of wormholes in sensor networks. In: WiSe'04, Proceeding of the 2004 ACM workshop on Wireless Security, pp. 51–60. ACM Press (2004)
22. Lu, L., Hussain, M.J., Luo, G., Han, Z.: Pworm: passive and real-time wormhole detection scheme for WSNs. Int. J. Distrib. Sens. Netw. **2015**, Article ID 356382, 16 pp. (2015)
23. Singh, R., Singh, J., Singh, R.: WRHT: a hybrid technique for detection of wormhole attack in wireless sensor networks. J. Mob. Inf. Syst. **2016**, Article ID 8354930, 13 pp. (2016)

Handover Latency Improvement and Packet Loss Reduction in Wireless Networks Using Scanning Algorithm

K. Aswathy, Parvathy Asok, T. Nandini and Lekshmi S. Nair

Abstract In the era of wireless technology, communication caters many advantages such as increased mobility, productivity and scalability. One challenge in Mobile IPv4 is getting continuous access to the Internet. In this paper, we propose a scanning algorithm that would scan for the nearest access point and establish the connection. After the connection establishment, the packet transmission starts. We understand that the mobile node is not stationary, and hence, the scanning algorithm runs in the background and finds the next nearest access point and establishes the connection with the new access point. Using this technique, we aim to reduce the handover latency to a good extent, thereby reducing the packet loss. This paper gives an analysis to scale down the handover latency and maximize the packet transmission.

Keywords Mobile node · Handover · Latency · Throughput
Access point

1 Introduction

In the world of mobile devices such as tablets, cell phones which makes human life easier, there are many challenges which can be enhanced such as seamless Internet connection, high-speed data transmission and power saving mechanisms. This

K. Aswathy (✉) · P. Asok · T. Nandini · L. S. Nair
Department of Computer Science and Engineering, Amrita School of Engineering, Amrita Vishwa Vidyapeetham, Amrita University, Amritapuri, India
e-mail: aswathyramesh9@gmail.com

P. Asok
e-mail: parvathyasok95@gmail.com

T. Nandini
e-mail: nandinithanu.264@gmail.com

L. S. Nair
e-mail: lekshmisn@am.amrita.edu

P. K. Sa et al. (eds.), *Recent Findings in Intelligent Computing Techniques*, Advances in Intelligent Systems and Computing 708,
https://doi.org/10.1007/978-981-10-8636-6_5

paper focuses mainly on the challenge—"seamless Internet connection". Seamless Internet connection means that the ongoing session should not come to an extremity even when the mobile node moves from one point of connection to another. In wireless networks, there is need for continuous Internet connectivity to support end-to-end TCP or UDP session [1].

The usual scenario involves many complications when a mobile node moves from one wireless network to another. The movement of mobile node from one wireless network to another network is called as handover. Handoff perturbs the quality of service precisely. Basically, handoff is of two types—inter-cell handover and intra-cell handover. When the source and the destination are present in distinct cells, then it is defined as inter-cell handover. This ensures that the transmission is continued as the user moves away from the source and towards the destination. In an intra-cell handover, the source and the destination are located in the same cell and only the channel is altered during the handover. The objective of this handover is to change the channel which has lot of disturbances (Fig. 1).

The movement of mobile node completely depends on the user. The time taken to detach from current network and attach to the nearest neighbouring network is called handover latency. While the course of implementing, the major issues which are to be resolved are as follows:

(1) During the handover latency, the node gets segregated from its detached network.
(2) Depending on the handover latency, packet loss will occur.
(3) All wireless networks must be mobility managed.

Generally, handoff [2] is divided into three phases—scanning, authentication and re-association. Handoff delay is the sum of delay aroused during scanning and re-authentication phase. Handoff latency can be branched into three delays:

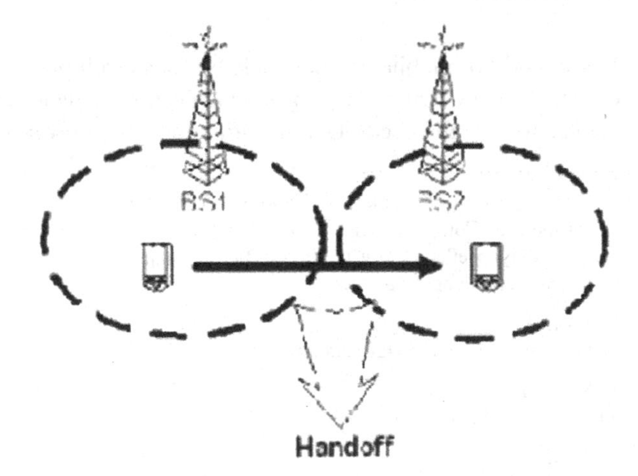

Fig. 1 Handoff of a mobile node

(1) Scanning delay—It is the time taken by a mobile node to find the nearest access points.
(2) Authentication delay—Authentication service is used by all stations to authorize the identity to station which they communicate.
(3) Re-association delay—It is the time taken to re-associate with the new access point.

In this paper, a handover latency scheme is proposed to overcome the above-mentioned challenges. Our work is mainly branched into two phases:

(1) The first phase is to curtail the handover latency of mobile node. This is accomplished by the following method. The mobile node scans for the nearest access point periodically by calculating the distance between the current node and the neighbouring access point as it moves. Finally, it gets connected with the access point which has the shortest distance with it and the packet transfer starts with that access point.
(2) The next phase is to augment the throughput of the packets transferred. Throughput can be defined as an average of successful message delivery over a communication channel. The maximum throughput is accomplished for distinct data rates which is calculated using a formula to enhance the accuracy of calculation.

2 Related Work

The authors in [3, 4] recommended scanning mechanism which uses GPS by which handover delay can be reduced. In this method, the scanning phase is completed just before the handover takes place. From their implementation, the handover delay can be reduced.

In the past few years, researchers have proposed a Hexagonal Cell Concept [5] to improve the handoff process. In this technique, if the calculated distance between a mobile node and the access point is greater than a predefined threshold level, then the handoff process is initiated.

In [6], Venkata proposed Fast Handoff By Avoiding Probe Wait (FHAP) where mobile nodes will not wait for probe responses. The mobile node preserves a priority list of channels and scans for new access points accordingly.

In [7], Debabrata Sarddar proposed a method to reduce the handoff using neighbour graph.

In [8], authors proposed a Distributed Handoff Mechanism for reducing the handover latency by which all the available channels are grouped and scanned by selecting neighbouring nodes separately.

In [9], authors proposed a Zone-Based Interleaved Scanning Handoff scheme.

In [10], the system focused on state route selection during the route discovery phase, during which the route is constructed. The paper adopts Received Signal Strength Indicator (RSSI)-based distance measurement.

In [11], the author has taken into consideration both cellular and D2D communication. Authors have suggested to serve communication using a master–slave technique and also consider the SNR and power to reorganize the mobile clusters created.

3 Proposed Method

The synopsis takes into consideration that in a wireless network scenario, many base stations are connected to access points which prevail in a certain region. Hence when a mobile node travels away from one network, it gets wide variety of other foreign networks to get connected with. This selection of a new network is the method proposed in this paper. The choice of selecting the network completely depends on the shortest distance factor which is calculated using the Euclidean distance formula. Hereby, considering the above-mentioned algorithm the steps involved in the implementation are as follows:

(1) Firstly, configure the mobile nodes by mentioning the followings:

- Routing protocols used: A routing protocol enumerates how the routers advertise with each other propagating the orientation that empower them to select the relevant routes between any two nodes. The most frequently used routing protocols are DSDV, AODV and DSR. In this paper, the method is implemented using the DSDV protocol.
- MAC type: The Media Access Control layer is the lower layer of the Data Link Layer and is perturbed with allocating the physical connection to the network. The feasible values for the MAC type are Mac/802 11, Mac/CSMA/CA, Mac/Sat, Mac/Sat/Unslotted Aloha, Mac/TDMA. The proposed method uses Mac/802 11.
- Channel type: A communication channel can either be a physical transmission medium or it can be a logical connection over a multiplexed medium and is used to transmit an information signal. The channel type among the node configuration parameters specifies which channel to be used among channel/wireless channel, channel/Sat. The proposed algorithm uses channel/wireless channel.
- Link layer type: Link Layer is the lowest layer in an Internet protocol network architecture that transmits the data among the contiguous network nodes or among the nodes of the same network. The available link layer types are LL, LL/Sat. LL is used in the proposed method.
- Antenna model: Antenna converts electric power to radio waves in which it can either radiate uniformly in all directions or in a specific direction. The

antenna models available in the node configuration are antenna/omni-antenna. The same antenna type is used in this paper.

- Radio propagation model: It specifies the behaviour of the radio wave when they travel or propagate from one point to another. The Two Ray Ground, Propagation/algorithm in this paper.

Propagation types available in the configuration are propagation/shadowing. Propagation/two-ray ground is used to implement the

- Network interface type: Interface between two protocol layers in a network architecture that rebound the probable values defined in the network interface type catalogue. The possible values include Ethernet/FDDI/Loopback/PPP/Slip/Token Ring.
- Number of mobile nodes: Specifies the total number of mobile nodes used in this scenario.

(2) Initialize the standby positions of the access points and the mobile nodes within the window size.
(3) Calculate the minimum distance between the access points and the mobile nodes as the time increases. This is calculated since the mobile nodes are in motion. Let (x1, y1) be the coordinates of the access point and (x2, y2) be the coordinates of the mobile node. Then, the shortest distance D between the access point and the mobile node can be calculated by the following equation:

$$D = \sqrt{(x2 - x1)^2 + (y2 - y1)^2} \quad (1)$$

(4) The TCP connection is established between the access point and the mobile node which has the shortest distance with it among all access points.
(5) After step 4, packet transmission starts.
(6) As the node moves away from the current access point step 3 repeats. This continues till the stop time is reached (Fig. 2).

Algorithm 1 shows the steps to find the nearest access point by finding the shortest distance between the node and the access point.

Algorithm 2 {Euclidean Distance} shows the steps to find the shortest distance between node and access point.

Algorithm 1
Procedure NEARESTACCESSPOINT (dist1, dist2)

1: time 0.1
2: now timecurr
3: Set(XN,YN) (Xcurr_node,Ycurr_node)
4: Set(XAP-1,YAP-2) (X curr-AP-2,Ycurr-AP-2)

Fig. 2 Proposed approach

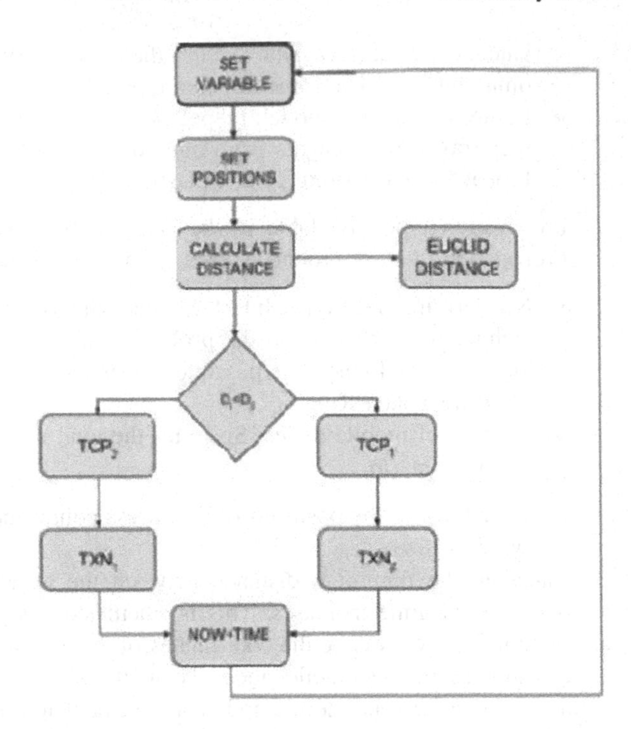

5: Set dist1 SHORTESTDISTANCE (Curr_node_position,ap_1_position)
6: Set dist2 SHORTESTDISTANCE (Curr_node_position,ap_2_position)
7: if dist 1 < dist2 AND node_scan_TYPE = ACTIVE
8: TCP1 established AND pkttransfer
9: Else
10: TCP2 established AND pkttransfer
11: For all now + time nearest_access_point()
12: end procedure

Algorithm 2 Euclidean Distance
Procedure SHORTESTDISTANCE(x1, y1, x2, y2)

1: *Let (x1,y1) be the coordinates of the first access point*
2: *Let (x2,y2) be the coordinates of the second access point*
3: *Then the shortest distance D can be calculated as:*

$$D = \sqrt{(x2 - x1)^2 + (y2 - y1)^2} \quad (1)$$

This method reduces the handover latency, and now, the throughput should be calculated using the AWK script. Throughput is the rate of successful message delivery over a communication channel.

The throughputs for the different data rates are found using the below-mentioned formula in AWK script to enhance the accuracy of calculation.

$$\text{Throughput} = \frac{\textit{size of packet received}}{\textit{start time} - \textit{stop time}} \times 0.008$$

This method reduces the handover latency, and now, the throughput should be calculated using the AWK script. Throughput is the rate of successful message delivery over a communication channel.

The throughputs for the different data rates are found using the below-mentioned formula in AWK script to enhance the accuracy of calculation.

Then, an Xgraph is plotted based on data rate values and throughput. On the x-axis, MAC/802 11 data rate is plotted, and on the y-axis, throughput is plotted. Using this Xgraph, we determine maximum throughput which will predict maximum efficiency.

4 Performance Evaluation

We determined the efficiency on an Xgraph using NS2 simulator. The graph is plotted based on data rate values and throughput. On the x-axis, MAC/802 11 data rate is plotted, and on the y-axis, throughput is plotted.

In the normal scenario, considering that the mobile node is already connected to an access point (packet transmission exists), the node starts its scanning phase only after its first packet drop or when the mobile node goes beyond the range of access point. This can be disadvantageous as the scanning phase creates a delay.

In our method, the node starts scanning for the best access point while simultaneously having a packet transmission with the current access point. The scanning of the access point is based on the shortest distance. Using AWK script, we calculate the throughput for its corresponding data rate. Figure 3 depicts the throughput calculation in our scenario.

Graphical representation shows that the throughput is in low range in the normal scenario, while on the other hand the throughput has gradually increased to a certain level where the maximum efficiency has reached. Table 1 shows the sample throughput which are generated using the AWK script.

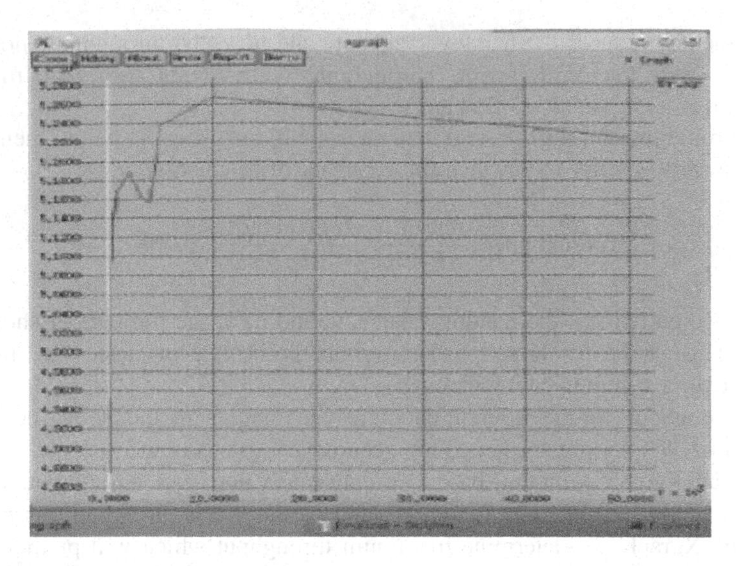

Fig. 3 Throughput calculation

Table 1 .

Data rate (Mb)	Throughput (Kbps)
100	4875.82
200	5032.11
300	5054.86
400	5133.98
500	5145.79
600	5139.14
700	5161.49
800	5148.88
900	5145.84
1000	5168.77
2000	5188.45
3000	5163.32
4000	5157.16
5000	5237.60
10000	5267.90
50000	5224.91

5 Conclusion and Future Scope

This paper gives a better perceptive of handoff and a concise summarization of various handover schemes. In this study, we have devised, analysed and evaluated an enhanced handover control scheme where a mobile node selects its new network based on the shortest distance between that mobile node and the access point. The performance evaluation of this method shows that handover latency of a mobile node can be scaled down to a good extent.

According to the above algorithm implemented, there has been a gradual efficient performance when compared to the normal scenario, and this is achieved by reducing the handover latency. The method has not taken into consideration the power consumption while the scanning algorithm is continuously being executed. The future scope of the proposed scheme is that it can also be enforced on a heterogeneous platform where modern gadgets can be utilized with enhanced data transmission techniques.

References

1. Jana, T.: Minimization of hand-off latency by area comparison method using GPS based map. Springer (2011)
2. A review on hand-off latency reducing techniques in IEEE 802.11 WLAN. IJCA (2014)
3. Reducing handover delay by pre selective scanning using GPS. IJDPS **1**(2) (2010)
4. Menodhan, T., Hayasaka, M., Miki, T.: Novel handover scheme for improving the performance of WLANs based on IEEE802.11 Apcc'06. In: Proceedings of Asia-Pacific Conference 2006, pp. 1–5 (2006)
5. Sarddar, D., Banerjee, J., Chakraborti, T., Sengupta, A., Naskar, M.K., Janat, T., Biswas, U.: Fast hand-off implementation using distance measurements between mobile station and APs. In: Students' Technology Symposium (TechSym), IEEE, pp. 81–86 (2011)
6. Chintala, V.M., Zeng, Q.A.: Novel MAC layer hand-off schemes for IEEE802.11 wireless LANs. In: Wireless Communications and Networking Conference, WCNC 2007. IEEE, pp. 4435–4440
7. Sarddar, D., Jana, T., Patra, T., Biswas, U., Naskar, M.K.: Fast hand-off mechanisms in WLANs based on neighbour graph information. In: 2010, 1st International Conference on Parallel Distributed and Grid Computing (PDGC), pp. 334–338 (2010)
8. Zhang, Z., Pazzi, R.W., Boukerche, A.: A fast MAC layer hand-off protocol for WiFi based wireless networks (2010)
9. Sarma, A., Gupta, R.K., Nandi, S.: A zone based interleaved scanning technique for fast handoff in IEEE 802.11 wireless networks pervasive systems, algorithms, and networks (ISPAN). In: 2009 10th International Symposium, pp. 232–237 (2009)
10. Rajesh, M., Vanishree, K., Sudharshan, T.S.B.: Stable route AODV routing protocol for mobile wireless sensor networks. In: 2015 International Conference on Computing and Network Communications, CoCoNet 2015
11. Giriraja, C.V., Ramesh, T.K.: SNR based master-slave dynamic device to device communication in underlay cellular networks. In: 2015 International Conference on Advances in Computing, Communications and Informatics, ICACCI 2015 (2015)

5 Conclusion and Future Scope

References

Optimized Location for On-Body Sink for PRPLC Protocol in WBAN

Arifa Anwar and D. Sridharan

Abstract Increased life expectancy and present-day habits have increased the number of patients with metabolic diseases. This makes even youngsters more dependent on continuous health care services. Wireless Body Area Network (WBAN) is a solution to this problem with mobile health care wherever you are, with required attention to fatal body symptoms. The implantation of on-body sink has been a subject of research, since the failure of this sink causes missing of important health information. This paper compares the performance of the routing protocol PRPLC by changing the sink to five different positions. The performance is compared with respect to three metrics, such as network lifetime, propagation delay and packet delivery ratio (PDR).

Keywords WBAN · Routing protocol · On-body sink · Lifetime
Delay · PDR

1 Introduction

An assembly formed by compact self-regulating and trivial weighted sensor nodes placed in, on or around the human body, is termed as a Wireless Body Area Network (WBAN). The IEEE 802.15.6 standard was specifically designed taking into consideration the rising need of mobile and autonomous health care systems without being coupled to hospital environments. BANs are used for medical and nonmedical purposes. The nonmedical purposes include clothing commercial electronics and entertainment gadgets [1]. The standard specifies either a one-hop or two-hop topology for communication in WBAN. Even when the maximum number of hops is restricted to be two, routing protocols in WBAN are an open research area due to the complex propagation conditions that arise within the human body

A. Anwar (✉) · D. Sridharan
Department of ECE, CEG Campus, Anna University, Guindy, Chennai 600025, India
e-mail: arifaanwar25@gmail.com

P. K. Sa et al. (eds.), *Recent Findings in Intelligent Computing Techniques*,
Advances in Intelligent Systems and Computing 708,
https://doi.org/10.1007/978-981-10-8636-6_6

and the minute range of the nodes used for human body communications meeting the strict specific absorption rate (SAR) defined by the medical standards.

The various types of routing protocols proposed for WBAN include temperature-based, cluster-based, cross-layer, cost-effective and QoS-based routing [2, 3]. The cost-effective routing protocols define a cost for each link based on their reliability status for communication in the next time slot. Probabilistic Routing with Postural Link Cost (PRPLC) is such an example. But various proposed routing protocols have their sink placement in various locations. For example, PRPLC and Distance Vector Routing with Postural Link Cost (DVRPLC) have their on-body sink located on the right ankle, whereas the protocols Stable Increased throughput Multi-hop Protocol for Link Efficiency (SIMPLE) and Adaptive Threshold-based Thermal-aware Energy-efficient Multi-hop proTocol (ATTEMPT) have their sinks located on the chest [2, 4, 5].

This paper compares the performance of the routing protocol PRPLC [2] by changing the sink positions. The on-body sink is placed at five different locations, such as right and left ankles, right arm, chest and waist. The comparison is done on the basis of overall network lifetime, sink node death, propagation delay and packet delivery ratio (PDR). The rest of this paper is arranged as follows. Section 2 explains the PRPLC protocol in detail. Sections 3 and 4 describe the simulation set-up and the results obtained from the comparison, respectively. Section 5 concludes the paper.

2 Probabilistic Routing with Postural Link Cost (PRPLC)

A total of seven nodes are arranged on the human body as two on the upper arms, two on the ankles, two on the thighs and one on the waist. PRPLC assigns a link likelihood factor (LLF), $P^t_{i,j}$, for a link $L_{i,j}$ between nodes i and j which quantifies the chances of the link to be connected over a discrete time slot t. For a link connected over a time slot t, $L_{i,j}$ is set as 1 and 0 otherwise. LLF is decided to be dynamically updated after the *t*th time slot as follows.

$$P^t_{i,j} = \begin{matrix} P^{t-1}_{i,j} + (1 - P^{t-1}_{i,j})\omega & \text{if Li, j is connected.} \\ P^{t-1}_{i,j}\omega & \text{if Li, j is disconnected.} \end{matrix} \quad (1)$$

Thus, the historical connectivity quality (HCQ) $\omega^{(i,j)}(t)$ determines the rate at which the LLF increases and decreases when the link is connected and disconnected, respectively. Thus, HCQ should capture the long-term history of the link, so that LLF should increase steadily and decrease slowly for historically good links and vice versa for historically poor links. The HCQ is defined as

$$\omega^{i,j}(t) = \sum_{n=t-Twindow}^{t} \frac{L_{i,j}^{n}}{Twindow}. \quad (2)$$

Twindow is the number of slots over which HCQ is averaged. The optimum Twindow size is between 7 and 14 [6]. All nodes retain and amend their LLF with all neighbouring nodes and sink through periodic *Hello* messages. PRPLC intends to select the most likely link to route the packet to avoid buffering and packet loss. When a node i wants to communicate to the sink s, it compares the LLF with its one-hop neighbours with the LLF of itself with the sink s. The comparison goes as follows if N is the set of all one-hop neighbours of i.

```
For all j ƐN
if P_{i,j}^t>P_{i,s}^t
forward the packet to node j
else
buffer the packet till link between i and s comes.
```

3 Simulation Parameters

The simulation parameters are selected according to the MICS specifications [7] and the experimentally found path loss properties. The simulation parameters used are enlisted in Table 1.

Table 1 Simulation parameters

Parameter	Value/Attribute
Simulation tool	Matlab
Simulation area	1.7 m
Maximum data collection rounds	2000
Data packet length	4096 bits
Frequency of operation	403 MHz
Attenuation factor	4.22
Energy for transmission	16.7 nJ per bit
Energy for reception	3.6 nJ per bit
Random scatter around mean	8.18 dB

4 Results and Discussion

The best position for an on-body sink is the one that provides a better lifetime and reliability with a lesser delay. Figures 1, 2 and 3 show the obtained results.

As is seen clearly from the results, although in a corner of the network, a sink on the left ankle optimizes the available network energy. Although the waist node is at the centre of the network and has direct communication with others, it dies off quickly than others rendering the network useless, whereas a higher LLF leads more nodes to transmit their packet through left ankle making it a better choice of sink.

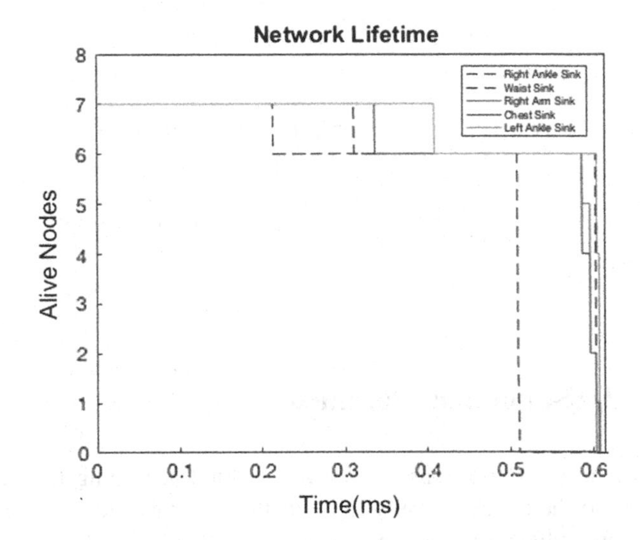

Fig. 1 Number of nodes still left with residual energy (*alive nodes*) as time progresses (*time*). This figure shows that a left ankle sink manages more number of alive nodes

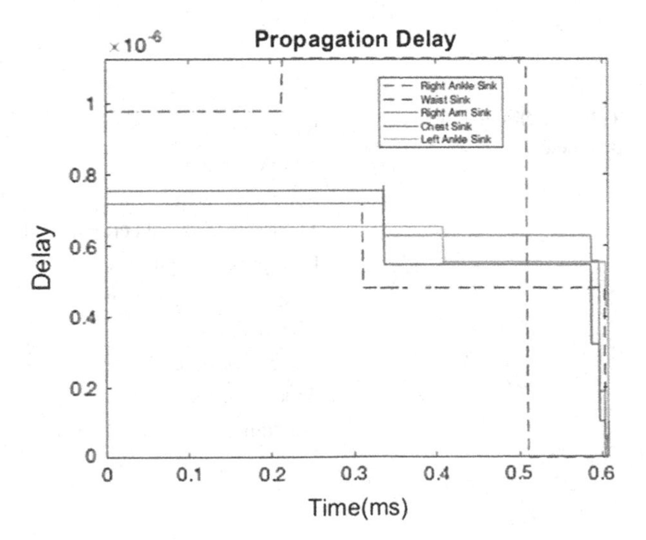

Fig. 2 Latency (*delay*) for sending the packets to on-body sink against total time taken (*time*). This figure shows that left ankle sink provides lesser delay

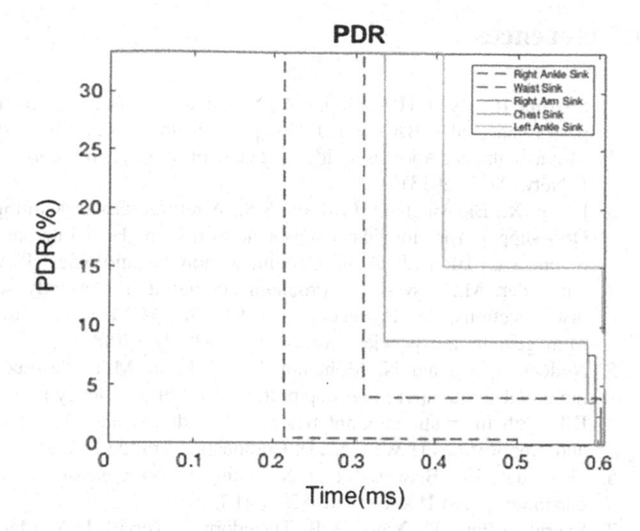

Fig. 3 Reliability score of different sink positions (*PDR*) against running time (*time*). This figure shows that the left ankle sink has a higher reliability

Propagation delay in PRPLC depends on the LLF of the available links. Higher LLF links towards the sink reduce the propagation delay for the packets to reach the sink. As observed in the network lifetime case, links towards the left ankle have higher chances and hence left ankle sink manages to receive the packets with lesser delay.

PDR is directly proportional to higher LLF in PRPLC protocol. Hence, the left ankle sink gets the higher score in getting the packets delivered reliably.

5 Conclusion

The paper deals with the optimum position of sink for the cost-effective routing protocol PRPLC. Five different positions are compared for the placement of on-body sink. The same sensor node is acting as the sink and sensor, and no separate node has been implemented as sink to avoid medical intricacies. From the results, it was found left ankle is an optimum position for the on-body sink in terms of network lifetime, latency and reliability. The links to the left ankle from other nodes have higher connection likelihood, and hence, left ankle is a better choice than right ankle as is used in the traditional PRPLC protocol. The delay for all the sink positions comes under the 20 ms deadline specified by the IEEE 802.15.6 standard [8, 9].

Declaration. The authors hereby declare that no human subjects or any such were involved in the study that requires approval from the ethical committee.

Acknowledgements This work is financially supported by the Visvesaraya PhD scheme for electronics and IT by MeitY, Government of India.

References

1. Sharma, R., Ryait, H.S., Gupta, A.K.: Impact of sink node position in the human body on the performance of WBAN. Int. J. Comput. Sci. Inf. Secur (IJCSIS) **14**(2) (2016)
2. Movassaghi, S., Abolhasan, M.: A review of routing protocols in wireless body area networks. J. Netw. **8**(3) (2013)
3. Liang, X., Balasingham, I., Byun, S.S.: A reinforcement learning based routing protocol with QoS support for biomedical sensor networks. In: First International Symposium on Applied Sciences on Biomedical and Communication Technologies (ISABEL'08), pp. 1–5, Oct 2008
4. Quwaider, M., Biswas, S.: Probabilistic routing in on-body sensor networks with postural disconnections. In: Proceedings of the 7th ACM International Symposium on Mobility Management and Wireless Access, pp. 149–158 (2009)
5. Nadeem, Q., Javaid, N., Mohammad, S.N., Khan, M.Y., Sarfraz, S., Gull, M.: SIMPLE: stable increased-throughput multi-hop protocol for link efficiency in wireless body area networks. In: IEEE 8th International Conference on Broadband and Wireless Computing, Communication and Applications (BWCCA'13), Compiegne, France (2013)
6. Quwaider, M., Biswas, S.: DTN routing in body sensor networks with dynamic postural partitioning. Ad Hoc Netw **8**, 824–841 (2010)
7. Sayrafian-Pour, K., Yang, W.B., Hagedorn, J., Terrill, J., Yazdandoost, K.Y.: A statistical path loss model for medical implant communication channels. In: 2009 IEEE 20th International Symposium on Personal, Indoor and Mobile Radio Communications (2009)
8. Yu, J., Park, L., Park, J., Cho, S., Keum, C.: CoR-MAC: contention over reservation MAC protocol for time-critical services in wireless body area sensor networks. Sensors **16**(5), 656 (2016)
9. Yang, J., Yoon, Y.: QoS of the BAN. Doc.: 15-07-0649-00-0ban. March 2007

A Machine Learning Approach to Find the Optimal Routes Through Analysis of GPS Traces of Mobile City Traffic

Shreya Ghosh, Abhisek Chowdhury and Soumya K. Ghosh

Abstract The rapid urbanization in developing countries has modernized people's lives in various aspects but also triggered many challenges, namely increasing carbon footprints/pollution, traffic congestion and high energy consumption. Traffic congestion is one of the major issues in any big city which has huge negative impacts, like wastage of productive time, longer travel time and more fuel consumption. In this paper, we aim to analyse GPS trajectories and analyse it to summarize the traffic flow patterns and detect probable traffic congestion. To have a feasible solution of the traffic congestion issue, we partition the complete region of interest (ROI) based on both traffic flow data and underlying structure of the road network. Our proposed framework combines various road features and GPS footprints, analyses the density of the traffic at each region, generates the road-segment graph along with the edge-weights and computes congestion ranks of the routes which in turn helps to identify optimal routes of a given source and destination point. Experimentation has been carried out using the GPS trajectories (T-drive data set of Microsoft) generated by 10,357 taxis covering 9 million kilometres and underlying road network extracted from OSM to show the effectiveness of the framework.

Keywords GPS trace · OpenStreetMap (OSM) · Classification · Traffic

1 Introduction

The staggering growth of urban areas and increased populations drive governments for the improvement of public transportation and services within cities to reduce

S. Ghosh (✉) · A. Chowdhury · S. K. Ghosh
Department of Computer Science and Engineering, Indian Institute of Technology, Kharagpur, India
e-mail: shreya.cst@gmail.com

A. Chowdhury
e-mail: abhisekchowdhury9@gmail.com

S. K. Ghosh
e-mail: skg@iitkgp.ac.in

P. K. Sa et al. (eds.), *Recent Findings in Intelligent Computing Techniques*, Advances in Intelligent Systems and Computing 708,
https://doi.org/10.1007/978-981-10-8636-6_7

carbon footprints and make a sustainable urbanization [1, 2]. As traffic congestion poses a major threat to all growing or developing urban areas, analysing traffic flows and detecting such issues in urban areas are strategically important for the improvement of people's lives, city operation systems and environment. Noticeably, with the availability of GPS data and advances in big data and growing computing and storage power, urban planning, capturing city-wide dynamics and intelligent traffic decisions have attracted fair research attentions in last few decades. These lead to various interesting applications in navigation systems and map services like route prediction, traffic analysis and efficient public transportation facilitating people's lives and serving the city [3, 4]. Also with the advent of emerging technologies like self-driving car, we need innovative technologies and an efficient and fault-tolerant traffic management system which can automatically and unobtrusively monitor traffic flow and detect any anomaly condition like traffic congestion and road blockage. Prior research on GPS traces analysis [5], traffic flow analysis and prediction has focused on the analysis of individual's movement patterns [4]. Few works have studied traffic flow conditions in a city region [6]. But, traffic flow prediction from individual level becomes difficult due to individual's life patterns and randomness of human movement nature. To this end, we have addressed the problem of traffic congestion detection from the city-wide vehicular movements and also analysed various road network features which were missing in the existing studies. We turn our attention to design a smart and automatic system that will detect the congestion in real time and subsequently manage it efficiently to ensure smooth traffic flow. To address the aforementioned traffic congestion problems, we propose a framework which involves (i) segmentation of road network and creating buffered regions of different road types, (ii) generating trajectory density function at each road type and analysing probability distribution of traffic flow, (iii) selection of traffic congestion features of a region and threshold depiction and (iv) build road-segment graph of the region and analyse congestion ranking of the paths. One of the real-life application scenario is rerouting of vehicles to avoid more traffic problems or even a recommendation system can be built which will notify the traffic condition of the major intersection points of a road network to carry out intelligent routing decisions. The remainder of the paper is organized as follows: Section 2 illustrates our proposed congestion detection and management framework; Section 3 describes the basic features of the visualization model and represents the obtained result; and finally, Sect. 4 concludes the paper with a highlight on the scope of future work.

2 Architecture of the Proposed Framework

In this section, we present the overall architecture of the proposed framework. Figure 1 depicts various modules of the framework. *GPS data pre-processing* module involves extracting road map of the region of interest (ROI) along with the OpenStreetMap (OSM) road features of the region. To analyse large-scale traffic data of any city region, a systematic storage method of GPS data with all contextual

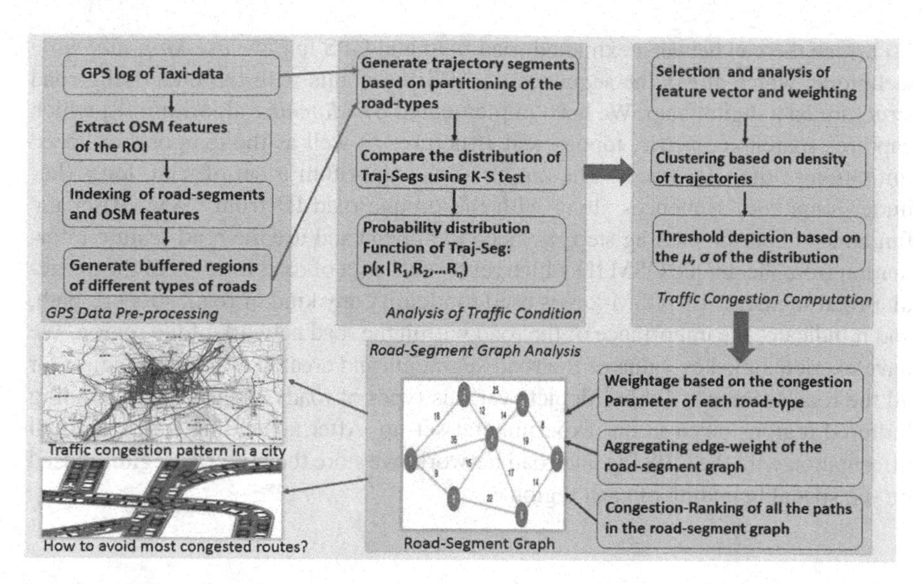

Fig. 1 Architecture of the framework

information and road network is required. To this end, we used a storage schema to efficiently access useful information for our application. The *Analysis of Traffic Condition* and *Traffic Congestion Computation* modules generate trajectory segments from the raw GPS log of the vehicles and compute the probability density function of the GPS footprints in each of the road types. Based on the computed density function of the historical GPS log, threshold parameter of the congestion in each type of roads is determined. In the next module, namely *Road-Segment Graph Analysis*, road graph is modelled and edge-weights are assigned. Based on the congestion ranking of each route, a probable less-congested path from a given source and destination can be identified. The framework also summarizes the traffic congestion pattern of the region of interest from the GPS footprints and road-feature set. The details of the modules are described in the following sections of the paper.

2.1 GPS Data Pre-processing

We aim to extract optimal routes analysing GPS traces of the city traffic. A typical GPS log consists of a series of timestamped latitude and longitude: $< lat_i, lon_i, t_i >$. To tackle the traffic congestion issues, the timestamped positions of traffic are not sufficient, underlying road network and road features play an important role. Therefore, we extract underlying road network of the region of interest from OpenStreetMap (OSM) [7]. From the GPS data log, we create the bounding box which covers all GPS traces and extract the road network (.shp file) from the OSM map.

To bridge the gap between extracted road map and GPS log, we use *Map-matching* technique, which aligns the sequence of GPS log points with the underlying road structure on a digital map. We have implemented *ST-Matching* algorithm [8] which captures spatial geometric, topological structures as well as the temporal or speed limitation of the GPS traces. The output of the algorithm is set of <lat, lon> (latitude, longitude) sequences along with the unique road-ID from OSM. After the fundamental pre-processing step, we need to append and use the road-feature information from the unique OSM ID which represents a set of road features of the particular road segment. The OSM-key is used to identify any kind of road, street or path, and it indicates the importance of the road within the road network of the region. We have extracted the key value of the road segments and created a buffered region for all the road segments. Table 1 depicts various types of roads and the corresponding buffered regions taken in our experimental set-up. After appending the contextual information with the GPS log and road network, we store the buffered regions based on the spatial bounding of each region.

Table 1 Road network feature and buffered regions

OSM-key/value	Road network feature	Buffered region
Motorway	Major divided highway generally having two or more running lanes and emergency hard shoulder	$l_w + 3.5$ km
Trunk	The most important roads in a country's system and may not have a divided highway	$l_w + 3$ km
Primary	A major highway which links large towns, in developed countries normally with two lanes	$l_w + 2.5$ km
Secondary	A highway which forms linkage in the national route network.	$l_w + 2$ km
Tertiary	It connects or links between smaller settlements, and local centres also connects minor streets to major roads	$l_w + 1$ km
Residential	Roads accessing or around residential areas, street or road generally used for local traffic within settlement	$l_w + 1$ km
Service	Access to a building, service station, beach, campsite, industrial estate, business park, etc.	$l_w + 0.5$ km

2.2 Analysis of Traffic Condition

Road segmentation (i.e. classification of various road segments in the data set to 'highway' or 'local') is followed by feature selection. Feature selection or extraction is one of the most important for selecting a subset of relevant features for the construction of a predictive model. In our problem, probable traffic congestion is determined by average velocity and number of vehicles in a particular bounding box or region or increased vehicle queuing. It has been considered that congestion may occur if the number of cars in a bounding box is greater than a particular threshold value and the average velocity of cars is less than a particular threshold value. Threshold values vary for different types of buffered regions like highways and locality region. The challenges are to predict or detect possible congestion from traffic flow. In addition, different types of buffered regions may have different parameters for traffic condition detection due to the different capacities of the regions. Traffic congestion (T_c) can be represented as in Eq. 1, where v, n, l_w are average velocity, number of vehicles and width of the road, respectively. α and β are normalizing constants.

$$T_c = \frac{\alpha}{v} + \frac{n}{l_w} \times \beta \quad (1)$$

To model traffic flow of a region, we need to summarize road conditions in different time interval. It is quite obvious that traffic scenario at morning(5.00–7.00 am) will significantly differ in peak times (9.00–11.00 am). Therefore, we need to model the time-series data $x = (x_0, x_1, \dots, x_n)$, where each x_i represents T_c value at t_i timestamp. We divide each day in equal partition of 1 h and carry out the analysis. After plotting the histogram of the above time-series data, we observe that Gaussian distribution can approximate the histograms. Using parametric method of distribution fitting, we estimate the μ and σ^2 of the distribution from the mean and standard deviation of the data set, where mean, $m = \frac{\sum T_c}{t_n}$; standard deviation, $s = \sqrt{\frac{1}{t_n - 1} \times \sum (T_c - m)}$. In this section, we demonstrate the process of analysing GPS log along with road features and summarize the traffic movements by defining a probability distribution for seven different regions (depicted in Table 1) of the road network.

2.3 Traffic Congestion Computation

For the traffic congestion computation, two sub-modules are required. First, we need to generate the normal traffic probability distribution from the historical GPS log; next, we need to carry out a classification algorithm for classifying the regions into congested and non-congested regions. Traffic congestion on a road segment means disrupting the normal traffic flow of the region. Hence, to detect such condition we depict congestion threshold parameter which will denote probable traffic bottleneck

of fluctuations in normal traffic conditions. We use *k-nearest neighbour* classifier for this learning. It is based on learning by analogy, i.e. it compares a given test tuple with other training tuples having similar feature set. The training tuples are described by a set of feature attributes. *'Closeness'* or *'Similarity'* is defined in terms of a distance or similarity metric. We have used Euclidean distance measure given two points or tuples, say, $X_1 = (x_{11}, x_{12}, x_{13} \dots, x_{1n})$ and $X_2 = (x_{21}, x_{22}, x_{23} \dots, x_{2n})$, is $dist(X_1, X_2) = \sqrt{\sum_{i=1}^{n} (x_{1i} - x_{2i})^2}$ where each entry of the tuple represents the feature values of the elements denoting the road traffic condition at different timestamp values. We use $k = 3$, where neighbourhood is defined based on the road network feature ranking in Table 1. Algorithm 1 provides an abstract view of the process, where based on the test data set (road segments), road-feature rank is extracted and neighbourhood is defined. Then, based on the k-NN classifier algorithm, we classify the test data set by comparing it with training tuples. Here, we specify p = 5, i.e. congestion ranking is carried out, where p = 1 denotes no congestion, and p = 5 denotes highest level congestion.

2.4 *Road-Segment Graph Analysis*

Road-Segment Graph: *RGraph* $= \{(V, E) | 1 < v_i < |V|, 1 < e_i < |E|\}$, where each node $v_i \in |V|$ denotes road intersection point of the underlying road network, and each edge of the graph $e_i \in |E|$ denotes existing roads between two or more intersecting points. Each node $v_i \in |V|$ stores information about the intersection point: [*ID*, *Name*, *CLevel*]. CLevel represents congestion level at a particular intersection point which is derived from the data distribution of the connected edges. It is observed that traffic congestion normally follows a regular pattern which can be detected from the traffic density function and incorporated in the *RGraph* structure. We use a cost function to feed in the traditional shortest path algorithm. One of the most popular and cost efficient methods to determine single source shortest path is Dijkstra Algorithm. Specifically, the cost function is

$$f(n) = g(n) + h(n) \tag{2}$$

The algorithm finds out the value of g(n), i.e. the cost of reaching from initial node (entry point) to n. A heuristic estimate (function h(n)) is carried out to reach from on to any goal node or exit points. In our approach, g(n) is calculated by aggregating edge-weights of road network and h(n) is determined by calculating traffic congestion data.

Algorithm 1: Traffic condition detection based on density analysis and congestion threshold parameter

Input: Traffic density function of various regions: $\mathcal{N}(\mu, \sigma|R)$, congestion threshold parameter: *Thres*
Output: Road-segment Graph: *RGraph*, Congestion level: *C*
initialization: Create *RGraph*(V, E) :V represents intersection of road-links, E: road segments of the network;
while $e_i \in E$ **do**
  r=ExtractRoadType(eId_i);
  for *all training data set available* **do**
    T=Extract $f(x|r-1, r, r+1)$;
  end
  for *all time intervals n in the traffic time series and* $t \in T$ **do**
    $p_i = f(x|n_i, r)$;
    $s = compareTraffic(p_i, f(x|n_i, t))$;
    if $s > Thresh$ **then**
      congest=1;
      C=(s-Thresh)/p;
      Append(C, *RGraph*);
    else
      congest=0;
      C=1 ;
      Append(C, *RGraph*);
    end
  end
end

3 Experimentation

3.1 Data set

Beijing T-drive Data Set: The data set contains the GPS trajectories of 10,357 taxis during the period of 2 February–8 February 2008 within Beijing. The data set contains huge amount of GPS points, about 15 million and a total distance covered by the trajectories reaches to 9 million kilometres [3, 4, 9].
OSM Map: OpenStreetMap stores physical features on the ground (e.g. roads or buildings) using OSM-tags attached to its basic data structures (its nodes, ways and relations). Each tag depicts a geographic attribute of the feature being represented by that specific node, way or relation [7].

3.2 Results and Discussion

After the data pre-processing steps and threshold depiction steps are completed, we partition the data set in 'Training Data Set' and the remaining as 'Test Data Set'. In our experiment, 1195 examples constituted the 'Training Data Set'. The learning model is trained using the concept of k-nearest neighbours with training examples from the 'Training Data Set'. Specifically, model is run for 25 iterations and accuracy of the model (in terms of % of examples correctly classified) is calculated. Overall accuracy of the model is 93%. Our visualization model, developed with PHP, Python, JQuery and AJAX, imitates traffic congestion on a predefined road map of the ROI. We have used OSM Leaflet API to simulate the map. On the back-end system, we have stored the data in PostgreSQL database and retrieve the data using spatial extension of psql. The model has salient three features: Feature I performs the pre-processing steps, including generation of bounding box and extracting road features and partitioning the road network into various buffered regions. Feature II depicts the level of congestion (traffic flow condition) in different regions of the network. In Feature III, the system recommends alternative routes based on the traffic flow condition on the road. Few visualization results are shown in Fig. 2.

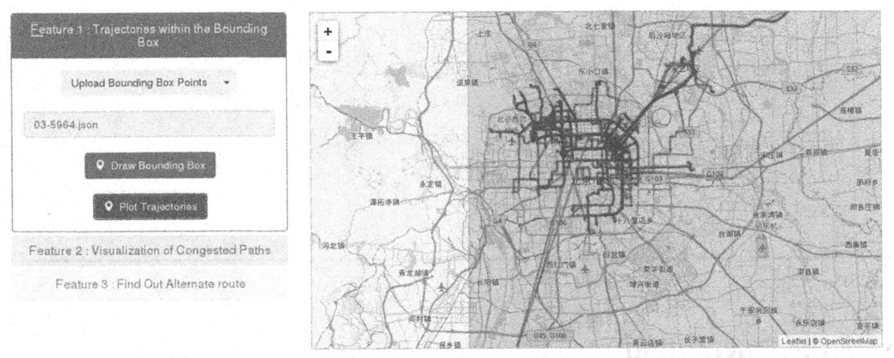

(a) Bounding Box/Segmentation Region for Congestion Detection

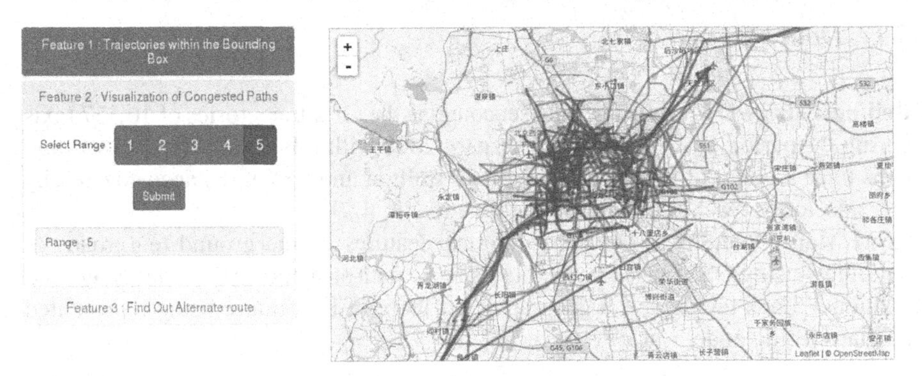

(b) Visualization of congestion Detection

Fig. 2 Snapshots of the visualization framework

4 Conclusion and Future Work

In this paper, we aim to detect optimal route in a city analysing the city-wide traffic flow. It involves segmentation of road network into different types of roads and analysing traffic condition in each of them. It models the regular traffic patterns in a city region from the mobile traffic data and detects anomalous conditions like fluctuation of road traffic from the data. Finally, the model classifies traffic condition in different categories of congestion level and few visualization results are shown. In future, we would like to build a real-time recommendation system, which can adaptively recommend the optimal route to the users. Further, we would like to enhance the accuracy and efficiency of the proposed model by using different indexing or storage schema and deploying advanced machine learning techniques.

References

1. Zheng, Y., et al.: Urban computing with taxicabs. In: Proceedings of the 13th International Conference on Ubiquitous Computing. ACM (2011)
2. Zheng, Y., et al.: Urban computing: concepts, methodologies, and applications. ACM Trans. Intell. Syst. Technol. (TIST) **5**(3), 38 (2014)
3. Yuan, J., Zheng, Y., Zhang, C., Xie, W., Xie, X., Sun, G., Huang, Y.: T-drive: driving directions based on taxi trajectories. In: Proceedings of the 18th SIGSPATIAL International Conference on Advances in Geographic Information Systems, GIS'10, New York, NY, USA, pp. 99–108. ACM (2010)
4. Zheng, Y., et al.: Learning transportation mode from raw gps data for geographic applications on the web. In: Proceedings of the 17th International Conference on World Wide Web. ACM (2008)
5. Ghosh, S., Ghosh, S.K.: THUMP: semantic analysis on trajectory traces to explore human movement pattern. In: Proceedings of the 25th International Conference Companion on World Wide Web 2016, Montreal, Canada, Apr 11, pp. 35–36. International World Wide Web Conferences Steering Committee
6. Hoang, M.X., Zheng, Y., Singh, A.K.: FCCF: forecasting citywide crowd flows based on big data. In: Proceedings of the 24th ACM SIGSPATIAL International Conference on Advances in Geographic Information Systems. ACM (2016)
7. https://www.openstreetmap.org/
8. Lou, Y., et al.: Map-matching for low-sampling-rate GPS trajectories. In: Proceedings of the 17th ACM SIGSPATIAL International Conference on Advances in Geographic Information Systems. ACM (2009)
9. Yuan, J., Zheng, Y., Xie, X., Sun, G.: Driving with knowledge from the physical world. In: The 17th ACM SIGKDD International Conference on Knowledge Discovery and Data Mining, KDD'11, New York, NY, USA. ACM (2011)

Conclusion and Future Work

References

Zigbee-Based Wearable Device for Elderly Health Monitoring with Fall Detection

Syed Yousuff, Sugandh Kumar Chaudary, N. P. Meghana, T. S. Ashwin and G. Ram Mohana Reddy

Abstract Health monitoring devices have flooded the market. But there are very few that cater specifically to the needs of elderly people. Continuously monitoring some critical health parameters like heart rate, body temperature can be lifesaving when the elderly is not physically monitored by a caretaker. An important difference between a general health tracking device and one meant specifically for the elderly is the pressing need in the latter to be able to detect a fall. In case of an elderly person or a critical patient, an unexpected fall, if not attended to within a very short time span, can lead to disastrous consequences including death. We present a solution in the form of a wearable device which, along with monitoring the critical health parameters of the elderly person, can also detect an event of a fall and alert the caretaker. We make use of a 3-axis accelerometer embedded into the wearable to collect acceleration data from the movements of the elderly. We have presented two algorithms for fall detections—one based on a threshold and the other based on a neural network and provided a detailed comparison of the two in terms of accuracy, performance, and robustness.

S. Yousuff (✉) · S. K. Chaudary · N. P. Meghana · T. S. Ashwin
G. Ram Mohana Reddy
Department of Information Technology, National Institute of Technology Karnataka, Surathkal, India
e-mail: yousuff145@gmail.com

S. K. Chaudary
e-mail: sugandh.kumar4@gmail.como

N. P. Meghana
e-mail: meghananpmegha@gmail.com

T. S. Ashwin
e-mail: ashwindixit9@gmail.com

G. Ram Mohana Reddy
e-mail: profgrmreddy@gmail.com

P. K. Sa et al. (eds.), *Recent Findings in Intelligent Computing Techniques*,
Advances in Intelligent Systems and Computing 708,
https://doi.org/10.1007/978-981-10-8636-6_8

Keywords Fall detection · Health monitoring · Wearable device · OpenHAB
Feature maps · Neural networks

1 Introduction

Elderly persons are at a constant risk of falling because of reduced physical strength and balance. There are several commercially available wearable devices for health monitoring in the market such as Mi-Band, Fitbit. But almost all of them are meant for fitness tracking and do not specifically cater to the needs of the elderly population. Several approaches have been proposed for fall detection. Mastorakis et al. proposed a Kinect-based fall detection system which uses a 3-D bounding box to measure velocity and thus requires no pre-knowledge of the scene (i.e., floor) [1]. Several other algorithms that make use of skeletal tracking feature of Kinect have also been proposed [2–4]. But the inherent limitation of this approach is that it requires the user to be in the line of sight of Kinect at all times. Other approaches try to overcome this limitation by using a network of low-resolution cameras to reconstruct a 3-D scene to detect falls [5–7]. But this method requires extensive computation and high installation cost. Lina Tong et al. [8] proposed a Hidden Markov Model-based method using a 3-axis accelerometer. Jin Wang et al. [9] proposed a Body Sensor Network that includes a Cardiotachometer along with a 3-axis accelerometer to detect falls. [10] proposes a signal processing technique for fall detection using RADAR. We propose a wearable device solution for fall detection that uses an accelerometer to gather acceleration data from the movements of the elderly. We have used two approaches for detecting a fall from the acceleration data—one based on a simple threshold and the other that uses an artificial neural network. We have compared the two approaches in terms of accuracy, computational requirement, and real-time performance.

Our key contribution in this paper is the proposed set of features that can be extracted from the acceleration stream of data to form a meaningful training set for a neural network. With these extracted features, we have shown that a trained Artificial neural network can provide both sensitivity and specificity of greater than 97%. Apart from this, we have proposed a novel method of using a Zigbee-based device that increases the battery life of the wearable device and unlike most Bluetooth-based devices in the market does not require constant pairing with a smartphone.

2 Methodology

A digital motion sensor consisting of a 3-axis accelerometer and a 3-axis gyroscope is embedded in a wearable device. The sensor readings are read through an I2C interface into an Atmega328 low-power microcontroller. The wearable also embeds a heart rate sensor and a body temperature sensor which are simultaneously sampled by the microcontroller. The microcontroller which is connected to an Xbee 2.0 Radio module transmits this data to another Xbee 2.0 Radio that is configured as a coordinator and connected to the central hub. A Raspberry Pi is used as a central hub that runs OpenHAB Home Automation Framework. The wearable device is powered using a 1000 mAh Li-Polymer battery. The hub is connected to the Internet and the OpenHAB framework provides for a user-friendly Web interface as well as an Android and IOS application through which the user can interact with the system. The system architecture is as shown in Fig. 1.

To distinguish a fall from other routine events, we identified six events that an average person is frequently exposed to. They are walking, running, sitting/standing, transition, idle vibrations, sudden motions/jerks. Hence, our dataset contains a total of seven classes (including fall). We collected 3-axis acceleration data from routine

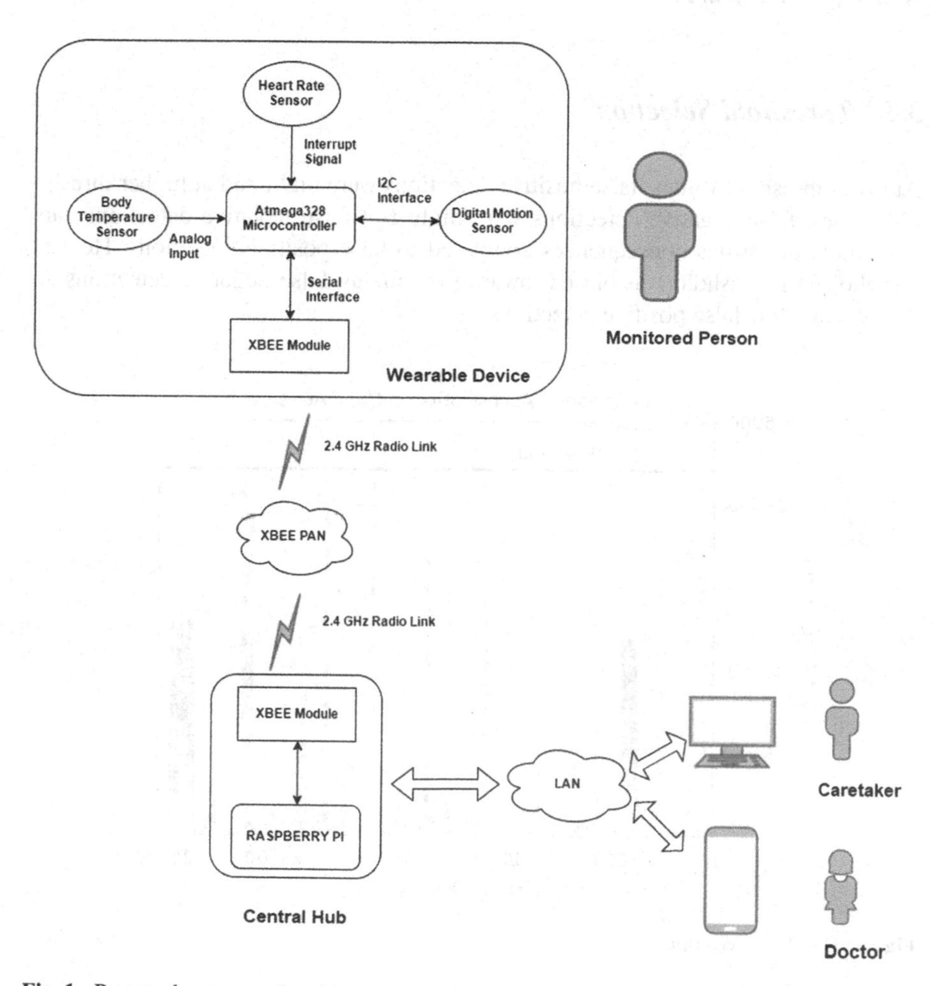

Fig. 1 Proposed system and architecture

actions of three different persons wearing the device in turn. Each sample was logged to a file along with a tag representing one of the above six activities that the person was performing. Those participating in the data collection were asked to mimic a natural fall in different ways. The data for the same was also collected and logged with a tag representing a fall.

3 Fall Detection Algorithm—Threshold Based

From the collected data, we observed that the net acceleration for all activities except a fall was well below a threshold value. This can be easily inferred from the acceleration plot shown in Fig. 2.

3.1 Threshold Selection

At lower threshold values, false positive detections were high, and at higher threshold values, false negative detections were high. But false negative detections can have more disastrous consequences compared to false positive detections. Hence, our choice of threshold was biased toward minimizing false negative detections at the cost of a few false positive detections.

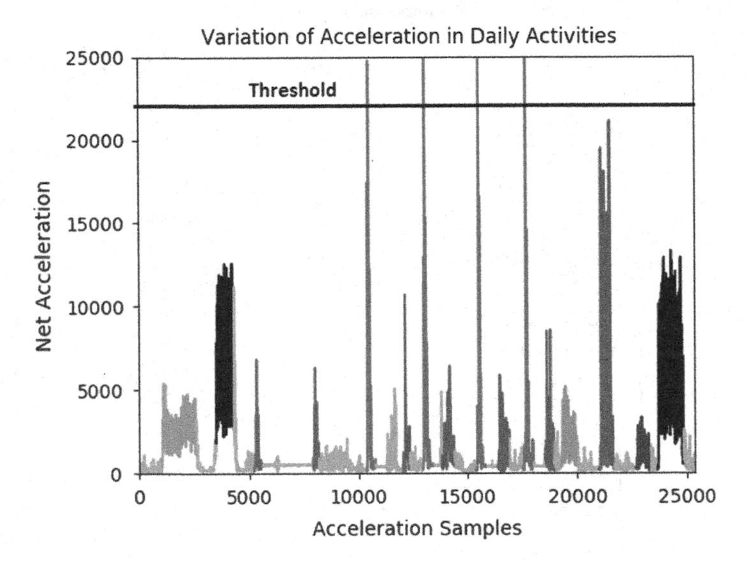

Fig. 2 Threshold detection

4 Fall Detection Algorithm—Artificial Neural Network

Though threshold method can perform fairly well in most circumstances, a well-designed machine learning approach can give better results since it can identify patterns in the sensor data rather than just comparing it with a threshold. But for a neural network to give good results, it is important to construct a meaningful feature set from the data. In the following section, we discuss the set of features that were extracted from the training data.

4.1 Feature Extraction

For each datapoint, we extracted features from a window of consecutive net acceleration values from the collected data. The window size was set to 300 as it was large enough to contain most events. The extracted features were

- **Maximum Difference**: During a fall, the maximum difference between two net accelerations within a window will tend to be high compared to the same during other events.
- **Maximum Slope**: During a fall, we noticed a spike in net acceleration which is characterized by a steep slope. Thus, maximum slope within a window tends to be high in the event of a fall.
- **Variance**: During walking, running, or vibrations, the net acceleration oscillates around a mean position resulting in low variance within a window. But in case of a fall, the mean acceleration is high and most of the neighboring points are far away from the mean resulting in high variance.
- **Maximum Acceleration**: This feature works as the threshold method since a fall is characterized by a higher maximum acceleration compared to other events.

4.2 Feature Maps

Four features were extracted from every sample window. The feature maps plotted for the training dataset are shown in Fig. 3.

5 Results and Analysis

5.1 Threshold Approach

With a good threshold, the method works fairly well in most circumstances. It is computationally highly inexpensive and can be easily implemented within a resource-constrained wearable device. The detection was accurate in case of all events except when the user performed sudden unexpected actions. Thus, the method tends to raise

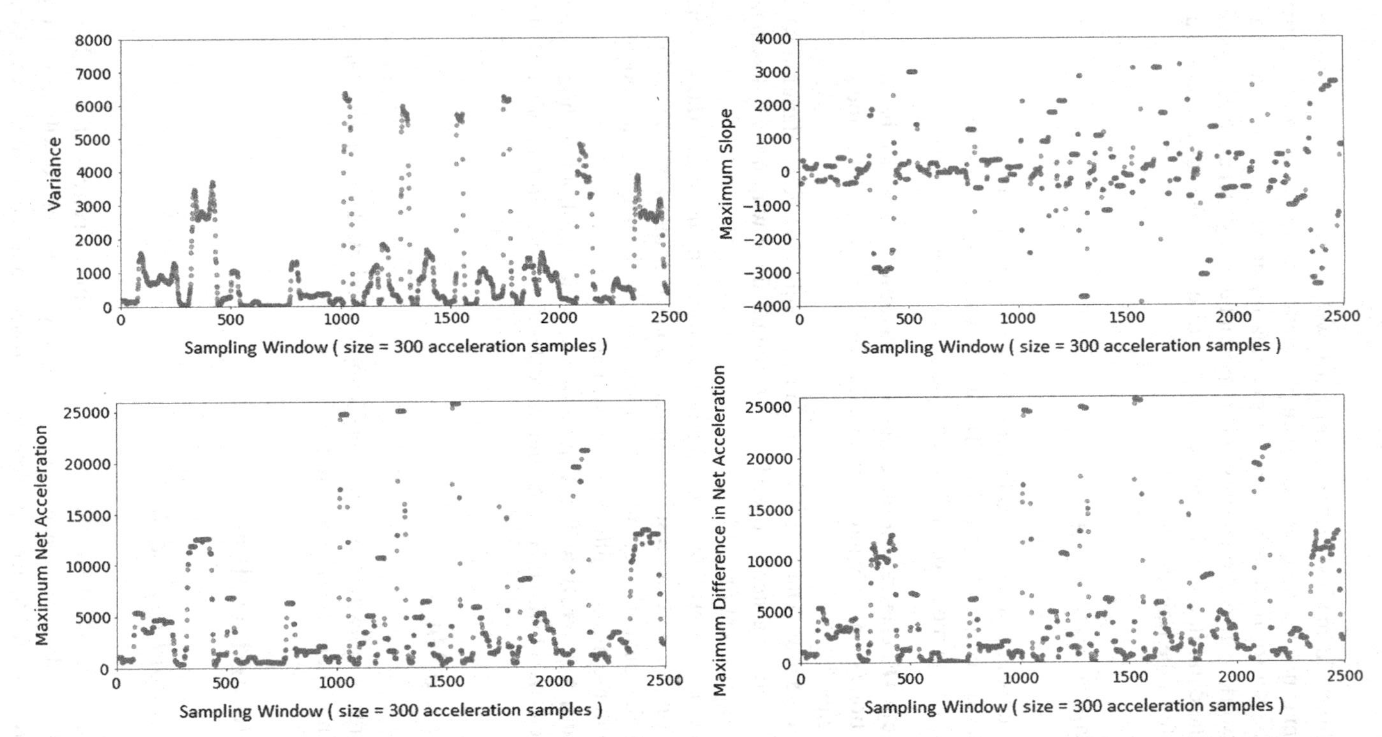

Fig. 3 Feature maps: (In clockwise direction from top left) Variance, Slope, Maximum Net Acceleration Difference and Maximum Net Acceleration

more false alarms. But this shortcoming can be overcomed by adding an alarm diffuser, a simple switch on the wearable device which the elderly can use to deactivate the false alarm.

5.2 *Neural Network Approach*

The NN is capable of identifying patterns in the data stream and accurately detecting a fall. The network is able to differentiate between a fall and all other events with both specificity and sensitivity greater than 97%.

But computation requirement is high with this approach as compared with the threshold approach. This is not a problem in our system as the acceleration data is being streamed to the central Zigbee hub and all computations are being performed there. The classifier being resource intensive runs on the central hub. Thus, the sensor data has to be streamed from the wearable device to the hub via the Zigbee link. The Zigbee link provides a maximum throughput of 250 kbps which is a bottleneck for data transfer. Also, a testing data point requires 300 sensor samples (window size). The sensor can sample at 200 Hz causing a delay of 1.5 s.

6 Conclusion and Future Work

We presented a novel design for a health monitoring system which is uniquely applicable to hospitals or old age homes. Our device is different from most existing devices in that it does not pair with a phone, but connects with a central hub, thus relieving the user from carrying a phone at all times. It also provides authorized access to caretakers through a user-friendly application. We presented a detailed comparison of two most commonly used approaches for fall detection in terms of accuracy, performance, resource requirements, and robustness. The power requirement of the wearable device can be reduced by efficient use of microcontroller sleep cycles. The accuracy can be improved by identifying and extracting more features from the dataset which can meaningfully distinguish between a fall and other events.

7 Declaration

Authors have obtained all ethical approvals from appropriate ethical committee and approval from the subjects involved in this study.

References

1. Mastorakis, G., Makris, D.: Fall detection system using Kinects infrared sensor. J. Real-Time Image Process. **9**(4), 635–646 (2014)
2. Nizam, Y., Jamil, M.M.A., Mohd, M.N.: A depth image approach to classify daily activities of human life for fall detection based on height and velocity of the subject. In: International Conference on Movement, Health and Exercise. Springer, Singapore (2016)
3. Nizam, Y., Mohd, M.N.H., Jamil, M.M.A.: Classification of human fall from activities of daily life using joint measurements. J. Telecommun. Electron. Comput. Eng. (JTEC) **8**(4), 145–149 (2016)
4. Kwolek, B., Kepski, M.: Fuzzy inference-based fall detection using kinect and body-worn accelerometer. Appl. Soft Comput. **40**, 305–318 (2016)
5. Skubic, M., et al.: Testing non-wearable fall detection methods in the homes of older adults. In: 2016 IEEE 38th Annual International Conference of the Engineering in Medicine and Biology Society (EMBC). IEEE (2016)
6. Debard, G., et al.: Camera-based fall detection using real-world versus simulated data: how far are we from the solution? J. Ambient Intell. Smart Environ. **8**(2), 149–168 (2016)
7. Senouci, B., et al.: Fast prototyping of a SoC-based smart-camera: a real-time fall detection case study. J. Real-Time Image Process. **12**(4), 649–662 (2016)
8. Tong, L., et al.: HMM-based human fall detection and prediction method using tri-axial accelerometer. IEEE Sens. J. **13**(5), 1849–1856 (2013)
9. Wang, J., et al.: An enhanced fall detection system for elderly person monitoring using consumer home networks. IEEE Trans. Consum. Electron. **60**(1), 23–29 (2014)
10. Amin, M.G., et al.: Radar signal processing for elderly fall detection: the future for in-home monitoring. IEEE Signal Process. Mag. **33**(2), 71–80 (2016)

A Colored Petri Net Based Modeling and Analysis of Packet Dropping Scenario for Wireless Network

Moitreyee Dasgupta and Sankhayan Choudhury

Abstract Losing data is one of the major concerns of wireless communication network. Corrective measure cannot be taken until the specific cause behind the activity is detected. Thus, packet dropping, manifestation of losing data, demands a detailed analysis for identification of the corresponding causes. In general, the packet dropping may occur due to the congestion, malicious activities, and random errors. In this work, the overall packet dropping scenario is modeled through Colored Petri Net to figure out and segregate the reasons for the same. The proposed two-layer CPN model expresses all possible situation of packet dropping. The lower layer analyzes the given scenario, while the top layer reflects the dropped packets segregated as per underlying causes. The model also provides the information about the affected zones and nodes that in turn lead the designer to take the proper corrective measure. The exhaustive simulation has been performed to validate the proposed model.

Keywords Packet drop analysis · Colored Petri Net · Anomaly detection

1 Introduction

Throughput is one of the measures of performance in a wireless network, and packet dropping is the activity that reduces the throughput. Packet drop happens due to many reasons. Wireless medium is a typical reason for such drop. In general, due to the behavior of the routing protocols (e.g., due to broadcasting nature of a protocol), always there will be a drop of packets and these are considered as valid drops. But exceptional situation with a substantive amount of packet dropping may occur due to congestion or some intended malicious activities. This phenomenon seriously affects the QoS, and thus, some recovery mechanism is needed to restore the QoS value at the desired level. The prerequisite for the said recovery mechanism is to apprehend

M. Dasgupta (✉) · S. Choudhury
Calcutta University, Technology Campus, Salt Lake City, Kolkata 700098, India
e-mail: moitreyee77@gmail.com

S. Choudhury
e-mail: sankhayan@gmail.com

P. K. Sa et al. (eds.), *Recent Findings in Intelligent Computing Techniques*,
Advances in Intelligent Systems and Computing 708,
https://doi.org/10.1007/978-981-10-8636-6_9

the cause of packet dropping. An analysis of such packet dropping environment is the way to find the causes, and accordingly, necessary corrective measure can be taken.

Always for a critical system, it is important to ensure that the solution to be deployed should be validated beforehand. The traditional way of checking correctness has some limitations. It requires correct annotations and proof of their correctness through human intervention. Moreover, any modification requires further re-evaluation and proof of the same. An alternative approach is to express the scenario in form of a formal model. The model can be verified using formal analysis such as behavioral verification.

Petri Net based modeling of the activities in a network is a well-known approach for discovering the insights through necessary analysis, and it is being used by the researchers for quite some time [1–3]. A method named TCPN for modeling and verification of reputation-based trust systems using Colored Petri Nets is presented for P2P network [1]. In [2], CPN has been applied to discover and analyze the vulnerabilities of session initiated protocol (SIP) against denial-of-service (DoS) attacks in Voice-over-IP (VoIP) scenario. CPN-based modeling also used to model and analyze Diffie–Hellman algorithm [3], stop-and-wait protocol [4], and many other diversified areas of application. A real prototype or implementation may suffer from the fact that some unexpected actions may happen under adverse situations in reality and the correctness of the system can be affected [5]. As an example, significant packet losses happen rarely in ideal cases, but due to substantive packet drop at a given time, the network may face catastrophic consequences. The modeling itself should cope with these unexpected behaviors and consequently will be able to offer the insights after necessary analysis. In this paper, an attempt has been made to model the packet dropping through Colored Petri Net. The proposed model is validated for all possible cases. The model is analyzed extensively and is able to provide an insight regarding the cause of packet dropping.

2 Background

Packet loss leads to one of the major performance issues of the network. It decreases throughput severely when the network experiences significant packet drops. The following three [13] causes are among the most common types a network may encounter [6]: (1) link congestion, which happened due to accumulation of network data on a device queue for transmission, failing which results in packet/data loss; (2) packet drop attack, which happened due to the launch of DoS attacks by a malicious user/device; and (3) drop due to protocol behavior, for example, due to expiration of packet lifetime (TTL = 0) or may be when also network nodes discarded the duplicate packets and packets received due to broadcasting and not relevant to the node. These drops do not affect the throughput due to valid protocol behavior. We have also categorized drops at receiver or sender node as a valid drop. Significant effort [7–11] has been made to solve the issue. As described before, unexpected packet drop mainly happens firstly due to congestion and secondly denial-of-service (DoS) attacks within the network. Both congestions and network security have got individual attention from the researchers for more than a decade. Based on the congestion control mechanism used, systems react immediately when it detects significant packet drops and takes measures such as the use of an alternative path for information flow or some data flow reduction mechanism to avoid data loss. Regular TCP reduces blindly its congestion window by half when a packet drop is detected without considering that the reason for packet drop may not be congestion rather a security threat to the system. The same is applicable for all DoS control security mechanisms too. The result is even more degradation of system throughput. In this paper, we have considered all the above-mentioned categories while modeling packet drop scenario of network. Between major two, while the first one is the result of congestion, the latter is due to DoS attack. There is no such solutions to the best of our knowledge which can handle both the causes together. In our proposed CPN model, we have identified the reasons behind packet dropping once it is detected by the system. Once done, any standard congestion control or security mechanism can be applied based on the identified result.

3 Proposed Model

Let us describe the scenario first. The entire deployment area is supposed to be divided into zones where each zone is marked with a unique color code. Nodes are mobile and hold the same color of the zone it belongs to. The packets generated by the nodes from a typical zone are marked with a specific color. The packets are moving from sender node to destination through intermediate nodes. While traveling, the packets will change the color with hop if required and always reflect the color of its last traversed node. An intermediate node may transfer or drop the packets. There is always a small number of packet drops (due to typical behavior of protocol), which is considered as valid. Sometimes, packet dropping may be exceptional and that may be due to congestion or some unwanted intentional dropping by a suspicious node. Each and every node in the network would maintain a data structure (table) locally. All intermediate nodes (routers) create an entry for each packet that it overhears whenever a packet gets drops. Every time it would detect a packet drop, it would make an entry of packet id (p_id) and its color and increased the dropped packet count by one corresponding to the node id (n_id) from which the packets get dropped. If no such entry exists, then the node will create one and set the packet drop count value to one. Discarding of packets at the source and/or receiver node is due to protocol behavior, and thus, we have categorized that as a valid drop. Once the dropped packets reached the threshold value either for the single node or a zone, it would analyze the dropped packets and notify to the zonal head the reason of the same. We have also assumed that neither the source node nor the receiver node discards the packets maliciously.

We have proposed a CPN-based multilayer model to represent the above scenario. One of the major reasons we have chosen CPN as our modeling tool is because of its features like support of datatypes as a high-level language construct. We have used the CPN datatypes (colors) to distinguish the tokens of different zones. This helps us to rightfully detect the affected zone from which the packets are getting dropped during transmission. Other versions of Petri Net do not support data types; thus, tokens (nodes and packets) are of same types only and cannot be separated from each other. The model is depicted in Figs. 1 and 2. The top layer (layer 1) of the model deals with the movement of the transmitted packets and initially tries to distinguish between valid and invalid drops. The lower layer (layer 2) examines the packets as marked invalid drop and finds out the cause behind the dropping through dropped packet analysis. In the top layer, there are eight places named SENDER, RECEIVER, INTERMEDIATE_NODE, DROP, VALID_DROP, INVALID_DROP, CONGESTION, and SUSPECIOUS_NODE. The top layer is associated with seven transitions among which transitions T_1 to T_5 and T_7 are the immediate transitions, while transition T_6 is a substitution transition for layer 2. The transition T_6 would be responsible for detecting the reason for unnatural packet drop. The SENDER place determines the initial status of the network. The number of tokens in this place indicates the number of packets generated by the sender nodes. Each packet is represented with the attributes (p_id, n_id, c, s_id, d_id, TTL) . Here, the associated color (c) of each

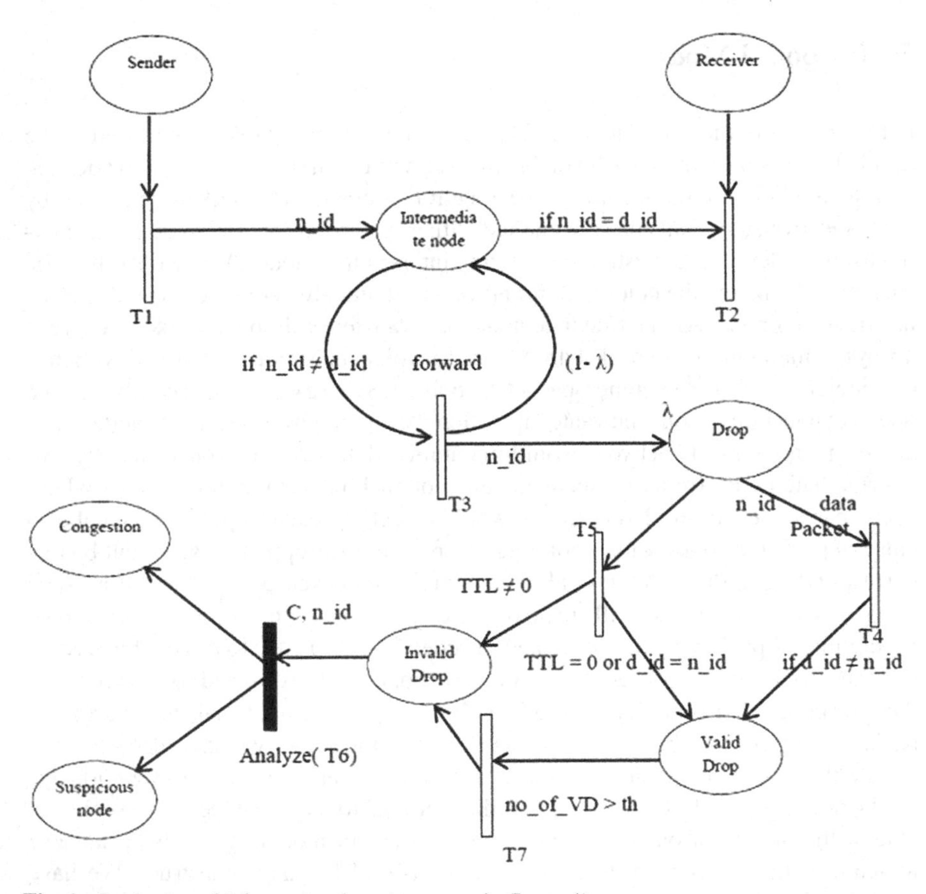

Fig. 1 Petri net model for packet dropping scenario (Layer 1)

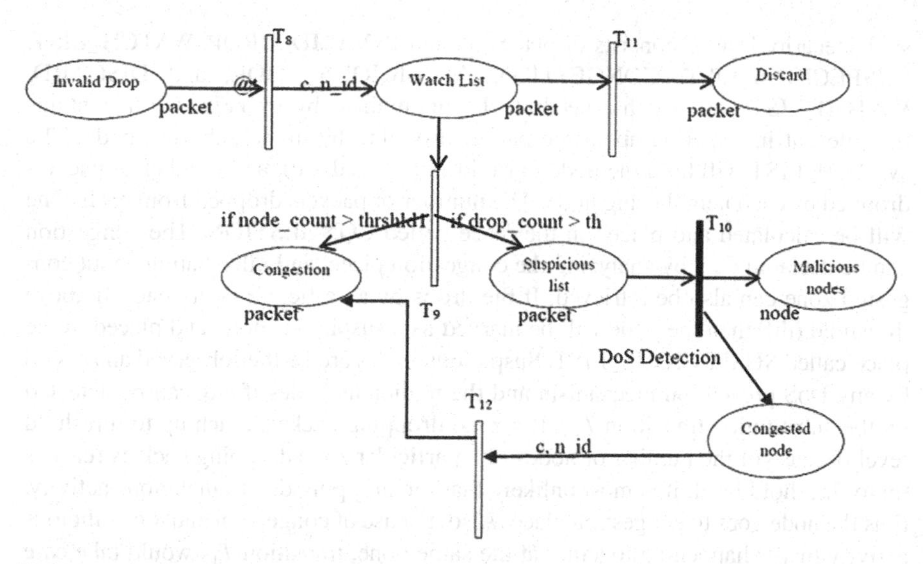

Fig. 2 Sub-net of analyzed substitute transition (Layer 2)

token is the color of the zone from where the sender initiates the packet. While forwarded by the intermediate nodes, color(c) changes with the corresponding zone color in which the node belongs to. The p_id, n_id, s_id, and d_id represent unique identification of packet, node, source, and destination, respectively, for the corresponding packets, while TTL (Time-to-Live) represents its standard representation. A packet would either be forwarded by all the intermediate nodes (represented by the place INTERMEDIATE_NODE), or it may eventually drop in between. Each packet drop phenomenon has been captured, and the dropped packet, i.e., the corresponding token, has been transferred to the place named DROP.

If the packet is dropped by the destination, we consider it as a valid drop as it happens according to the protocol behavior. Also, the drop due to TTL counter (when TTL hits 0) is considered as valid drop and is modeled subsequently. All neighboring nodes (but not a node within the selected path) overhear a data packet due to wireless broadcast transmission and obviously drop the packet. All the above-said activities are modeled, and the corresponding tokens are transferred to a place named VALID_DROP. The proposed model considers any further drops exclusive of those discussed above as invalid drop. These could be due to either congestion or malicious activity. The packets generated due to invalid drops are represented by the tokens of the place INVALID_DROP. These tokens have been analyzed, categorized, and placed in the appropriate places named CONGESTION and SUSPECIOUS_NODE. The analysis part has been taken care by layer 2, and the detailed diagram is depicted in Fig. 2. To analyze all possible scenarios with, congestion and malicious activity (attack), we have four possible cases: (1) no-congestion and no-attack scenario, (2) no-attack but with congestion scenario, (3) no-congestion but with attack scenario, and 4) with attack and with congestion scenario. To accommodate the above-said scenario, layer 2 consists of places named INVALID_DROP, WATCH_LIST, SUSPECIOUS_LIST, CONGESTION, MALICIOUS_NODE, and DISCARD. WATCH_LIST is a list that needs to be maintained by every node. It contains the relevant information about the packet dropping by its neighboring node. The WATCH_LIST will hold the node id (n_id), color code (c), and number of packets dropped by each neighboring node. The number of packets dropped from each zone will be calculated and placed in the place called CONGESTION. The congestion can be declared only by analyzing the congestion place, and information about congested zone can also be retrieved. If the drops by a particular node reached above threshold (th), then the node will be marked as a suspicious node and placed in the place called SUSPECIOUS_LIST. Suspicious nodes are further observed and tested by any DoS prevention mechanism and the malicious nodes if any can be detected by the substitution transition T_{10}. If a zone dropping packets reach up to threshold level because of the number of nodes of a particular zone dropping packets reaches up to threshold level, it is most unlikely that the dropping due to malicious activity, thus the node goes to congestion place. Also, in case of congestion and the malicious activity that is happening together at the same zone, transition T_{12} would take care of the same. The arc is associated with T_{10}, and the place MALICIOUS_NODE is assigned higher priority for the same. Other transitions associated with layer 2 are transition T_8 and T_9 responsible for categorizing between congestion and malicious activity.

4 Formal Definition

Our proposed model can be defines as 8-tuple model
$\{\sum, P, T, A, C, G, E, Var\}$.
• $\sum$ *is a finite set of non – emptytypes, called color sets.*
$\sum = \{V, I, B, G, Y, O, R, P, BL, W, PACKET, NODE, ZONE\}$;
•*A finite set of places* $P = P_{TOP}\ U\ P_{ANALYZE}$ *where*
P_{TOP} = { SENDER, RECEIVER, INTERMEDIATE_NODE, DROP, VALID_DROP, INVALID_DROP, CONGESTION and SUSPECIOUS_NODE } *is the set of TOP layer places;*
$P_{ANALYZE}$ = { INVALID_DROP, WATCH_LIST, SUSPECIOUS_LIST, CONGESTION, MALICIOUS_NODE and DISCARD } *is the set of layer 2 places;*
•*A finite set of transitions* $T = T_{TOP}\ U\ T_{ANALYZE}$, *where*
$T_{TOP} = \{T_1, T_2, T_3, T_4, T_5, T_6, T_7\}$ *is the set of layer1 transitions;*
$T_{ANALYZE} = \{T_8, T_9, T_{10}, T_{11}, T_{12}\}$ *is the set of layer2 transitions;*
•*A is a finite set of arcs such that* :
$P \cap T = P \cap A = T \cap A = \phi, A \subseteq P \times T \subseteq T \times P.$
• *C is a color function. It is defined from P into* $\sum$ *such that* :
$\forall$ p $\in$ P, *there exist one or more elements in* $\sum$ *such that for every p in P there is at least one element in* $\sum$ *and p.*
• *G is a guard function. It is defined from T into expressions such that:*

$$\forall t \in T : [Type(G(t)) = BoolType(Var(G(t))) \in \sum].$$

• *E is an arc expression function. It is defined from A into expressions.*

5 Simulation Using CPN Tools

We have simulated the proposed model using CPN tool [12]. The simulation of the model can detect errors, but does not ensure the absence of errors. In order to do that, we ensure that our tests cover all possible execution paths. We have generated full state space of our model which confirms the same. State space is a directed graph with a set of nodes and arcs. The nodes represent all reachable states from the initial marking, and arcs represent the state transitions. State space comprises of all the possible executions of the modeled system, thus detecting the vulnerable states if any. Full state-space analysis has been performed for the proposed model, and the corresponding reachability has been generated to ensure no error has been occurred for all possible execution paths of the proposed CPN model. For the sake of implementation, we have divided the network deployment area into nine different zones such as zone1, zone2,.., zone9 with color assigned as violet, indigo, blue, green, yellow, orange, red, pink, and black, respectively. Also, we have assumed that the packet generation is evenly distributed over all deployment zones. We have all four cases discussed before, under which the network behavior is analyzed. Simulation is performed, considering 99 packets that are generated by the sender and thus traveling through the randomly chosen intermediate nodes to the receiver nodes. The nodes are distributed to different zones. We consider that all the packets generated by the node for transmission are in SENDER place initially. If it reaches the destination successfully, packets are placed in the RECEIVER place. If it drops due to network behavior, then packets are either of VALID_DROP, CONGESTION, SUDPECIOUS_LIST place. The SUSPICIOUS_LIST place is further investigated to get the corresponding node information which is responsible for dropping the packets placed in SUSPICIOUS_LIST.

Case 1: No-Congestion, No-Attack Scenario—No-congestion and no-attack scenario has been implemented to verify and validate the proposed model. We have set the packet drop probability same as valid drop probability, i.e., 0.002. This means that there will be no congestion in the network if the model works properly. We set the attack probability (0.001) very low as well to ensure no malicious activity. We have generated the full state space for case 1, and according to the report generated, there is no-dead transition. For case 1, we initialized the network with 99 packets. The result holds the fact true that in the normal network scenario only less than 1% of the packets get dropped. And we categorized this type of packet drop as valid packet drop as it is nothing to do with congestion or malicious activity within the network.

Case 2: Modeling Congestion—We have considered the state-space analysis after adding congestion to our CPN model. This has been done by changing the initial marking for the model. We have added more number of tokens to few color zones to do the same. The simulation results show that out of 99 packets, 54 packets reached the destination thus undropped, 9 packets get dropped as a valid drop, and 36 packets are shifted to the congested zone; among them, black and indigo zones experience maximum packet drops (8 each) followed by red and blue zones (6 each). Rest of the zones are not congested as they drop negligible amount of packets. No token is placed in SUSPICIOUS_NODE, so no malicious activity is identified within the network. The final marking after the simulation is $M_f = [0, 54, 9, 36, 0]$.

Case 3: Modeling Attack—Next we have launched malicious activity by adding intruder. The partial model for the same has been designed which cannot be shown here due to lack of space. To do the same, we have increased the probability of packet dropping by the transition named intruder. The intruder launching DoS attacks by dropping packets. The probability used for dropping packets by the intruder is 0.33. We have included a place called INTRUDER to our CPN model as an extension and then checked the performance of the model in the presence of the intruder. We have simulated this scenario for the same 99 packets, and the results show that out of 99, again 42 get delivered, while 2 packets shifted to the place VALID_NODE, thus representing valid drop, whereas rest 30 packets shifted to the place named SUSPECIOUS_List, and corresponding 2 tokens placed to SUSPICIOUS_NODE, signifies corresponding nodes responsible for malicious activity much to our expectations. Also, 25 packets placed at the CONGESTION place. This indicates that with high attack probability, chances of network congestion increase. The final marking after the simulation is Mf = [0, 42, 2, 25, 30].

Case 4: Modeling Both Congestion and Attack—Finally, the model has been simulated in the presence of both congestion and malicious activity. For no-attack scenario, the probability for generating false packets by the intruder was 0.01. For these cases (cases 1 and 2), there was no packet drops by the intruder. But when the attack took place (cases 2 and 4), the probability of dropping packet is 0.33 for both congestion and packet drop probability. So, in the presence of both congestion and attack, out of 99, 27 packets gets undropped, 32 packets dropped because of congestion, and 37 packets got marked as a reason for suspicious activity, while rest 3 packets dropped as a valid drop. Final marking after simulation of case 4 is $M_f = [0, 37, 3, 32, 27]$.

6 Performance Analysis

We have generated full state space along with the corresponding reachability graph for our model. The state-space statistics gives reachability graph with 1099 nodes and 2314 arcs, therefore ensuring the finiteness of our model. No packets in the place named CONGESTION as well as SUSPECIOUS_NODE along with the generation of full state space for no-congestion, no-attack scenario (case 1) signify the expected outcome after finite number of execution steps proves the functional correctness of the model. There is a list of dead marking investigated with query tool to ensure that those are for the final model output states only. Other properties generated from the state-space report like no-dead transition instances, no-live transition instances, and no-infinite occurrence sequence clarify that the proposed model has finite occurrence sequence to reach to the o/p state, and thus, it is a terminating model. Simulation result of case 1 shows that the model takes 1217 steps to complete the execution for 99 packets and reaches its final expected state.

No-dead transition signifies the absence of deadlock. With the help of the query, it has been detected that the list of generated dead markings represents the expected model behavior. The 11 dead markings [815,579, 532, 284, 1085,...] signify our expected output states only; i.e., tokens are in either RECEIVER, VALID_DROP, CONGESTION, SUSPECIOUS_NODE. This justifies correctness of our model. Event packets generated by source nodes and transmitted through a set of intermediate nodes are either delivered to the destination or dropped with all possible sequence of events. No unexpected behavior is identified.

All places of the proposed model have an upper bound N that holds for all reachable marking and are called N-safe model. Here, N is the total number of tokens generated by sender nodes of the network. Once bounded, it can generate the full state space of the model and thus it ensures the proof of correctness. Live signifies each reachable marking enables occurrence sequences containing all transitions. In context of our model, here no-live transition instances justified the system behavior, all packets reached to its destination or dropped signify that the system has reached its final reachable marking Mf= [0, 98, 1, 0, 0], and thus, no transition is enabled at that point. Fairness properties are only relevant if there are infinite firing sequences (IFS), otherwise CPN Tools report: "no-infinite occurrence sequences." No-infinite occurrence sequence, i.e., cycles in a state space that once entered can never be left, and within which no progress is made with respect to the purpose of the system (Table 1).

7 Conclusion

The main focus of our work is to figure out the inherent cause of packet drop within the network out of many possible reasons as discussed in the literature. Hence, the result is getting a clear direction to resolve the problem. The beauty of our model is

Table 1 Packet drop statistics for all four cases for 99 packets

Cases	Drop Prob for congestion	Drop prob for intruder	No of valid drop	Received packet	Drop due to congestion	Drop by the suspicious node
Case 1	0.002	0.01	02	97	0	0
Case 2	0.33	0.01	09	54	36	0
Case 3	0.002	0.3	07	74	0	18
Case 4	0.33	0.3	03	37	32	27

that we can have a detailed insight of packet drops, i.e., percentage of drop, the cause behind the same (due to congestion and/or malicious activity), and also the affected network zone(s). This can be done by examining only two places of our proposed model named CONGESTION and SUSPECIOUS_NODE by exploiting the property of Colored Petri Net. Another major advantage is that any standard prevention mechanism can act locally to repair the damage. Once the node and its zone information are available, the recovery mechanism can act locally. Our proposed method definitely made the process easy as the affected zone would be declared by the model, and therefore, the correction mechanism could only focus on the affected zone rather than the taking entire network in its consideration. We would also like to come up with different repair models for the same which would be able to work reactively depending on our current model outcome as future work.

References

1. Bidgoly, A.J., Ladani, B.T.: Trust modeling and verification using colored petri nets. In: 8th International ISC Conference on Information Security and Cryptology (ISCISC) (2011)
2. Liu, L.: Uncovering SIP vulnerabilities to DoS attacks using coloured petri nets. In: International Joint Conference of IEEE TrustCom (2011)
3. Long, S.: Analysis of concurrent security protocols using colored petri nets. In: International Conference on Networking and Digital Society (2009)
4. Li-li, W., Xiao-jing M.: Modeling and verification of colored petri net in stop and wait protocol. In: International Conference On Computer Design And Appliations (2010)
5. Jensen, K.: The practitioner's guide to Coloured Petri nets. In: 24th International Conference on Applications and theory of Petri nets (2003)
6. http://www.annese.com/blog/packet-loss-and-network-security
7. Chen, W et al.: Joint QoS provisioning and congestion control for multi-hop wireless networks. In: EURASIP Journal on Wireless Communications and Networking (2016)
8. Lee, K.Y., Cho, K., Lee, B.: A congestion control algorithm in IEEE 802.11 wireless LAN for throughput enhancements. In: IEEE Digest of Technical Papers International Conference on Consumer Electronics (2007)
9. Staehle, B. et al.: Intra-mesh congestion control for IEEE 802.11s wireless mesh networks. in: 21st IEEE International Conference on Computer Communications and Networks (2012)
10. Masri, A.E., et al.: Neighborhood-aware and overhead-free congestion control for IEEE 802.11 wireless mesh networks. J. IEEE Trans. Wirel. Commun. 13(10) 5878–5892 (2014)
11. Shen, Z.M., Thomas, J.P.: Security and QoS self-optimization in mobile Ad Hoc networks. J. IEEE Trans. Mob. Comput. **7**(9), 1138–1151 (2008)
12. http://cpntools.org/documentation/start
13. Haywood, R., Peng, X.-H.: On packet loss Performance under varying network conditions with path diversity. In: ACM International Conference on Advance Infocomm Technology (2008)

Hole Detection in Wireless Sensor Network: A Review

Smita Das and Mrinal Kanti DebBarma

Abstract Nowadays wireless sensor networks (WSNs) have massive relevance from environmental observation to endangered species recovery, habitat monitoring to home automation, waste management to wine production, medical science to military applications. While organizing the sensor nodes in a WSN, covering the area-to-be-monitored is a tricky job and this quality is compromised in the presence of holes. Hole may be defined as an area in WSN around which a sensor is incapable of sensing or transmitting data. As holes can cause permanent or temporary interruption in sensing or in communicating task, therefore detection of holes in a coverage area is an essential job. In this paper, a detail literature review is done on hole detection, categorization, characteristics, and their effect on sensor network's performance on the basis of the most recent literature.

Keywords Wireless sensor networks · Sensor nodes · Coverage
Hole detection · Hole healing · Boundary detection

1 Introduction

In recent years, microelectromechanical systems (MEMS) technology and wireless communication technology together have drawn global attention due to the invention of tiny, economical wireless sensor nodes. These nodes are well organized to sense, evaluate, and collect data from a particular environment, although they have limited amount of power, processing speed, and memory. From mid-1990s exploration in the field of WSN started and among all the research topics, coverage has got the highest preference since last few years. Coverage quantifies how well an area is sensed by the nodes they are deployed into. In the

S. Das (✉) · M. K. DebBarma
Department of Computer Science & Engineering, NIT Agartala, Tripura 799046, India
e-mail: smitadas.nita@gmail.com

M. K. DebBarma
e-mail: mkdb06@gmail.com

P. K. Sa et al. (eds.), *Recent Findings in Intelligent Computing Techniques*,
Advances in Intelligent Systems and Computing 708,
https://doi.org/10.1007/978-981-10-8636-6_10

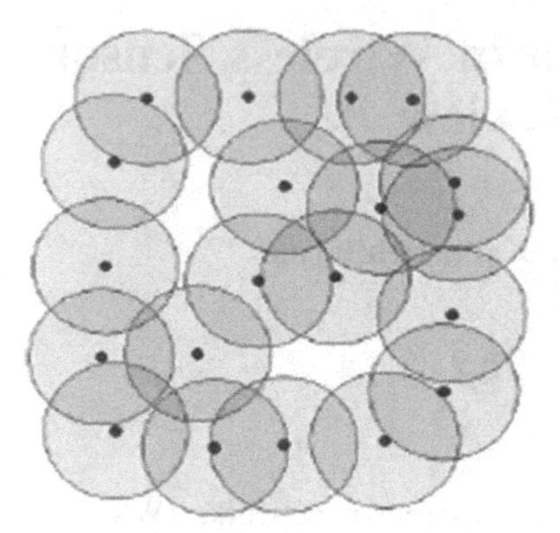

Fig. 1 Coverage hole

coverage problem, each sensor has to wrap up some subregion and summing up these entire covered subregions one can have a totally covered region in the WSN. Therefore, it is clear that the random deployment of the sensors in the target area cannot promise an optimal solution at the very first attempt. While discussing the deployment strategies, it is assumed that the sensing field will entirely be covered with sensors. But if we see practically just like Fig. 1, there may be several coverage holes, which are the areas not being covered by any sensor. These coverage holes may be created due to several reasons, for example, when sensors are thrown into a battlefield through some in-flight arrangements, the node deployment becomes random in nature as well as unstructured. Therefore, a few gaps may be created unintentionally in the sensing field which results hole in the coverage area. As a result, a few nodes might fail to sense data or communicate with other nodes and finally performance of sensor network may degrade. The rest of the paper is organized as follows: In Sect. 2, we have identified a few reasons behind the creation of holes and discussed why hole detection is important. In Sect. 3, we discussed the categorization of various types of holes found in sensor network. In Sect. 4, there is an elaborate description about hole detection and healing algorithms, especially on coverage hole detection, with respect to various recent literatures. Summarizations of different algorithms are done in Sect. 5. And finally, we have concluded in Sect. 6.

2 Overview of Hole

The prime job of a sensor node is to sense and communicate with other nodes in the network, but when a sensor fails to do so in an area, then that area in the network produces a hole. There are several causes [1] behind the creation of holes, such as:

- Drainage of power—a sensor node is meaningless without its power source. If the power source is over-used for sensing or communication purpose, then the

node will run out of power and hence causing a hole in the network. Generally, coverage and routing holes are created due to power exhaustion.
- Adverse environment—an adverse environment (e.g., fire in forest) can destroy the nodes and hole may be created in the network. Routing holes are created under such condition.
- Incorrect design—while designing the network, if there is some ambiguity in the topology, then it will lead to the formation of the coverage hole.
- Obstacle—due to some obstacle in the sensing field routing or jamming holes can be formed.
- Node replacement—in place of a faulty node, if a new node is placed, then the route for communication may change which results in routing hole in the network.

Holes are the reason behind the performance deterioration of sensor network. If there is some hole in the network, then communication becomes weak between the nodes due to the fact that, sensed data is routed along the boundary of the hole repeatedly. Therefore, detection of holes is very essential as follows:

- Identifies whether a node is fully operational or not.
- Guarantees elongated network lifetime.
- Provides sufficient quality of service in network coverage by identifying whether each point in sensing field has the compulsory quantity of coverage or not.
- With the help of hole detection, we can assess whether extra nodes are required in the region of interest (ROI) or not which results speedy covering the holes.
- Detection of holes prevents data loss in the network. Also, it helps to identify any substitute communication passage to normalize flow of data.

3 Category of Holes in WSN

In 2005, Ahmed et al. [2] have classified network holes in the following four categories: *Coverage holes* (shown in Fig. 1) [3] are produced due to random deployment of nodes, drainage of battery and faulty network topology. *Routing holes* (shown in Fig. 2—i) [4] occur due to adverse environmental conditions, obstacle present in the sensing field or replacement of an old node with a new one. If the radio frequency used for communication between sensor nodes is blocked by a jammer with some high-frequency signals then *Jamming holes* (shown in Fig. 2—iv, v and vi) arise [5]. This jamming can be either intentional or unintentional. The last category (shown in Fig. 2—ii and iii) [6, 7] of hole defined by Ahmed et al. is the *Sink/black holes/Wormholes*. Denial of service attacks can originate *Sink/black* holes [8]. In this case, an alternate route suggestion toward sink is provided by an opponent to mislead the neighboring nodes. Wormhole [9] is also initiated through denial of service that comes under *malicious holes* category.

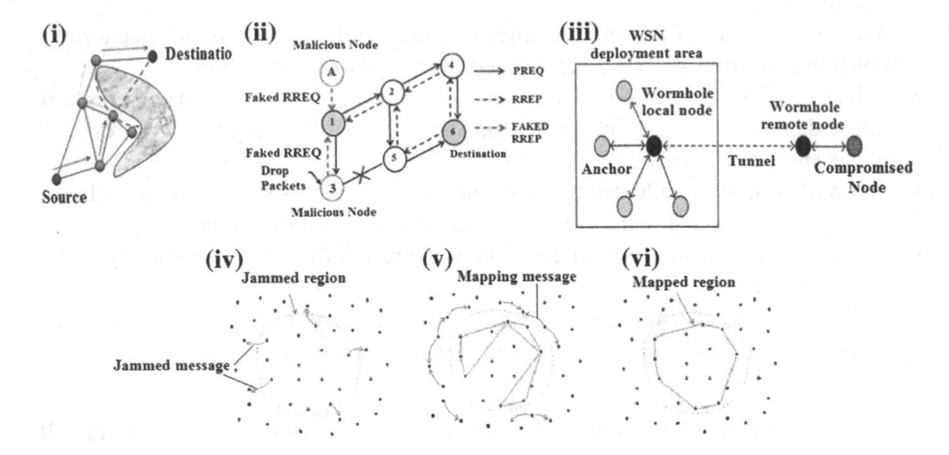

Fig. 2 Different types of holes in sensor network (i) Routing Holes: GPSR—Greedy Perimeter Stateless Routing for WSN (ii) Black Hole attack problem (iii) Wormhole attack to location and neighbor discovery (iv), (v), and (vi) Jamming Holes: Overview of Collaborative mapping of nodes in a jammed region in the network

In 2013, Jabeur et al. [10] introduced PLMS (physical/logical/malicious/semantic), a cause-based taxonomy. *Physical holes* occur due to the limited capacity of processing, overuse of energy or inappropriate sensor nodes. Coverage holes and Routing holes fall into this category. Cluster-based approach, where a sensor node cannot be sustained from its neighboring nodes, initiates *Logical holes.* Jamming hole, sinkhole, and wormhole belong from the category of *Malicious holes* and they can occur when a few sensors in the network behaves abnormally. *Semantic holes* are caused due to processing and routing of data. The authors in [7] also suggested that further categorization of holes can be done on the basis of mobility (static or moving), lifetime (temporary or permanent), purpose (intentional or accidental), and affected function (functional or non-functional).

4 Algorithms for Hole Detection

In this section, we are going to discuss various hole detection algorithms, particularly, coverage hole detection, as numerous works are done on this topic. In **2005**, Ghrist and Muhammad [11] have proposed a centralized algorithm that detects holes via homology without prior knowledge of sensor locations. This algorithm detects only single-level coverage holes but unable to detect boundary holes. In **2008**, Li and Hunter [12] proposed one distributed algorithm named 3MeSH (Triangle Mesh Self-Healing) that detects the existence of holes and also recover them. This static approach can recover large holes produced by accidental node failure or topology changes in networks. Recovery of trivial holes is also possible depending upon the availability of node location information and the distances between nodes. Kanno et al. [13] in **2009** proposed a distributed method that determines the number of holes along with their location in a non-planar sensor network with a given communication graph. From this non-planar graph, a planar graph is obtained and further divided into subgraphs with the help of 'partition network' algorithm. If a subgraph contains no holes, it is eliminated from the list; otherwise, graph is further divided into subgraphs until the holes are adequately bounded. In **2010**, Yang and Fei [14] proposed a hole detection and adaptive geographical routing (HDAR) algorithm to detect holes and to deal with local minimum problem. If the angle between two adjacent edges of a node is greater than 120°, then hole detection algorithm is initiated. If for some node the value of hole detection ratio is greater than a pre-defined threshold, then it is on the boundary. Yan et al. [15] in **2011** used the concepts of Rips complex and Cech complex to discover coverage holes and classify coverage holes to be triangular or non-triangular. This is based on topological approach. A distributed algorithm with only connectivity information was proposed for non-triangular holes detection. In **2012**, Zeadally et al. [16] proposed a hop-based approach to find holes in sensor networks. There are three phases, namely information collection phase, path construction phase, and finally path checking phase. If the communication path of x-hop neighbors of a node is broken, then it is boundary node. Algorithm works for a node of degree of 7 or higher. Senouci et al. [17], in **2013,** proposed a hole detection and healing (HEAL) algorithm. This allows a local healing using hole detection Algorithm (HDA) for detecting the holes and another boundary detection algorithm (BDA) is proposed to identify the boundary holes within the RoI. Moreover, they have used a virtual force-based hole healing algorithm. This algorithm relocates only the adequate nodes within the shortest range.

In the most recent literatures, Ghosh et al. [18] in **2014** has proposed two novel distributed algorithms as DVHD (distance vector hole determination) and GCHD (Gaussian curvature-based hole determination). DVHD uses Bellman–Ford algorithm to calculate the shortest distance path between a pair of nodes in a weighted Delaunay graph. If the distance is less than k, which is a constant greater than number of nodes in the graph, the nodes are treated to be in the same boundary. Otherwise, nodes are in different boundaries and thus resulting holes. GCHD uses Gauss–Bonnet theorem to calculate the distributed curvature to detect the no. of holes.

Li and Zhang [19] in **2015** proposed a novel algorithm to detect coverage holes using 'empty circle property' by forming Delaunay triangulation of the network. If empty circle radius R_c is greater than sensing radius R_s, there exists some hole. The holes are further clustered by connecting the center of empty circles of each Delaunay triangle with its neighbor by a line segment.

In **2016**, Zhao et al. [20] proposed an algorithm which has two phases namely: distributed sector cover scanning (DSCS), that is, used to identify the nodes on hole borders and the outer boundary of WSN and directional walk (DW) that can locate the coverage holes based on the boundary nodes identified with DSCS.

Sahoo et al. [21] in **2016** proposed a distributed coverage hole detection (DCHD) algorithm to detect the bounded or non-bounded coverage holes in the sensor network. This method uses critical intersection point (CIP) to resolve the faults of the perimeter-based coverage hole detection by reducing the time complexity of coverage hole detection. At first, each sensor finds out its CIP set and then verifies if any point belongs to a covered points (CP) set or not. Finally, each sensor in the clock-wise direction collaborates with its one-hop neighbors and unites to its CIP to detect the occurrence of a coverage hole.

Again in **2016**, Beghdad and Lamraoui [22] proposed an algorithm is based on connected independent sets (BDCIS) and is divided into three steps. At first, each node collects connectivity information by sending and receiving messages toward its neighbors and constructs its one-hop neighbors' graph only. In the second step, independent sets (IS) of cardinality α are established with the help of the minimum or maximum id in the graph G_i, which will be the first element of the IS1. Then by removing all the neighbors of this node from G_i, another node having the minimum id among the remaining nodes is chosen and this process continues to build all other possible ISs. Finally, the independent sets are connected in order to search for the closed path to detect holes based on some rules.

In **2017**, Amgoth and Jana [23] proposed an algorithm having two phases namely: coverage hole detection (CHD) and coverage restoration (CR). In CHD, each sensor node separately detects hole by updating certain information with its neighbor nodes. For this information update, a sensor node searches for cells with their coordinates inside its sensing range and then covering the maximum sensing range R^*. For CR, a sensor node with comparatively high residual energy is given priority to cover up the hole closer to it by increasing its sensing range up to a maximum limit.

5 Summarization

In the previous section, we have done an extensive review of the most recent literature on coverage hole. In this section, we are going to summarize the literature in a tabular format for the interest of the research community. In Table 1, the final summarization is provided which is primarily carried out with the Distributed Algorithm. Only [11] is from centralized type of algorithm. From this, we can suggest that distributed algorithms for coverage hole detection are much more efficient than centralized one.

Also, we have found that maximum of these algorithms follows computational geometry-based approach and topology-based approach. A few also follow other approaches like virtual force based, perimeter based, or mathematical model based. Hence, computational geometry-based approach or graph-based approach may be treated to be the best approach to use for algorithm design. In the table, we have put the algorithms in chronological order according to their year of publishing. Besides that, from all algorithms we have taken a few fields, given in the table, to highlight

Table 1 Summarization of coverage hole detection on the basis of distributed algorithm

Year and ref. no.	Pre-requisites	Key features	Hole detection	Disadvantage	Simulator used
Computation geometry-based approach					
2005 [11][a]	Connectivity information	Homology	Single-level detection	Cannot detect boundary holes.	MATLAB
2008 [12]	Node location information	Distance between nodes, recovery of trivial holes.	Large hole detection	Failure in trivial hole recovery	MATLAB
2009 [13]	Communication Graph	Homology, location of holes	Local holes	Failure in multiple hole detection	MATLAB
2010 [14]	Angle between two adjacent edges of node	Hole detection ratio, adaptive geographical routing	Local holes	Not scalable	Easim 3D
2011 [15]	Connectivity information	Rips complex, Cech complex, detection of non-triangular holes	Local holes	Cannot detect boundary holes.	MATLAB
2016 [20]	Node location information	Directional walk of holes	Outer boundary of holes	Complex calculations are required	MATLAB
2016 [22]	Connectivity information	Connected independent set, energy efficiency	Local holes	Communication overhead	MATLAB
Topology-based approach					
2012 [16]	Connectivity information	X-hop neighbor matrix, node of degree 7 or higher	Local holes	–	GTSNetS
2014 [18]	Node location and incident angle	Empty circle property, distributed curvature	Local holes	Communication overhead	MATLAB
2015 [19]	Node location information	Empty circle property, clustering of holes	Triangular and non-triangular holes	Increased computational complexity	MATLAB

(continued)

Table 1 (continued)

Year and ref. no.	Pre-requisites	Key features	Hole detection	Disadvantage	Simulator used
Virtual force-based approach					
2013 [17]	Node location information	Design and evaluation of holes, hole healing	Local holes	Cannot detect boundary holes.	Ns-2
Perimeter-based approach					
2016 [21]	Connectivity information	Critical intersection point, clustering of holes	Non-bounded holes	–	Ns-2
Mathematical model-based approach					
2017 [23]	Neighbor node coordinates	Hole restoration, Energy efficiency	Local	Not scalable	Dev C ++ and MATLAB

[a]***Coverage Hole detection on the basis of Centralized Algorithm***

each algorithm more specifically. The fields are chosen as follows: if any *pre-requisites* or criteria are required before applying the algorithm, the main characteristics, i.e., the *key features* of each algorithm, the level of *detection of hole* and its type, if there is any *disadvantage* of the given algorithm and finally the *simulator* required to test the effectiveness of the algorithm. On the basis of these fields, we can have the observation that node location information and connectivity information are two most important prerequisites for algorithm. Also, we have noticed that most of the algorithms face the drawback of failure of detection of holes that fall in boundary locations. Therefore, further research can be done on detection of holes which fall in boundary location. Finally, we can conclude this summarization saying that till date MATLAB may be the best simulator to be used to test the proposed coverage hole detection algorithm.

6 Conclusion

Unlike other networks, WSN is very much application specific and for supporting those multidisciplinary applications, deployment of the nodes should explicitly be defined. While designing the coverage of the sensing field, quality can be jeopardized at the occurrence of hole. Hence, in this paper, we have wrapped up different types of holes and their characteristics, cause of creation of particular hole and reason behind their detection. Among different types of network holes, coverage holes are treated to be the most important to detect, as they play a key position in QoS assurance in WSN. Also, we have shown an elaborate review with respect to the extremely recent literature of coverage hole detection in wireless sensor network. In addition to that, we have reviewed special issues from the available coverage hole detection algorithms from different angles like approach, features, prerequisites, types of algorithm.

References

1. Bhardwaj, R., Sharma, H.: Holes in wireless sensor networks. Int. J. Comput. Sci. Inform. **2** (4), 62–63 (2012)
2. Ahmed, N., Kanhere, S.S., Jha, S.: The holes problem in wireless sensor networks: a survey. ACM SIGMOBILE Mob. Comput. Commun. Rev. New York USA **9**(2), 4–18 (2005)
3. Ramazani, S., Kanno, J., Selmic, R.R., Brust, M.R.: Topological and combinatorial coverage hole detection in coordinate-free wireless sensor networks. Int. J. Sens. Netw. **21**(1), 40–52 (2016)
4. Karp, B., Kung, H.T.: GPSR: Greedy perimeter stateless routing for wireless networks. In: Proceedings of the 6th Annual International Conference on Mobile Computing and Networking. ACM, (2000)
5. Wood, A.D., Stankovic, J.A., Son, S.H.: JAM: a jammed-area mapping service for sensor networks. In: 24th IEEE Real Time System Symposium (RTSS'03), pp. 286–298 (2003)

6. Vangili, A., Thangadurai, K.: Detection of black hole attack in mobile ad-hoc networks using ant colony optimization–simulation analysis. Indian J. Sci. Technol. **8**(13) (2015)
7. García-Otero, M., and Población-Hernández, A.: Secure neighbor discovery in wireless sensor networks using range-free localization techniques. Int. J. Distrib Sens Netw (2012)
8. Karlof, C and Wagner, D.: Secure routing in wireless sensor networks: attacks and countermeasures. In Proceedings of the 1st IEEE International Workshop (SNPA '03) (2003)
9. Hu, Y.C., Perrig, A., Johnson, D.B.: Wormhole detection in wireless adhoc networks. Technical Report TR01-384, Department of Computer Science, Rice University (2002)
10. Jabeur, N., Sahli, N., Khan, I.M.: Survey on sensor holes?: a cause-effect-solution perspective. Proced. Comput. Sci. **19**, 1074–1080 (2013)
11. Ghrist, R., Muhammad, A.: Coverage and hole-detection in sensor networks via homology. In: Fourth International Symposium on Information Processing in Sensor Networks, IPSN 2005. IEEE (2005)
12. Li, X., Hunter, D.K.: Distributed coordinate-free hole recovery, In: IEEE International Conference on Communications Workshops, ICC Workshops' 08. IEEE (2008)
13. Kanno, J., Selmic, R.R., Phoha, V.: Detecting coverage holes in wireless sensor networks. In: 17th Mediterranean Conference on Control and Automation, MED'09, pp. 452–457 (2009)
14. Yang, J., Fei, Z.: HDAR: hole detection and adaptive geographic routing for ad hoc networks. In: Proceedings of the 19th International Conference on Computer Communications and Networks (ICCCN '10), pp. 1–6 (2010)
15. Yan, F., Martins, P., Decreusefond, L.: Connectivity-based distributed coverage hole detection in wireless sensor networks. IEEE Global Communications Conference GLOBECOM, pp. 1–6 (2011)
16. Khan, I.M., Jabeur, N., Zeadally, S.: Hop-based approach for holes and boundary detection in wireless sensor networks. IET Wirel Sens Syst **2**(4), 328–337 (2012)
17. Senouci, M.R., Mellouk, A., Assnoune, K.: Localized movement-assisted sensor deployment algorithm for hole detection and healing. IEEE Trans Parallel Distrib. Syst. 1–11 (2013)
18. Ghosh, P., Gao, J., Gasparri, A., Krishnamachari, B.: Distributed hole detection algorithms for wireless sensor networks. In: 11th International Conference on. Mobile Ad Hoc and Sensor Systems (MASS 2014). IEEE (2014)
19. Li, W., Zhang, W.: Coverage hole and boundary nodes detection in wireless sensor networks. J Netw. Comput. Appl. **48**, 35–43 (2015)
20. Zhao, L.H., Liu, W., Lei, H., Zhang, R., Tan, Q.: Detecting boundary nodes and coverage holes in wireless sensor networks. Mob. Inf. Syst. **2016,** 16, Article ID 8310296. https://doi.org/10.1155/2016/8310296
21. Kumar Sahoo, P., Chiang, M.J., Wu, S.L.: An efficient distributed coverage hole detection protocol for wireless sensor networks. Sensors **16**(3), 386 (2016)
22. Beghdad, R., Lamraoui, A.: Boundary and holes recognition in wireless sensor networks. J. Innov. Digit. Ecosyst. **3**(1), 1–14 (2016)
23. Amgoth, T., Jana, P.K.: Coverage hole detection and restoration algorithm for wireless sensor networks. Peer-to-Peer Netw. Appl. **10**(1), 66–78 (2017)

A Two-Tailed Chain Topology in Wireless Sensor Networks for Efficient Monitoring of Food Grain Storage

Alekha Kumar Mishra, Arun Kumar, Asis Kumar Tripathy and Tapan Kumar Das

Abstract With 17.5% of world population, food safety and security is of primary concern in India. Despite increase in food grain production in last decade, there is heavy loss of food grains during transit and storage. This is due to lack of adequate infrastructures and technologies to monitor adverse environmental conditions that result in storage losses. In this paper, we propose a topology for wireless sensor networks called two-tailed chain topology that is suitable for efficiently monitoring of environmental factors in food grain depots. The proposed topology is designed by considering all infrastructure constraints of a food grain depot. Results show that the proposed topology is better than traditional topologies in terms of detection accuracy and coverage.

Keywords Food grain monitoring · Environmental factors · Wireless sensor networks · WSN topology · Two-tailed chain

1 Introduction

Wireless sensor networks (WSNs) have been proved to be the most effective technology for monitoring and data acquisition in a wide range of applications [1]. Food security is one of the primary concerns in populated country like India. More than 20% of the food grains are wasted in storage due to lack of proper monitoring and controlled infrastructure. The environmental factors such as temperature, amount of moisture, and light of a food grain storage influence the lifetime of food grains

A. K. Mishra (✉)
School of Computer Science and Engineering, VIT University, Vellore, India
e-mail: alekha.mishra@vit.ac.in

A. Kumar
Department of Electrical & Computer Engineering, National University of Singapore, Singapore, Singapore

A. K. Tripathy · T. K. Das
School of Information Technology and Engineering, VIT University, Vellore, India

P. K. Sa et al. (eds.), *Recent Findings in Intelligent Computing Techniques*,
Advances in Intelligent Systems and Computing 708,
https://doi.org/10.1007/978-981-10-8636-6_11

[2, 3]. Traditional or manual depot monitoring process has the disadvantage of limited reachability to the food grains in the depot. WSNs play a significant role in monitoring and controlling of environmental factors in various applications [4]. WSNs can be deployed with various topologies such as mesh, grid, cluster to monitor the environmental factors [5]. However, these topologies are not efficient for monitoring in food grain storages due to various accessibility and structural constraints. In this paper, we propose a topology for WSNs that is suitable for efficient monitoring of environmental factors in a food grain depot. Directional antennas are used in sensors instead of omnidirectional antenna to improve the reachability of the proposed topology. The proposed topology is designed by considering food grain storage infrastructure and accessibility constraints. It is observed from the result that the proposed topology is efficient compared to traditional topologies in terms of detection accuracy and coverage.

2 Survey of WSN Topologies

This section provides a quick overview of various topologies commonly used in WSN. A WSN topology must address the issues such as coverage, connectivity, energy consumption [6]. The star topology is the simplest and efficient for small networks. It is a single-hop system, where all sensor communicate directly. The overall power consumption is lower in star topology compared to others. However, the communication range of the network is limited. In a network of tree topology, the base station (BS) acts as the root. The BS is connected with relay nodes situated at more than one level down from the BS in the tree hierarchy. The tree topology may reduce power consumption of the network and can easily extend the network communication range. However, in tree topology, it is required that all nodes are time synchronized [5].

In mesh topology, each node is connected to all the nodes within its communication range. It commonly adopts multi-hop communication to send sensed data to the BS. A mesh network is highly fault tolerant because of having multiple paths to the BS or to other nodes. Grid-based network divides a network into a rectangular grid, where sensors are deployed at grid points [7]. Since, ideal placement strategy such as grid cannot be achieved in practice, sometimes a number of nodes instead of one are placed on a square grid within a unit area. Cluster-based topology divides the entire network into number of clusters [8, 9]. A cluster consists of a set of nodes, where a node is designated as cluster head (CH). The nodes other than CH are responsible for sensing data and sent to the CH. The CH coordinates the cluster, collects sensed data from all members, and sends it to the BS. It is reported that inter-cluster communication via bridge nodes (non-CH nodes) consumes lesser energy than direct communication between cluster heads. The advantage of cluster-based topology is it minimizes number of messages sent over the network. Sometimes it is suitable to use a hybrid topology [8] by combining one or more previously discussed topologies. Clustering tree forms a hierarchical structure among the clusters formed in the network [10]. To minimize energy consumption of a CH, a number of non-CH nodes are selected as the relay node to share the load of CHs. Therefore, clustering-tree topology achieves better network lifetime. A star-mesh topology has the simplicity and low-power feature of star topology and the extendibility of mesh network. In a clustered mesh topology, deployment region is divided it into rectangular cells of equal size [7]. Each of this cell is considered as a cluster. An inter-CH mesh is formed among all the CHs of the network to forward sensed data.

3 Issues in Food Grain Depots

The foremost issue is the lack of improper infrastructure with the farmers. Since most of the regions of food generation in India are humid (70–98%), the traditional storage methods cannot preserve the quality of the primary food grains such as wheat and rice throughout the year. The second issue is the lack of technology and facilities to monitor environmental factors such as temperature, moisture, humidity, light that highly affect the quality of food grains stored in large depots [2, 3]. The third and most important issue is the irregularities in manual inspection to be performed by officials to check and monitor the quality of food grains. The above-mentioned issues result in wastage of food grains during storage. Therefore, an automated and efficient monitoring system is required to keep track of quality of food grain in the depots to reduce the wastage.

A WSN is a desirable technology for this kind of applications. The most important step in establishing a WSN infrastructure is the designing of topology for the given application. To design or select a suitable and efficient WSN topology for a given application, it is required to understand the characteristics of its infrastructure. The standards regarding food grain depots that are followed by major food corporations such as Food Corporation of India (FCI) is enlisted in the following subsections.

3.1 Infrastructure Details of Food Grain Depot

The general structure of the depots is doom in shape. The standard specifications of the conventional type of depots are as follows:

- Area of the depot (center to center) = 125.55 m × 21.80 m.
- Area of the depot (outside to outside) = 126.01 m × 22.26 m.
- Depot height = 5.5–6.25 m.
- No. of compartments in 5000 Metric Ton depot = 3.
- Length of each compartment = 41.85 m.
- Capacity of one compartment = 1670 M.T.
- No. of stacks in each compartment = 12.
- Size of stacks = 6.10 m × 9.15 m.

- No of ventilators used = approx. 130 (on the top and bottom side of long walls).
- Area requirement for 5,000 M.T. capacity = 2.98 Acres (approx.).
- Area requirement for 10,000 M.T. capacity = 4.70 Acres.
- Moisture content range for rice: 12.2–13.8%
- Moisture content range for wheat: 12.8–13.9%

Galvanized Iron Corrugated (GIC) silos are suitable for storage of food grains throughout the year [11]. However, GIC silos are expensive compared to standard grain depot. Cover and plinth (CAP) storages are commonly practiced by the farmers to store the food grain in addition to depot storage. The Fig. 1 shows a standard food grain depot structure from inside.

4 Suitability of Existing Topologies

In this section, we discuss the feasibility of commonly practiced topology of WSN for food grain depot. A typical large-scale WSN consists of one or more sinks (or BSs), and tens to thousands of sensor nodes that organized themselves into a multi-hop wireless network deployed over a geographical region of interest. It is clear from Fig. 1 that it is impossible to deploy sensor nodes in the middle of the depot due to vehicle and personnel movement. According to the practice of officials, it would hinder the process of bulk loading and unloading of food grain, if sensors are deployed in between the food grain bags. Due to lack of support of pillars inside the depots area where sensor nodes could be fixed, only option left is to deploy the sensors on the inside wall of the depot.

Based on the above deployment constraints, existing topologies such as star, tree, mesh, grid-based, and cluster-based are unsuitable for this infrastructure. This is because, in these topologies, the nodes are deployed inside the region of interest. It is reported in the literature [12] that above topologies assume that the omnidirectional

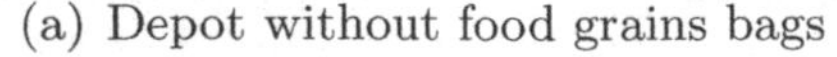

(a) Depot without food grains bags (b) Depot with food grains bags

Fig. 1 Internal structure of food grain depot in FCIs

antennas are used for communication. It is also reported that directional antennas are more suitable for indoor applications compared to omnidirectional antennas. Using star topology for food grain depot, we cannot guarantee that the gateway node lies in the communication range of sensing nodes. Since cluster-based topology assumes that sensors are deployed densely near the region of interest, it difficult to use this topology for food grain depot. Grid-based topology is suitable up to some extent provided that the inside wall of the depot is considered as the deployment plane of the grid. However, the grid-based topology uses sensors with omnidirectional antenna, which limits the sensing range to few meters near the wall. The tree topology is unsuitable because the sensors may require more number of relaying node due to above infrastructure constraint. In this paper, we propose a two-tailed chain topology suitable for food grain depot. The following sections discuss the proposed topology in detail.

5 Proposed Work

The two-tailed chain topology consists of a sequence of sensors that are linked with each other in the form of a chain. The sensors are equipped with directional antenna instead of omnidirectional antenna for communication. The range for reliable communication is considered within 40 m. In this topology, sensors are deployed on the inside wall of the depot. The sensors present at the end of the chain are called tail of the chain. The sensor present at the top of the chain is called head of the chain. In addition, one special relaying node is associated with the chain of sensors. The job of the relaying node is to collect and forward the data from head node to the BS. It is assumed that the relaying node is present in the communication range of the BS. The direction of the communication antenna of each sensor is faced toward the geographical location of head node as shown in the figure.

5.1 Energy-Efficient Data Collection Mechanism

In each round of data collection, the tail initiates the data collection process. The tail node sends its data to the next node toward the head of the chain. Each intermediate node in the chain receives the data from a node from tail end, aggregates its data with the received value, and forward to a node in the chain toward head node. Each node waits for data from the lower end for a specific period of time. If data is not received within a given time due to failure, the node sends an additional failure status bit along with failed node ID with its data to the node in the chain toward head node. Once the aggregated data reaches the head node, it is forward to BS via the relay node.

6 Analysis

In this section, we analyze the efficiency of the proposed topology for food grain depot compared to existing ones in WSN.

6.1 *Number of Sensor Nodes Required*

The proposed topology requires minimum number of sensors to collect the environmental data compared to existing ones. This is because, in this topology the sensors are deployed in the form of a chain on the inside wall of the depot. The topologies such as star, tree, cluster-based required that the sensors to be deployed inside the region which is quite difficult in a food grain depot. Even if it would be possible to deploy inside the region, these topologies may require more number of sensor nodes for communication because most of the time the region of interest is occupied by grain bags. Therefore, sensors with omnidirectional antenna would hardly get a clear line of sight for communication. Due to data communication across the chain of sensors using directional antenna, the proposed topology is able to achieve clear range of communication using lesser number of nodes which can be depicted from Fig. 2.

6.2 *Energy Consumption*

The data communication mechanism in the proposed topology uses aggregation at each intermediate node, which reduces the overhead of the head node to transmit all data messages to the BS. In each round, a single-aggregated data message is sent to BS by the head via relay node. Therefore, none of the nodes in the network is overloaded by the task of transmitting higher number of messages.

6.3 *Node Failure*

Upon failure of a node, it is informed to the BS via a failure status bit and node ID by its predecessor in the chain. However, connectivity may be retained by reorganization of the nodes in the network, only if the coverage of nodes across the chain is greater or equal to two. The increasing the node coverage would also increase the number of nodes in the network. Therefore, a trade-off between size of the network and node coverage is required to be considered during implementation.

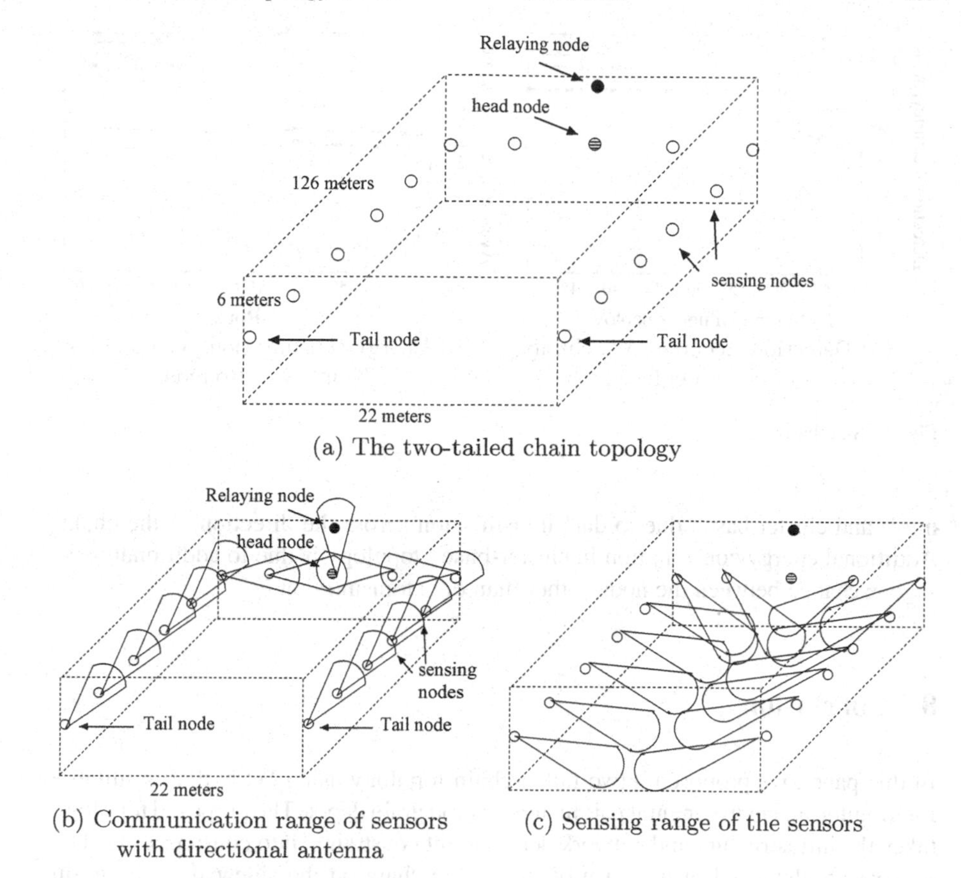

(a) The two-tailed chain topology

(b) Communication range of sensors with directional antenna

(c) Sensing range of the sensors

Fig. 2 Proposed two-tailed chain topology

7 Simulation and Results

The performance of proposed topology is verified for food grain depot environment using simulation. The parameters considered are detection accuracy and energy consumption. The moisture and temperature are varied in the simulation via physical process modules. The existing mesh and cluster-based topology are simulated by overriding deployment constraints.

The comparison of detection accuracy versus number of nodes deployed is shown in Fig. 3a. It is observed that the proposed topology is able to detect almost all environmental parameter variations with lesser number of nodes compared to mesh and cluster-based topology. This is because of the chain topology that is successfully able to sense data at various locations inside the food grain depot. The comparison of energy consumption versus number of rounds is shown in Fig. 3b. It is observed from the result that the energy consumption per node in each round is almost remains constant. The energy consumption in the proposed topology is marginally lesser than

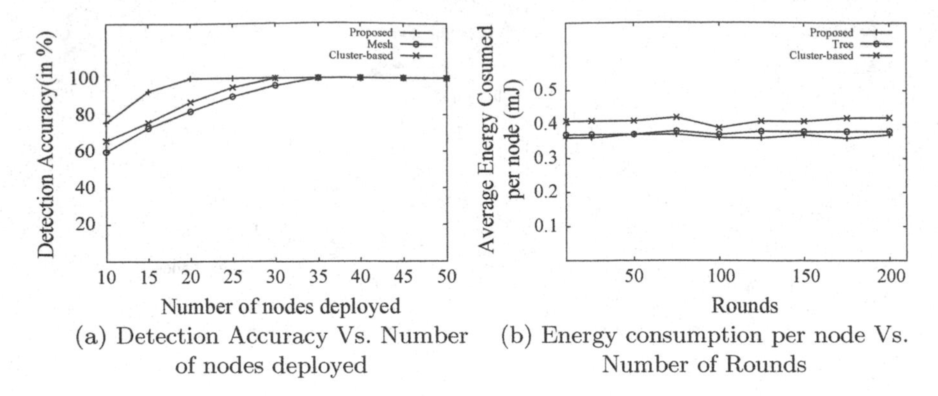

(a) Detection Accuracy Vs. Number of nodes deployed

(b) Energy consumption per node Vs. Number of Rounds

Fig. 3 Simulation results

mesh and cluster-based due to data transmission across the direction of the chain. Additional energy consumption in cluster-based topology is due to additional message exchange between the nodes other than data transmission.

8 Conclusion

In this paper, we proposed a two-tailed chain topology using WSN that is suitable for monitoring environmental parameters in food grain depot. The proposed topology takes the infrastructure and network deployment constraints into consideration. The sensors are deployed in the form of two-tailed chain on the sidewall of the grain depot. The directional antenna is used communication among the nodes inside the food grain depot to achieve clear line of sight among the sensors. The simulation result shows that the proposed strategy is able to achieve accurate detection with lesser number of nodes compared to mesh and cluster-based topology.

References

1. Tanwar, S., Kumar, N., Joel, J.P.C.: Rodrigues. A systematic review on heterogeneous routing protocols for wireless sensor network. J. Netw. Comput. Appl. **53**, 39–56 (2015)
2. Shakuntala, M.N., Shadaksharaswamy, M.: Foods Facts and Principles, 3rd edn. New Age International (2008)
3. Sawant, A.A., Patil, S.C., Kalse, S.B., Thakor, N.J.: Effect of temperature, relative humidity and moisture content on germination percentage of wheat stored in different storage structures. AgricEngInt CIGR J. **14**(2), 01–14 (2012)
4. Carlos-Mancilla, M., Lpez-Mellado, E., Siller, M.: Wireless sensor networks formation: approaches and techniques. J. Sens. **2016**(2016), 1–18 (2016)
5. Ceclio, J., Furtado, P.: Wireless sensors in heterogeneous networked systems. In: Chapter Wireless Sensor Networks: Concepts and Components, pp. 5–25. Springer (2014)

6. Fan, G., Jin, S.: Coverage problem in wireless sensor network: a survey. J. Netw. **5**(9), 1033–1040 (2010)
7. Alsemairi, S., Younis, M.: Forming a cluster-mesh topology to boost base-station anonymity in wireless sensor networks. In: Proceedings of IEEE Wireless Communications and Networking Conference, WCNC, vol. 2016, pp. 01–06 (2016)
8. Aziz, A.A., Sekercioglu, Y.A., Fitzpatrick, P., Ivanovich, M.: A survey on distributed topology control techniques for extending the lifetime of battery powered wireless sensor networks. IEEE Commun. Surv. Tutor **15**(1), 121–144 (2013)
9. Santos, A.C., Duhamel, C., Belisrio, L.S.: Heuristics for designing multi-sink clustered WSN topologies. Eng. Appl. Artif. Intell. **50**(2016), 20–31 (2016)
10. Hong, Z., Wang, R., Li, X.: A clustering-tree topology control based on the energy forecast for heterogeneous wireless sensor networks. 68 IEEE/CAA J. Autom. Sin. **3**(1), 68–77 (2016)
11. Deshpande, N., Shaligram, A.D., Botre, B.A., Bindal, S., Sadistap, S.S.: Embedded E-nose application to sense the food grain storage condition. In: Proceedings of International Conference of Computational Intelligence and Communication Networks, pp. 608–611, Nov 2010
12. Yu, Z., Teng, J., Bai, X., Xuan, D., Jia, W.: Connected coverage in wireless networks with directional antennas. In: Proceedings of IEEE INFOCOM, pp. 2264–2272, Apr 2011

Designing an Adaptive Vehicular Density-Based Traffic Signal Controlling System

Monika J. Deshmukh and Chirag N. Modi

Abstract This paper proposes an approach for solving vehicular traffic congestion problem by automating a traffic signal based on traffic density. It avoids traffic jams in complex environmental conditions such as sunny, rainy, cloudy days, sunrise, sunset, or nighttime. We design a control algorithm for complex environmental conditions, which can automate a traffic signal with minimal computation cost and maximum accuracy. It uses dark channel prior approach to remove the impact of weather and light, followed by foreground extraction with the help of Gaussian mixture model. Based on the extracted foreground, threshold-based control algorithm diverts the traffic.

Keywords Computer vision · Intelligent systems · Traffic control system
Vehicle detection · Background subtraction

1 Introduction

Due to the ever-increasing traffic demands, road transportation faces the problem of traffic congestion. This results in loss of travel time and costs huge societal and economical loss. Nowadays, smart and effective road traffic control systems have become a major focus. There are different approaches for adaptive traffic light control which have been investigated. Statistics shows that with adaptive traffic control systems (ATCSs), traffic congestions and the degree of air pollution can be greatly reduced. Intrusive methods [1] with devices such as bending plate, pneumatic road tube, piezoelectric circuits, and inductive loops are having limitations such as requirement of road excavation, high installation and maintenance cost,

M. J. Deshmukh (✉) · C. N. Modi
Department of Computer Science and Engineering, National Institute of Technology Goa, Ponda, India
e-mail: monikadeshmukh72@gmail.com

C. N. Modi
e-mail: cnmodi@nitgoa.ac.in

P. K. Sa et al. (eds.), *Recent Findings in Intelligent Computing Techniques*,
Advances in Intelligent Systems and Computing 708,
https://doi.org/10.1007/978-981-10-8636-6_12

large calibration, pneumatic road tube, low performance and cannot do lane monitoring. To address these problems, we propose a design of traffic signaling system which overcomes the traffic jams and reduces the waiting time at road intersections. It uses dark channel prior (DCP) approach [2] to remove haze and noise present in the video. To improve the video quality in nighttime, an improved DCP algorithm [1] is used. For foreground detection, the Gaussian mixture model (GMM) [3] is used. For control algorithm, threshold-based progressive area expanding approach is proposed, which makes decision of green time based on area covered by vehicles.

The rest of this paper is organized as follows: Sect. 2 discusses existing approaches in the area of vehicular traffic monitoring. Section 3 presents the proposed design of density-based adaptive traffic signal controlling system in detail. Finally, Sect. 4 concludes our research work followed by references.

2 Existing Traffic Monitoring and Signal Processing Systems

The existing approaches can be classified into sensor based and vision based. The Sydney Coordinated Adaptive Traffic System [4, 5] is deployed using inductive loops beneath the road. Split Cycle Offset Optimization Technique [4] is based on a centralized architecture and scheduling algorithm. Predefined detection points are installed on the road. It requires high cost of maintenance. Schutter [6] has derived an approximate model which describes lengths of queues as a continuous time function. Here, optimization is used over a fixed number of switch-overs of lights. Tubaishat et al. [7] have proposed a three-tier architecture based on wireless sensor network which collects data using sensor network. It is deployed in and outside of road on every intersection. Mirchandani et al. [8] have proposed an architecture which calculates the flow of traffic and optimally controls the flow. Wenjie et al. [9] have designed a wireless sensor network-based method for dynamic vehicle detection and traffic signal controlling system. The sensor-based technologies have high maintenance and installation cost. In contrast to this, the vision-based systems collect useful information regarding the traffic.

InSync [4] traffic monitoring system uses camera as a data collector, which measures the traffic flow as well as allows the live monitoring through the Internet. It is based on finite-state machine in which detection zone is divided into segments and a number of vehicles and waiting time is calculated. Wu et al. [8] have used histogram extension (HE) for lightning condition and gray-level differential value method for extracting moving objects. Tian et al. [10] have used virtual line-based sensors to detect the vehicles. Sabri et al. [11] have presented a method for real-time detection and tracking of moving targets. Siyal et al. [12] have proposed a method to extract accurate vehicle queue parameters using image processing and neural networks. Harini et al. [13] have used blob tracking algorithm with Kalman filter.

It uses GMM as a segmenting algorithm. Zhang et al. [14] have proposed a vision-based vehicle queue detection method which uses segment-based architecture. It uses edge detection and binary thresholding technique.

3 Proposed Approach

The objective of this paper is to design an automated traffic controlling system which can accept a real-time data from the cameras installed on road intersections and can calculate an area covered by vehicles with their waiting time for each route. Based on these parameters, it should prioritize the green time allocation to give sequence of traffic signals optimally. To achieve above objectives, we use dark channel prior (DCP) dehazing approach [1, 2] and background subtraction with GMM [3].

As shown in Fig. 1, a generalized model consists of three modules, viz. *data collector* (e.g., cameras), *controller*, and *centralized server*. Cameras collect the data about road traffic. Controller acts as processing infrastructure. Controller dynamically decides timing and sends it to a signal controller based on the calculated density. Centralized server collects data from controller, removes noise, and extracts foreground. It depends on density of objects and gives the results to controller.

The proposed approach uses progressive area expanding approach in which the initial area is considered from stop line of each road. It then defines regions of interest (ROIs), which are limited to the road surface only. The initial ROI is approximately equal to 15 s of green signal time (refer Fig. 2). When an area occupied by vehicles is more than threshold, i.e., 75%, it increases the ROI approximately by a vehicle size and repeats until defined threshold is satisfied, which depends on traffic conditions, timing on a particular intersection (refer Figs. 3, 4, 5, and 6).

The proposed approach is divided into three modules, viz. *video preprocessing*, *object detection*, and *control algorithm*.

Fig. 1 Generalized system

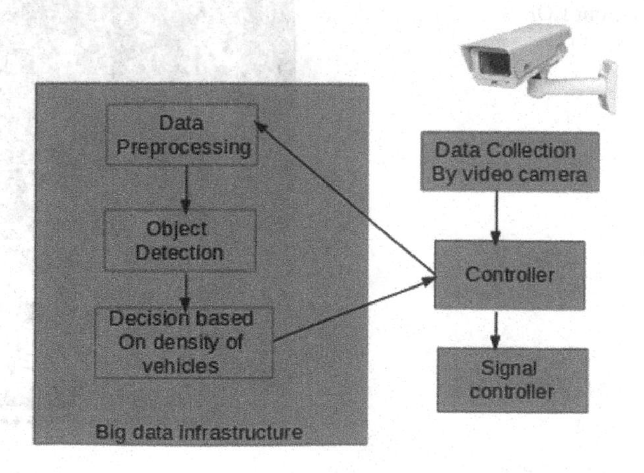

Fig. 2 Initial ROI defined

Fig. 3 Vehicles in initial ROI

Fig. 4 Extracted foreground from ROI

Fig. 5 Expanded ROI

Fig. 6 Extracted foreground for ROI

3.1 Video Preprocessing

As shown in Fig. 7, the video captured from four roadside cameras is converted into frames, i.e., 10fps. After conversion into frames and by checking the pixel value, the images are classified as either daytime or nighttime image based on the mean of intensities of each pixel in the image. If mean of pixels is greater than threshold, then it is considered as a daytime; otherwise, it is considered as a nighttime image. Based on this classification, the preprocessing takes place. For nighttime images, we use an improved DCP [1]; otherwise, it checks the RGB pixel values. If all RGB values are high and no value tends to zero, then the image is considered as haze-free image. This image is converted into gray scale and sent to the object detection. If any value tends to zero that implies haze is present in the

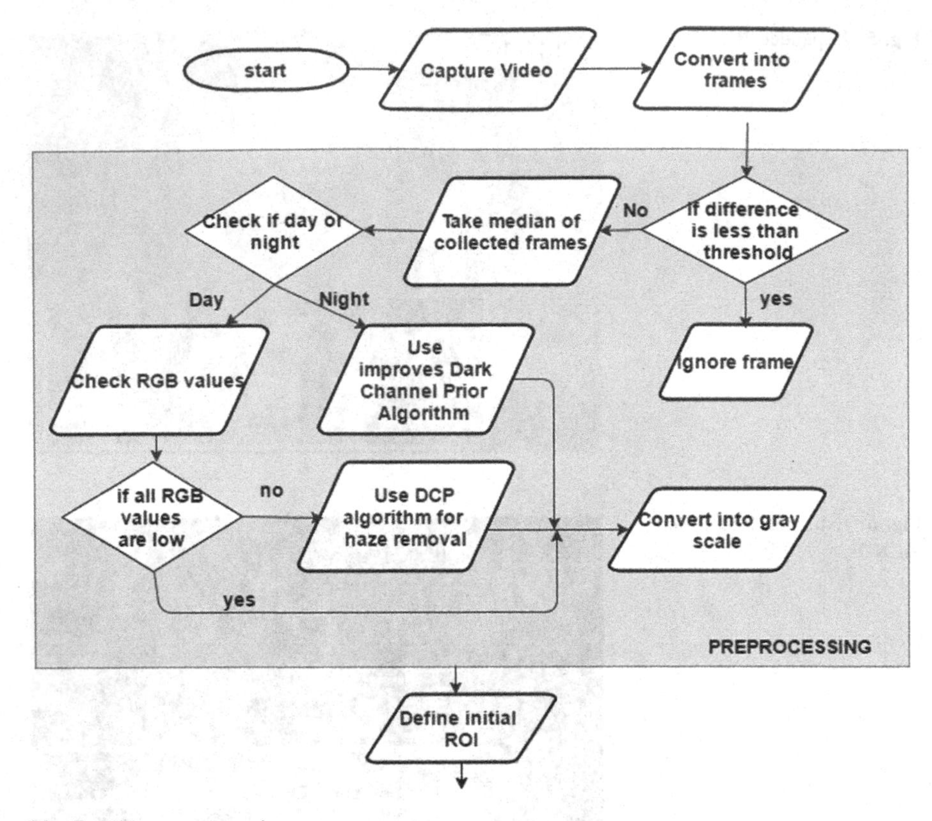

Fig. 7 Video preprocessing

image and processed using DCP haze removal technique [2]. It gives the haze-free output with initial ROI (refer Fig. 2).

3.2 *Object Detection*

As shown in Fig. 8, it defines region of interest (ROI) on starting of road for approximately 15 s. The gray image is fed to GMM for foreground detection. GMM gives an area occupied by each vehicle. As preprocessing module removes the effects of environment, it gives good results with less complexity.

3.3 *Control Algorithm*

It takes the detected foreground as input and calculates an area occupied by vehicle (refer Fig. 9). If an area occupied is greater than 75%, it increments that area by

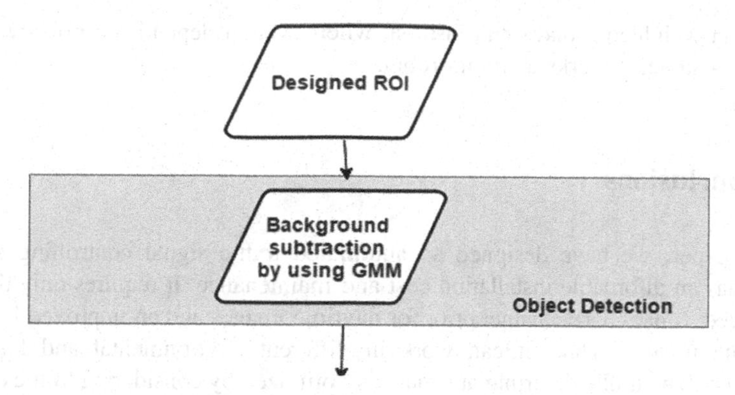

Fig. 8 Object detection approach

approximate vehicle size and timer by 5 s. The decision about giving priority to green signal is taken based on the density as well as waiting time. For giving priorities, it considers the following conditions: The waiting time should not exceed 200 s, and green signal time should not exceed 100 s. When green time is over, initialize area, priority, and time for corresponding thread and process the other

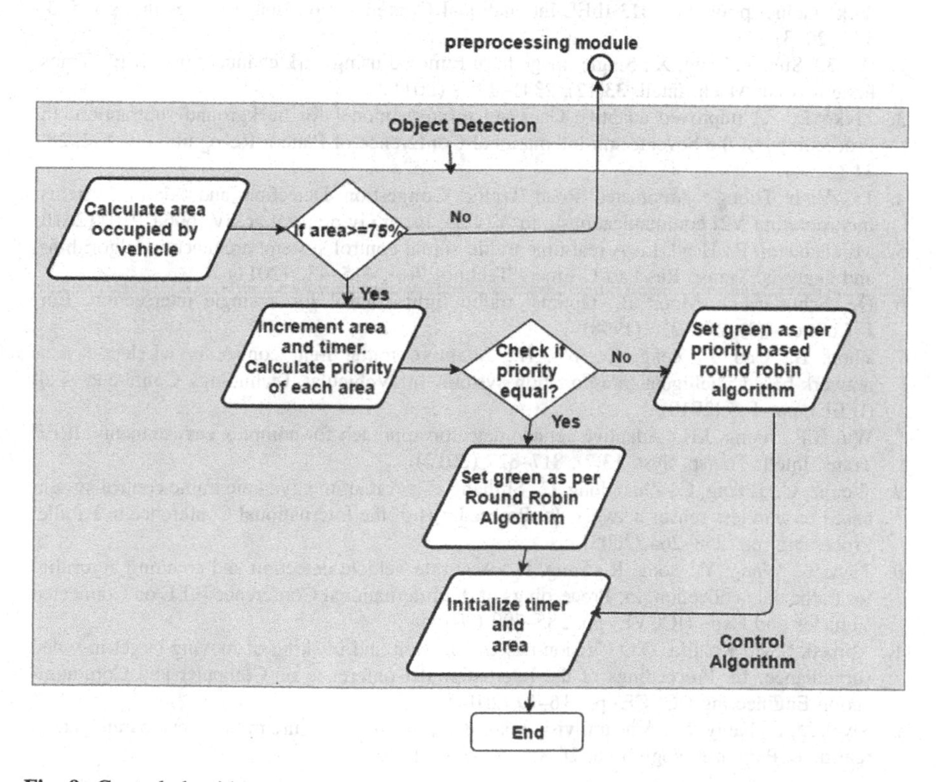

Fig. 9 Control algorithm

threads in weighted round robin fashion, where weight depends on priority. If the priority is same, it works as round robin.

4 Conclusions

In this paper, we have designed an automated traffic signal controlling system which has an affordable installation cost and maintenance. It requires only the live video feed. It uses dark channel prior for daytime images and an improved DCP for nighttime images. Thus, it can work in different environmental and lightning conditions. The traffic diverting at signals is optimized by considering traffic density and waiting time. The proposed approach offers an efficient and effective traffic signaling system with an affordable computation cost. Thus, it is very encouraging.

References

1. Jiang, X., Yao, H., Zhang, S., Lu, X., Zeng, W.: Night video enhancement using improved dark channel prior. In: 2013 IEEE International Conference on Image Processing, pp. 553–557 (2013)
2. He, K., Sun, J., Tang, X.: Single image haze removal using dark channel prior. IEEE Trans. Pattern Anal. Mach. Intell. **33**(12), 2341–2353 (2011)
3. Zivkovic, Z.: Improved adaptive Gaussian mixture model for background subtraction. In: Proceedings of the Seventeenth International Conference of Pattern Recognition, vol. 2, 28–31 (2004)
4. Ta, Vinh Thong.: Automated Road Traffic Congestion Detection and Alarm Systems: Incorporating V2I communications into ATCSs. In: arXiv preprint arXiv:1606.01010 (2016)
5. Mirchandani, P., Head, L.: A real-time traffic signal control system: architecture, algorithms, and analysis. Transp Res Part C Emerg Technol **9**(6), 415–432 (2011)
6. De, Schutter, De Moor, B.: Optimal traffic light control for a single intersection. Eur. J. Control **4**(3), 260–276 (1998)
7. Zhou, B., Cao, J., Zeng, X., Wu, H.: Adaptive traffic light control in wireless sensor network-based intelligent transportation system. In: Vehicular Technology Conference Fall (IEEE), pp. 1–5 (2010)
8. Wu, B.F., Juang, J.H.: Adaptive vehicle detector approach for complex environments. IEEE Trans. Intell. Transp. Syst. **13**(2), 817–827 (2012)
9. Wenjie, C., Lifeng, C., Zhanglong, C., Shiliang, T.: A realtime dynamic traffic control system based on wireless sensor network. In: Proceedings of the International Conference in Parallel Processing, pp. 258–264 (2005)
10. Tian, Y., Wang, Y., Song, R., Song, H.: Accurate vehicle detection and counting algorithm for traffic data collection. In: Proceedings of the International Conference IEEE on Connected Vehicles and Expo (ICCVE) pp. 285–290 (2015)
11. Ahmed, S.M., Khalifa, O.O.: Vision-based detection and tracking of moving target in video surveillance. In: Proceedings of the International Conference on Computer and Communication Engineering (ICCCE) pp. 16–19 (2014)
12. Siyal, M.Y., Fathy, M.: A neural-vision based approach to measure traffic queue parameters in real-time. Pattern Recogn. Lett. **20**(8), 761–770 (1999)

13. Veeraraghavan, H., Masoud, O., Papanikolopoulos, N.P.: Computer vision algorithms for intersection monitoring. IEEE Trans. Intell. Transp. Syst. **4**(2), 78–89 (2003)
14. Zhang, J., Liu, Z.: A vision-based road surveillance system using improved background subtraction and region growing approach. In: Eighth ACIS International Conference on Software Engineering, Artificial Intelligence, Networking, and Parallel/Distributed Computing, vol. 3, pp. 819–822 (2007)

Moving Object Detection and Tracking in Traffic Surveillance Video Sequences

Pranjali Gajbhiye, Naveen Cheggoju and Vishal R. Satpute

Abstract In the field of traffic video surveillance systems, multiple object detection and tracking has a vital role to play. Various algorithms based on image processing techniques have been used to detect and track objects in video surveillance systems. So it is required to develop an algorithm which is computationally fast and robust to noisy environment. In this paper, variance-based method for multiple object detection and tracking of vehicles in traffic surveillance is proposed and compared with the existing method. This new method is based on five-frame background subtraction, five-frame differencing, and variance calculation for object detection and tracking. To evaluate the proposed method, it is compared with the existing methods such as standard mean shift method. The experimental results show that this method gives better accuracy and takes comparable computational time when compared with mean shift method, which is required in traffic surveillance systems. The comparative shows that the proposed method gives about 99.57% accuracy in detection whereas mean shift gives 88.88% accuracy.

Keywords Multiple object tracking · Background subtraction
Five-frame differencing · Variance method

P. Gajbhiye (✉) · N. Cheggoju · V. R. Satpute
Department of Electronics and Communication Engineering,
Visvesvaraya National Institute of Technology, Nagpur, India
e-mail: gajbhiyepranjali@gmail.com

N. Cheggoju
e-mail: cheggojunaveen@students.vnit.ac.in

V. R. Satpute
e-mail: vrsatpute@ece.vnit.ac.in

P. K. Sa et al. (eds.), *Recent Findings in Intelligent Computing Techniques*,
Advances in Intelligent Systems and Computing 708,
https://doi.org/10.1007/978-981-10-8636-6_13

1 Introduction

Moving object detection and tracking in the video sequences is one of the important works in the field of computer vision. The purpose of moving object detection and tracking is to check the presence of object and track it in the moving video sequence. The main part in object tracking is to get the relationship of the object in consecutive frames. The commonly used methods for tracking are the Gaussian mixture model (GMM), mean shift method, frame differencing with centroid method, and the Kalman filter method. There exist a lot of parameters which affect the detection of moving object such as illumination changes, shadow of moving objects, background movement, and the movement of camera [1]. The applications of the object tracking include traffic surveillance, object detection and tracking, face tracking, object tracking, image recognition, image processing in medical field.

To build a higher level vision-based intelligence system, an accurate and efficient tracking capability is essential. The main objective of object tracking in such systems is to analyze the object details of the consecutive frames in the video sequence. When performing the object detection, the main aspects to be considered are the random motion of the objects in the video sequence and the process of the video capturing phenomenon. The detection can be very difficult if the motion of the object is relatively faster than the frame rate. Considering previous work mentioned in [2], one can say that the object detection and tracking is a difficult process indulging a lot of problems even for a fixed camera network. These problems include object segmentation and performance improvements in moving object tracking. In this paper, the above-mentioned problems are taken into consideration and contributions are made to overcome the problems in object tracking. One of the important algorithms for tracking object is mean shift. It is an iterative process to get weighted average value of data points with shifting. This method is usually used in clustering and computer vision task such as image segmentation and object tracking [3, 4]. Disadvantages of the mean shift algorithm [5, 6] include inaccurate false tracking and its non-adaptability to the changes in object size or shape. Although this method is computationally very efficient, it fails in handling the occlusions caused by multiple objects.

The main objectives of the paper are as follows:

- To improve the object segmentation algorithm in the video sequence.
- To analyze methods for single and multiple object tracking.

The glimpses of the next sections are as follows: The proposed variance-based algorithm for multiple object detection and tracking is discussed in detail in Sect. 2. Experimental results and comparison between mean shift method and the proposed algorithm are discussed in Sects. 3 and 4 concludes the paper.

2 Proposed Variance-Based Method for Multiple Object Detection

In this paper, a new variance-based approach is presented for multiple object detection. Here, standard five-frame differencing method and five-frame background subtraction method are used for separating the object from its background. The resulting frames are further binarized, and morphological operation is used for removing unwanted pixels and noise from the binarized frames. After binarization, we are using variance-based approach for moving object detection and tracking.

The flowchart shown in Fig. 2 explains the complete flow of the proposed algorithm. The video capturing and converting it into grayscale format is the first step toward video processing. Now to detect the moving pixels from the video frame, we are using five-frame background subtraction method and frame difference method. The separate outputs of five-frame difference and background subtraction method are shown in Figs. 1 and 3 respectively. For getting perfect counter of moving pixels, we are using ORing operation between these two outputs. Morphological operation is done for the removal of unwanted noise pixels which came due to illumination changes or camera motion. At this stage, we are getting perfect binary image (after applying thresholding). Variance calculation leads to give proper bounding box around detected moving pixel. Each step is discussed in detail in the next subsections.

2.1 *Five-Frame Difference Method*

To overcome the disadvantage associated with the two-frame and three-frame difference [7], we have chosen five-frame difference method which is shown in Fig. 1. Here, five consecutive frames are taken namely I_{n-1}, I_n, I_{n+1}, I_{n+2}, and I_{n+3}. Four differential images (I_A, I_B, I_C, and I_D) are obtained from I_{n-1} and I_n, I_n and I_{n+1}, I_{n+1} and I_{n+2}, and I_{n+2} and I_{n+3}. The frames I_E, I_F, and I_G are obtained by logical OR operation between I_A and I_B, I_B and I_C, and I_C and I_D, respectively. The frames I_H and I_I are obtained by logical OR operation between frames I_E and I_F, and I_F and I_G, respectively. The resultant frame I_J is obtained by logical OR operation between frames I_H and I_I.

2.2 *Five-Frame Background Subtraction Method*

Background subtraction is widely used process for moving pixel extraction from video frames. In background subtraction, the current frame is subtracted from reference frame (followed by morphological operation) for getting moving pixels. But for accuracy purpose, here we are using five-frame background subtraction. Here, five consecutive frames are taken namely I_n, I_{n+1}, I_{n+2}, I_{n+3}, and I_{n+4}

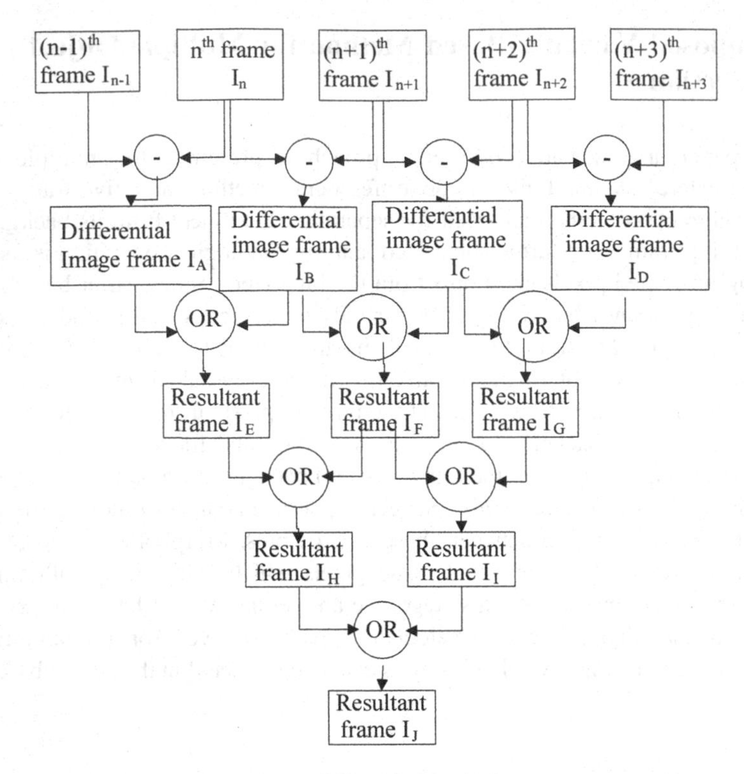

Fig. 1 Schematic diagram of five-frame differencing method

(where n = 1). Four differential images (I_K, I_L, I_M, and I_N) are obtained from I_{n+1} and I_n, I_{n+2} and I_n, I_{n+3} and I_n, and I_{n+4} and I_n. The frames I_O, I_P, and I_Q are obtained by logical OR operation between I_K and I_L, I_L and I_M, and I_M and I_N, respectively. The frames I_R and I_S are obtained by logical OR operation between frames I_O and I_P, and I_P and I_Q, respectively. The resultant frame I_T is obtained by logical OR operation between frames I_R and I_S.

2.3 *Binarization and Morphological Operation*

Binarization is the technique for converting a grayscale image into binary image, i.e., black and white. For binarization process, a threshold value is needed to be calculated. Here, a global threshold is calculated which will vary from frame to frame depending upon the mean (u) and standard deviation (σ). The formula for calculating threshold (T) is given in (1), (2), and (3), respectively:

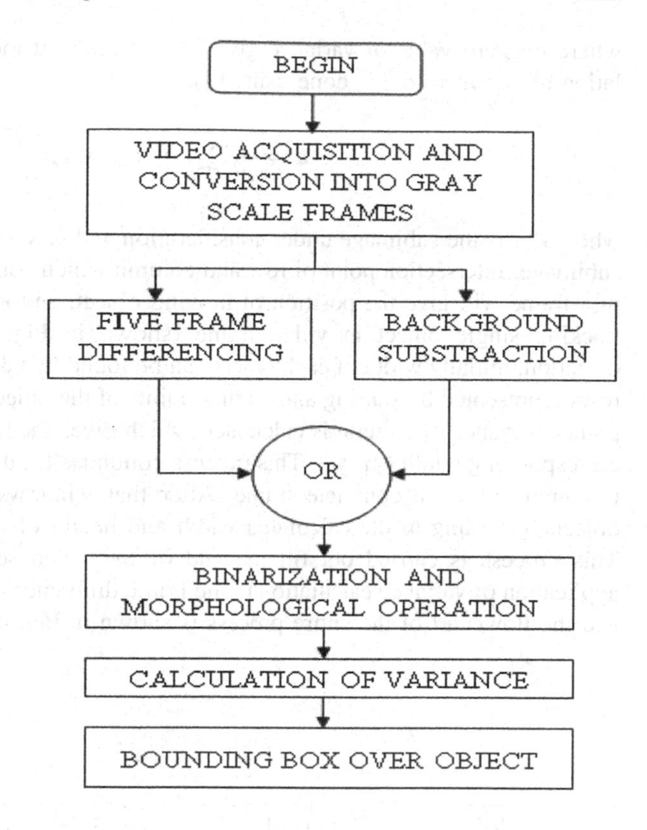

Fig. 2 Steps of variance-based algorithm for multiple object detection

$$= \frac{1}{r*c} \sum_{i=1}^{r} \sum_{j=1}^{c} \mathrm{I_C}(\mathrm{i},\mathrm{j}) \tag{1}$$

$$\sigma = \sqrt{\frac{1}{r*c} \sum_{i=1}^{r} \sum_{j=1}^{c} (\mathrm{I_C}(\mathrm{i},\mathrm{j}) - u)^2} \tag{2}$$

$$\mathrm{T} = 0.05 * \sigma \tag{3}$$

where r is the number of rows and c is the number of columns. To remove small unwanted pixels and noises from images, morphological operations such as opening, closing, erosion, dilation, and morphological filters are used.

2.4 Variance Calculation

In this paper, a new variance-based method is used for finding the location of moving objects. Variance [8] shows the variation in the image intensity in frames, where nonzero value of variance gives the location of moving object. The calculation of variance (σ^2) is done using (4).

$$\sigma^2 = \frac{1}{n^2} \sum_{x-1}^{m} \sum_{y-1}^{n} I_{sub}^2(x, y) \tag{4}$$

where I_{sub} is the subimage under consideration and $m \times n$ represents the size of the subimage. Intersection point of row and column which contains maximum variance in a frame will give the position of moving object, and it is useful in finding and tracking single object in video frame (shown in Fig. 5). For multiple object detection, initially width of each object can be found by taking values of variance in rows represented by starting and ending points of the object (x_1, x_2). Between these points, variance of column is calculated which gives the height of the object in the corresponding width (y_1, y_2). This process continues till the calculation of variance is completed for a complete frame. After that windows are formed around the objects according to the calculated width and height of the corresponding object. This process is carried out till the end of the video sequence. The process of application of variance calculation to the frame difference output is shown in Fig. 2, and the flowchart of the entire process is shown in Fig. 4.

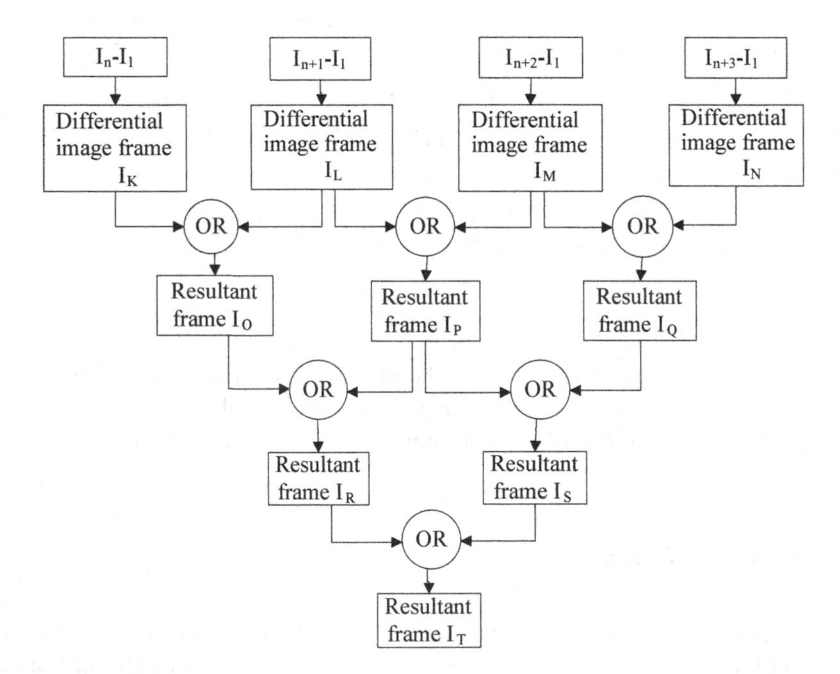

Fig. 3 Schematic diagram of five-frame background subtraction method

3 Simulation Results and Discussion

The proposed method has been tested on various video databases which are captured at different locations. A sample database of videos with various sizes and textures used for testing the algorithm is given in Table 1. The software simulation tool used for the evaluation of the algorithm is MATLAB which is executed on the hardware configuration of Intel Core (TM) i7 processor with 3.40 GHz clock frequency and 32 GB of RAM. True–false analysis is carried out on the algorithm to evaluate its speed and accuracy. From Fig. 6. it is clearly evident that moving object detection done successfully without noise for the sampled frame.

Fig. 4 Flowchart of calculation of variance

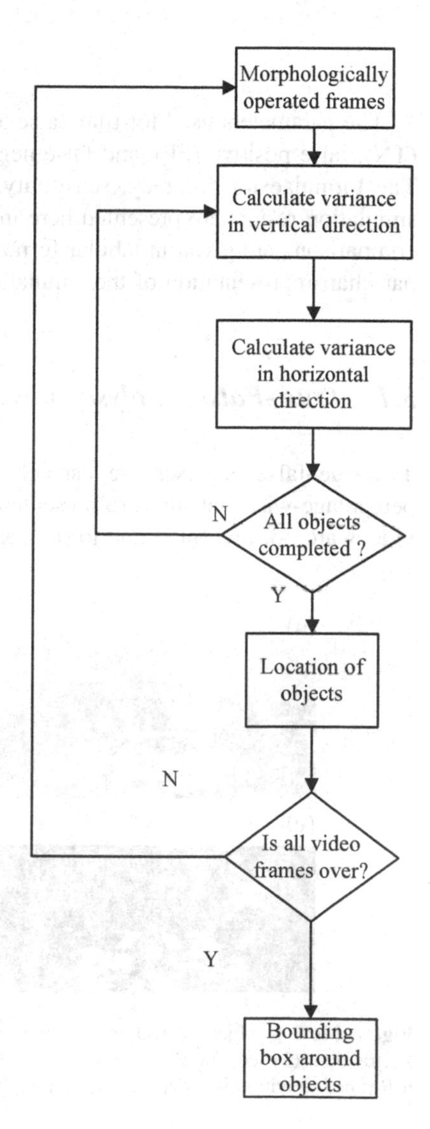

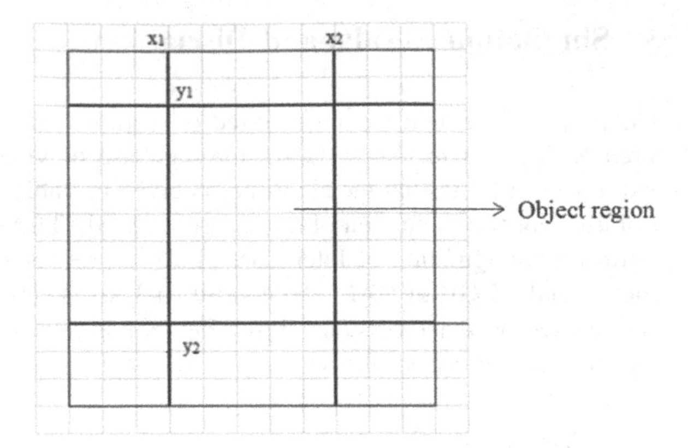

Fig. 5 Location of object pixels found by variance method

The parameters used for true–false analysis are true positive (TP), true negative (TN), false positive (FP), and false negative (FN) which are explained in Table 2. The formulae for accuracy, sensitivity, and specificity are given in Table 3. The simulation results are presented here in two categories: (i) True–false analysis and comparisons are given in tabular format, and (ii) the reconstructed images and the bar chart representation of the simulations are presented.

3.1 True–False Analysis and Algorithm Comparison

The true–false analyses are carried out on the proposed algorithm, and their percentage-wise statistics are presented in Table 1. In this table, the results for all videos are not presented due to the space constraint. For example, in Table 1, the

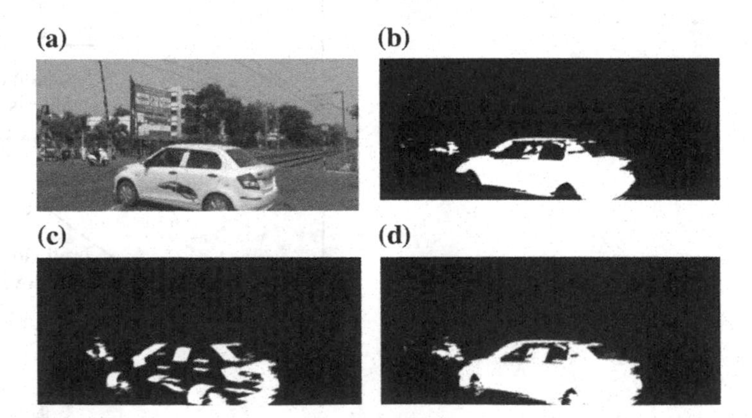

Fig. 6 Moving object detection. **a** Original image. **b** Detected moving pixel background subtraction (313ed frame). **c** Five-frame difference moving pixel detection (313ed frame). **d** Refined moving pixel detection after morphological operations and ORing between (**b**) and (**c**)

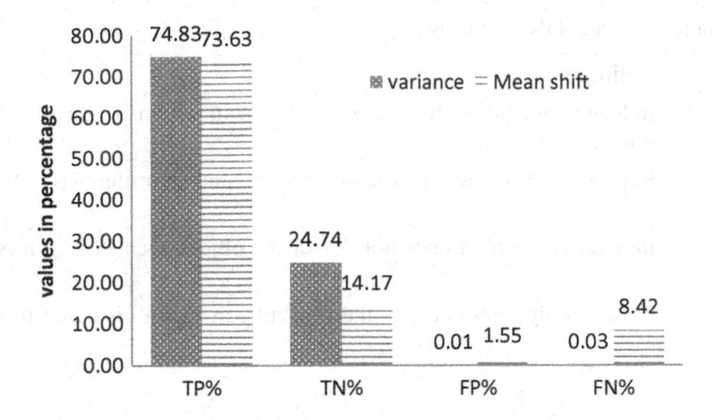

Fig. 7 Bar chart—comparative analysis of variance-based method

Table 1 Comparative analysis between mean shift and variance

Video name (mp4)	Resolution	No of frames	Method	% tp	% tn	% fp	% fn	A	Se	Sp
test9	640 * 350	246	M	76	15	2	3	91	0.7	1.0
			V	64	35	0	0	100	1.0	1.0
trial4	1920 * 1080	764	M	88	8	0	2	96	0.9	1.0
			V	95	4	0	0	100	1.0	1.0
trial5_1	640 * 350	477	M	74	18	0	8	92	0.8	1.0
			V	78	20	0	0	100	1.0	1.0
Manish_traffic	1920 * 1080	273	M	28	5	1	17	79	0.6	0.9
			V	21	78	0	0	100	1.0	1.0
manish2_traffic	1920 * 1080	759	M	97	0	2	3	97	0.9	0.1
			V	99	0	0	0	100	1.0	1.0
mamish6-1traffic	1920 * 1080	979	M	89	2	1	4	91	0.9	0.7
			V	95	4	0	0	100	1.0	1.0
manish7_1_traffic	1920 * 1080	762	M	75	16	1	3	92	0.9	0.9
			V	81	18	0	0	99	1.0	0.9
manish17_1	1920 * 1080	405	M	77	22	0	5	100	0.9	0.0
			V	72	27	0	0	100	1.0	1.0
car_track	1920 * 1080	187	M	50	4	9	13	55	0.7	0.3
			V	63	37	0	0	100	1.0	1.0
manish_26	1920 * 1080	994	M	76	2	0	1	78	0.9	1.0
			V	75	24	0	0	100	1.0	1.0

Note M—Mean shift, V—variance

statistics of the video "manish2_traffic.mp4" can be explained as follows: 99.47% of frames have objects in them and are detected properly, 0.45% of frames do not have object and are not detected, 0.00% of the frames have object but the algorithm does not detect, and 0.00% frames do not have object but it is detected. The proposed algorithm is compared with the existing algorithm such as mean shift, and

Table 2 Parameter for true–false analysis

Parameters	Definition
True positive (TP)	Indicates number of frames of the video in which object id is there, and it got detected correctly
True negative (TN)	Represents object is not there in the video and algorithm has failed to detect it
False positive (FP)	Indicates algorithm does not detect the object even though it is there
False negative (FN)	Indicates algorithm detects the object even though it is not there

Table 3 Formulae of accuracy, sensitivity, and specificity

Parameters	Definition	Formula
Sensitivity (Se)	It measures the proportion of positives that are correctly identified. It is also called true positive rate	$\frac{TP}{TP+FN}$
Specificity (Sp)	It measures the proportion of negatives that are correctly identified. It is also called true negative rate	$\frac{TN}{TN+FP}$
Accuracy (A)	The degree to which the result of a calculation confirms to the correct value or standard	$\frac{TP+TN}{Totalframes}$

the average results are given in Table 1. Mean shift algorithm is found out to be the nearest competitors to the proposed algorithm. Hence, this method is taken for comparison with the variance-based method.

3.2 Reconstructed Images and the Bar Chart Representation of Simulated Results

This subsection contains some experimental results for multiple object detection and tracking using variance-based method. In Fig. 8, the results of 161st frame of "highway_traffic.mp4" are presented in which (e) represents original image and (f) represents multiple object detected frame of variance-based detection. The white boxes shown in Fig. 9 represent that the object is found and the algorithm is able to track the object. For showing the continuity of tracking, frame #370, frame #398, frame #419, and frame #450 are shown in subfigures (g), (h), (i), and (j). From the series of (g) to (j), it can be clearly observed that all objects in those frames are tracked without being skipped in any of the frame. From the above resultant frames, it can be clearly observed that single object detection and multiple object detection are done perfectly. This represents the robustness of the proposed algorithm in continuous detection and tracking. Figure 7 represents the bar chart in which the

Fig. 8 Multiple object detection. **e** Original frame (161st of highway_traffic.mp4). **f** Detected frame using variance-based method

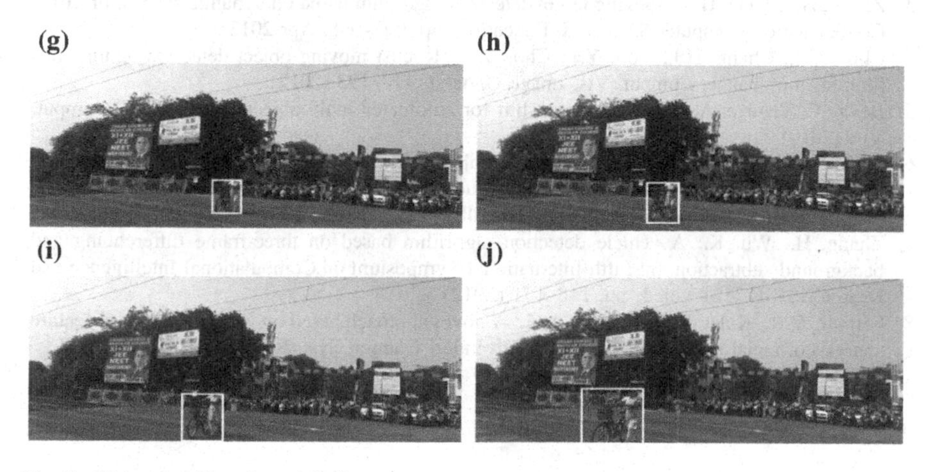

Fig. 9 Object tracking from (g)–(j)

proposed algorithm is compared with mean shift method. From the bar chart, it is very clear that the proposed method is outperforming the other compared methods in terms of percentage detection. It can be seen that FP and FN values constitute less than 0.01% of overall detection; this shows that the algorithm is adaptable to the sudden changes occurring in the frame.

4 Conclusion

In this paper, variance-based approach for multiple object detection and tracking is proposed. The proposed method is compared with mean shift method. Both the algorithms are tested on different types of videos. From results section, it is clearly evident that the detection accuracy of the proposed method is far better than the existing algorithms. In terms of numbers, the proposed algorithm has the detection accuracy of 100% and the mean shift has 88.88% accuracy. For the comparison of these, numbers clearly indicate the robustness of the proposed method. Hence, it can be concluded that variance-based approach is giving good result over the other methods. The experimental results prove the feasibility and usefulness of the proposed method.

References

1. Piccardi, M.: Background subtraction techniques: a review. Syst. Man Cybern. **4**, 3099–3104 (2004)
2. Hu, W., Tan, T., Wang, L., Maybank, S.: A survey on visual surveillance of object motion and behaviours. IEEE Trans. Syst. Man Cybern. Part C (Appl. Rev.) **34**, 334–352 (2004)
3. Zhang, H., Zhang, H.: A moving target detection algorithm based on dynamic scenes. In: IEEE Conference on Computer Science & Education, pp. 995–998, Apr 2013
4. Choi, J.M., Chang, H.J., Yoo, Y.J., Choi, J.Y.: Robust moving object detection against fast illumination change. Comput. Vis. Image Underst. 179–193 (2012)
5. Beyan C, Temizel A.: Adaptive mean-shift for automated multi object tracking. IET Comput. Vis. (2010)
6. Santosh, H., Krishna Mohan P.G.: Multiple objects tracking using extended kalman filter, gmm and mean shift algorithm—a comparative Study. In: Advanced Communication Control and Computing Technologies (ICACCCT), pp. 1484–1488 (2014)
7. Zhang, H., Wu, K.: A vehicle detection algorithm based on three-frame differencing and background subtraction. In: Fifth International Symposium on Computational Intelligence and Design (ISCID '12), vol. 1, pp. 148–151 (2012)
8. Satpute, V.R., Kulat, K.D., Keskar, A.G.: A novel approach based on variance for local feature analysis of facial images. IEEE Recent Adv Intell Comput Syst (RAICS) 210–215 (2011)

Dynamic Distance-Based Cluster Head Election for Maximizing Efficiency in Wireless Sensor Networks Using Artificial Neural Networks

A. K. P. Kovendan, R. Divya and D. Sridharan

Abstract The development of ICT (information and communication) technologies has paved a very great path for sophisticated life through numerous smart gadgets. Such an advancement in hardware technology paved pathway for the evolution of wireless sensor networks (WSNs). WSN has grasped the attention of a number of researchers in the recent era due to its potential in wide areas forced by scalable and adaptable nature of such tiny sensor nodes. Though it can facilitate number of critical and potential applications, it still suffers a number of challenges like energy efficiency, coverage. Artificial neural networks (ANNs) have been a proven technology in WSN for the development of localization-based frameworks. In this paper, a comprehensive approach is being proposed for incorporating ANN in determining the distance parameter for electing cluster head by modifying the existing LEACH protocols to attain energy efficiency in order to ensure the effective utilization of available power resource in WSN. A dynamic algorithm is proposed for cluster head election as it utilizes distance factor achieved from the received signal strength. Artificial neural network is being incorporated to utilize its effectiveness of faster computation without compromising the computational cost. The performance effectiveness of the proposed algorithm has been evaluated using comparative study with existing algorithms.

Keywords WSNs (Wireless sensor networks) · ANNs (Artificial neural networks) · Localization · Cluster head · Energy efficiency · LEACH

A. K. P. Kovendan (✉) · R. Divya · D. Sridharan
Department of ECE, CEG campus, Anna University, Guindy, Chennai 600032, Tamil Nadu, India
e-mail: kpkvendan@gmail.com

P. K. Sa et al. (eds.), *Recent Findings in Intelligent Computing Techniques*, Advances in Intelligent Systems and Computing 708,
https://doi.org/10.1007/978-981-10-8636-6_14

1 Introduction

Rapid technological advancement in the domain of electronics, communication technology especially wireless communication and mechanical systems has flashed the light for WSN technology. Wireless sensor network or wireless sensor and actuator networks are a collection of spatially distributed tiny sensor node capable of communicating each other without any external means of network elements and thus forming an autonomous network. Such a network is capable of adapting various topologies and can effectively monitor environmental parameters or physical conditions for enabling an effective monitoring system that can be applied for any potential or critical cause [1]. A very few of such cause are disaster management, biomedical health monitoring, military operations, smart grids, etc. Artificial neural network (ANN) is a computational approach [2–4]. This approach uses a collection of large data modeled in a way that a biological brain is processing data. Such type of modeling is being used in WSN for localization of the sensor nodes. Localization is a critical factor in WSN especially when incorporated in disaster monitoring or military applications, etc. [5] Nodes in WSN are powered by small AA sized battery as it operates at ultra-low-power radio frequency range. In most of the cases, it becomes harder in order to physically access the deployed nodes, and it becomes mandatory to utilize the available power effectively to retain the network in normal operating state [6]. There are a number of routing algorithms that have been proposed for maximizing the reliability and energy efficiency; still there exist a wider opportunities for researchers to further enhance the technology. Based on the recent researches and survey on WSN's routing platform, the classification of such protocols can be listed under four main groups, namely geographical-based routing, data-based routing, cluster-based routing, and hybrid routing [7, 8]. Among the available protocols, cluster-based protocols have been incorporated in maximum because of its low-complex computation and energy efficient nature. Among various available protocols for cluster-based routing in WSN, LEACH is the most predominant one.

LEACH is an abbreviated form of low energy adaptive clustering hierarchy. LEACH is a time division multiple access (TDMA)-based medium access control (MAC) protocol which incorporates clustering and routing of WSN to reduce the energy consumption in WSN [9–12]. LEACH is one of the most popular hierarchical protocols for maximizing the lifetime of WSN by attaining energy efficiency in the data aggregation through election of cluster heads/base station by randomized election of cluster heads. In LEACH, sensor nodes will self-organize thus forming a cluster and the protocols instead perform self-organizing and re-clustering in each energy rounds to achieve maximized lifetime of network node and also in data aggregation process by transmitting compressed to cluster head. There are many protocols that have been proposed to overcome the disadvantages of existing LEACH as an advancement and are named as A-LEACH (angled LEACH), LEACH-B (balance LEACH), LEACH-C (centralized LEACH), LEACH-E (energy LEACH), LEACH-F (fixed no. of cluster low energy adaptive clustering hierarchy), M- LEACH (multi-hop LEACH) and V-LEACH (vice cluster level low energy adaptive clustering hierarchy) [12]. The comparison of available LEACH protocols and the comprehensive study has been given. ANN-based WSN for indoor localization has been presented, and various other works symbolize the adaptability and performance enhancement of WSN through ANN incorporation which is better explained [13–15].

Organization of the rest of the paper is as follows. Section 2 explains the ANN-based WSN available methodology and protocols. Section 3 describes the proposed system model and the obtained results along with appropriate comments on the obtained results. Section 4 summarizes the results as conclusion and necessary future research directions and advancements needed.

2 Literature Review on ANN-Based Networking with WSN

The traditional LEACH protocol first elects the cluster head at its setup phase and then advertises the neighbor nodes. The non-head of the cluster joins the cluster based on the received signal strength. This protocol fails to note the distance of the node from sink, and also, the cluster head itself has to be chosen by considering the distance from sink at setup phase. The distance parameter is being considered because the node energy consumption profile variation mainly depends on the distance factor [14]. Artificial neural networks were biological-induced system where the working of human brain is being considered as a working model. The number of layers and the number of data sets have no restriction and can be processed with lesser complexity in ANN for obtaining optimized results [13, 15]. ANN-based localization and security conflicts in WSN paved the path for introducing distance-based cluster head election in a dynamic fashion on each energy rounds to optimize the energy usage profile in WSN data aggregation strategy. The distance parameter can be obtained from the signal strength itself instead of incorporating additional localization techniques, which provides the path for reliable data delivery as the analogy states, and signal strength will be more only in low attenuated and shortest path by distance. Such an additional factor is being considered for electing cluster head in the setup phase. Clustered architecture of WSN is shown in the Fig. 1 [16]. The ANN-based computation utilizes less cost of energy compared to the sensor-based processing. The node level cluster head election is based on the energy level available with the sensor nodes at that particular energy. But, the ANN-based election considers the received signal strength and performs the computation within the neural network and thus involves lesser energy cost profile [17]. The ANN-based data aggregation has resulted in a better reliability and optimized sensor power management.

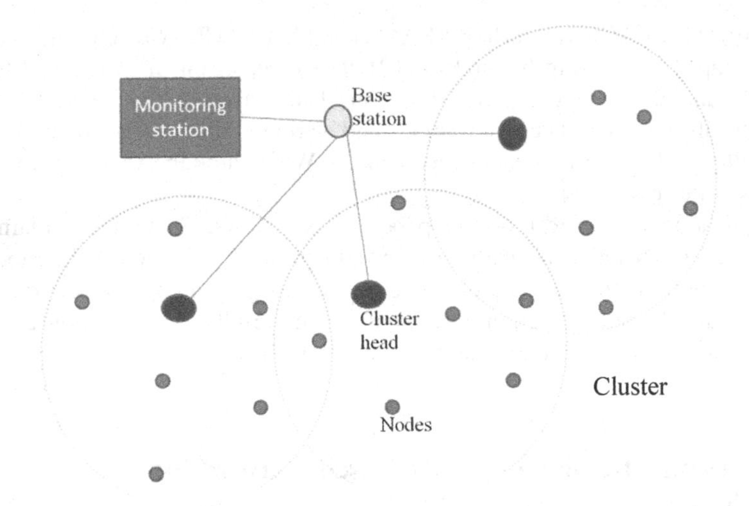

Fig. 1 WSN clustered architecture

3 Proposed System Model

The key factor that has to be considered with WSN protocol design is energy efficiency. The efficiency in power consumption can be resulted by optimized usage of energy profile. To optimize the power consumption, the cost on energy has to be reduced without compromising the system performance. The initial energy will be a fixed amount at the initial setup phase of the network and to be taken as E_o. Let the distance be taken as d between nodes. Neural networks have been used for computation, and the computational energy is conserved as compared to the traditional approach. The maximum energy factor is being used in LEACH for electing the cluster head and in the modified version of LEACH with ANN with distance-based cluster head is the proposed algorithm and that uses minimum distances node with more energy as the cluster head. The artificial neural approach shown in Fig. 2 is used for electing the cluster heads.

3.1 Algorithm for Dynamic Cluster Head Election-Based on Shortest Distance Factor Using ANN

3.1.1 Assumptions

The initial energy of each node was assumed to be maximum and equal among all the nodes in the random area of about 200 sq.m. Thirty nodes were created in total number for measuring the performance of the proposed network, and the nodes are

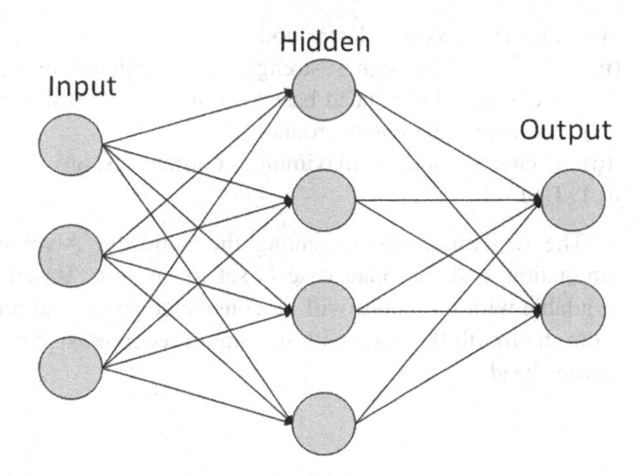

Fig. 2 Artificial neural network

deployed in a random fashion. The probability for each node to become as cluster head is set to 0.05. A threshold distance has been considered based on the coverage of sensor nodes as 90 m beyond which no transmission going to be happened. The path loss factor considered is free space propagation model, and the occurrence of multi-path is also considered. Constant data packet and energy consumption is considered among nodes provided for the same distance factor among nodes in transmission range.

3.1.2 Algorithm

In the proposed algorithm, the cluster head election is split into energy rounds and numbered. In each energy rounds, the setup and manage phase exists as considered in LEACH protocol. Since the amount of energy is same and the nodes level consumptions are assumed to be same, LEACH-C is applied for first three energy rounds.

(a) Start.
(b) Check for the number of rounds and proceed.
(c) If the number is more than three, go to step d else go to step e.
(d) Apply LEACH-C and go to step n.
(e) Check whether the nodes (NN) in cluster are alive.
(f) If the NN is non-zero, go to step g else go to step p.
(g) Determine the number of clusters to be formed.
(h) Artificial neural network is being applied to find the maximum energy level E_{max} among the nodes.
(i) Check whether obtained E_{max} is maximum.
(j) If yes, go to step k else go to step l.
(k) Node elected as cluster head based on received signal strength and proceed to step m.

(l) Functions as the cluster node.
(m) The received signal strength is calculated among all the cluster nodes including cluster head based on the threshold distance.
(n) Increment the energy round.
(o) If energy round is maximum, continue else go to step e.
(p) End.

The maximum energy among the nodes in ANN is considered to be 1 on simulation, and the other case is set to be zero. Based on the remaining energy available with the nodes will be considered in second and third energy round and from the fourth the node with maximum received signal strength will be elected as cluster head.

3.2 *Energy and Throughput Profile for Proposed System with Number of Nodes*

This section summarizes the proposed system performance with the graph obtained by MATLAB simulation. The performance measure of the proposed system is being compared with the existing LEACH-C. Figure 3a briefs the cost of energy profile with the increasing number of nodes. The same has been obtained for LEACH-C along with the proposed model, and the same has compared to show that the proposed system consumes lesser energy compared to existing LEACH-C. Figure 3b describes the throughput profile for the proposed system and the existing LEACH-C. This study also concludes that the proposed system is a better performing as the success ratio is more in comparison.

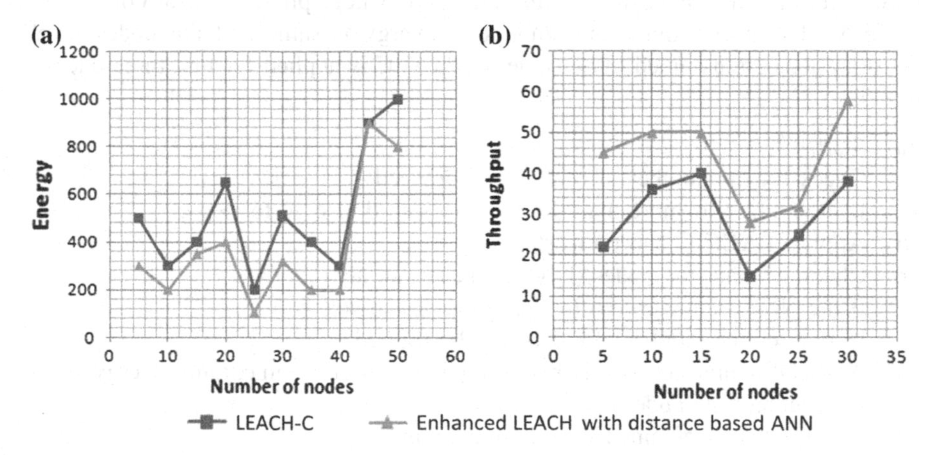

Fig. 3 **a** No. of nodes versus energy. **b** No. of nodes versus throughput

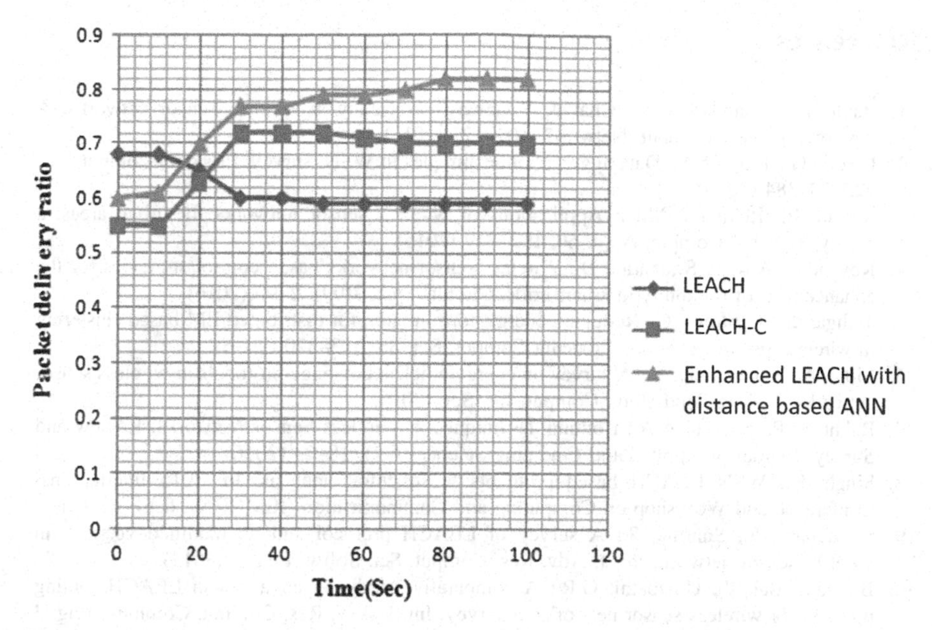

Fig. 4 Packet delivery ratio (PDR) versus simulation time measured in sec (T (s))

Figure 4 briefs the system performance as the measure of PDR (packet delivery ratio). PDR is better for LEACH when the number of packets increased with increasing simulation time PDR of LEACH drops, but the proposed system has better performance in PDR compared to LEACH and LEACH-C.

4 Conclusion and Future Scope

A dynamic algorithm for electing cluster head based on received signal strength computed through artificial neural networks for wireless sensor networks has been designed, and the same has been compared with the existing LEACH algorithm. Thus, energy efficiency attained from the proposed system is evident from the results shown which clearly describes the increased lifetime of the network. The future extension of this work will incorporate the cloud computing to overcome the coverage issues and facilitating the remote monitoring ability at ease of access. Priority queue can be incorporated to maximize the system reliability when applied in critical applications.

References

1. Rault, T., Bouabdallah, A., Challal, Y.: Energy efficiency in wireless sensor networks: A top-down survey. Comput. Netw. **67**, 104–122 (2014)
2. Dhand, G., Tyagi, S.S.: Data aggregation techniques in WSN: survey. Procedia Comput. Sci. **92**, 378–384 (2016)
3. Rashid, B., Rehmani, M.H.: Applications of wireless sensor networks for urban areas: a survey. J. Netw. Comput. Appl. **60**, 192–219 (2016)
4. Kovendan, A.K.P., Sridharan, D.: Wireless sensor networks based control strategies for the enhancement of reliability in smart grids. Circuits Syst. **7**(09), 2499 (2016)
5. Tsiligaridis, J., Flores, C.: Reducing energy consumption for distributed EM-based clustering in wireless sensor networks. Procedia Comput. Sci. **83**, 313–320 (2016)
6. More, A., Raisinghani, V.: A survey on energy efficient coverage protocols in wireless sensor networks. J. King Saud Univ.-Comput. Inf. Sci. (2016)
7. Bakht, M.P., Shaikh, A.A.: Routing Techniques in Wireless Sensor Networks: Review and Survey. Journal of Applied and Emerging Sciences **6**(1), 18–23 (2016)
8. Singh, K.: WSN LEACH based protocols: a structural analysis. In: 2015 International Conference and Workshop on Computing and Communication (IEMCON). IEEE (2015)
9. Choudhary, S., Sharma, S.: A survey of LEACH protocol and its modified versions in wireless sensor network. Int. J. Adv. Res. Comput. Sci. Softw. Eng. 1 (2014)
10. Braman, Alakesh, Umapathi, G.R.: A comparative study on advances in LEACH routing protocol for wireless sensor networks: a survey. Int. J. Adv. Res. Comput. Commun. Eng. **3** (2), 5683–5690 (2014)
11. Razaque, A., et al.: H-LEACH: hybrid-low energy adaptive clustering hierarchy for wireless sensor networks. In: 2016 IEEE Long Island Systems, Applications and Technology Conference (LISAT). IEEE (2016)
12. Bakr, B.A., Lilien, L.T.: Comparison by simulation of energy consumption and WSN lifetime for LEACH and LEACH-SM. Procedia Comput. Sci. **34**, 180–187 (2014)
13. Farid, Z., et al.: Hybrid indoor-based WLAN-WSN localization scheme for improving accuracy based on artificial neural network. Mob. Inf. Syst. 2016 (2016)
14. Assaf Ahmad, El, et al.: Robust ANNs-Based WSN localization in the presence of anisotropic signal attenuation. IEEE Wirel. Commun. Lett. **5**(5), 504–507 (2016)
15. Payal, A., Rai, C.S., Reddy, B.V.R.: Artificial neural networks for developing localization framework in wireless sensor networks. In: 2014 International Conference on Data Mining and Intelligent Computing (ICDMIC). IEEE (2014)
16. Barbancho, J., et al. Using artificial intelligence in wireless sensor routing protocols. In: International Conference on Knowledge-Based and Intelligent Information and Engineering Systems. Springer, Berlin, Heidelberg (2006)
17. Serpen, Gursel, Li, Jiakai, Liu, Linqian: Ai-wsn: adaptive and intelligent wireless sensor network. Procedia Comput. Sci. **20**, 406–413 (2013)

Implementation of FPGA-Based Network Synchronization Using IEEE 1588 Precision Time Protocol (PTP)

Atul Chavan, Sayyad Nagurvalli, Manoj Jain and Shashikant Chaudhari

Abstract Synchronization is an essential prerequisite for all wired and wireless networks to operate. It is fundamental requirement for data integrity, and without it data will be affected by errors and networks will have outages. The IEEE 1588 Precision Time Protocol (PTP) enables precise synchronization of clocks via packet networks with accuracy down to the nanoseconds range. PTP-based solution can be used in various heterogeneous systems like industrial automation, RADAR, Telecom networks. FPGA-based implementation of PTP eliminates Ethernet latency and jitter issues through hardware Timestamping. It gives precise and accurate synchronization in comparison with Operating System-based Timestamping solutions. Accuracy in the range of 10–100 ns is achievable using FPGA-based platform. This paper explains basic mechanism of PTP protocol and advantages of FPGA-based platform for its implementation. It provides detail implementation of PTP on FPGA platform, design flow, testing methodology and its results.

Keywords PTP · FPGA · GPS · VHDL · QUARTUS · NIOS

1 Introduction

The modern network involves different devices distributed across the network. These devices will be capturing the events which need to be accurately Timestamped for proper managing, securing, planning, and debugging a network. Time provides the only frame of reference between all devices on the network, and thus, time synchronization across the devices in the network is very crucial for a robust network. Precise time is crucial in variety of network services. Telecommunication networks, financial networks, industrial automation, electrical power grids, etc, rely on precision time synchronization for operational efficiency of the devices distributed across the network and performance of the overall network.

A. Chavan (✉) · S. Nagurvalli · M. Jain · S. Chaudhari
Central Research Laboratory, Bharat Electronics Ltd, Bangalore, India
e-mail: atulbabanchavan@bel.co.in

P. K. Sa et al. (eds.), *Recent Findings in Intelligent Computing Techniques*,
Advances in Intelligent Systems and Computing 708,
https://doi.org/10.1007/978-981-10-8636-6_15

In industrial automation and other time critical networks where high synchronization accuracy (in sub-microsecond range) is required, the IEEE 1588v2 PTP [1] is the most popular protocol used widely. It does not need a separate networking infrastructure since it uses the existing Ethernet-based Local Area Networks (LAN) for time synchronization. Master clock and Slave clocks are distributed across the network. Master clock is usually synchronized to a higher reference source such as Global Positioning System (GPS) [2]. Master clock sends this timing information to the Slave clocks through the exchange of the PTP messages across the network.

As title of the paper talks about Field Programmable Gate Array (FPGA) [3]-based platform, some introduction about this is given below. FPGAs are re-programmable silicon chips. Using pre-built logic blocks and programmable routing resources, it can be configured to implement custom hardware functionality. FPGAs are parallel processing devices. FPGA configuration is generally specified using Hardware Description Language (HDL) like VHDL or Verilog. In current work, Proprietary hardware (FPGA based) is used for Synchronization solution. Altera FPGA, NIOS soft processor and tools like Quartus [4], NIOS IDE are used in this process.

1.1 IEEE1588 Precision Time Protocol (PTP)

Sync, Follow-Up, Delay Request, and *Delay Response* are the PTP packets required for Master Slave synchronization.

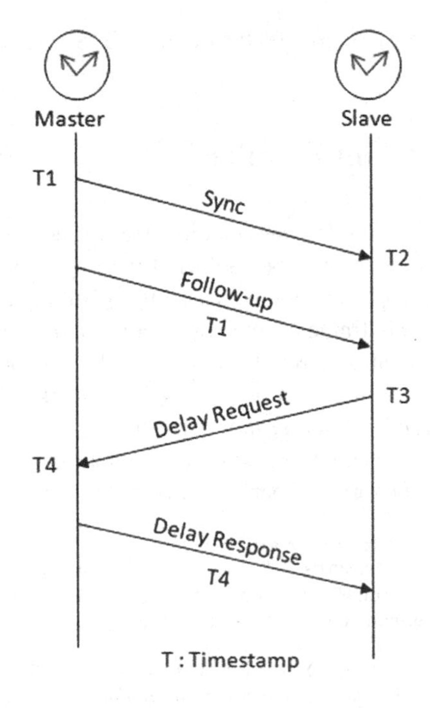

Fig. 1 PTP message flow

Sync packet will be sent periodically by Master to the Slave. And subsequent packet flow will happen following the *Sync* packet as shown in Fig. 1. Whenever PTP packets will be sent or received on Ethernet ports, Timestamp T1 and T4 will be taken at Master and similarly T2 and T3 will be taken at Slave. To synchronize Slave with Master, offset between Slave and Master should be calculated at Slave. The Slave requires 4 Timestamp values namely T1, T2, T3, and T4 to calculate offset. For calculation of offset at Slave, T1 and T4 value should be sent to Slave. Job of *Follow Up* and *Delay Response* packet is to send T1 and T4 value, respectively, to the Slave. The offset and propagation delay is calculated using following formula:

$$\text{Offset} = ((T2 - T1) - (T4 - T3)) / 2$$
$$\text{Propagation Delay} = ((T2 - T1) + (T4 - T3)) / 2$$

Slave will correct its clock using the offset and propagation delay value calculated as mentioned in the above formula. The precision of the result depends on the precision of the Timestamps taken by the system.

2 System Description

2.1 FPGA-Based Implementation of PTP Protocol

PTP protocol is implemented on Altera-based FPGA. Operating system is not used in this process. NIOS soft processor is used for writing PTP protocol stack while hardware Timestamping is implemented in VHDL. The major advantage of implementing Timestamp mechanism in VHDL is its precision.

As shown in Fig. 2, if Timestamp engine detects Sync or Delay Request message, it notes the Timestamp value of there occurrence. This Timestamp value buffered into Timestamp buffer and used latter for assigning to the corresponding PTP message. Real-time clock counts oscillator ticks which are used to keep track of seconds to nanoseconds. In NIOS processor, offset will be calculated between Master and Slave using PTP protocol. This offset will be provided to real-time clock to correct its time. Time of the system is based on real-time clock. Timestamp Engine runs in VHDL; hence, it runs in no time. Number of clock cycles required for packet traversing from the Timestamp Engine till going out from Ethernet PHY chip is known. So there is no jitter added by this process, and hence, precision of synchronization is high using FPGA-based implementation.

2.2 Follow_Up PTP Packet Is Not Required in FPGA-Based Implementation

Timestamp T1 will be taken when *Sync* packet is about to leave Ethernet port of Master. Timestamp T1 can be sent in *Sync* packet or it can be sent in *Follow_Up*

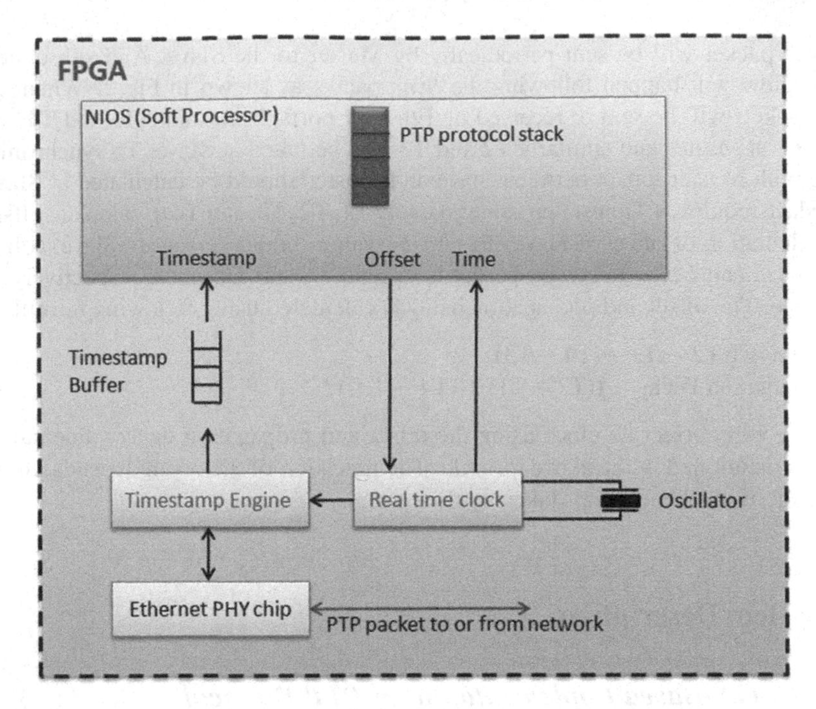

Fig. 2 Functional block diagram of FPGA-based implementation

packet. It is totally implementation dependent. Depending on the implementation platform, accuracy of the Timestamp (T1) may affect if the T1 tried to send in the same *Sync* packet. The job of *Follow_Up* packet is to send the captured Timestamp value of T1 to the Slave.

In FPGA-based implementation, when *Sync* packet comes to Timestamp engine, Timestamp T1 will be captured and immediately captured value of T1 will be filled in the *Sync* packet. This filling of Timestamp value will happen in VHDL only (Hardware-based Timestamp). In VHDL, it is easy to know how many number of clock cycles are required to fill the Timestamp value in the *Sync* packet. This processing delay is always fixed and can be compensated easily. In this process, jitter will not be added in capturing and sending Timestamp value. Due to high precision of getting Timestamp value and zero jitter, there is no need of *Follow_Up* packet. PTP message flow without *Follow_Up* packet is shown in Fig. 3.

In OS-based implementation or software-based approach of Timestamping, *Follow_Up* packet is required, as jitter will be added by protocol stack depending on the implementation of the protocol.

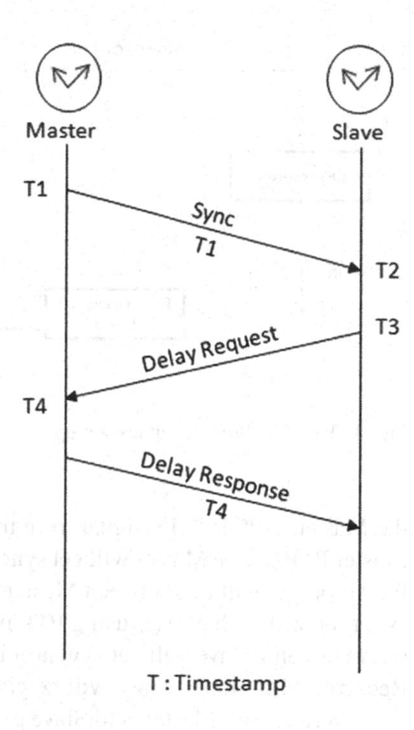

Fig. 3 PTP message flow without Follow_Up packet

2.3 High Precision of Accuracy Can Be Achieved Using VHDL/VERILOG on FPGA-Based Platform

Oscillator used in the PTP board is the heart of the synchronization mechanism. Oscillator generates number of clock cycles per second depending on its operational frequency. A free running counter is used to track the number of cycles generated by the oscillator. For example, oscillator with frequency 32.768 MHz will generate 32768000 cycles per second. In this case, when free running counter completes 32768000 cycles, it is understood that 1 s is completed. On this basic phenomenon, time will be calculated. In VHDL implementation, single cycle error in the code can be removed easily using Signal Tap Logic Analyzer [4] tool of Altera. In case of oscillator with 32.768 MHz frequency, if single cycle error happens in the coding implementation, then it gives 30 ns of penalty in the accuracy. Thus, FPGA-based platform gives control for nanosecond level of accuracy in the implementation.

3 Test Methodology and Results

Test set up for FPGA-based synchronization is as shown in Fig. 4. Highly precise 1 PPS (Pulse Per Second) signal generated by GPS is provided as a reference to

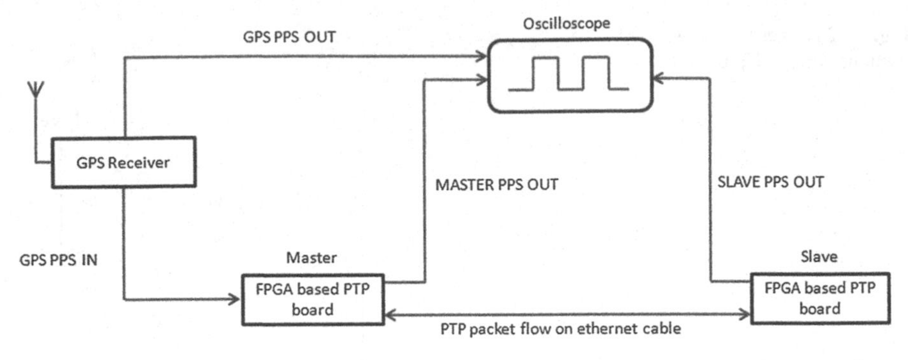

Fig. 4 PTP synchronization test set up

the Master. GPS PPS IN signal from the GPS Receiver will be given as input to the Master PTP board. Master will get synchronized with GPS using GPS PPS IN signal. PTP protocol will run between Master and Slave over Ethernet cable. Slave will get synchronized with Master using PTP protocol. Finally, Master will be synchronized with GPS and Slave will get synchronized with Master. PPS OUT signal from GPS Receiver, Master, and Slave will be given to oscilloscope for measurement.

GPS Receiver, Master, and Slave give PPS OUT signal at each second. As shown in Fig. 5, rising edges from this three signals are captured on the oscilloscope. Width

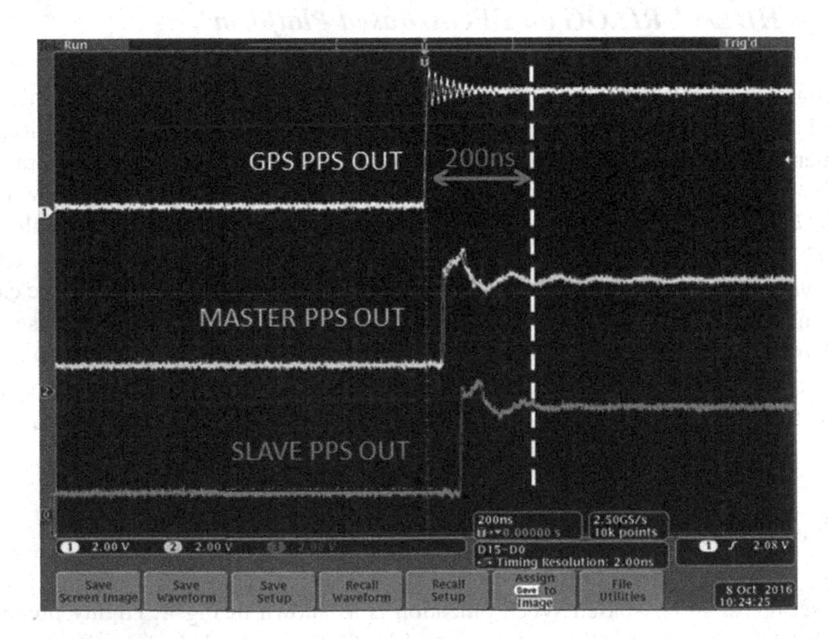

Fig. 5 Screen shot of the synchronization accuracy

of grid adjusted on oscilloscope is 200 ns. In reference to GPS signal, synchronization between Master and Slave is achieved with less than 100 ns of accuracy. The screen shot of the oscilloscope is as shown in Fig. 5.

4 Conclusion

Accuracy in hardware-based Timestamping, coding implementation of PTP protocol and hardware platform on which IEEE 1588 PTP protocol has to be implemented plays a vital role in achieving the precise level of network synchronization (in nanoseconds). It is proved that using FPGA-based platform less than 100 ns of accuracy is achievable between Master and Slave PTP board. In FPGA-based implementation, it is easy to debug the code for nanosecond level of accuracy and Follow_Up message of PTP protocol is not required to be implemented.

References

1. IEEE Standard for a Precision Clock Synchronization Protocol for Networked Measurement and Control Systems. IEEE Std 1588-2008 (Revision of IEEE Std 1588-2002)
2. Misra, P., Enge, P.: Global Positioning System: Signals, Measurements and Performance, 2nd edn
3. Farooq, U., Marrakchi, Z., Mehrez, H.: An overview. In: Tree-based Heterogeneous FPGA Architectures, FPGA architectures (2012)
4. QUARTUS, NIOS, SignalTap Logic Analyzer Tool. https://www.altera.com

4 Conclusion

References

A Cloud-Based Intelligent Transportation System's Framework to Support Advanced Traveler Information Systems

Swathi Lanka and S. K. Jena

Abstract In vehicular networks, each vehicle is constrained with limited computation and resources, which may result in low storage and data processing capability which challenges the rapid growth of intelligent transportation systems (ITSs). However, migrating traditional ITS into cloud-based ITS will help to resolve the challenges by availing cloud services to share among multiple vehicles, roadside units (RSUs), and traffic centers. Advanced traveler information systems (ATISs) are a well-established research field in ITS, whose objective is to collect the data sent by the vehicles as their periodic beacon messages and process those data to get useful information. In this paper, a generic cloud-based ITS framework has been proposed, which can accommodate multiple ITS administrators and cloud service providers on the top of a hierarchical cloud-based VANET architecture. The results of the analysis give us some guidelines in designing efficient ATIS protocols for a cloud-based ITS. To the best of our knowledge, no framework has been designed till date for a systematic design of vehicular data processing applications on cloud-based ITS, and our work is the first approach toward this objective.

Keywords Data center · Cloud · Traveler information · Computing resources

1 Introduction

Vehicular ad hoc network (VANET) is one of the key enabling technologies which can provide communications among vehicles, and between a vehicle and infrastructure network like the Internet [1]. VANET is a component of intelligent transportation systems (ITSs). ITS is a distributed information system built on top of a VANET, which can bring noticeable improvements in transportation systems

S. Lanka (✉) · S. K. Jena
C.S.E Department, National Institute of Technology, Rourkela, India
e-mail: swathivanet@gmail.com

P. K. Sa et al. (eds.), *Recent Findings in Intelligent Computing Techniques*,
Advances in Intelligent Systems and Computing 708,
https://doi.org/10.1007/978-981-10-8636-6_16

toward decreasing congestion and improving safety and traveler convenience [2]. ITS aims to provide relevant travel and timely information to road authorities and users as well [3]. However, due to the resource constraints of a VANET, the large computation and storage demands of ITS are becoming a significant challenge that hinders the rapid development of vehicular networks. A VANET is formed with a set of onboard unit (OBU)-equipped vehicles, which have limited sensing, computing, and communication capabilities through which it makes wireless communication (IEEE 802.11p) among themselves and with a set of roadside units (RSUs). These RSUs are interconnected by dedicated high-speed communication channels, and they are again connected to a set of back-end servers through the Internet. Resource limitations arise from various components of a VANET. The computing and communication capabilities of an OBU are very limited. Although RSU has much higher computing resources than the OBUs, its resources are still inadequate to process a large volume of data collected from a large number of vehicles over a long period of time. The back-end servers in a VANET usually have sufficient computing and storage capabilities. A data server at the back end of the VANET can manage and store data for a specific zone of the city covering multiple RSU areas [4, 5]. In order to overcome these resource limitations of the traditional ITS, the concept of cloud-based VANET has been proposed very recently. In the cloud-based vehicular network architecture, clouds can be formed at three stages, viz. (1) vehicular clouds which are formed among a set of neighboring vehicles, (2) roadside cloud—which is formed among the RSUs, and (3) central cloud—which can be a public cloud and/or private cloud connected to the VANET through the Internet. In this hierarchical and hybrid VANET cloud computing structure, the computation tasks of an ITS can be decomposed into components, the component tasks can be partitioned according to their characteristics, and each partition can be assigned to a specific type of cloud. A task can be characterized depending upon its resource demands, data requirements, functional characteristics, security requirements, etc. For instance, the resources of a vehicular cloud could be used to implement an intelligent parking service or to download a large data file from the Internet into a vehicle, and a roadside cloud can provide distributed storage service that could be used by video surveillance applications.

2 Advanced Traveler Information Systems (ATISs)

ATIS is one of the subcomponents of ITS and also widely deployed application area. With the rapid growth of Internet and wireless technologies in the recent years, the scope of increasing ATIS applicability is also enlarged. ATIS is designed to provide advanced traveler information about road traffic, driving behavior, road surface condition, accurate roadmaps and accidents to road users, traffic police, transportation authorities, and hospitals.

3 Cloud-Based Intelligent Transportation Systems (CITSs)

In vehicular networks, normally each vehicle is limited by individual resources including computation, storage, and bandwidth. Due to low-cost and small-size hardware, each vehicle is constrained with limited computation and resources, which may result in low data processing capability. On the other hand, some emerging transportation applications and traffic management systems require a high computation and large storage capability. This situation necessitates a promising solution which should allow sharing the computation and storage resources among the vehicles, roadside units, and also between vehicles and roadside units. This motivates the concept of designing cloud-based ITS. The traditional ITS is benefited in many ways when cloud concept has incorporated into ITS to become CITS [6]. In CITS, nearby vehicles will be formed as a vehicle cloud and connected with nearest RSUs. In this setup, vehicles connected in the cloud can share the information, bandwidth, storage resources. Through cloud computing, single vehicle can represent its neighboring vehicles and forward all vehicles' data into nearest RSU [7]. Therefore, it may reduce the number of data entries into RSU and also one RSU can get multiple vehicles' information within a less time. In CITS, RSU cloud forms with multiple nearby RSUs perform computation by sharing appropriate algorithms and techniques among the RSUs. It may be advantageous by reducing time delay and processing complexity. CITS also allows multiple RSUs to connect with back-end servers virtually; therefore, multitasking is possible to reduce the computation time.

4 A Generic Framework for Cloud-Based ITS

4.1 A Novel Framework for CITS

We propose a novel framework for cloud-based ITS and also address the security goals for CITS. With the support of cloud computing technologies, it will go far beyond basic ITS functionality, addressing issues such as infinite system scalability, reducing the establishment cost, and minimizing the total cost of ownership. Our model represented in Fig. 1 has six main components such as ITS administrator, vehicles, clouds (public or private), roadside units, the data center, and the private key generator. In addition, Fig. 5 demonstrates the information flow of components in detail.

Vehicles or End users can subscribe to multiple ITS for different services. A vehicle can generate vehicular service requests like vehicular safety traffic management services, path enquiry, weather information, social networking to an ITS under which it is registered.

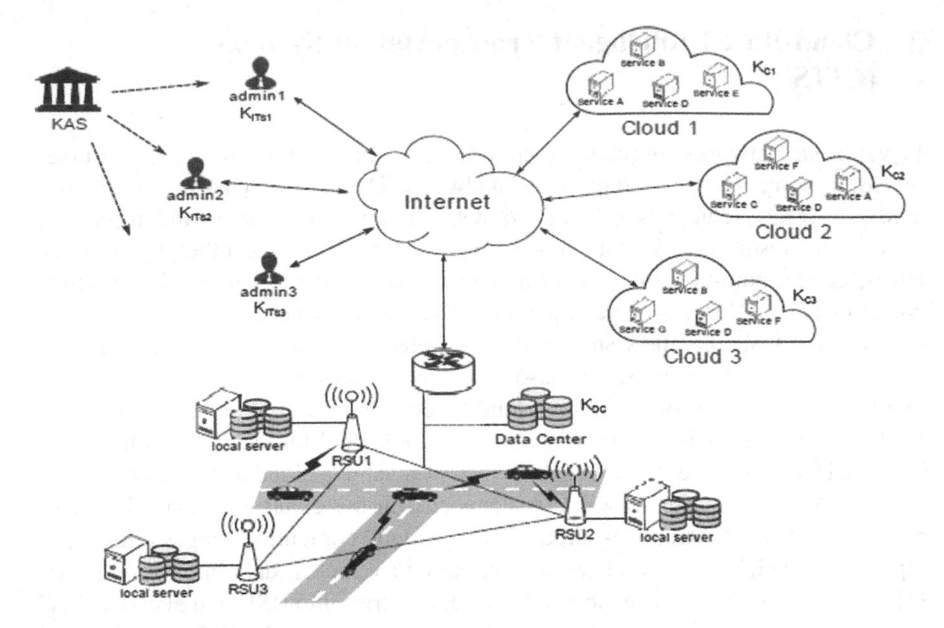

Fig. 1 Proposed novel framework for CITS

Roadside Unit (RSU) is responsible for bringing all the ITS services like providing Internet access and information on weather or traffic to the end users. Each RSU has its own local servers to perform small computations. These are semi-trusted entities which can only be trusted for timely delivery of messages. In our model, the RSU is also responsible for rerouting the vehicle service request packets to the appropriate cloud from which ITS rents the requested service type and forwarding the request reply message back to the vehicle.

Data Center (DC) is deployed to store the periodic beacon messages about road condition, traffic conditions, etc., produced by the end users. These periodic beacon messages are used to provide various services to end users and ITS administrators.

In our proposed model, a city can have many ITSs managed by **ITS administrators** and will share the same VANET infrastructure which tends to reduce the establishment and maintenance costs. An ITS can rent different clouds or deploy its private cloud to do the required computations. ITS administrators are responsible for providing information regarding traffic conditions, free parking slots, driving misbehaviors etc., and various services like weather forecasting, the Internet, to end users as well as administrative services like traffic management and road authorities. Often, ITS administrators send a data request message as surveillance request, real-time traffic condition data, remote vehicle diagnosis, etc. to the cloud over the Internet.

Cloud Service Providers (CSPs) are responsible for providing different ITS services. In our model, a cloud is accountable to address both ITS and vehicle requests for a service and can ask DC for stored beacon messages whenever needed for processing the requests. It also maintains a local copy of beacon messages to minimize communication cost.

Key Authentication Server (KAS) is a city-wide trusted entity responsible for generating and providing the private key corresponding to the identity of all entities of the network. The transmission of private keys occurs over a secure channel on verifying the identity of the requesting entity.

4.2 Information Flow Diagram Description

In this section, we are describing how the information and vehicle requests are processed among the entities as shown in Fig. 2. Mainly, three sets of communication are taking place, all of which are described below.

Vehicle Request Processing: The vehicle request packet includes the sender's identity, the receiver's ITS identity, service type as packet header, and message 'm' and nonce 'n' as encrypted payload. Here, each vehicle uses the following protocol for requesting a service from both ITS and RSU (Fig. 3).

1. First, vehicle sends an encrypted message to ITS administrator for registration. After receiving it, ITS administrator generates and sends a vehicle authentic key as a reply to vehicle. In addition, it also forwards the same key to nearest RSU. Second, vehicle sends a service request to RSU along with periodic beacon messages and waits for acknowledgment.

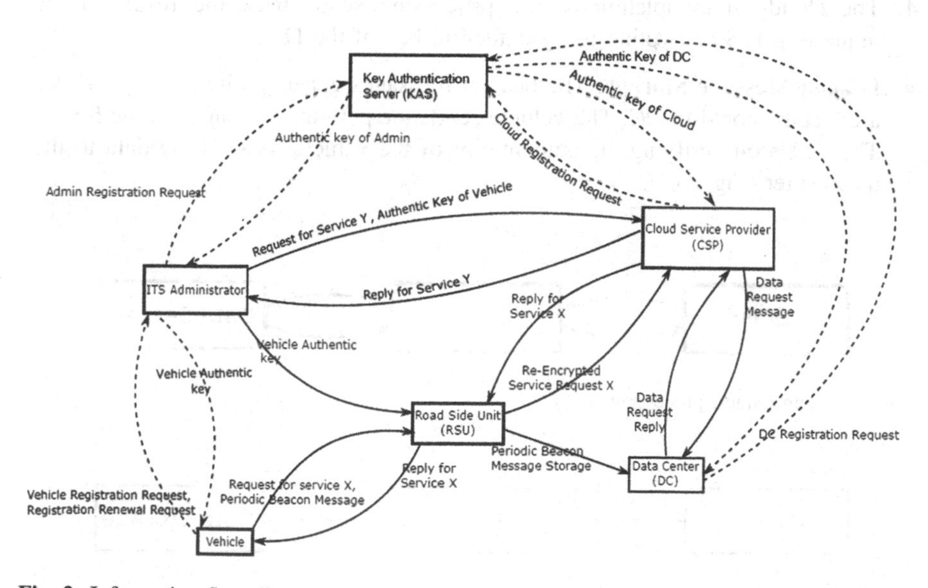

Fig. 2 Information flow diagram

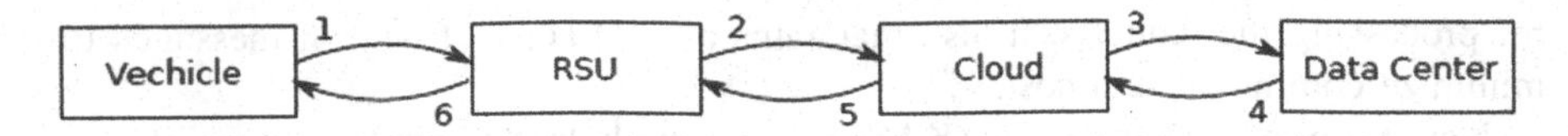

Fig. 3 Vehicle request processing

2. The RSU on verifying the authenticity of the message routes the request packets according to the routing table after the re-encryption.
3. The cloud on receiving the request message from RSU will perform some operations or may request the data center for historical data if required.
4. The data center provides all the data requested by the cloud on verifying the identity of the cloud.
5. On completion of the processing, the cloud then encrypts the response packets using public key of vehicle and sends it to the RSU.
6. The RSU then forwards the response message to the vehicle, later the vehicle can view the message by calling Decrypt function.

ITS Admin Request Processing: The administrator can request any service from the cloud anytime as shown in Fig. 4. The admin follows the following protocol to request a service from the cloud:

1. The administrator requests cloud for any service, encrypting the request packets using the authentic key of the cloud.
2. The cloud performs computation based on the service request and may retrieve the data from DC if required.
3. The DC verifies the identity of the cloud and then forwards the requested data back to the cloud.
4. The cloud on completion of the processing sends back the result to the requesting ITS encrypting using authentic key of the ITS.

- **Beacon Message Storage**: The beacon messages generated by all the vehicles are locally stored in DC. The vehicle sends the periodic messages to the RSUs. The RSUs on verifying the authenticity of the vehicle forward the data to the data center (Fig. 5).

Fig. 4 Admin request processing

Fig. 5 Beacon message storage

5 Challenges of CITS

Transparency in process execution by CSP cannot be achieved because of lack of vehicle–CSP direct interaction. RSUs and clouds can compromise at any instant of the time, since they are **Semi-trusted entities**. Hence, high-security mechanisms must be used. **Communication overhead** may be increased due to decentralized process. **Transmission delay** occurs due to improper Internet facilities. If any vehicle requests for service, its data will be stored in DC and the data will be retrieved by a specific cloud. Due to this phenomenon, **Privacy** of vehicle data cannot be achieved. **Service delay** may occur due to improper Internet services and decentralized execution.

6 Conclusion

In this paper, we have discussed about cloud ITS, which allows to share data, computation, and applications among multiple vehicles, RSUs and clouds may help to give faster response to vehicle user. This piece of research proposed a novel framework for CITS which aims to provide real-time, economic, secure, and on-demand services to customers through the associated clouds. In addition, information flow description of the framework is also described to show how the information and vehicle requests are processed among the entities by considering three sets of communication. As a result, the analysis gives us some guidelines in designing efficient ATIS protocols for a cloud-based ITS.

References

1. Hartenstein, H., Laberteaux, L.P.: A tutorial survey on vehicular ad hoc networks. IEEE Commun. Mag. **46**(6) (2008)
2. Atkinson, R., et al.: Explaining International IT Application Leadership: Intelligent Transportation Systems (2010)
3. Sen, R., Raman, B.: Intelligent transport systems for indian cities. NSDR (2012)
4. Pham, D.-T., et al.: A constructive intelligent transportation system for urban traffic network in developing countries via GPS data from multiple transportation modes. In: 2015 IEEE 18th International Conference on Intelligent Transportation Systems (ITSC). IEEE (2015)
5. Chen, C., et al.: iBOAT: isolation-based online anomalous trajectory detection. IEEE Trans. Intell. Transp. Syst. **14**(2), 806–818 (2013)
6. Yu, R., et al.: Toward cloud-based vehicular networks with efficient resource management. IEEE Netw. **27**(5), 48–55 (2013)
7. Wu, G.Y., Da Li, X.: Developing vehicular data cloud services in the IoT environment. IEEE Trans. Ind. Inform. **10**(2), 1587–1595 (2014)

Part II
Communication Systems, Antenna Research, and Cognitive Radio, Signal Processing

Statistical-Based Dynamic Radio Resource Allocation for Sectored Cells in MIMO-OFDMA System

Suneeta V. Budihal and Sneha Shiralikar

Abstract The paper provides framework for statistical-based dynamic radio resource allocation for sectored cells in MIMO-OFDMA systems. The availability of multiple antennas in the present wireless equipment provides higher data rate. In the multicellular system, the users suffer from severe interference at the boundary of the cell termed as Inter-Cell Interference (ICI). The sharing of common radio resources by the users at the edge leads to ICI. In order to mitigate the ICI at the edge, the resources need to be dynamically categorized, managed and allocated to the users. The statistical-based dynamic allocation of resources like subcarriers and the corresponding power is carried out, for the categorized users. The simulation results enhance overall system throughput, spectral efficiency at increased resource utilization.

Keywords Radio resource management · Frequency reuse · ICI

1 Introduction

LTE system employs Multiple-Input Multiple-Output (MIMO) Orthogonal Frequency Division Multiple Access (OFDMA) as it accommodates multiple users in the same channel. Conventional wireless communication system faces the problem of limited availability of spectrum and is utilized inefficiently. Hence, the present design of radio networks must provide continuous service for a wider area by economically using available spectrum. In practical scenarios, all users do not attain minimum required data rates due to the interference from adjacent cells, near the boundary of a cell. The users residing at edges of radio cell suffer a lot of ICI

S. V. Budihal (✉) · S. Shiralikar
Department of ECE, BVBCET, Hubballi 580031, India
e-mail: suneeta_vb@bvb.edu
URL: http://www.bvb.edu

S. Shiralikar
e-mail: sneha.shiralikar19@gmail.com

P. K. Sa et al. (eds.), *Recent Findings in Intelligent Computing Techniques*,
Advances in Intelligent Systems and Computing 708,
https://doi.org/10.1007/978-981-10-8636-6_17

compared to the users near base station. This leads to overall reduced cell throughput and spectral efficiency. Among the existing interference mitigation techniques, interference avoidance scheme is preferred in the proposed scheme.

Various authors have worked on interference avoidance techniques and proposed many schemes for radio resource allocation. Frequency reuse techniques are proposed in this direction such that the available frequency needs to be planned and managed in order to avoid interference. In [1], the research is made on several frequency reuse schemes among which partial frequency reuse-3 is suggested under specific traffic conditions. Few literatures [2, 3] utilize soft frequency reuse scheme for avoiding interference but fails to attain maximum capacity along with user satisfaction. Authors in [1] implement an opportunistic scheduler which allocates subcarriers dynamically but fails to maintain the performance as the radius of center region is increased. Further, many researches [4] focused on providing power dynamically by devising a parameter called power ratio. Most of the researches [5–7] are based on the performance of various soft frequency reuse schemes for different traffic loads and diverse power ratios. In interference avoidance scheme, resource allocation to the best channel gains user results [8] in a biased distribution of resources. In this paper, a framework for sectored frequency reuse scheme for a multicellular layout is proposed. To achieve improved throughput and user satisfaction at increased resource utilization, we need dynamic resource allocation algorithms. The further sections of the paper discuss the proposed methodology. Following are the contributions of paper. In the paper authors,

1. Proposed a statistical-based dynamic categorization of the users in a hexagonal cell layout as edge and center users.
2. Proposed an algorithm to dynamically allocate the radio resources to edge and center users based on statistical approach in LTE downlink system.

The paper is organized such that the Sect. 2 covers the system model. The Sect. 3 highlights the proposed frequency planning, Sect. 4 provides resource allocation methodology. Sect. 5 explains the results with the analysis, and final Sect. 6 provides conclusion.

2 System Model

The overall network is assumed to be consisting of several hexagonal coverage areas which are adjacently placed and are called as cells. Each cell consists of a base station and a number of users which seek for the resources. The study concentrates on transmission of signal from base station to user equipment usually termed as downlink. The measured strength of the signal at the user during the downlink transmission is considered as Signal-to-Interference-plus-Noise Ratio (SINR) [9]. The signal undergoes degradation during transmission due to multiple path loss, signal fading, and interferences from the adjacent users. Equipment noise is another factor contributing to the degradation. The SINR is computed by the following relation.

$$S_k = \frac{Pt_b H_{bk} Ga_{bk}}{\sigma_N^2 + \sum_{z \epsilon Z} Pt_z H_{bz} Ga_{bz}} \tag{1}$$

where Pt_b is the transmitted power from serving base station, H_{bk} is the channel gain, and Ga_{bk} is path loss component from the user k and the base station b. The term σ_N^2 used indicates the noise power. The user is denoted as k ranging from 1 to K in a cell. The subcarriers are denoted as n ranging from 1 to N, total available number of subcarriers. The set Z presents adjacent BS which are using common subcarriers of the spectrum as user k. In the study, H_{bk} and H_{bz} are considered as unity. Using the SINR of each user in the radio cell, compute the capacity of user k on subcarrier n. This is computed for a targeted Bit Error Rate (BER) using the subcarrier spacing. A scaling factor of Θ is computed using BER value and calculated as,

$$\Theta = \frac{1.5}{\ln(5BER)} \tag{2}$$

Therefore, the capacity computation using subcarrier spacing Δf is as follows:

$$C_{k,n} = \Delta f log_2(1 + \Theta S_{k,n}) \tag{3}$$

3 Frequency Planning

A geographical area is sectioned into a number of hexagonal cellular regions of radius R. Hexagonal configuration is preferred, as circular configuration leads to overlapping and is complex for analysis. The hexagonal cell radius is divided into three equal sectors each covering 1/3rd of cell area. The sectorization of cells is repeated in all the cells in such a way that sharing of the same radio resources in adjacent sectors is avoided. In order to distinguish users as center and edge users, a user bifurcation technique is used. The bifurcation technique is an interval-based statistical approach to find a near optimal threshold. The input to this method is a dataset consisting of user's position from the base station. The approach follows the following steps.

1. Compute Mean (M), Variance (V), Standard Deviation (SD) for the given dataset.
2. Subtract 1 from the sample size to find the Degrees of Freedom (df).
3. Find the α value considering the certain % of confidence.
4. Using α and *df* find the critical value using t-distribution table.
5. Compute tolerance limit to set bounds.
6. Obtain effective center radius.
7. end if (radius > threshold).

Once the center region is computed, the overall bandwidth B (MHz) is divided proportionally to center users and edge users. The division is same for rest of the cells in the sector for a particular instant. The ratio of proportion for center and edge is determined and denoted by the term ξ. This fraction is further used to compute external bandwidth N_{ext} and internal bandwidth N_{int}.

As the values of N_{int} and N_{ext} are in terms of MHz and this quantity cannot be directly used for subcarrier allocation, there is a need to find out the actual number of subcarriers using subcarrier spacing as per LTE standards. Number of available subcarriers for center as well as for edge region using N_{int} and N_{ext} is computed. Sub_{int} and Sub_{ext} are the subcarriers allocated to center and edge region, respectively. These parameters are computed using subcarrier spacing Δf which is 15 KHz in the proposed framework used in LTE system. Now the obtained Sub_{int} and Sub_{ext} are integer values which can be used for subcarrier allocation. The subcarriers allocated to edge region is divided into three equal parts for three sectors: f_1, f_2 and f_3 as shown in Fig. 1.

After successful division of bandwidth using the center radius, designer has an account of subcarriers being allocated to users. The bifurcation method has the potential to yield a thresholding radius which is bound to have coverage area for center users below 50%. But in the considered system model, the coverage area is restricted for center users as above 60%. The coverage area of center and edge is restricted to be 6 : 4 or above to avoid overlapping of resources as depicted in Fig. 1. The overall process is depicted in the Fig. 2. To avoid interference in sectored frequency reuse scheme, set up common center region to all the cells of a sector. Though the area of the center region may vary every instant, care is taken that it remains common to all the cells at that particular instant. Here, do not restrict or reserve frequency bands for center or edge but the ratio of bandwidth sharing in all the cells will remain common for that instant hence we consider this scheme as a dynamic allocation of bandwidth.

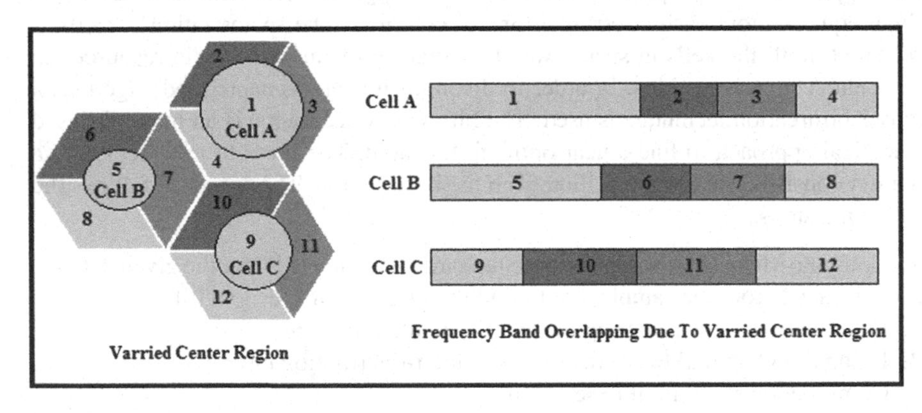

Fig. 1 Cellular layout with users randomly distributed in the cell area with three cells and users are categorized with varying sizes

Fig. 2 Algorithm describing the overall process

Algorithm 1 Process of sectorization and respective resource allocation

```
Input: total_users,radius
load_network_parameters ()
generate_network_cells_&_users ()
for cell=1 : total_cells do
  for user=1 : total_users do
    calculate_SINR_per_user ()
  end for
  do
     calculate_center_radius.
  end do (while center_radius < (0.65×radius))
  for user=1 : total_users do
    if user_distance ≤ center_radius then
      user= center_user
      allot_subcarriers_&_power_per_user ()
      caluclate_capacity_per_user ()
    else
      user = edge_user
      if edge_user_distance ≤ (sector1 || sector2 || sector3)
      then
        allot_subcarriers_&_power_per_user ()
        caluclate_capacity_per_user ()
      end if
    end if
  end for
  calulate_throughput_per_cell ()
  calulate_user_satisfaction_per_cell ()
  calulate_spectral_efficiency_per_cell ()
end for
```

4 Resource Allocation

In a cellular network, users at the edge experience lower SINR as compared users at center, and as a result they pull down the overall throughput. These users require comparatively more power to accomplish the same level of transmission as center users. Hence, a solution that provides better transmission rates without sacrificing the transmission power and the fairness index is proposed in the paper. The constraints need to be set which take care of the allocation of subcarriers minimizing the total power requirement. The following approach deals with dynamic power and dynamic subcarrier allocation mechanisms. Optimization variables both $x_{k,n}$ and $P_{k,n}$ are considered together in the formulation of optimization problem. Each subcarrier n is allocated to at maximum one user k during time t, which is indicated by $x_{k,n}$ · $x_{k,n}$ is set as unity if subcarrier is allocated to user k and set to nil if n is not assigned to k at instant t.

It is subject to the three constraints that are mentioned as follows. Maximize throughput $\Gamma^{p}_{k,n}$

$$\Gamma^{p}_{k,n} = \max \sum_{l=1}^{N} \sum_{k=1}^{K} x_{k,n} \hat{C}^{p}_{k,n} \tag{4}$$

such that,

$$\sum_{j} x_{k,n} > 1 \quad \forall n \quad (ALLOC) \tag{5}$$

$$\sum_{j} P_{k,n} < P_{total} \quad \forall n \quad (POWER) \tag{6}$$

$$\sum_{n} F\left(\frac{P_{k,n}, h_{k,n}}{\sigma_N^2}\right) x_{k,n} \leq \Gamma \quad \forall k \quad (FAIR) \tag{7}$$

Here, function F describes the number of subcarriers n for user k with a transmit power of $P_{k,n}$. First constraint (ALLOC) assures the allocation of at least one subcarrier to one user at a time. The next constraint (POWER) guarantees that the addition of assigned power to the subcarrier equals to or less than the maximum allowable transmission power P_{total}. The third constraint (FAIR) implies minimum throughput has to be maximized Γ per user terminal per downlink phase. Hence, the resource allocation is divided mainly into subcarrier allocation and power allocation.

4.1 Subcarrier Allocation

In the previous study, it is found that the subcarrier allocation technique is universally known to all base stations (BS), such that if subcarrier is allocated to the middle region in a radio cell, then the neighboring cells are allocated with subcarriers w.r.t. that subcarrier along with the gain of the subcarrier. The centralized algorithms need the data of the complete network topology and channel gain of every downlink transmission, The distributed algorithm requires information of SINRs of those receivers which are local to the single reference cell. The centralized algorithms require global knowledge of all channel gains and increases complexity. SINR of every user is the most important parameter used in this allocation algorithm. Each user's SINR is scaled to a parameter such that the user with highest SINR provides least value while the user with lowest SINR provides highest value. This is done so that the user with least SINR ends up with higher number of subcarriers as the user is an edge user. The allocation continues until all the subcarriers reserved for the respective sector are completely exhausted.

4.2 Power Allocation

A method which cooperates with the subcarrier allocation is implied to the network. We have used a power amplification factor in order to make use of available power efficiently. The ratio of number of subcarriers allocated to internal to the external region is termed as power amplification factor φ. The center region transmit power P_{int} is completely reserved for the center users while the transmit power for edge region P_{ext} is divided among the sectors using the amplification factor. Lagrange multiplier technique is used to compute the optimal power with the total power constraint. The transmitted power over a subcarrier n to a user k, denoted as $P_{k,n}$ where i is range of subcarriers ranging from ($i = 1, 2, 3, \ldots n$) and N is the total available subcarriers. The Lagrange function is expressed as,

$$J\left(P_1, \ldots P_n\right) = \frac{1}{N}\sum_{i=1}^{N}\frac{u_i}{2}\exp\left(-\frac{w_i}{2}G_iP_i\right) + \lambda\left(\sum_{i=1}^{N}P_i - K \cdot P_t\right) \quad (8)$$

where u_i and w_i are obtained by the M-QAM modulation scheme applied to the subcarrier and they are mathematically defined as follows:

$$u_i = \frac{2\left(\sqrt{M} - 1\right)}{\sqrt{M} \cdot log2\sqrt{M}} = \frac{2\left(\sqrt{2^{v_i}} - 1\right)}{\sqrt{2^{v_i}}log2\sqrt{2^{v_i}}} \quad (9)$$

$$w_i = \frac{3}{M - 1} = \frac{3}{2^{v_i} - 1} \quad (10)$$

For special subcarrier i, the value v_i is known. The term v_i represents the subcarriers allocated to the user k in the proposed algorithm. Substituting the above u_i and w_i values in equation and final value for power allocated by the ith subcarrier is,

$$P_i = \begin{cases} \frac{-2}{w_i \cdot G_i}\ln\frac{4K\lambda}{u_i \cdot w_i \cdot G_i} & G_i > 0 \\ 0 & \text{others} \end{cases}$$

The final power allocated is always less than the total power allocated. This power obtained is considered while calculating the final throughput, and the resulting impact on the system is discussed in the results section.

Table 1 Simulation parameters for the system model

System specifications	Values
Number of user	100
Cell radius	500 m
Cell bandwidth	10 MHz
Carrier frequency	2 GHz
Spacing between subcarrier	15 kHz

5 Discussion of Results

The proposed dynamic radio resource allocation strategy is applied for a multi-cell OFDMA downlink system. The standard system specifications considered during simulation are listed in Table 1. As shown in Fig. 3, the simulation is carried out for a multi-cell system over 10 adjacent radio cells, with optimal radius R. The radius is optimized such that available radio resources are utilized optimally and achieve maximum satisfaction level. Each cell user is served by a multiple antenna base station, i.e., $1, 2, \ldots 10$, located at the middle of a radio cell. The overall bandwidth B is divided into three equal parts in a cellular layout. Figs. 1 and 3 represent the random categorization of the users and optimal categorization of the users. Figure 4 presents the throughput attained by users in one sector cell after optimal allocation of subcarriers. The throughput is summation of capacities achieved by all users in respective cell. The maximum capacity achieved in the proposed scheme is around 180 Mbps while the maximum achieved in existing scheme is around 135 Mbps. However, the performance is constant in existing scheme, while the proposed scheme shows better performance when the range of users is from 70 to 100. The existing systems provide highest system throughput for around 80 users, and the performance drops when users are increased further. In the proposed technique, the performance drops

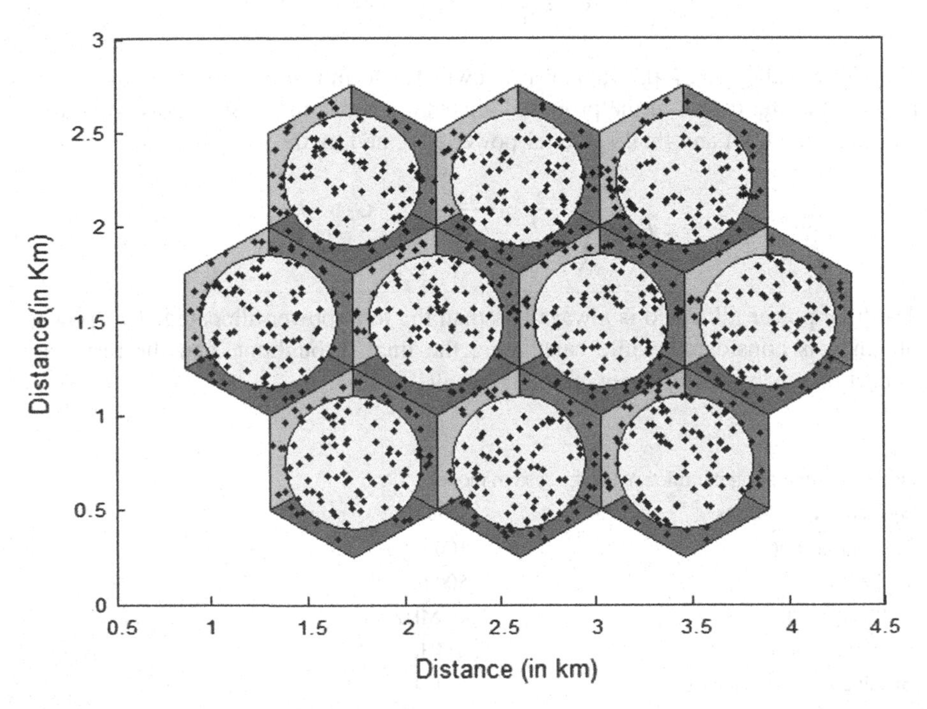

Fig. 3 Cell layout with users randomly located in the entire area in ten cells and the users are categorized optimally as cell center and edge users using statistical model

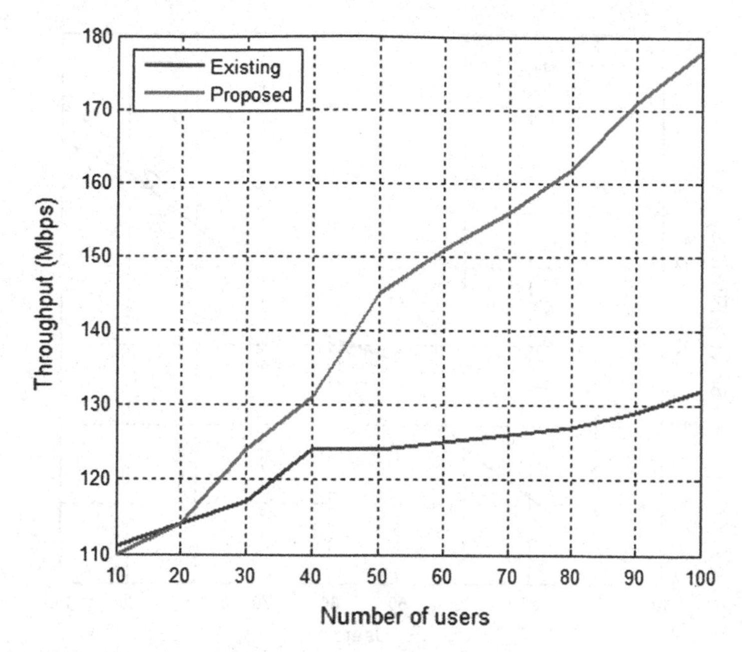

Fig. 4 Plot is for throughput achieved in the cell when the proposed statistical-based dynamic resource allocation scheme is applied

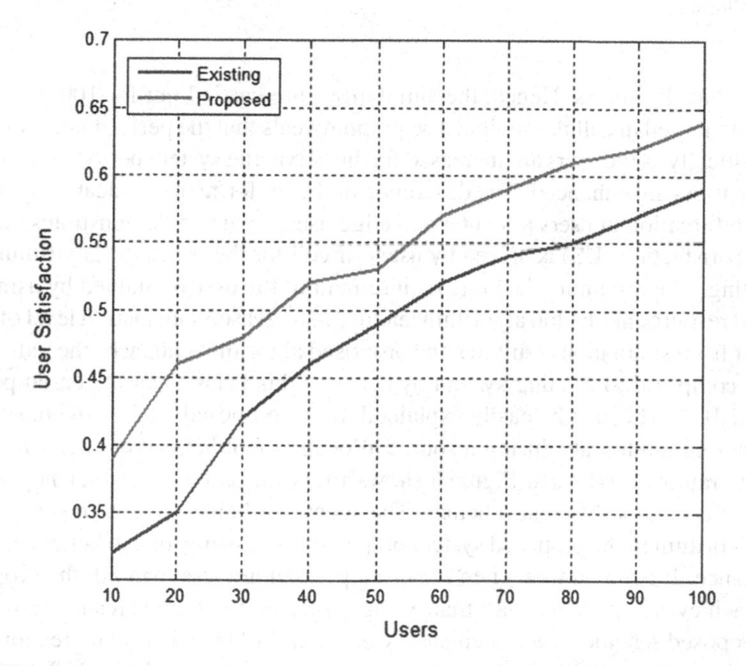

Fig. 5 Plot depicts the user satisfaction achieved in the cell when optimized resource allocation scheme is employed

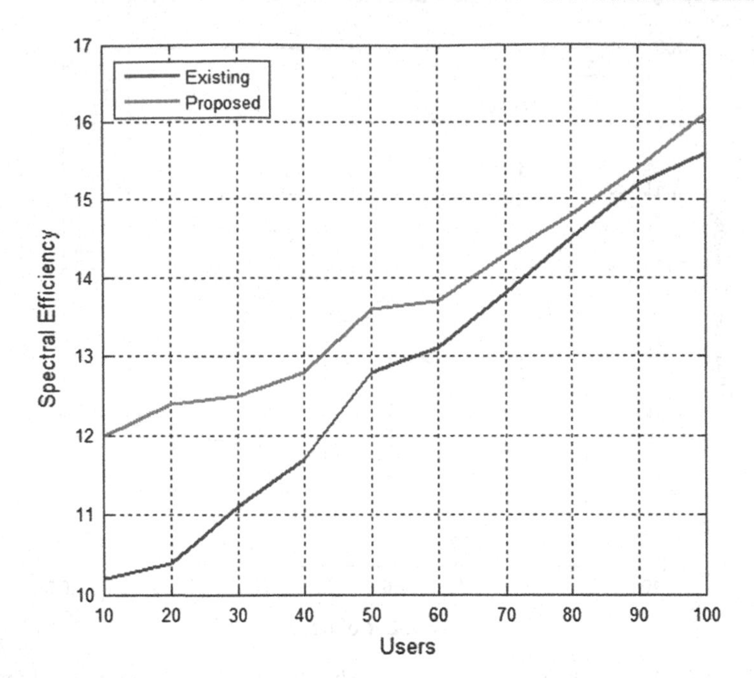

Fig. 6 Plot depicts achieved spectral efficiency by varying users traffic in cell with optimized resource allocation

for more than 100 users. Hence, the simulations are carried out for 100 users randomly distributed in cellular region. The graph reveals that the performance does not drop drastically as the users are increased further. Existing system performance drops when the users are enhanced. The difference in the performance is clearly due to the optimal bifurcation of users as center and edge users. Figure 5 demonstrates the plot of User Satisfaction (US) achieved by users of cell for the proposed algorithm over the existing. The minimum data rate requirement of the user is attained by using the proposed resource allocation algorithm leading to increased satisfaction level of user. It is clear from simulation result that the proposed algorithm enhances the individual users as compared to existing system by 0.1, and this leads to an increased performance of 10%. This can be easily explained as the proposed algorithm considers a predefined minimum rate during resource allocation that helps more users to fulfill their data rate requirements. Figure 6 shows the average spectral efficiency versus number of users, for different schemes. The spectral efficiency due to resource allocation algorithm in the proposed system outperforms existing, due to better cell edge performance. It is found that when the users per cell are less than 30, the proposed scheme achieves improved results than existing system. Simulation results reveal that in the proposed scheme, spectral efficiency of around 13 bits/Hz is obtained for users ranging from 10 to 40. Similarly, a slight increase is maintained up to 50 users and then from users ranging from 60 to 100 we find the efficiency gradually increasing and drops afterward.

6 Conclusion

The paper provides framework for statistical-based dynamic radio resource allocation for sectored cells in MIMO-OFDMA systems. Users in the cell are sectored and categorized based on statistical model. The proposed frequency plan provides the user separation and resource management such that systems do not suffer from severe interference at the edge of the cell. Hence, the common radio resources are not shared by the users at the edge. The statistical-based allocation of resources like subcarriers and the corresponding power is carried out for the users. The simulation results show that there is an enhancement in the overall cell throughput, spectral efficiency and user satisfaction by the statistical-based dynamic radio resource allocation.

References

1. Kivanc, D., Li, G., Liu, H.: Computationally efficient bandwidth allocation and power control for OFDMA. IEEE Trans. Wirel. Commun. **2**(6), 1150–1158 (2003)
2. Zhang, H., Chu, X., Guo, W., Wang, S.: Performance evaluation of frequency planning schemes in OFDMA-based networks. IEEE Commun. Mag. **53**(3), 158–164 (2008)
3. Bohge, M., Gross, J., Wolisz, A.: The potential of dynamic power and sub-carrier assignments in multi-user OFDM-FDMAa cells. In: GLOBECOM'05. IEEE Global Telecommunications Conference, vol. 5, pp. 2932–2936 (2005)
4. Ali, S.H., Leung, V.C.M.: Dynamic frequency allocation in fractional frequency reused OFDMA networks. IEEE Trans. Wirel. Commun. **8**(8), 4286–4295 (2009)
5. Lim, J.H., Badlishah, R., Jusoh M.: Dynamic frequency allocation in fractional frequency reused OFDMA networks. In: Second IEEE International Conference on Electronic Design (ICED), vol. 8, no. 8, pp. 527–532 (2014)
6. Ullah, R., Fisal, N., Safdar, H., Maqbool, W., Khalid, Z., Khan, A.S.: Voronoi cell geometry based dynamic fractional frequency reuse for OFDMA cellular networks. In: IEEE International Conference on Signal and Image Processing Applications (ICSIPA), pp. 435–440 (2013)
7. Biagioni, A., Fantacci, R., Marabissi, D., Tarchi, D.: Adaptive subcarrier allocation schemes for wireless OFDMA systems in WiMAX networks. IEEE J. Sel. Areas Commun. **27**(2), 217–225 (2009)
8. Aliu, O.G., Mehta, M., Imran, M.A., Karandikar, Abhay, Evans, Barry: A new cellular-automata-based fractional frequency reuse scheme. IEEE Trans. Veh. Technol. **64**(4), 1535–1547 (2015)
9. Sneha S., Budihal, S.V.: Sector based resource allocation for MIMO-OFDMA System. Int. J. Technol. Sci. **9**(2), 57–60 (2016). ISSN: (Online) 2350–1111, (Print) 2350–1103

Classification of Protein Sequences by Means of an Ensemble Classifier with an Improved Feature Selection Strategy

Aditya Sriram, Mounica Sanapala, Ronak Patel and Nagamma Patil

Abstract With decreasing cost of biological sequencing, the influx of new sequences into biological databases such as NCBI, SwissProt, UniProt is increasing at an ever-growing pace. Annotating these newly sequenced proteins will aid in ground breaking discoveries for developing novel drugs and potential therapies for diseases. Previous work in this field has harnessed the high computational power of modern machines to achieve good prediction quality but at the cost of high dimensionality. To address this disparity, we propose a novel word segmentation-based feature selection strategy to classify protein sequences using a highly condensed feature set. Using an incremental classifier selection strategy was seen to yield better results than all existing methods. The antioxidant protein data curated in the previous work was used in order to facilitate a level ground for evaluation and comparison of results. The proposed method was found to outperform all existing works on this data with an accuracy of 95%.

Keywords Proteomics · Knowledge base · Word segmentation · Entropy Incremental classifier selection (ics) · Ensemble classifier · Radial basis function (rbf) network

---

A. Sriram (✉) · M. Sanapala · R. Patel
National Institute of Technology Karnataka, Surathkal, Mangalore 575025, Karnataka, India
e-mail: adityasriram1995@gmail.com

M. Sanapala
e-mail: smounica9@gmail.com

R. Patel
e-mail: ronak1395@gmail.com

N. Patil
e-mail: nagammapatil@nitk.ac.in
URL: http://www.nitk.ac.in/

P. K. Sa et al. (eds.), *Recent Findings in Intelligent Computing Techniques*, Advances in Intelligent Systems and Computing 708,
https://doi.org/10.1007/978-981-10-8636-6_18

1 Introduction

Over the past few years, with advances in technology and reduction in sequencing costs, there has been a large amount of data influx into public databases which need to be annotated [1]. A lot of work has been done on genomic data due to its high prediction quality in diseases and discovery of new active sites. As a result, the field of protein analysis was highly neglected due to the comparatively lower prediction quality. It is only in recent times that the focus has been slowly shifting toward analyzing proteins as scientists believe it holds the key to a lot of important discoveries. One such family is the antioxidant protein.

The antioxidant proteins play a vital role in slowing the progression of various diseases including cancer and DNA-induced diseases [2]. It also helps preserve food texture. Hence, accurately predicting them goes a long way in obtaining useful information regarding the physiological processes behind these diseases and food deteriorations, thereby providing a basis for developing drugs and extending the food shelf life.

Proteins are typically short sequences composed of 20 different amino acids, thereby making the analysis complex and challenging as compared to genomic data which is longer and have just four bases [3]. The domain knowledge known thus far in proteins is very scarce due to the low annotation numbers.

The aim of this work is to harness the potential of modern computational power and employ improved feature selection strategies to achieve a higher prediction quality for proteins and paving the way for future discoveries.

2 Related Work

In spite of the importance, very few computational methods have been proposed. Since the traditional experimental inferences are time-consuming, it is important to develop more computational methods. To this end, Feng et al. [4] used an amino acid composition and dipeptide composition-based feature extraction and trained the model using a Naive Bayes (NB) classifier. Zhang et al. [5] in their work realized a g-gap dipeptide composition that went on extract a few more discriminant and informative features as compared to [4], thereby increasing the model accuracy. They later extended their work [6] to accommodate higher tier correlation residues between adjoining amino acids in protein sequences which greatly improved prediction quality. They went on to use an ICS strategy and multiple feature extraction methods which complement each other to extract valuable information. This work yielded an accuracy of 94% following the feature selection done using the Relief-Incremental Feature Selection (IFS) ranking algorithm. Although it manages to address most of the shortcomings of the earlier approaches, we aim to adopt an improved feature selection approach to achieve excellent prediction quality on an extremely condensed feature space.

3 Methodology

Protein sequences are analogous to Chinese text; hence, we apply text-based feature extraction strategies peculiar to proteins. The work done by Yang et al. [7] was incorporated along with a modified segmentation algorithm to select a highly informative and discriminant feature set.

Figure 1 depicts the block diagram of the proposed methodology. The idea is to convert the textual information of each protein sequence in the input data into a more comprehensible numerical feature vector for further analysis.

3.1 Building Knowledge Base

K-mer counting [8] is a simple yet effective solution to most problems in biological analysis. Through experimentation, it was found that for k lengths beyond four, the computational complexity increases and the features extracted degrade the prediction quality. Hence, we extract the k-mers of lengths 1 to 4 and select the 20 amino acids (monomers), top 100 dimers, top 100 trimers, and top 100 4-mers to accumulate a total of 320 features. The feature ranking criteria adopted are:

1. **Frequency**—Records the frequency of k-mers according to their appearance across all sequences in the corpus [7]. Let $f_{t,s}$ be the frequency of k-mer t in sequence s, w_t be the weight of t, and N be the size of the training set. The weight is given by

$$w_t = \sum_{s=1}^{N} f_{t,s} \tag{1}$$

2. **tf-idf value**—tf-idf is calculated for a k-mer in a single sequence [7]. It is a product of tf (term frequency) and idf (inverse document frequency) values. Let $w_{t,s}$ be the tf-idf value for a k-mer t in sequence s, and n_t be the number of sequences

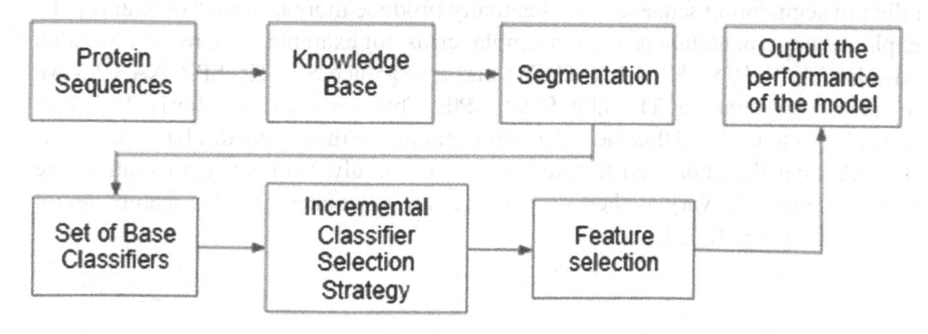

Fig. 1 Work flow of the proposed protein predictor

in which t appears. The tf-idf score is given by

$$w_{t,s} = f_{t,s} \times log\frac{N}{n_t} \tag{2}$$

The weight of t, w_t, is defined as the maximum value of $w_{t,s}$:

$$w_t = \max_{s \in T} w_{t,s} \tag{3}$$

3. **Entropy value**: It is an information theoretic idea which specifies the amount of information contained in a k-mer. The higher the entropy value, more the information contained in it. For simple computational purposes, it was refined in [7] as follows:

$$w_{t,s} = log(f_{t,s} + 1.0) * (1 + \frac{1}{logN} \sum_{s \in T} [\frac{f_{t,s}}{n_t} log\frac{f_{t,s}}{n_t}]) \tag{4}$$

and the maximum value of $w_{t,s}$ is assigned to the weight of t as Eq. (3).

Unlike the work in [6], the 320 features extracted using the above three ranking criteria are not aggregated together. They are used to generate independent feature spaces. This is to facilitate a comparative study between each of the three criteria for feature extraction and selection.

It is to be noted that the frequency criterion selects frequently occurring k-mers in the corpus, while the other two criteria select particularly infrequent words [9].

3.2 Segmentation

The algorithm in [7] has been improvised by using a sliding window-based segmentation method controlled by the weight of the k-mer encountered in each window. This method was found to be more effective than the one used in [7]. The improvised algorithm is given below in Algorithm 1. We also put our model to test if the effect of segmenting sequences does actually produce more informative features. To explain the segmentation process in simple terms, for example, we have a knowledge base B={A, E, P, S, AP, AAA, TPTS} and a sequence S = TPTSPPPAAAPE, we manage to segment S as TPTS|P|P|P|AAA|P|E. Thus, the feature vector is {0, 1, 4, 0, 0, 1, 1}. It is to be noted that the order of the features in the knowledge base matters as it would alter the numerical feature vector accordingly. Without segmentation, the feature values will vary as their occurrences are calculated, and the feature vector thus becomes {3, 1, 5, 1, 1, 1, 1}.

Algorithm 1 Modified segmentation algorithm

```
 1: procedure SEGMENTATION(KB, S)
 2:   N ← StringLength
 3:   maxLen ← 4                              ▷ Maximum length of kmer
 4:   maxWeight ← 0                           ▷ Used for sliding the window
 5:   first ← 0                               ▷ Start of sliding window
 6:   last ← first + 4                        ▷ End of sliding window
 7:   while first < N do
 8:     if last > N then
 9:       last ← N
10:     if kmer(fromfirsttolast)inKB then
11:       if kmerWeight > maxWeight then
12:         maxWeight ← kmerWeight
13:       first ← last
14:       last ← first + maxLen
15:     else
16:       last ← last − 1                     ▷ To shift sliding window to 1 position left
17:   return segmentedsequence
```

3.3 Incremental Classifier Selection Strategy

Once the feature space is generated by the numerical feature vectors corresponding to each sequence, the model is trained on six base classifiers, viz RBF network [10], random forest (RF) [11], J48 decision tree (J48) [11], Naive Bayes (NB) multinomial [12], support vector machines (LibSVM) [11], and sequential minimal optimization (SMO) [11]. An ICS is then used to find the ensemble classifier which gives the best results.

3.4 Feature Selection

The Relief-IFS feature ranking algorithm [6] is then used to rank the features and further condense the feature space by selecting the optimal feature subset. The ensemble classifier is then applied to this feature subset to achieve the highest accuracy.

4 Results and Analysis

The dataset used in our experiments is an antioxidant protein dataset curated by Zhang et al. [6]. It is accessible at http://antioxidant.weka.cc/.

The benchmark training dataset has 200 protein sequences and two classes, namely antioxidant and non-antioxidant, out of which 100 sequences are antioxidant and the other 100 are non-antioxidant.

The proposed methodology was applied on the dataset, and the results are as follows:

Table 1 depicts the results obtained using the entropy ranking technique. The feature spaces obtained with and without segmentation were trained on six base classifiers, and it was found that RBF network is predicted with a good accuracy of 92% for the feature space obtained using the segmentation. The table also goes on to show the positive effect of segmentation in feature selection.

In a similar fashion, the accuracies were obtained using frequency and tf-idf ranking technique as well. The sliding window-based segmentation method works by comparing the weight of each adjoining k-mer and would ensure that smaller k-mers are selected due to their higher weights for the frequency technique, thus explaining the low accuracy. The tf-idf scores tend to be very high for k-mers which occur very frequently in a single sequence. This would also mean a lot of informative features would be lost in the process. Hence, it is observed from all the results obtained that the entropy measure with segmentation works best yielding an accuracy of 92% using the RBF network classifier.

Table 2 depicts the ICS strategy based on the results obtained from Table 1.

The base classifiers are grouped one by one into the classifier subset, and the ensemble classifier is then trained on the model. Based on the results obtained, it turns out that the subset featuring RBF network, RF, NB multinomial, and LibSVM

Table 1 Results using entropy

Classifier	No. of features	Accuracy without segmentation (%)	Accuracy with segmentation (%)
RBF network	320	79	**92**
Random forest	320	89.5	88.5
NB multinomial	320	77.5	88.5
LibSVM	320	87	88.5
SMO	320	77	84
J48 Decision tree	320	79	75.5

Table 2 Incremental classifier selection strategy

Classifier	Accuracy (%)
RBF network	92
RBF net + RF	92.5
RBF net + RF + NB multinomial	92.6
RBF Net + RF + NB multinomial + LibSVM	**93.5**
RBF net + RF + NB multi + LibSVM + SMO	90.5
RBF net + RF + NB multi + LibSVM + SMO + J48	93

Table 3 Comparison with existing work

Method	Accuracy (%)	No. of features	Sensitivity	Specificity
Zhang et al.	94	152	0.95	0.93
Proposed method	95	125	0.95	0.95

is predicted with an accuracy of 93.5% as compared to 92.5% using the ensemble classifier in [6].

The Relief-IFS-based feature selection algorithm was applied on this 320 feature subset using the ensemble classifier to rank the features and further condense the space. At the end of this feature selection process, a total of 125 optimal features were selected giving an accuracy of 95% using the ensemble classifier.

Table 3 depicts the comparison of our proposed methodology with the existing work, and it was found that our method performs better giving an accuracy of 95% with a more condensed feature space as compared to the existing model.

5 Conclusion and Future Work

The proposed methodology was found to perform exceptionally well as compared to the previous works on the antioxidant dataset. It was a challenging task to work with proteomic data as the domain knowledge is very scarce; hence, the multiple feature selection technique employed in this work can be a very efficient step toward training a classifier model with a highly reduced feature space. This method also establishes the use of text categorization methods to model protein sequences in an effective and comprehensible manner. The ensemble classifier was also able to work well by using the different decision boundaries of the base classifiers to combine the prediction results.

We also look to extend our work to the parallel platform using Hadoop on a large corpus of sequences to reduce the computational complexity and in turn generalize our work by deploying it on various other proteomic datasets.

References

1. Patil, N., Toshniwal, D., Garg, K.: Effective framework for protein structure prediction. Int. J. Func. Inform. Pers. Med. **4**(1), 69–79 (2012)
2. Valko, M., Rhodes, C.J., Moncola, J., Izakovic, M., Mazur, M.: Free radicals, metals and antioxidants in oxidative stress-induced cancer. Chem.-Biol. Interact. **160**, 1–40 (2006). https://doi.org/10.1016/j.cbi.2005.12.009 PMID: 16430879
3. DNA, RNA and Protein: The Central Dogma. http://science-explained.com/theory/dna-rna-and-protein

4. Feng, P.M., Lin, H., Chen, W.: Identification of antioxidants from sequence information using naïve Bayes. Comput. Math. Methods Med. 2013: 567529 (2013). https://doi.org/10.1155/2013/567529 PMID: 24062796
5. Zhang, L.N., Zhang, C.J., Gao, R., Yang, R.T.: Incorporating g-gap dipeptide composition and position specific scoring matrix for identifying antioxidant proteins. In: 28th IEEE Canadian Conference on Electrical and Computer Engineering, Halifax, Canada (2015)
6. Zhang, L., Zhang, C., Gao, R., Yang, R., Song, Q.: Sequence based prediction of antioxidant proteins using a classifier selection strategy. PLoS ONE **11**(9), e0163274 (2016). https://doi.org/10.1371/journal.pone.0163274
7. Yang, Y., Lu, B.-L., Yang, W.-Y.: Classification of protein sequences based on word segmentation methods. In: 2007 IEEE Symposium on Computational Intelligence in Bioinformatics and Computational Biology (2007)
8. K-mer Counting—A 2014 Recap. http://homolog.us/blogs/kmer-counting-a-2014-recap/
9. TF-IDF and Log Entropy Model. http://stats.stackexchange.com/difference-between-log-entropy-model-and-tf-idf-model
10. Radial Basis Function Network Tutorial. http://mccormickml.com/2013/08/radial-basis-function-network-rbfn-tutorial/
11. Machine Learning Algorithms for Classification. www.cs.princeton.edu/picasso-minicourse.html
12. Naive Bayes Multinomial Text Classification made easy document. http://nlp.stanford.edu/htmledition/naive-bayes-text-classification-1.html/

Spectrum Aware-Based Distributed Handover Algorithm in Cognitive Radio Network

Shefali Modi and Mahendra Kumar Murmu

Abstract Cognitive radio network (CRN) has evolved to explore the unused spectrum by secondary users (SUs) where spectrum license is held by primary user (PU). CRN has two important features such as node mobility and spectrum mobility. The node mobility allows SUs to move freely in the network, whereas spectrum mobility is due to the random appearance of PU in channels. SU node may be dynamically disconnected due to both of the reasons and rejoins if the spectrum is free from PU in opportunistic manner. Therefore, in CRN, it is very difficult to ensure connectivity, quality of service (QoS), bandwidth efficiency, and throughput. To address this problem, we propose a distributed spectrum handover algorithm for CRN. The spectrum handover refers to the procedure that when the current channel of a SU is no longer available, the SU needs to break off the on-going transmission, vacate that channel, and determine a newly available channel to continue the transmission. Our algorithm is based on awareness of the channel characterization where SU node uses prediction method to find available channel list to proceed possible spectrum switching. Our algorithm is a kind a proactive handover algorithm.

Keywords Primary user · Secondary user · Handover · QoS
Cognitive radio network

1 Introduction

Increasing number of wireless nodes and mobile users has led to the uneven exhaustion of radio spectrum. Reports of FCC [1], highlights the inefficient utilization of spectrum band. The need to efficiently utilize spectrum has led to the

S. Modi · M. K. Murmu (✉)
Computer Engineering Department, NIT Kurukshetra, Kurukshetra 136119, Haryana, India
e-mail: mkmurmunitkkr@gmail.com

S. Modi
e-mail: shefalimodi08@gmail.com

P. K. Sa et al. (eds.), *Recent Findings in Intelligent Computing Techniques*,
Advances in Intelligent Systems and Computing 708,
https://doi.org/10.1007/978-981-10-8636-6_19

origin of cognitive radios. Cognitive radios (CR) [2] are software defined radios that increase spectrum utilization by opportunistically accessing the unused spectrum. Thus, forming an opportunistic network [3]. Users in cognitive radio are secondary users (SU) or unlicensed users and primary users (PU) or licensed users [4]. A cognitive radio networks (CRNs) can be viewed as a collection of SU nodes. In CRN, SU works on the principles to observe, analyze, learn, and adapt. The spectrum holes identified at each SU node is termed as local channel set (*LCS*). From *LCS*, some channels are used as control channels which discover the neighbor node [5]. The channel set is termed as common control channel (*CCC*). Once PU appears in the channel, SU starts observing for suitable spectrum holes, i.e., *LCS*, and reconfigures to available alternative channel. This process of changing the frequency of operation is known as spectrum handover in CRN.

Broadly, spectrum handover scheme can be viewed in two ways, namely never spectrum handover (NSHO) and always spectrum handover (ASHO). In NSHO, waiting is preferred over switching. This technique may introduce huge delay if PU activity is for long duration. However, frequent and short duration of PU is beneficial. ASHO scheme can be viewed in terms of proactive spectrum handover (PSHO) and reactive spectrum handover (RSHO) [6]. In PSHO [7, 8, 9], SU makes decision about the next channel prior to the interruption by PU and leaves the current channel before PU appears. This decision is based upon some prediction which is the consequence of learning skills of SU. PSHO scheme introduces less delay in the handover process. However, the efficacy of PSHO depends upon the exactness of prediction [9]. Prediction error can cause SU to switch to a channel that may become occupied in between the switching time and thus causing data collision or data loss. RSHO [10, 11] scheme is an observe and react approach. This scheme introduces more sensing delay as sensing starts after interruption.

Handover initiation is highly dependent on behavioral activity of PU. PU prediction may reduce sensing efforts of SU. However, PU prediction is a difficult task. Interested readers may refer [12] for various prediction models. The prediction of PU may assist SU to take handover decisions. Spectrum handover process introduces some delay in transmission process; this extra delay may degrade quality of service (QoS) [13]. During a communication process, multiple handovers may be observed. However, low handover rate is preferable [13]. Spectrum handover comes with many challenges as follows:

i. Deal with imperfect sensing and prediction errors.
ii. Selection of the best target channel.
iii. Reducing overall sensing delay by reducing the number of channels to be sensed in order to find the required channel.

In literature, most of the research work has been conducted on spectrum handoff and less work is done on spectrum handover in CRAHN. In [7, 8, 10, 11, 14], spectrum handoff is considered. While in [6, 9, 13, 15–17], spectrum handover has been explored, most of the work consider single pair of SUs. Spectrum handover is the extension of spectrum handoff in terms of resumption of transmission from target channel once handoff is initiated. Less work is done in the area of spectrum handover. Hence, there is a need of more exploration in this area.

In this paper, we designed a scheme that is hybrid of ASHO and NSHO with intelligent switching decision ability. Good switching decision helps in reducing unnecessary delay due to handover and thus increases spectrum utilization by SU. Besides switching decision, we also focused on selection of next channel. Rather than random selection of next channel, selecting an efficient channel will result in better spectrum utilization. For this, we characterized channels on the basis of bandwidth, channel noise, and idle duration and then select next channel.

The rest of the paper is organized as follows. In Sect. 2, system model is described. Section 3 includes the algorithm description. Simulation and numerical results are specified in Sect. 4. Finally, we conclude the paper in Sect. 5.

2 System Model

For our work, we considered a CRN with M independent licensed channels and N SUs. Each channel is characterized by a unique *id*, *bandwidth*, and *channel noise*. PUs are considered with exponential ON/OFF traffic model. Every PU is defined with their PU activity pattern, transmission range and number of arrivals and departures on licensed channel.

A channel can have two states in case of exponential ON/OFF traffic model. Either a channel can be in ON (*Busy*) state or it can be in OFF (*Idle*) state. *Busy* state of a channel denotes its occupancy by PU while the *idle* state represents the absence of PU at the channel. In exponential ON/OFF model, transition from one state to another is guided by a birth-death process. If α is the death rate then duration of *busy* state follows an exponential distribution with mean α^{-1} and if β is birth rate then *idle* period follows an exponential distribution with mean β^{-1}. SUs are operating in overlay mode of spectrum sharing. For our work following assumptions are considered:

$$CHP\,cost = T_S + T_C + \mu + \delta \tag{1}$$

where *CHP cost* (channel handover process cost) includes sensing time (T_S), next channel selection time (T_C), switching delay (μ), *a*nd negotiation delay (δ).

$$C_i = \frac{T_{idle}}{T_{idle} + \mu} \times c \tag{2}$$

where C_i is effective channel capacity, T_{idle} denotes the idle duration of channel, μ is the switching delay assumed above, and c is normalized channel capacity.

During the process of handover, switching decision plays an important role. Switching decision refers to decide whether to switch to another channel or to wait on current channel for the PU to finish. In our work, we evaluated *CHP* cost which

```
Switching decision (Output)
  begin
  | If (CHP cost < T_on)
  | |     return switch decision = true;
  | else
  | └─    return switch decision =false;
  end
  └─
```

Fig. 1 Algorithm for switching decision

includes the delay in selection of next channel with the time required to sense the candidate channel. The cost also includes the switching delay and the time required for negotiation on CCC between transmitter and receiver before resuming of transmission. This cost forms the basis for taking switching decision for handover. Among all the factors in *CHP cost*, total time to select candidate channel is a controllable factor. A quick and right selection of next channel reduces further delays.

3 Proposed Algorithm

3.1 Switching Algorithm with Description

The switching algorithm is for taking switching decision (shown in Fig. 1). If the switching decision is true, then switching process or channel handover process is initiated while the false decision restricts the switching process and waiting on current channel for PU to finish is preferred over switching. Decision is evaluated by comparing CHP cost with the PU activity ON duration at current channel. If cost to handover is less than the *busy* duration, then it is better to initiate handover process rather to wait for PU to finish. Similarly, if PU activity is for short duration and delay introduced is less than the delay in case of handover then it is better to wait rather to initiate handover process. In this way, this algorithm guides to take switching decision.

3.2 Next Channel Selection Algorithm with Description

The channel selection algorithm helps in deciding the next channel in case of handover (shown in Fig. 2.) For next channel selection, several factors are considered like channel bandwidth, idle duration, noise, and effective channel capacity. Decision of next channel is taken by evaluating each channel until the channel that fulfills all criteria is found. At first, the idle duration of channel is compared with the packet size. If idle duration is large enough to accommodate the packet

```
Current channel (Input)
next channel (Output)
   const  MaxChannels = 11;
   var    channel: 0..MaxChannels;
   begin
      channel = 0;
      repeat
         channel = channel + 1;
         if ( T_OFF >= packet size )
              Calculate interference at that channel;
              if ( interference >  threshold )
                    Discover other channel;
              else
                 Calculate effective channel capacity (C_i);
                 if ( C_i > = C_R )
                      return next channel;
      Until channel = MaxChannels
   end
```

Fig. 2 Algorithm for next channel selection

Fig. 3 Number of handover for different traffic load

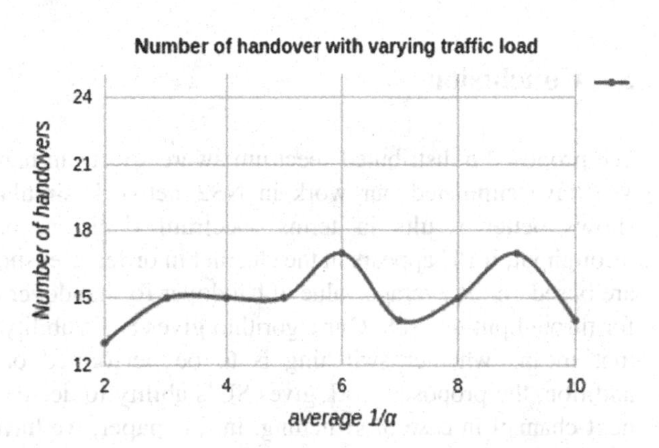

completely then the interference at that channel is evaluated. Further if interference is less than the threshold value, then effective channel capacity is evaluated. If this is greater than the required capacity then the channel is selected as the next channel.

4 Simulation Results

Figure 3 shows the handover rate of SUs for channel if PU appears in the channel. The number of handover does not vary large with increased duration of PU activity. This shows the efficiency of channel selection and switching decision. The channel

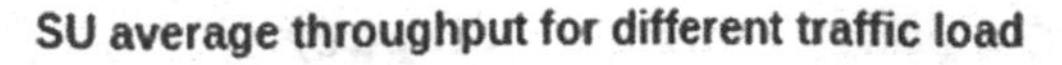

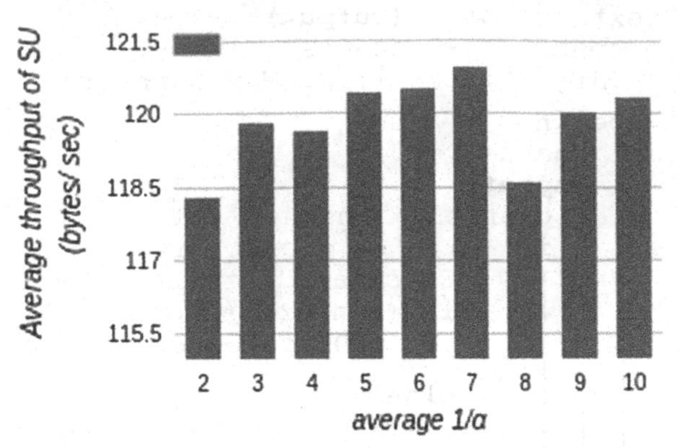

Fig. 4 SU average throughput for different traffic load

switching decision takes right decision according to PU activity duration and this decision helps in reducing unessential handover. Figure 4 shows the effect of PU activity duration over SU throughput. The results show better throughput of SUs in CRN.

5 Conclusion

We proposed a distributed spectrum aware spectrum handover algorithm for CRN. We have simulated our work in NS2 network simulation tools. The algorithm shows better results in terms spectrum decision, bandwidth utilization, and throughput, if PU appears in the channel in order to ensure QoS in CRN. Our results are based on the average value of handover for handover rate and SU's performance for throughput in CRN. Our algorithm gives SU's ability to take switching decision that means whether switching is to be performed or waiting is beneficial. In addition, the proposed work gives SU's ability to decide for appropriate channel as next channel in case of switching. In this paper, we have considered only a single pair of SUs. Our future scope is to extend to *n* SUs in cognitive radio network.

References

1. Powell, C.: Commissioners Copps, Martin and Adelstein, FCC, Notice of proposed rule-making and order, ET Docket no. 03–322 (2003)
2. Akyildiz, I.F., Lee, W.Y., Vuran, M.C., Mohanty, S.: NeXt generation/dynamic spectrum access/cognitive radio wireless networks: a survey. J. Comput. Netw. **50**(13), 2127–2159 (2006)
3. Alam, S.S., Marcenaro, L., Regazzoni C.: Opportunistic spectrum sensing and transmissions. CRIMTS. IGI Global, pp. 1–28 (2013)

4. Akyildiz, I.F., Lee, W.Y., Chowdhury, K.R.: CRAHNs: Cognitive radio Ad Hoc Networks. J. Ad Hoc Netw. **7**(5), 1–27 (2009)
5. Khan, A.A., Husain, R.M., Yasir, S.: Neighbor discovery in traditional wireless networks and cognitive radio networks: Basics, taxonomy, challenges and future research directions. J. Netw. Comput. Appl. **52**, 173–190 (2015)
6. Ma, B., Xie, X.: Spectrum handover mechanism based on channel scheduling in cognitive radio networks. In: ARECWAC, pp. 408–413 (2011)
7. Song, Y., Xie, J.: ProSpect: a proactive spectrum handoff framework for cognitive radio ad hoc networks without common control channel. IEEE Trans. Mob. Comput. **11**(7), 1127–1139 (2012)
8. Nejatian, S., Syed-Yusof, S.K., Latif, N.M.A., Asadpour, V.: Proactive integrated handoff management in cr-manets: a conceptual model. In: IEEE ISWTA, pp. 32–37 (2012)
9. Ma, B., Xie, X., Liao, X.: PSHO-HF-PM: an efficient proactive spectrum handover mechanism in cognitive radio networks. J. Wirel. Pers. Commun. **79**(3), 1679–1701 (2014)
10. Wang, C.W., Wang, L.C., Adachi, F.: Modeling and analysis for reactive-decision spectrum handoff in cognitive radio networks. In: IEEE Globecom, pp. 1–6 (2010)
11. Wang, C.W., Wang, L.C.: Analysis of reactive spectrum handoff in cognitive radio networks. J. Sel. Areas Commun. **30**(10), 2016–2028 (2012)
12. Yasir, S., Husain, R.M.: Primary radio user activity models for cognitive radio networks: a survey. J. Netw. Comput. Appl. **43**, 1–16 (2014)
13. Kahraman, B., Buzluca, F.: An efficient and adaptive channel handover procedure for cognitive radio networks. J. Wirel. Commun. Mob. Comput. **15**(3), 442–458 (2015)
14. Wang, L.C., Wang, C.W.: Spectrum handoff for cognitive radio networks: reactive-sensing or proactive-sensing?. In: IEEE IPCCC, pp. 343–348 (2008)
15. Kahraman, B., Buzluca, F.: A Novel channel handover strategy to improve the throughput in cognitive radio networks. In: IEEE IWCMC, pp. 107–112(2011)
16. Mardeni, R., Anuar K., Hafidzoh, M., Alias, M.Y., Mohamad, H., Ramli, N.: Efficient handover algorithm using fuzzy logic underlay power sharing for cognitive radio wireless network. In: IEEE ISWTA, pp. 53–56 (2013)
17. Lu, D., Huang, X., Zhang, W., Fan, J.: Interference-aware spectrum handover for cognitive radio networks. J. Wirel. Commun. Mob. Comput. **14**(11), 1009–1112 (2014)

A Reputation-Based Trust Management Model in Multi-hop Cognitive Radio Networks

Trupil Limbasiya, Debasis Das and Ram Narayan Yadav

Abstract Cognitive radio network (CRN) is made of numerous intellectual users with the capabilities of sensing and sharing in underutilized spectrum called cognitive users (CUs). The spectrum allocated to licensed users also called primary users (PUs) is not fully utilized. The cognitive radio came into the picture to fulfill spectrum's demand by enriching unproductive bandwidth. However, the set of spectrum available to CUs varies time-to-time and space due to dynamics of PUs activities. Therefore, CUs' spectrum allocation changes from the PUs and the resource to both customers distributing the licensed spectrum. To realize the concept of CRN, each CU is required to send its sensing outcome and individual information to the central cognitive user base station (CUB) in centralized CRN. But CUs' dynamic environment and effortless compromise make the CUB susceptible in data falsification and PU emulation attack, which will deceive its global decision making and degrades the system performance. For this reason, an efficient trust management scheme is required to manage the CRN properly. In this paper, we design a trust-based method for centralized cognitive radio networks to achieve reliable practice of the system. Results present that the suggested can expose the malevolent behavior and make available transparency in the architecture.

Keywords Attack · Cognitive radio networks · Integrity · Trust

T. Limbasiya (✉) · R. N. Yadav
NIIT University, Neemrana, Rajasthan, India
e-mail: limbasiyatrupil@gmail.com

R. N. Yadav
e-mail: narayanram.1988@gmail.com

D. Das
Birla Institute of Technology & Science, Pilani, Goa, India
e-mail: deba16@gmail.com

P. K. Sa et al. (eds.), *Recent Findings in Intelligent Computing Techniques*,
Advances in Intelligent Systems and Computing 708,
https://doi.org/10.1007/978-981-10-8636-6_20

1 Introduction

Cognitive radio networks (CRNs) can practice this exiguous spectrum efficiently and envision as developing engineering sciences that can carry into effect the future interests. In order to rectify the concern of spectrum scarcity, a new technology is developed, which is known as "cognitive radio" [1] and it has capability to sense the environment. The customer, who wishes to use license mediums opportunistically, are known as a cognitive user (CU). CUs may use unused channel for their transmissions with nominal intervention to licensed customer (preliminary user). The spectrum sharing scheme has been separated into two main tracks: overlay and underlay [2]. In the overlay technique, CUs access temporary idle channels of primary systems facilitated by the outstanding spectrum sensing and signal processing methods when PUs are not practicing their linked spectrum. In the underlay system, PUs and CUs transfer at the same point where CUs have lower priority as compared to PUs to access the spectrum. To limit the intervention caused by CUs to PUs, it is controlled by the concerned authority that indicates the maximum endurable interference threshold. In this paper, we are taking overlay system of spectrum sharing into the consideration in which CUs are allowed to access the licensed medium whenever it is not utilized by PU as shown in Fig. 1.

To realize the CRN, each CU is required to send its spectrum sensing outcomes to the node which decides, manages, and allocates the channels to CUs in CRN [3]. To ensure the connectivity and robustness of CRN, CUs need to share their available channels and other control information with neighbor [4]. To minimize the sensing error due to hidden node, shadowing, and fading, cooperative spectrum sensing mechanism can be employed that reduces the interference to PU transmissions caused by CU transmissions and improves the system overall performance. Due to open nature of CRN, the decision-making outcomes of the central node may misled

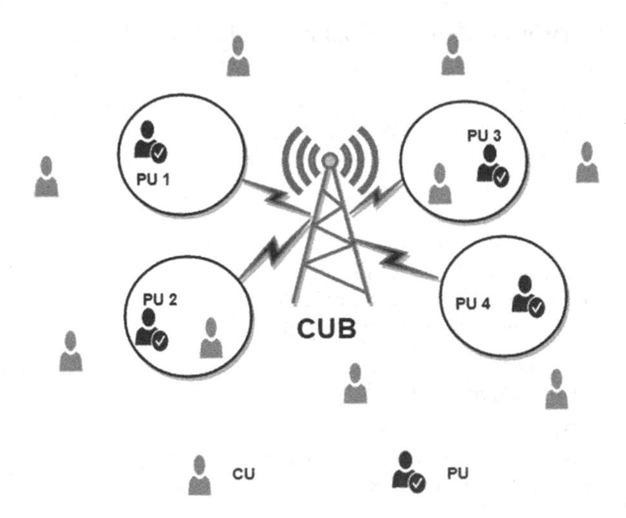

Fig. 1 A centralized structure of cognitive radio network

when SUs intentionally send false data. In CR ad-hoc network, an SU can communicate with other SUs provisionally and each node must have CR capability. Recently, scientists have concentrated on spectrum sensing [5] and sharing problems [6]. Security issues have been discussed on cognitive radio ad-hoc network [7–10]. However, in the case of centralized CRN, CUs send their individual sensed data to one centralized cognitive user base station (CUB) which decides and manages the cognitive cycle.

Sometimes, CUB's decision may mislead when corresponding CUs send falsified or forged data intentionally. This affects the CUB decisions and may reduce the system performance such as false alarm attack and miss detection attack. Due to CUs' dynamic environment and easy compromise, it makes the CUB vulnerable to data falsification and PU emulation attack, which will misled its global decision making and degrades the system performance. So, the open nature of CRN presents various security challenges such as sensing data authentication, PU emulation attacks, data falsification in realizing the CRNs [11]. Accordingly, an effective security mechanism is required to address various security problems present in centralized CRN.

In this paper, we propose a new trust management mechanism for centralized CRNs to address the sensing data falsification and authentication. In wireless systems, the centralized CUB gets charge of the foundation and trust instrument administration. The base station calculates reputation focused on the behavior of CUs and then assigns incentive and punishment values in order to monitor the network. In our trust scheme, the base station can recognize the genuine CUs and the malicious CUs in view of the long-haul conduct records of their activities.

Paper Organization: In Sect. 2, we present a brief survey regarding various CRN schemes. In Sect. 3, we suggest a trust management model for CRN. In Sect. 4, we perform an experiment focused on the proposed protocols. In Sect. 5, we analyze the proposed system in terms of security. Finally, we conclude in Sect. 6. References are at the end.

2 Background and Related Work

A hopeful solution is a cognitive radio (CR) for the restricted radio spectrum point which has already practiced in vehicular network environments. Frequently, CR technology can support cognitive users (CUs) because they do not have a pre-elucidated range for wireless interfaces to employ the original users PUs' permitted spectrum. The provision is that CUs should not discontinue PUs' standard communications. In a different way, CUs should have the knowledge for recognizing existence of PU. Furthermore, they should furnish the privileged spectrum for PUs whenever it is demanded by the same. PUs have no obligation to switch their actual foundation to serve CUs. CR can be practiced to decrease the spectrum drawbacks in vehicle ad-hoc networks additionally [1]. In [12], authors proposed an efficient trust management model for centralized cognitive radio network to facilitate proper functioning of cooperative tasks in CRN. Pravin et al. [11] addressed security challenges to achieve security communication based on trust-based authentication.

A collaborative spectrum sensing protocol has been presented in [13] based on trust management which also focuses on energy efficiency. To achieve benefits of cooperative spectrum, they assessed integrity of the reports sent by CUs and also required to identify the presence of misbehaving or malicious CUs in network. In [14], Orumwense et al. studied the effect of malicious customers on the energy efficiency of CRNs. Mousavifar et al. [15] analyzed the effect of sensing results using trust-based mechanism. Quantum-resistance authentication framework has been proposed in [16] for centralized cognitive and discussed authentication issues based on automated verification of Internet security etiquettes and applications, provided formal authentication with the help of BAN logic, and suggested the resistance of the propounded framework against multiple vulnerabilities.

3 Proposed Model

We have suggested a trust management etiquette focused on cognitive radio networks in this section which is employed in determining the correct information regarding band sensing. Here, we have practiced the concept of asymmetric key cryptosystems for secure data transmission from one node to other node/centralized server or centralized server to other nodes. A node generates a pair of keys, namely public key (K_{Pub}) and private key (K_{Pri}). All nodes share their K_{Pub} with all other nodes of the network framework to obtain some level of security in the cognitive radio networks. If any node is interested in sending some important information on sensing, then a sender can transmit packets to the recipient after employing the receiver's public key (K_{Pub}) as per suggested protocol. A node will be terminated for various kinds of services if the dislike value ($\Re$) is more than fixed threshold value of T_T.

3.1 Proposed Algorithms

We propose two protocols through which we can provide better trustworthiness result regarding nodes. First algorithm is helped in understanding the secure sensed data transmission to the CUB, and second is to decide reputation of various customers. Customers play a crucial role in the CRN by suggesting correct information to the CUB. However, there are some malicious users and they can disturb the system by informing false/modified sensing intelligence which affects faithful users. We can provide trust to various customers regarding the same concern with the help of the following protocols. We have fixed the trusted factor (T_F) of all nodes as a one initially ($T_F = 1$).

3.1.1 Conserved Intelligence Etiquette

Users can transmit various types of information to CUB in a protected manner using this etiquette in order to compute reputation of a specified customer, and thus, users can fill confidence regarding the system.

1. Firstly, a CUB computes distance between the CUB and different participated nodes in data dissemination procedures based on Euclidean distance mapping in two dimensions as follows:

$$\therefore \quad d_{ab} = \sqrt{|(a_1 - b_1)^2 + (a_2 - b_2)^2|}$$

2. If d_{ab} belongs to the transmission span of a node, then only it will check the dislike value ($\mathfrak{R}$) of a certain node and it is a significant value for all different nodes. $\mathfrak{R}$ is calculated by performing verification steps of a node as either a fallacious node or an intermediator and/or an original sender by CUB, and it shares $\mathfrak{R}$ of various customers regularly within the network system.
3. In the condition of an intermediator or an original sender, a node will calculate a time difference between a received datagram and sent packet by enumerating $\Delta T \leq T_2 - T_1$, where T_1 is a computing time stamp at the sender side, T_2 is a time stamp at a recipient side, and ΔT is a maximum allowed time limit to accept a packet legally. In a case of more time difference rather than ΔT, it will be considered as an unsatisfied result and it will terminate the packet forwarding process at that moment. Otherwise, it will continue to *Step-4*.

4. The sender/forwarder will encrypt sensing data (D_s) with the help of K_{Pub} of the reception by performing an Ex-OR operation as follows, and $\{A, T_1\}$ will be transmitted to the receiver.

$$\therefore \quad A = E_{K_{Pub}}(D_s \oplus T_1)$$

5. After the experiment of a time validation, it will check the value of A by calculating the operation as follows:

$$\therefore \quad D_{K_{Pri}}(A) = D_{K_{Pri}}(D_s \oplus T_1)$$

$$\therefore \quad D_{K_{Pri}}(A) = D_s \oplus T_1$$

If the recipient can obtain valid information after calculating D_s, then we can trust on received packet and the sender/intermediator in this secure data dissemination. On the other hand, we can say that there was an occasion of happening information modification and this event will be marked as unsatisfied.

3.1.2 Reputation Etiquette

In this algorithm, the CUB calculates the dislike value ($\mathfrak{R}$) of different users which have sent packets with D_s. Hence, $\mathfrak{R}$ is computed based on different credentials to provide accurate outcome of $\mathfrak{R}$ to a precise customer. And it is computed and focused on various activities of a node. The following protocol will be useful in calculating $\mathfrak{R}$.

1. First of all, CUB verifies the availability of all nodes by computing d_{ab}.
2. The CUB finds participation of any node even though that node was unavailable during the communication, then CUB increases ρ by one value.
3. The CUB checks overall packet communication period and sets a specific value (ΔT) for the same. If a node delivers any packet beyond the ΔT, then it will raise the value of τ by one point directly.
4. The CUB has K_{Pub} of all customers which is beneficial in the enumerating process of data retrieving from different nodes. If any legitimate/malicious customer has forged any kind of information, then it will be identified during this step. At the same moment, the CUB will increment ω by one worth.

If any illegal activity has been found by the CUB, then it will enumerate ($\mathfrak{R}$) and as well as the trusted factor (T_F) of a precise user as follows, and it will be communicated on the regular basis in the interest of the customers to be aware of spiteful users within the system.

$$\therefore \quad \mathfrak{R} = \rho + \tau + \omega$$

$$\therefore \quad T_F = T_F - \frac{1}{\rho}$$

4 Performance Evaluation

We have experimented a scenario of CRN by using NS2 with 10 customers totally in which 1 customer has been marked as a malicious user ($\mathcal{A}$) initially and various spiteful activities have been carried out by $\mathcal{A}$ to interrupt the regular mechanism. We have executed the outlined method with different ΔT (by fixing 0.80). A faithful node does not performed any malicious activity during the experiment and thus, T_F is remained as 1 as shown in Fig. 2. At the same point, an attacker node has executed some illegal actions (by delaying 0.02 s each packet) during the experimental process. And we can identify that a malicious user has been crystallized as shown in Fig. 3 after performing a certain number of time. However, a proposed mechanism

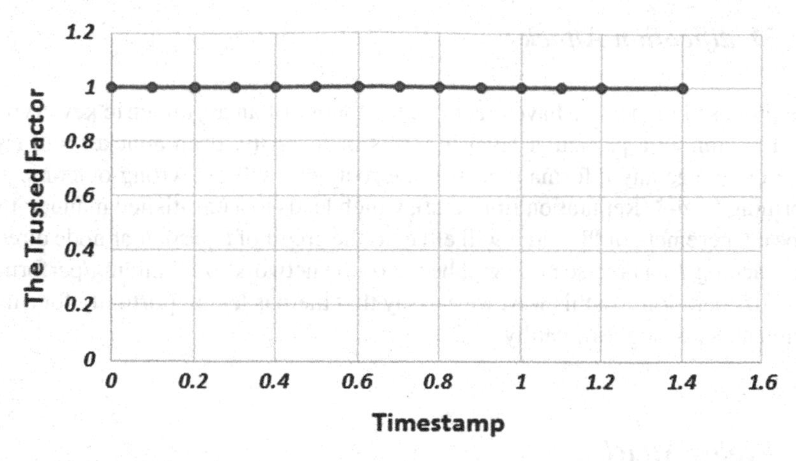

Fig. 2 The trusted factor for a truthful node

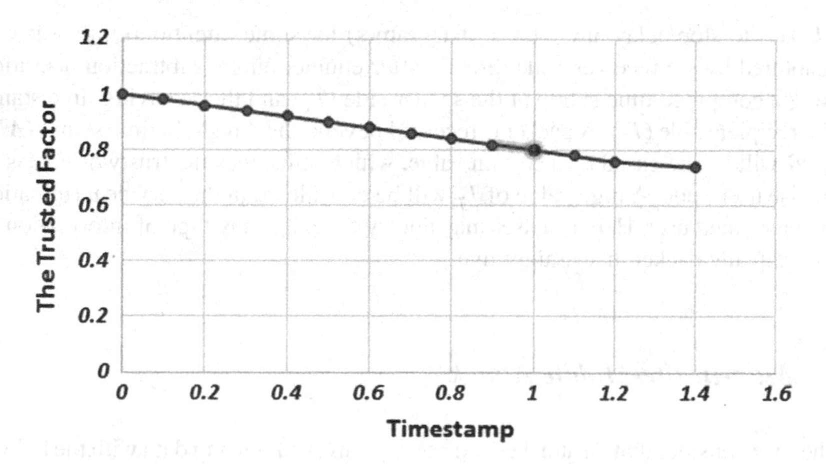

Fig. 3 The trusted factor for a spiteful node

administrator can decide the value of ΔT as well as cryptographic algorithm based on the requirement of the protection level within the framework.

5 Security Observations

In this section, we have done analysis of the suggested protocol with respect to data integrity and availability within a reasonable period of time and explained how a recommended algorithm can withstand against various attacks easily.

5.1 Modification Attack

In the proposed model, we have practiced the theory of an asymmetric key cryptography to maintain a particular level in terms of integrity. If an attacker will try to update or change any information, then the recipient will get wrong outcome after performing *Step-3* (Reputation Etiquette), which leads to a unsatisfied manner; there will be an increment of $\mathfrak{R}$, and it will effect to the credit of a particular node directly. After reaching T_t, a precise node will be out of the network for obtaining/performing variety of operations. In this way, we can say that there is less opportunity for a modification attack straightforwardly.

5.2 Replay Attack

If $\mathcal{A}$ tries to stop/delay any type of datagram(s) for some intention(s), then it will be captured by the receiver node directly after enumerating a subtraction operation among a computed time stamp at the sender side (T_1) and the receiving time stamp at the recipient side (T_2). A packet is received beyond the threshold time stamp (ΔT), then $\mathfrak{R}$ will be increased by a certain value, which influences the trustworthiness of a precise user node. A high value of T_T will be considered in the service termination of a determined user. Hence, a user may not try to update any type of information or delay/stop any packet in a straight line.

5.3 Man-in-the-Middle Attack

Hither, we consider that an attacker can access conveyed sensing data with the help of numerous schemes. In such circumstances, we have practiced the concept of public key cryptography and attackers have to cognizant regarding K_{Pri} of the recipient then only they can obtain the significant information about spectrum sensing else they cannot understand communicated data from the sender to the receiver. Generally, there are two keys in public key cryptography in which K_{Pub} can be accessible to users which are part of the networks and K_{Pri} is available to the owner of the key individually.

6 Conclusion

We have explained a trust management system in CRNs with different types of users (primary and cognitive). In this paper, we have presented algorithms for cognitive radio networks focused on trust management system, which can be employed to perform different cooperative functionalists such as cooperative, spectrum sensing, spectrum decision, spectrum assignment. Our proposed method satisfies different necessary demands for an example easy node identification within a span, confidentiality, integrity, and conclusive availability which are more useful to communicate in a variety of situation. In addition, the proposed algorithms can withstand against a replay attack, a man-in-the-middle attack, and a modification attack. Through simulation, we have demonstrated the effectiveness of the proposed etiquettes in terms of malicious node identification.

Acknowledgements We are thankful to NIIT University for providing research laboratory and funding support to carry out this research work.

References

1. Haykin, S.: Cognitive radio: brain-empowered wireless communications. IEEE J. Sel. Areas Commun. **23**(2), 201–220 (2005)
2. Goldsmith, A., Jafar, S.A., Maric, I., Srinivasa, S.: Breaking spectrum gridlock with cognitive radios: an information theoretic perspective. Proc. IEEE **97**(5), 894–914 (2009)
3. Yadav, R.N., Misra, R.: κ-Channel connected topology control algorithm for cognitive radio networks. In: 2016 8th International Conference on Communication Systems and Networks (COMSNETS), pp. 1–8. IEEE (2016)
4. Yadav, R.N., Misra, R.: κ-channel connectivity in cognitive radio networks for bounded treewidth graphs. Int. J. Commun. Syst. (2017)
5. Cabric, D., Mishra, S.M., Brodersen, R.W.: Implementation issues in spectrum sensing for cognitive radios. In: Conference Record of the Thirty-Eighth Asilomar Conference on Signals, systems and computers, vol. 1, pp. 772–776 (2004)
6. Zhao, Q., Swami, A.: A survey of dynamic spectrum access: signal processing and networking perspectives. In: IEEE International Conference on Acoustics, Speech and Signal Processing, 2007. ICASSP 2007, vol. 4, pp. 1349–1352 (2007)
7. Wei, Z., Tang, H., Yu, F.R., Wang, M., Mason, P.: Security enhancements for mobile ad hoc networks with trust management using uncertain reasoning. IEEE Trans. Veh. Technol. **63**(9), 4647–4658 (2014)
8. Vosoughi, A., Cavallaro, J.R., Marshall, A.: A cooperative spectrum sensing scheme for cognitive radio ad hoc networks based on gossip and trust. In: IEEE Global Conference on Signal and Information Processing (GlobalSIP), pp. 1175–1179 (2014)
9. Vosoughi, A., Cavallaro, J.R., Marshall, A.: Trust-aware consensus-inspired distributed cooperative spectrum sensing for cognitive radio Ad Hoc networks. IEEE Trans. Cogn. Commun. Netw. **2**(1), 24–37 (2016)
10. Gupta, A., Hussain, M.: Distributed cooperative algorithm to mitigate hello flood attack in cognitive radio Ad hoc networks (CRAHNs). In: Proceedings of the First International Conference on Computational Intelligence and Informatics, pp. 255–263 (2017)
11. Parvin, S., Han, S., Hussain, F.K., Al Faruque, M.A.: Trust based security for cognitive radio networks. In: Proceedings of the 12th ACM International Conference on Information Integration and Web-based Applications & Services, pp. 743–748 (2010)
12. Parvin, S., Han, S., Gao, L., Hussain, F., Chang, E.: Towards trust establishment for spectrum selection in cognitive radio networks. In: 24th IEEE International Conference on Advanced Information Networking and Applications, pp. 579–583 (2010)
13. Mousavifar, S.A., Leung, C.: Energy efficient collaborative spectrum sensing based on trust management in cognitive radio networks. IEEE Trans. Wirel. Commun. **14**(4), 1927–1939 (2015)

14. Orumwense, E.F., Afullo, T.J., Srivastava, V.M.: Effects of malicious users on the energy efficiency of cognitive radio networks. In: Proceedings of the Southern Africa Telecommunications Networks and Applications Conference (SATNAC), Hermanus, South Africa, pp. 431–435 (2015)
15. Mousavifar, S.A., Leung, C.: Transient analysis for a trust-based cognitive radio collaborative spectrum sensing scheme. IEEE Wirel. Commun. Lett. **4**(4), 377–380 (2015)
16. Bakhtiari Chehelcheshmeh, S., Hosseinzadeh, M.: Quantumresistance authentication in centralized cognitive radio networks. Secur. Commun. Netw. **9**(10), 11581172 (2016)

A Study: Various NP-Hard Problems in VLSI and the Need for Biologically Inspired Heuristics

Lalin L. Laudis, Shilpa Shyam, V. Suresh and Ajay Kumar

Abstract As engineering endures to dominate the planet, its application is pervasive in almost all arenas. Thus, the problems associated with engineering optimization takes up a hike. Certain problems may be categorized as single objective optimization (SOO), a problem which has a single criterion to be optimized. In contrary, all modern problems may be classified as multi-objective optimization (MOO) a problem which has two or more objectives to be optimized. The MOO gives the fortuity for selecting the desired solution for the Pareto font where human becomes a decision maker (DM). In extension, most problems are non-deterministic polynomial in nature (NP-hard problems) wherein the obtained solution always converges near to the exact solution. Consequently, algorithms come into picture for solving these NP-hard problems. In this research work, various NP-hard problems which are in association with VLSI design akin wirelength minimization, area minimization, dead space minimization and so forth are delineated and the significant role of biologically inspired algorithms is discussed in a concise manner. The results are compared with other optimization algorithms. The results apparently protrude that bio-inspired algorithms which produce promising results comparatively with other optimization algorithms dominate the arena of algorithms whilst solving problems which are indeed NP-hard. An attempt to analyse the problems with SOO and MOO is attempted which is indeed intriguing.

Keywords Problems in VLSI · Optimization · Bio-inspired algorithms
Combinatorial optimization · NP-hard problems in VLSI

L. L. Laudis (✉) · S. Shyam · V. Suresh · A. Kumar
School of Electronics and Communication Engineering, Mar Ephraem College of Engineering and Technology, Marthandam, India
e-mail: lalin@marephraem.edu.in

P. K. Sa et al. (eds.), *Recent Findings in Intelligent Computing Techniques*,
Advances in Intelligent Systems and Computing 708,
https://doi.org/10.1007/978-981-10-8636-6_21

1 Introduction

'Let a computer smear—with the right kind of quantum randomness—and you create, in effect, a 'parallel' machine with an astronomical number of processors … All you have to do is be sure that when you collapse the system, you choose the version that happened to find the needle in the mathematical haystack'.—from Quarantine, a 1992 science fiction novel by Greg Egan.

From the evolution of engineering, problems tend to become more complex than usual that their absolute values find so rigid to converge. Such problems can be formulated under non-deterministic polynomial hard problem (NP-hard). These problems can have a precise solution verified in polynomial time. For example, the Travelling Salesman Problem (TSP) which is considered as a NP-hard problem can be tested in polynomial time for a given tour of definite length. However, it is factual that no heuristics are being formulated till then that could solve the NP-hard problem in polynomial time [1]. The fact that we reason NP-hard problem as non-deterministic polynomial problem is that if we assume a non-deterministic machine, which could predict perfectly at each point, then it becomes obvious that it could solve the given problem in polynomial time, which in turn is the same as verifying the solution. It is, indeed, interesting that P = NP is an open question in theoretical computer science. It has also been enlisted in the seven most important problems across the field of mathematics depicted by Delvin (2005). Nevertheless, it is usually believed that $P \neq NP$ [1]. Moreover, when design problems in VLSI are taken into due consideration, they may by classified under NP-hard [2, 3]. The authors of [2, 3] attempt to solve the floor planning issue and cell placement problem as NP-hard problems. This article delineates a clear picture about the NP-hard problems specifically concentrated on problems related to VLSI design. In extension, the necessity of biologically inspired algorithms and their efficiency in solving these problems concerned with VLSI design is also elaborated. Hybrid methodologies of biological process also have a profound impact as bio-inspired algorithms in solving problems which are NP-hard [4]. Section 2 attempts to outline the definition of NP-hard and gives a brief info about the problem formulations. Section 3 portrays the various problems in VLSI design which is supposed to be tagged under NP-hard problem and their mathematical formulations. Section 4 introduces bio-inspired heuristics and discusses a few successful bio-inspired algorithms which perhaps have a scope in computing and optimization. As a consolidated summary of the obtained results, Sect. 5 tabulates and analyses the results of various bio-inspired algorithms for VLSI design. The analysis of the so obtained results is classified whilst considering the design problems as single objective optimization (SOO) problems as well as multi-objective optimization (MOO) problems. Finally, the paper concludes by justifying the necessity for bio-inspired heuristics in solving NP-hard problems in VLSI design and also claims with solid proof from simulations that bio-inspired heuristics dominate other algorithms for solving NP-hard problems related to VLSI design.

2 NP-Hard Problems

NP-hard problems may be demarcated in simple terms as: a decision problem with a property of, if the answer holds a positive response, then there holds a proof that can be checked in polynomial time. The intriguing fact is that NP is the set of decision problems where the observer can verify a positive response if he envisages the solution.

In electronics, the circuit satisfiability problem [5] is classified as NP. Consider the following problem associated with P_{opt}: standard decision problem (SDP) instance: $I \in E_P$ and a number x. The output is '*Positive*' if there is an $n^0 \in N(I)$ provided that $G(I, n^0) \leq x$; '*Negative*' otherwise. In specific, if $N(I) = \emptyset$, then the output is '*Negative*'. To portray the NP-hardness of an optimization problem using a classical approach, we attempt to construct a standard decision problem [6] that agrees to the actual problem P_{opt} and justifies the NP-hardness of the previously portrayed problem. This scheme is termed as the *standard scheme* for justification of the NP-hardness of an optimization problem [6]. *To solve: Step (1):* solve the optimization problem P_{opt}. *Step (2):* Given n^*, calculate $F(I, n^*)$, if ($N(I) \neq \emptyset$). *Step (3)*: Compare $F(I, n^*)$and a number x. Return 'positive' if $F(I, n^*) \leq x$, otherwise return 'negative'. Now, when step (1) and step (2) can be compiled in a polynomial time, then the SDP is solvable in polynomial. Moreover, if the SDP is justified to be NP-hard and step (2) can likely be performed in polynomial time, then step (1) can unlikely be performed in polynomial time unless P = NP. This fact concludes that the problem is NP-hard [6] (Fig. 1).

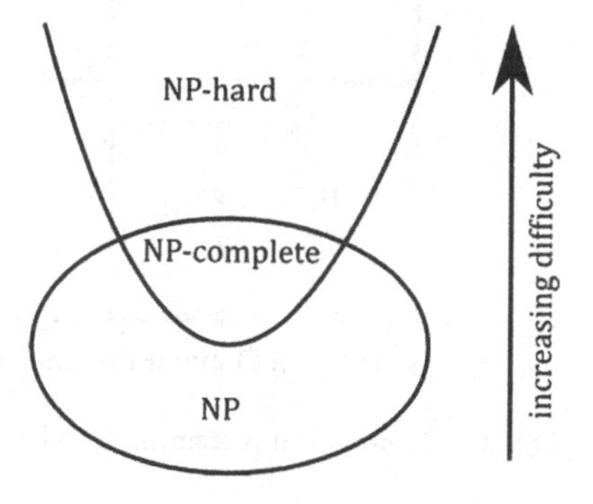

Fig. 1 Hierachy of problem with respect to solving difficulty

3 NP-Hard Problems in VLSI

Since the evolution of VLSI, the complexity of circuits is in par relative to the rate of lump of transistors in a die. Hence, it is evidently of significant importance to cogitate the problems in VLSI circuits, as the problems have to be optimized with multiple trade-offs. In general, the objectives of the problems are consolidated into one and solved as a single objective optimization (SOO) problem [7]. Nevertheless, in spite of the cumulative complexity researchers are paying attention to solve the problems with equal importance to several criteria as a multi-objective optimization (MOO) problem [8, 9]. Few problems are taken into the arena of NP-hard and elaborated in this section. These problems could be individually solved as a SOO problem and also in extension could be solved as considering MOO problem. However, when considering as MOO problem, the complexity increases and also the DM has the choice of selection of solution from the Pareto font.

3.1 *Area Minimization*

The efficiency of an ideal floor plan lies in expending the available area to the extreme. Given, a number of components 'n_i' and available area 'a_i', the criteria are that 'n_i' is placed optimally within the available 'a_i'. Consider a floor plan with 'a_i' and the number of blocks that has to be placed in the floor plan is 'n_i' where n = 7. Now, Fig. 2a protrudes the placement without optimality whereas Fig. 2b with optimality.

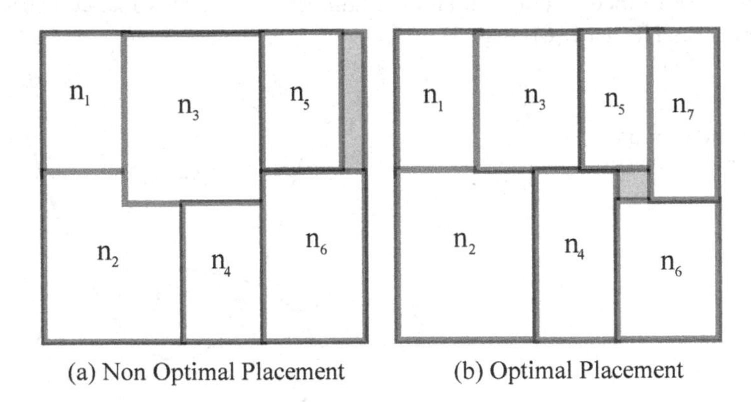

Fig. 2 **a** Non-optimal placement. **b** Optimal placement

3.2 *Wirelength Minimization*

The wirelength minimization problem is discussed in detail with a metaheuristics approach to solve it in [11, 12]. The authors provide a modified version of simulating annealing (SA) to solve the wirelength minimization problem. A theorem in [13] elicits that when a tree T is given which has vertices of degree 4 to the maximum, the problem falls under NP-complete when the deceive factor is T which has or not a layout with unit-length edges. The layout of a typical VLSI circuit has to be described in the form of data structure. Thus, the wirelength that interconnects the various blocks also has to be depicted in data structure. There are literatures that depicts the representation of interconnects and the floor planning in VLSI circuits. [14] proposes a well-known and considerably significant data structure based on B*tree representation (Fig. 3).

In [15], the author depicts the methodologies that educe a placement having minimum wirelength from a placement which is consistently relative. The problem is formulated as linear program and represented as

$$\text{Min} \sum W(N)\, x^{n+} - x^{n-} + y^{n+} - y^{n-} \tag{1}$$

The ultimate aim is to find an optimal routing of wires across the components of an integrated circuit.

3.3 *Dead Space Minimization*

The dead space of a floor plan is the area which is not occupied by any modules [10]. In a physical design of VLSI circuit, dead space is created when the horizontal size of the blocks is not in par accordance. In addition, the same is created when the

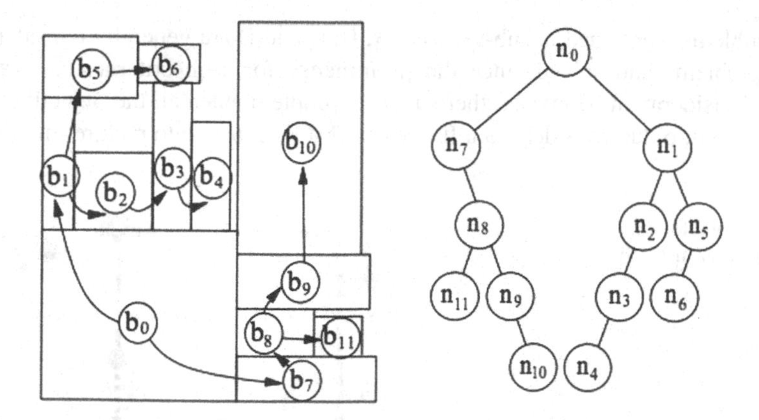

Fig. 3 An admissible placement and its corresponding B* tree representation as on [14]

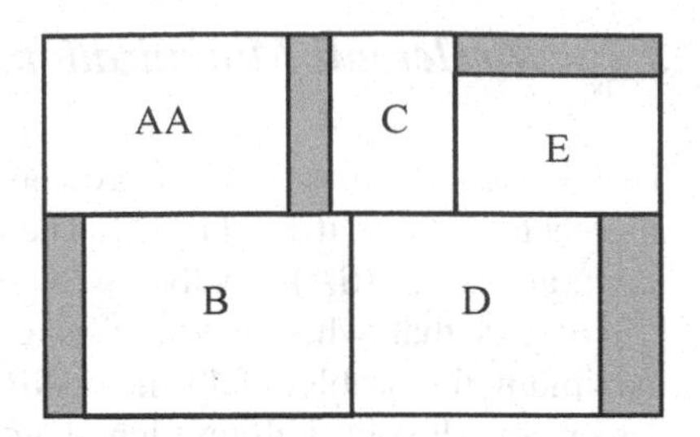

Fig. 4 A simple floor plan depicting dead space (grey shaded portion)

perpendicular size of two modules or sub-floor plans is alike [10]. In Fig. 2, the modules A, B, C are unlikely identical forming a dead space between them. For an efficient floor planning of VLSI circuit, the dead space has to be considerably reduced to minimum. Hence, this problem can be classified under optimization problem (Fig. 4).

3.4 Via Minimization

The via minimization problem is also put into the class of NP-hard problems [16]. For a metal net which is connected, a vertical connection gives rise to a layer change. We tend to abate the employment of these vias as vias to decrease the electrical reliability, performance of the chip and also decrease the manufacturing yield considerably. In the general case, the via minimization problem is NP-hard. In practice, however, via minimization is often either ignored or de-emphasized in routing tools and comes as an afterthought problem [16] (Fig. 5).

3.5 Multi-objective Problems

The problems mentioned in sub-sections A, B, C and D are generally considered as SOO problems and solved since the prominence for each criteria changes with design consideration. However, there arises a problem when all the objectives have to be taken into due consideration. It is where MOO comes into picture. In [17], the

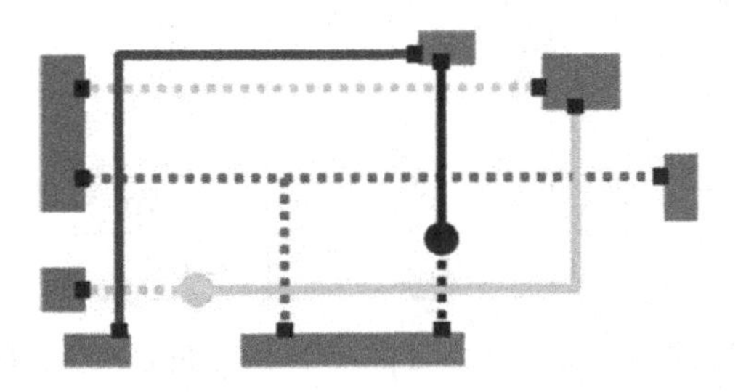

Fig. 5 A simple via representation in VLSI circuits

authors attempted to employ simulated annealing (SA) algorithm for MOO problems wherein the problems in VLSI were also considered. In addition, the MOO gives the fortuity for the DM to handpick the solution as per the DM's requirements. Also, multi-objective tools were also developed for solving optimization problems in VLSI design [18, 19].

4 Bio-Inspired Heuristics

4.1 Genetic Algorithm

GA is indeed an influential algorithm in solving optimization problems. The GA mimics the natural selection process in biology and refines at each generation (iteration). GA, generally, selects a random individual from the group of population and tends to modify and refines them. GA selects the individual solution in a stochastic manner and employs the solutions as parents to produce a children solution set for the consecutive generations. Also, the GA undergoes mutation (sudden changes in the chromosomes) and crossover operation which further refines the fitness of the solution, thereby converging the solution to the best as much as can. In addition, the GA has been successfully employed in solving combinatorial problems in VLSI design. The VLSI routing problem was successfully solved by using GA, and it reported high efficiency [20, 21]. As an essence, in the evolution of GA, it is been customized and has taken several dimensions and deformation to solve problems related to VLSI [20].

4.2 Particle Swarm Optimization

Particle swarm optimization (PSO) is also an intriguing bio-inspired algorithm which is used in a very large extent for optimization problems. In extension, PSO has also been successfully employed to solve problems related to VLSI design [22, 23]. The PSO algorithm's mechanism runs by having a group of population generally termed as swarm and also the candidate solutions which may be labelled as particles. The working is that these particles are moved in the search space with some constrains. The swarms in the population select the best known position and tend to move. The highlight is that when the swarm finds a best solution it converges to migrate to the best solution thereby refining its position and converging towards the very best position. The process is re-iterated, and it is perhaps believed to extract a favourable solution The PSO algorithm in *zero*th iteration and the converged solution set after '*N*th' iteration is revealed in Fig. 6.

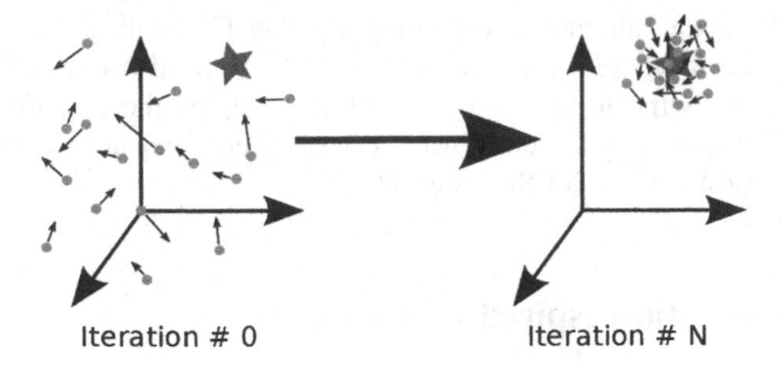

Fig. 6 PSO in iteration '0' and convergence in iteration 'N'

4.3 Ant Colony Optimization

Ant colony optimization (ACO) is also perhaps a well-known bio-inspired algorithm for solving optimization problems. It mimics the biological behaviour of ant. The ACO may also be classified under *swarm intelligence* method, and it is a probabilistic method of solving problem. The inspiration of ACO incepts from the foraging attribute of an ant colony. The ants are allegedly reported to communicate indirectly just by chemical pheromone trails. This helps them for optimizing a path from their prey to their habitat. This attribute of ant colonies is studied and imparted to solve discrete optimization problems [24]. ACO has been actively employed in solving problems related to VLSI floor planning. It is evident from that ACO, with MCNC benchmarks, has obtained impressive results in VLSI floor planning. Besides, ACO has also been employed in multi-objective optimization of operational amplifier [25].

4.4 Clonal Selection (CS) Algorithm

The clonal selection (CS) algorithm takes its inspiration from the theory of acquired immunity and thus falls under the same category of bio-inspired algorithms. The algorithm describes how B and T lymphocytes get adhered with their antigens and their response characteristics to antigens over a given time which is termed as affinity maturation. The CS algorithm has close proximity with *Darwin's* theory of natural selection where selection is based on affinity of antigen–antibody interaction and the reproduction is just due to the act of cell division. Moreover, the variation takes its inspiration by somatic hyper-mutation. CS algorithm is allegedly in close association with hill climbing techniques and GAs. CS algorithm has its profound impact in solving problems related to VLSI that too problems concentrated on floor planning issues [26]. The multi-objective approach is also taken into account for solving optimization problems [27].

5 Results and Tabulation

5.1 Simulation Platform Details

The algorithms mentioned in Sects. 4.1, 4.2, 4.3 and 4.4 were implemented in C/C ++ programming language. The platform utilized for simulation is Microsoft Visual Studio 10 Professional. The system held an OS MS Windows 7. The hardware specification was 64-bit OS. The processor was Intel i3 with one TB HDD and 4 GB RAM. The processor speed was 2.2 GHz clock speed.

5.2 Benchmark Circuits

The benchmark circuit used for testing the coded algorithm was Microelectronic Centre for North Carolina benchmarks. It is usually referred in literatures as MCNC benchmark. This benchmark cluster was unconfined for research, and it is quite often cited as MCNC benchmarks. These benchmarks are maintained by North Carolina's Microelectronics, Computing, and Networking Center, but are now located at the CAD Benchmarking Laboratory (CBL) at North Carolina State University. The standards pack the representation and attributes of certain pre-defined circuits which are given in Table 1. The benchmark circuits vary in size from 9 to 49 modules.

5.3 As SOO Problem

See Tables 2, 3 and 4.

Table 1 MCNC benchmark circuit details

Benchmarks	Modules	Nets	I/O Pads	Pins
apte	9	97	73	287
xerox	10	203	2	698
hp	11	83	45	309
ami33	33	123	42	522
ami49	49	408	22	953

Table 2 Dead space comparison

Benchmarks	GA	PSO	ACO	CSA	TCG	O-Tree	e-Otree
apte	7.43	7.21	7.34	7.13	7.90	7.75	8.10
xerox	5.22	5.10	5.09	5.12	5.50	5.67	5.87
hp	8.48	8.20	8.26	8.45	8.82	8.30	8.49
ami33	10.83	10.50	10.77	10.71	10.98	10.82	10.91
ami49	12.18	12.02	12.22	12.28	12.49	12.52	12.32

Table 3 Area comparison

Benchmarks	GA	PSO	ACO	CSA	TCG	O-Tree	e-Otree
apte	46.90	46.92	46.9	46.01	48.92	47.22	46.98
xerox	20	19.55	19.8	19.02	19.82	20.02	20.21
hp	9.03	9.10	9.18	9.15	10.94	9.98	9.16
ami33	1.22	1.28	1.20	1.21	1.80	1.98	1.29
ami49	37.50	37.01	37.2	37.08	36.77	37.70	37.73

Table 4 Wirelength comparison

Benchmarks	GA	PSO	ACO	CSA	TCG	O-Tree	e-Otree
apte	197	270	247	294	379	331	339
xerox	229	280	399	367	387	389	370
hp	69.36	62.27	152	150.87	159	162	150.1
ami33	45.25	48.48	52.23	48.87	52.22	50.14	51.12
ami49	622	680	768	688	690	689	710

5.4 As MOO Problem

See Tables 5 and 6.

Table 5 As MOO problem with wirelength and area as objective

Benchmarks	GA		PSO		ACO		CSA	
	A	WL	A	WL	A	WL	A	WL
apte	46.7	192	47.4	265	47.18	242	48.5	289
xerox	20.2	220	20.2	28.9	20.3	379	20.4	381
hp	9.85	68.3	9.65	60.3	9.46	149	9.60	149
ami33	1.29	46.2	1.29	58.4	1.25	58.3	1.25	48.10
ami49	39.5	612	40.6	693	37.5	788	38.2	690

Table 6 As MOO problem with wirelength and area as objective

Benchmarks	TCG		O-Tree		e-Otree	
	A	WL	A	WL	A	WL
apte	51.5	378	52.9	321	52	321
xerox	21.4	385	21.4	380	22.4	381
hp	9.50	152	9.50	153	9.50	152
ami33	1.84	50	1.98	51	1.70	52
ami49	38.2	663	39.6	689	39.6	703

6 Conclusion

This paper attempted to protrude the noteworthy role of biologically inspired algorithm in solving problems which are classified under NP-hard, in VLSI domain. The simulation results and the discussions delineate a flawless picture of the importance of employing bio-inspired algorithms for VLSI-related problems. Moreover, the multi-objective approaches for the proposed algorithm also convey the factual importance of the same. Mathematically modelling the problems in VLSI is an area of interest for future research.

References

1. Kendall, G., Parkes, A.J., Spoerer, K.: A survey of NP-complete puzzles. ICGA J. **31**(1), 13–34 (2008)
2. Binas, J., Indiveri, G., Pfeiffer M.: Spiking analog VLSI neuron assemblies as constraint satisfaction problem solvers. In 2016 IEEE International Symposium on Circuits and Systems (ISCAS), pp. 2094–2097. IEEE (2016)
3. Chen, X., Lin, G., Chen, J., Zhu, W.: An adaptive hybrid genetic algorithm for VLSI standard cell placement problem. In: 2016 3rd International Conference on Information Science and Control Engineering (ICISCE), pp. 163–167. IEEE (2016)
4. Kureichik, V., Kureichik, V. Jr., Zaruba, D.: Hybrid bioinspired search for schematic design. In: Proceedings of the First International Scientific Conference "Intelligent Information Technologies for Industry" (IITI'16), pp. 249–255. Springer International Publishing (2016)
5. Glaßer, C., Jonsson, P., Martin, B.: Circuit satisfiability and constraint satisfaction around Skolem Arithmetic. In: Conference on Computability in Europe, pp. 323–332. Springer International Publishing (2016)
6. Cheng, T.C.E., Shafransky, Y., Ng, C.T.: An alternative approach for proving the NP-hardness of optimization problems. Eur. J. Oper. Res. **248**(1), 52–58 (2016)
7. Deb, K., Sindhya, K., Hakanen, J.: Multi-objective optimization. In: Decision Sciences: Theory and Practice, pp. 145–184. CRC Press (2016)
8. Bhuvaneswari, M.C.: Application of evolutionary algorithms for multi-objective optimization in VLSI and embedded systems. Springer, Berlin, Germany (2015)
9. Guo, W., Liu, G., Chen, G., Peng, S.: A hybrid multi-objective PSO algorithm with local search strategy for VLSI partitioning. Front. Comput. Sci. **8**(2), 203–216 (2014)
10. Lichen, Z., Runping, Y., Meixue, C., Xiaomin, J., Xuanxiang, L., Shimin, D.: An efficient simulated annealing based VLSI floorplanning algorithm for slicing structure. In: 2012 International Conference on Computer Science & Service System (CSSS), pp. 326–330. IEEE (2012)
11. Laudis, L.L., Anand, S., Sinha, A.K.: Modified SA algorithm for wirelength minimization in VLSI circuits. In: 2015 International Conference on Circuit, Power and Computing Technologies (ICCPCT), pp. 1–6. IEEE (2015)
12. Laudis, Lalin L., Sinha, A.K.: Metaheuristic approach for VLSI 3D-Floorplanning. Int. J. Sci. Res. **2**(18), 202–203 (2013)
13. Bhatt, Sandeep N., Cosmadakis, Stavros S.: The complexity of minimizing wire lengths in VLSI layouts. Inf. Process. Lett. **25**(4), 263–267 (1987)
14. Chang, Y.-C., Chang, Y.-W., Wu, G.-M., Wu, S.-W.: B*-Trees: a new representation for non-slicing floorplans. In: Proceedings of the 37th Annual Design Automation Conference, pp. 458–463. ACM (2000)

15. Vygen, J.: Platzierung im VLSI Design und ein zweidimensionales zerlegungsproblem. Dissertation, University of Bonn (1996)
16. Liers, F., Tim N., Pardella, G.: Via Minimization in VLSI Chip Design
17. Laudis, L.L.: A study of various multi objective techniques in simulated annealing. Int. J. Eng. Res. Technol. (ESRSA Publications) **3**(2) (2014)
18. Anand, S., Saravanasankar, S., Subbaraj, P.: A multiobjective optimization tool for very large scale integrated nonslicing floorplanning. Int. J. Circuit Theory Appl. **41**(9), 904–923 (2013)
19. Subbaraj, P., Saravanasankar, S., Anand, S.: Multi-objective optimization in VLSI floorplanning. In: Control, Computation and Information Systems, pp. 65–72. Springer, Berlin, Heidelberg (2011)
20. Lienig, J.: A parallel genetic algorithm for performance-driven VLSI routing. IEEE Trans. Evol. Comput. **1**(1), 29–39 (1997)
21. Lienig, J., Thulasiraman, K.: A genetic algorithm for channel routing in VLSI circuits. Evol. Comput. **1**(4), 293–311 (1993)
22. Singh, R.B., Baghel, A.S., Agarwal, A.: A review on VLSI floorplanning optimization using metaheuristic algorithms. In: International Conference on Electrical, Electronics, and Optimization Techniques (ICEEOT), pp. 4198–4202. IEEE (2016)
23. Kaur, A., Gill, S.S.: Hybrid swarm intelligence for VLSI floorplan. In: 2016 International Conference on Computing, Communication and Automation (ICCCA), pp. 224–229. IEEE (2016)
24. Blum, C.: Ant colony optimization: introduction and recent trends. Phys. Life Rev. **2**(4), 353–373 (2005)
25. Chiang, C.-W.: Ant colony optimization for VLSI floorplanning with clustering constraints. J. Chin. Inst. Ind. Eng. **26**(6), 440–448 (2009)
26. Bachir, Benhala, Ali, Ahaitouf, Abdellah, Mechaqrane: Multiobjective optimization of an operational amplifier by the ant colony optimisation algorithm. Electr. Electron. Eng. **2**(4), 230–235 (2012)
27. Abdullah, D.M., Abdullah, W.M., Babu, N.M., Bhuiyan, M.M.I., Nabi, K.M., Rahman, M.S.: VLSI floorplanning design using clonal selection algorithm. In: 2013 International Conference on Informatics, Electronics & Vision (ICIEV), pp. 1–6. IEEE (2013)

ESMGB: Efficient and Secure Modified Group-Based AKA Protocol for Maritime Communication Networks

Shubham Gupta, Balu L. Parne and Narendra S. Chaudhari

Abstract Maritime communication network is one of the applications in wireless networks to establish the communication among the vessels and overcome the communication complexities between them. However, the existing maritime network is susceptible to various known attacks. To mitigate the congenital vulnerabilities, the efficient authentication protocol for secure group (EAPSG) communication was proposed but it fails to overcome all the security weakness. In this paper, we show that the EAPSG protocol suffers from redirection attack, DoS attack, and high bandwidth consumption. Later, we illustrate the countermeasure of identified attacks. We modified the EAPSG protocol and propose efficient and secure modified group-based (ESMGB) AKA protocol for the maritime communication network. The formal verification of the proposed protocol is performed using AVISPA tool. The analysis shows that the protocol successfully avoids all the identified attacks. The quantitative analysis of the proposed protocol shows less communication and computation overhead compared to existing protocols.

Keywords Maritime communication · Group authentication · Network security Authentication and key agreement

S. Gupta (✉) · B. L. Parne (✉) · N. S. Chaudhari
Department of Computer Science and Engineering, Visvesvaraya National Institute of Technology (VNIT), Nagpur 440010, Maharashtra, India
e-mail: guptashubham396@gmail.com

B. L. Parne
e-mail: baluparne@gmail.com

N. S. Chaudhari
e-mail: nsc0183@yahoo.com

N. S. Chaudhari
Department of Computer Science and Engineering, Indian Institute of Technology (IIT), Indore 453552, Madhya Pradesh, India

P. K. Sa et al. (eds.), *Recent Findings in Intelligent Computing Techniques*, Advances in Intelligent Systems and Computing 708,
https://doi.org/10.1007/978-981-10-8636-6_22

1 Introduction

In the sea area, an enormous amount of information such as videos and images is transmitted from vessels to the maritime administrative authority on land via non-3GPP network (WiMAX, WLAN). Simultaneously, the maritime authority also broadcasts the commands, confidential documents, and multimedia data to the vessels. In the past few years, various incidents have been addressed regarding pirates violation and vessel hijacking in the sea area [1, 2]. Hence, it needs to design a wideband communication scheme for vessels at low cost so that secure maritime communication activities can be established.

Due to high data rate and large coverage area, Worldwide Interoperability for Microwave Access (WiMAX) has been admired as a suitable technology to fascinate the increasing demand of traffic information at sea area [3–7]. However, it is very difficult to design a pervasive communication network at sea due to the costly surface and long coastline. Although, researcher has proposed the Extensible Authentication Protocol-Authentication and Key Agreement (EAP-AKA) [8] and other similar authentication schemes such as EAP-TTLS [9], EAP-LEAP [10], EAP-PEAP [11], EAP-SPEKE [12] to authenticate the devices at WiMAX network on land. But all these protocols are vulnerable to various kind of attacks such as denial-of-service (DoS) attack, redirection attack, and impersonation attack. Moreover, these protocols do not support the group authentication mechanism and consume high bandwidth during the authentication process. However, EG-AKA protocol [13] authenticates the group-based devices but vulnerable to similar kind of attacks as above protocols. Therefore, it is impossible to adopt these protocols for group-based maritime communication.

To design the AKA protocol for the maritime scenario, we consider a group of vessels that generates the traffic information and transmits it to the maritime authority on land. The group of vessels is required to be authenticated by the server whenever they enter in the coverage area of infostations (15 Km). Figure 1 shows the network architecture of maritime communication scenario. In this architecture, a group of vessels travel from one point to another point and need to access the WiMAX infostation which is connected to the Access Network Gateway (ASN-GW). The ASN-GW achieves the functionalities of authentication, authorization, and accounting (AAA) server and hosts the AAA clients. The traffic information is transmitted between infostation and ASN-GW that connects to the proxy AAA server (PAAA). The PAAA server resides in WiMAX connectivity service network (CSN) that obtains the mobility anchoring and traffic accounting. The PAAA works as an intermediary to transfer the information between infostation and Home AAA (HAAA). The HAAA connects to content server in non-3GPP network [14].

Recently, EAPSG protocol [14] (extension of EG-AKA) has been proposed for the maritime scenario that supports group-based authentication and defeats most of the attacks. But, EAPSG protocol is vulnerable to DoS attack and redirection attack. In addition, the protocol consumes high bandwidth during the authentication of vessels and PAAA. Hence, these protocols are not suitable for resource-constrained ves-

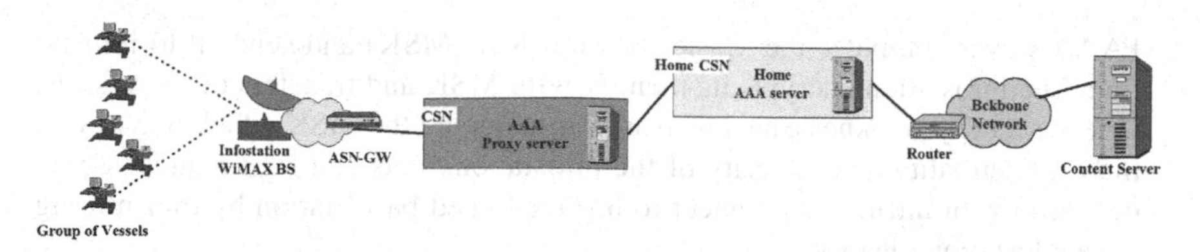

Fig. 1 Architecture of maritime communication network

sels/devices. On the basis of possible countermeasures of EAPSG protocol, we propose the efficient and secure modified group-based (ESMGB) AKA protocol. The formal verification of the proposed protocol is carried out by Automated Validation of Internet Security Protocols and Applications (AVISPA) tool. The quantitative analysis shows that the proposed protocol has less overhead at network compared to other existing AKA protocols.

The organization of this paper is as follows. Section 2 illustrates the security vulnerabilities in EAPSG protocol. The ESMGB-AKA protocol is proposed in Sect. 3. To overcome the security issues of EAPSG protocol, the security analysis and countermeasures are proposed in Sect. 4. The formal verification and performance analysis of the proposed protocol are shown in Sect. 5. Finally, Sect. 6 concludes the paper.

2 Related Work

Many AKA protocols have been proposed by the researchers to communicate the mobile devices in non-3GPP network to 3GPP network on land, but very few of them are acceptable for the maritime scenario. In this section, we discuss the EG-AKA protocol and its extension EAPSG protocol. Jiang et al. [13] proposed group-based EG-AKA protocol to authenticate mass devices in non-3GPP network. The EG-AKA protocol mutually authenticates each device at MME and maintains the confidentiality and data integrity. Moreover, the protocol preserves the privacy of devices and secures against most of the attacks but fails to defeat DoS attack and redirection attack. In addition, the protocol suffers from high computation overhead due to asymmetric key cryptosystem. Yang et al. [14] extend the application of EG-AKA protocol in special maritime communication networks and propose EAPSG protocol that mutually authenticates group of vessels at PAAA server connected to the infostations. The authentication process of the protocol is very similar to the EG-AKA protocol. Hence, the protocol also suffers from similar kind of attacks and problems.

- **Vulnerable to Redirection Attack**: The EAPSG protocol is penetrable to redirection attack as the transmitted information between PAAA and vessels obtains by an adversary in a decayed network. To authenticate the identity of infostation,

PAAA server computes the Master Session Key (MSK) and sends it to infostation. The infostation encrypts its identity with MSK and transfers to the vessels. Due to security weakness, an adversary can compute the MSK which may reveal the confidentiality and integrity of the infostation. Hence, a legitimate vessel is decoyed by an intruder to connect to his/her forged base station by transmitting false infostation identity.

- **Vulnerable to DoS Attack**: EAPSG protocol is vulnerable to DoS attack as an adversary inserts a false registration attempt that disrupts the synchronization of the initial value ($IV_{VG_{1-1}}$). It is assumed that the $IV_{VG_{1-1}}$ is persistently maintained between PAAA and vessels. There are many elements which lead to the inconsistency between the initial value of PAAA and vessels. For instance, if an adversary injects a false registration attempt, it makes MAC'_{PAAA} inconsistent at vessel as the value of $IV_{VG_{1-1}}$ is not matched. If the legitimate vessel tries to authenticate PAAA, it will always end in failure since MAC'_{PAAA} is not verified. Hence, all the time vessels attempt to be authenticated at PAAA but the PAAA will be observed as unauthorized; i.e., vessel suffers from the DoS attack.
- **Bandwidth Consumption at Network**: The protocol adopts public key cryptosystem-based Elliptic Curve Diffie–Hellman (ECDH) to authenticate each vessel in the group that generates high bandwidth consumption. The public key cryptosystem is a costly infrastructure that is not suitable for resource-constrained devices. Moreover, it generates the high computation overhead at the network that does not suit for the maritime communication network.

To overcome all the above-mentioned issues of EAPSG protocol, we propose the ESMGB-AKA protocol for the maritime communication network. The proposed protocol mutually authenticates each vessel and PAAA. The protocol obtains all the security requirements and defeats the identified attacks as redirection attack and DoS attack. In addition, the protocol adopts the symmetric key operations while authenticating the group of vessels and overcomes the problem of high bandwidth consumption from the network.

3 Proposed ESMGB-AKA Protocol

In this section, we propose the ESMGB-AKA protocol for mass vessels in a special maritime communication network. In the proposed AKA protocol, the first vessel executes a full authentication process and obtains the authentication vectors for the remaining vessels at HAAA. PAAA stores the authentication vectors from HAAA to authenticate each vessel in the group. The explanation of the proposed AKA protocol is as follows.

3.1 Group Formation and Initialization

A large number of vessels form groups based on similar application, region, behavior, and characteristics. Each vessel has a permanent identity ($PID_{VG_{i-k}}$) and installed by the service provider to allow each vessel for registration in maritime network. Meanwhile, each vessel in the group has a pre-shared secret key ($K_{VG_{i-k}}$) with HAAA. The service provider also provides a group key (GK_i) and identity (ID_{G_i}) to each group. Table 1 represents the notations used in the proposed AKA protocol. In addition, an index table [14] at HAAA is formed that contains the group (G_i), group identity (ID_{G_i}), and vessel identity ($ID_{VG_{i-k}}$). The proposed protocol is explained in two phases as shown in Fig. 2.

3.2 Registration Phase

In the registration phase, the first vessel ($V_{G_{1-1}}$) from the group G_1 communicates to HAAA and generates the authentication vectors for the authentication process of remaining vessels. The communication channel between PAAA and HAAA is secure [14]. The registration phase executes the following steps:

Step-1: An access request is sent by a vessel ($V_{G_{1-1}}$) to the infostation.

Step-2: To obtain the identity of $V_{G_{1-1}}$, the request identity message is sent from infostation to vessel.

Step-3: After receiving the request identity message, the $V_{G_{1-1}}$ computes the temporary identity ($TID_{VG_{1-1}}$) as

$$TID_{VG_{1-1}} = f^{1}_{K_{VG_{1-1}}}(PID_{VG_{1-1}}) \tag{1}$$

Table 1 Protocol notations and their acronyms

Notation	Definition
R_x	Random number computed by x
PID_{V_x}/TID_{V_x}	Permanent/Temporary Identity of x
GK_i	Group key of the ith group
GTK_{Gi}	Group Temporary Key of the ith group
MAC_x	Message Authentication Code calculated by x
$AUTH_x$	Authentication Token computed by x
f_1	Temporary Identity Generation Function
f_2	Generation Function of MAC
f_3	Generation Function of Group Temporary Key
f_4	Generation Function of Sharing Key

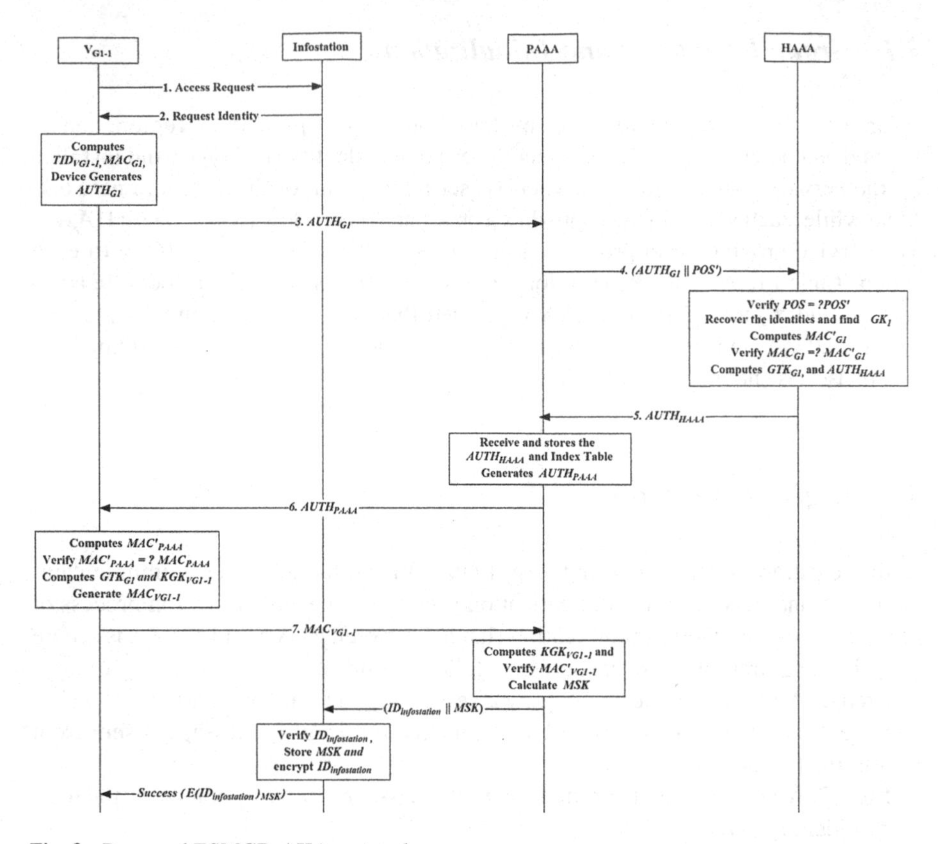

Fig. 2 Proposed ESMGB-AKA protocol

Then $V_{G_{1-1}}$ computes the MAC_{G_1} as

$$MAC_{G_1} = f^2_{K_{VG_{1-1}}}(ID_{G1} \parallel TID_{VG_{1-1}} \parallel R_{VG_{1-1}} \parallel ID_{infostation}) \tag{2}$$

And, computes the authentication code $AUTH_{G_1}$ as

$AUTH_{G_1} = (ID_{G1} \parallel TID_{VG_{1-1}} \parallel R_{VG_{1-1}} \parallel MAC_{G_1} \parallel POS \parallel ID_{infostation})$

Step-4: $V_{G_{1-1}}$ transfers the $AUTH_{G_1}$ to PAAA server over infostation. Then, the PAAA server attached its POS' with $AUTH_{G_1}$ and forwards to the HAAA.

Step-5: When HAAA receives the authentication request message, HAAA verifies whether the $POS = ?POS'$. If it is verified, HAAA generates the MAC'_{G_1} using $K_{VG_{1-1}}$ as Eq. 2 and verifies $MAC_{G_1} = ?MAC'_{G_1}$. If the authentication is successful, HAAA computes the $PID_{VG_{1-1}}$. HAAA also computes the respective group key GK_1 to generate the GTK_{G_1} as

$$GTK_{G_1} = f^3_{GK_1}(ID_{G1} \parallel R_{HAAA} \parallel AMF) \tag{3}$$

Then, HAAA generates $AUTH_{HAAA}$ as

$$AUTH_{HAAA} = (R_{VG_{1-1}} \parallel R_{HAAA} \parallel AMF \parallel GTK_{G_1} \parallel ID_{infostation})$$

HAAA transfers the computed $AUTH_{HAAA}$ with G_1's information collected in the index table to PAAA. PAAA server accepts and stores this information with $ID_{infostation}$, GTK_{G_1}, R_{HAAA} for the future use.

3.3 Authentication and Key Agreement Phase

After registration phase, the group authentication and key agreement phase of the protocol are executed.

Step-6: After receiving the $AUTH_{HAAA}$, the PAAA server selects R_{PAAA} and computes the MAC_{PAAA} for the mutual authentication with $V_{G_{1-1}}$ as

$$MAC_{PAAA} = f^2_{GTK_{G_1}}(ID_{G1} \parallel TID_{VG_{1-1}} \parallel R_{VG_{1-1}} \parallel R_{PAAA} \parallel R_{HAAA} \parallel AMF) \quad (4)$$

It computes the authentication code $AUTH_{PAAA}$ and transfers to $V_{G_{1-1}}$ as $AUTH_{PAAA} = (MAC_{PAAA} \parallel R_{VG_{1-1}} \parallel R_{PAAA} \parallel R_{HAAA} \parallel AMF)$

Step-7: $V_{G_{1-1}}$ generates the GTK_{G_1} (as Eq. 3). Then, it computes and verifies MAC'_{PAAA} (as Eq. 4). If $MAC'_{PAAA} \neq MAC_{PAAA}$, then $V_{G_{1-1}}$ sends an authentication failure response to PAAA server. Otherwise, PAAA server authenticated at $V_{G_{1-1}}$. Further, $V_{G_{1-1}}$ generates the key generation key $KGK_{VG_{1-1}}$ as

$$KGK_{VG_{1-1}} = f^4_{GTK_{G_1}}(ID_{G1} \parallel TID_{VG_{1-1}} \parallel R_{VG_{1-1}} \parallel R_{PAAA}) \quad (5)$$

To authenticate $V_{G_{1-1}}$ at PAAA server, $V_{G_{1-1}}$ computes the $MAC_{VG_{1-1}} = f^2_{KGK_{VG_{1-1}}}$ $(ID_{G1} \parallel TID_{VG_{1-1}} \parallel R_{VG_{1-1}} \parallel R_{PAAA})$ using $KGK_{VG_{1-1}}$ and transfers to PAAA.

Step-8: After receiving $MAC_{VG_{1-1}}$ from $V_{G_{1-1}}$, PAAA server generates the $KGK_{VG_{1-1}}$ (as Eq. 5) and computes $MAC'_{VG_{1-1}}$. Then, PAAA server verifies whether $MAC'_{VG_{1-1}} = MAC_{VG_{1-1}}$. It sends a successful authentication message if the verification passes otherwise an authentication failure response is transmitted to the vessel. To authenticate the identity of infostation at $V_{G_{1-1}}$ and infostation, PAAA server computes the MSK as

$$MSK = h(ID_{G1} \parallel R_{VG_{1-1}} \parallel R_{HAAA} \parallel R_{PAAA}) \quad (6)$$

Step-9: Then, PAAA server sends the stored $ID_{infostation}$ as $(ID_{infostation} || MSK)$ to infostation. The infostation authenticates its identity and stores the MSK after a successful authentication. Further, it transmits an encrypted $ID_{infostation}$ with MSK to $V_{G_{1-1}}$.

Step-10: Finally, $V_{G_{1-1}}$ decrypts and authenticates the $ID_{infostation}$ using MSK (as Eq. 6). It verifies whether $ID_{infostation}$ received from infostation in Step-9 matches with $ID_{infostation}$ used in Step-3. If the verification is successful, the authentication and key agreement phase for the first vessel is completed.

When the other vessels ($V_{G_{1-2}}$, $V_{G_{1-3}}$,......$V_{G_{1-n}}$) in group intend to access the network, PAAA server performs the mutual authentication with these vessels using GTK_{G_1} computed in the registration phase. The following steps are executed in this procedure.

Step-1: Vessel ($V_{G_{1-2}}$) accepts the request identity message from infostation to provide access of the infostations.

Step-2: Then, $V_{G_{1-2}}$ computes the $TID_{VG_{1-2}}$ (as Eq. 1) and authentication code $AUTH_{VG_{1-2}} = (ID_{G1} \parallel TID_{VG_{1-1}} \parallel R_{VG_{1-1}} \parallel ID_{infostation})$. It is not required for $V_{G_{1-2}}$ to transfer MAC_{G_1} because the PAAA server verifies $V_{G_{1-2}}$ explicitly without the knowledge of HAAA.

Step-3: After receiving the $AUTH_{VG_{1-2}}$ from $V_{G_{1-2}}$, PAAA server executes the $ID_{infostation}$, GTK_{G_1}, R_{HAAA}, and index table of G_1. To mutually authenticate the vessels and PAAA server, procedure from Step-6 to Step-10 is executed until the complete verification of each vessel in the group. Hence, a full authentication and key agreement of each vessel in the group is realized.

4 Formal Verification and Security Analysis of Proposed Protocol

The proposed ESMGB-AKA protocol mutually authenticates the entities, maintains the data integrity, provides the confidentiality, and preserves the privacy of vessels in the group. Moreover, the protocol defeats the impersonation attack and replay attack as the EG-AKA and EAPSG protocols. This section illustrates the formal verification and security analysis of the proposed protocol.

4.1 Formal Verification Using AVISPA Tool

In this subsection, the formal verification of proposed protocol is carried out using AVISPA tool [15, 16]. AVISPA accommodates automatic security analysis and verification such as On-the-Fly-Model-checker (OFMC) and SAT-based Model-Checker (SATMC). The protocol is coded in High-Level Protocol Specifications Language (HLPSL) to verify its security properties under AVISPA. The fundamental objective of the protocol is to achieve the mutual authentication and key agreement between the group of vessels and communication entities. Moreover, it is required to achieve

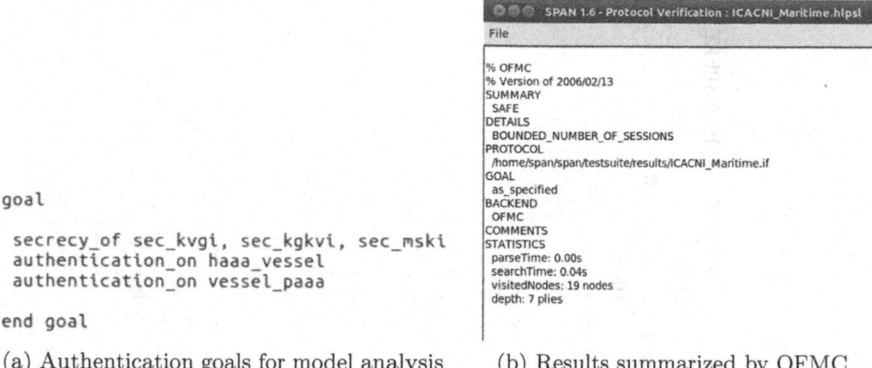

(a) Authentication goals for model analysis (b) Results summarized by OFMC

Fig. 3 Analysis of proposed protocol using AVISPA

secrecy of pre-shared keys ($K_{VG_{i-k}}$, GTK_{G_1}) in the authentication process. We verify the proposed protocol using OFMC. To verify the proposed protocol, the authentication goals are shown in Fig. 3a. In addition, the Fig. 3b shows the output of the model analysis results. From these results, it is observed that the proposed protocol achieves the security goals and defeats all the attacks.

4.2 Security Analysis of Proposed Protocol

In this subsection, the security analysis shows that the proposed protocol successfully avoids the problems of EG-AKA and EAPSG protocols.

- **Resistance to Redirection Attack**: To overcome redirection attack, we need to restrict an adversary to access the MSK while transmitting the identity of infostation to vessels. The proposed protocol computes the *MSK* to encrypt the identity of the infostation, thereby confining the network from redirection attack. The $R_{VG_{1-1}}$, R_{HAAA}, R_{PAAA} are random distinct numbers in each formation of the session key. Hence, an adversary fails to generate the valid *MSK*. In the proposed protocol, the PAAA server transmits $ID_{infostation}$ with session key *MSK* to infostation. The infostation verifies its identity and stores the *MSK*. Then, it sends an encrypted $ID_{infostation}$ with *MSK* to each vessel in the group. Hence, the proposed protocol preserves the privacy of $ID_{infostation}$.
- **Resistance to DoS Attack**: In the proposed protocol, PAAA server computes the MAC_{PAAA} (as Eq. 4) using GTK_{G_1} and the distinct random number of communication entities. The vessels also generate the GTK_{G_1} and authenticate the MAC'_{PAAA}. Different from EG-AKA and EAPSG protocols, we are not considering the initial values (IVs) to compute the *MAC*. Hence, the PAAA server and vessels mutually authenticate each other by verifying their *MAC*.

Table 2 Security parameters comparison in various AKA protocols

Security Parameters	AKA Protocols							
	ESMGB-AKA	EAPSG [14]	EG-AKA [13]	EAP-AKA [8]	EAP-TTLS [9]	EAP-PEAP [11]	EAP-LEAP [10]	EAP-SPEKE [12]
Cryptosystem	Symmetrical	Asymmetrical	Asymmetrical	Symmetrical	Asymmetrical	Asymmetrical	Asymmetrical	Asymmetrical
Bandwidth consumption	Low	High	High	Low	High	High	High	High
Privacy protection	Yes	Yes	Yes	No	Yes	Yes	No	No
Resistance to replay attack	Yes	Yes	Yes	Yes	Yes	Yes	Yes	Yes
Resistance to impersonation attack	Yes	Yes	Yes	No	No	No	No	No
Resistance to MiTM attack	Yes	Yes	Yes	No	Yes	Yes	Yes	Yes
Resistance to redirection attack	Yes	No	No	No	No	No	No	No
Resistance to DoS attack	Yes	No	No	No	No	No	No	No
Provide group authentication	Yes	Yes	Yes	No	No	No	No	No

- **Resistance to Bandwidth Consumption**: The proposed protocol adopts symmetric key operations during the registration and key agreement phase. Each vessel in the group has a pre-shared secret key ($K_{VG_{i-k}}$) with HAAA. Each vessel and PAAA mutually authenticate each other using symmetric key GTK_{G_1}. Moreover, the $ID_{infostation}$ is also verified at $V_{G_{1-1}}$ using symmetric key. Hence, the lightweight framework suits for resource-constrained vessels of the maritime communication network. Also, it reduces the computation overhead from the network.

In Table 2, we show the comparative analysis of various AKA protocols for non-3GPP network on the basis of various security parameters. From illustrated results, it can be observed that the proposed protocol achieves the extensive security performance. The protocol overcomes the issues observed in EAPSG and other existing protocols. In addition, the proposed protocol mandates the group key (GK_i) update whenever the vessels join or leave the group.

5 Performance Analysis of Proposed Protocol

In this section, we present the communication and computation overhead to analyze the performance of the proposed protocol.

- **Communication Overhead**: To analyze the communication overhead of the AKA protocols, we consider that there are n number of vessels in m groups. The signaling overhead of AKA protocols is shown in Table 3, where co_{first} and $co_{remaining}$ represent the communication overhead incurred by first vessel and each remaining vessel during the authentication process, respectively. In EG-AKA and EAPSG protocols, seven signaling messages are required to authenticate the first vessel. The remaining vessels of the group are verified in five message interactions. These protocols perform the asymmetric key cryptosystem to authenticate vessels. Although the proposed protocol requires the same number of signaling messages, the authentication process is based on the symmetric key operations. Hence, the communication overhead of the proposed protocol is reduced compared to EG-AKA and EAPSG protocols as shown in Fig. 4a.
- **Computation Overhead** To analyze the computation overhead of the AKA protocols, we consider the consumed time of cryptographic functions [17] as: T_{Hash}= 0.067ms; T_{mul}= 0.612ms; and T_{aes} (encryption/decryption)= 0.161 ms. Table 4 represents the computation overhead of EG-AKA, EAPSG, and proposed ESMGB-AKA protocols. The EG-AKA and EAPSG protocols perform the mutual authentication and key agreement using multiplication operations over an elliptic curve. However, the key operations of proposed protocol are based on the symmetric key cryptosystem; hence, it generates the less computation overhead as shown in Fig. 4b. We consider the average value of EG-AKA and EAPSG protocols as they have same computation overhead.

Table 3 Comparison of communication overhead of group-based AKA protocols

AKA Protocols	Number of bits		
	CO_{first}	$CO_{remaining}$	$CO_{overall}$
EG-AKA [13]	3200	1728	$1728(n-m)+3200*m$
EAPSG [14]	3328	1728	$1728(n-m)+3328*m$
ESMGB-AKA	2776	1456	$1456(n-m)+2776*m$

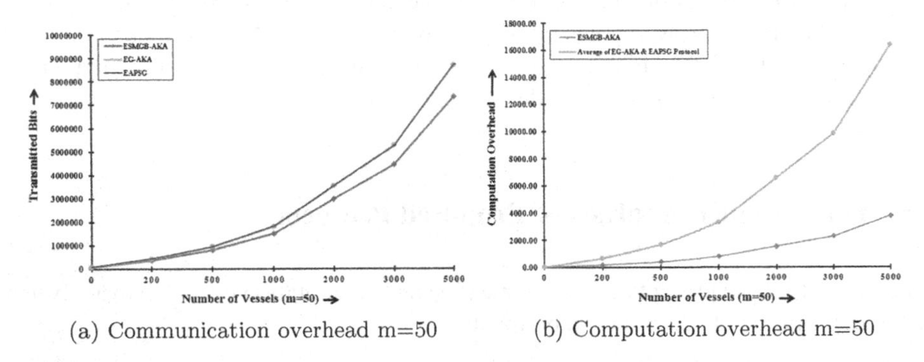

(a) Communication overhead m=50 (b) Computation overhead m=50

Fig. 4 Analysis of communication and computation overhead

Table 4 Comparison of computation overhead of AKA protocols

AKA Protocols	Number of bits		
	MTC Devices	Network	Total (in ms)
EG-AKA [13]	$(2T_{mul}+3T_{hash}+2T_{aes})*n$	$(2T_{mul}+2T_{hash}+T_{aes})*n+(T_{hash}+T_{aes})*m$	$(4T_{mul}+5T_{hash}+3T_{aes})*n+(T_{hash}+T_{aes})*m$
EAPSG [14]	$(2T_{mul}+3T_{hash}+2T_{aes})*n$	$(2T_{mul}+2T_{hash}+T_{aes})*n+(T_{hash}+T_{aes})*m$	$(4T_{mul}+5T_{hash}+3T_{aes})*n+(T_{hash}+T_{aes})*m$
ESMGB-AKA	$(2T_{hash}+2T_{aes})*n+T_{hash}$	$(2T_{hash}+T_{aes})*n+(T_{hash}+T_{aes})*m$	$(4T_{hash}+3T_{aes})*n+(T_{hash}+T_{aes})*m+T_{hash}$

6 Conclusion and Future Work

In this paper, we perform the security analysis of non-3GPP-based EAPSG protocol in maritime scenario. The security analysis shows that the EAPSG protocol is vulnerable to redirection attack and DoS attack. Also, the protocol executes the asymmetric key cryptosystem. Therefore, it suffers from high bandwidth consumption. Later, we address the countermeasure for the above-mentioned problems. We propose the

ESMGB-AKA protocol from the proposed countermeasures. To analyze the security parameters and secret keys, the proposed protocol is formally verified using AVISPA tool. The formal security analysis shows that the protocol is safe and avoids the identified attacks. The protocol is also compared with other existing group-based AKA protocols on the basis of communication and computation overhead. The graphical representation shows that the protocol has improved communication and computation overhead. In addition, the high computational consumption and communication complexity are observed during the handover. Hence, we will exploit our research in the handover of group-based AKA protocols in future.

References

1. MI News Network. Maritime Piracy Affected Areas around the World. http://www.marineinsight.com/marine-piracy-marine/10-maritime-piracy-affected-areas-around-the-world/ (2016). Accessed 21-July-2016
2. Yang, T., Liang, H., Cheng, N., Shen, X.: Towards video packets store-carry-and-forward scheduling in maritime wideband communication. In: 2013 IEEE Global Communications Conference (GLOBECOM), pp. 4032–4037. IEEE (2013)
3. Hoang, V.D., Ma, M., Miura, R., Fujise, M.: A novel way for handover in maritime wimax mesh network. In: 2007 7th International Conference on ITS Telecommunications, pp. 1–4. IEEE (2007)
4. Baldini, G., Karanasios, S., Allen, D., Vergari, F.: Survey of wireless communication technologies for public safety. IEEE Commun. Surv. Tutor. **16**(2), 619–641 (2014)
5. Verma, R., Prakash, A., Agrawal, A., Naik, K., Tripathi, R., Alsabaan, M., Khalifa, T., Abdelkader, T., Abogharaf, A.: Machine-to-machine (M2M) communications: a survey. J. Netw. Comput. Appl. **66**, 83–105 (2016)
6. Ghavimi, F., Chen, H.-H.: M2m communications in 3G pp lte/lte-a networks: architectures, service requirements, challenges, and applications. IEEE Commun. Surv. Tutor. **17**(2), 525–549 (2015)
7. Barki, A., Bouabdallah, A., Gharout, S., Traoré, J.: M2m security: challenges and solutions. IEEE Commun. Surv. Tutor. **18**(2), 1241–1254 (2015)
8. Arkko, J., Haverinen, H.: Extensible Authentication Protocol Method for 3rd Generation Authentication and Key Agreement (EAP-AKA) (2006)
9. Funk, P., Blake-Wilson, S.: Extensible Authentication Protocol Tunneled Transport Layer Security Authenticated Protocol Version 0 (EAP-TTLSv0) (2008)
10. Yuan, G., Zhu, K., Fang, N.-S., Wu, G.-X.: Research and application of 802.1 x/eap-peap. Comput. Eng. Des. **10**, 34 (2006)
11. Palekar, A., Simon, D., Zorn, G., Josefsson, S.: protected EAP protocol (PEAP). Work in Progress, vol. 6 (2004)
12. Dantu, R., Clothier, G., Atri, A.: Eap methods for wireless networks. Comput. Stand. Interfaces **29**(3), 289–301 (2007)
13. Jiang, R., Lai, C., Luo, J., Wang, X., Wang, H.: Eap-based group authentication and key agreement protocol for machine-type communications. Int. J Distrib. Sens. Netw. **2013** (2013)
14. Yang, T., Lai, C., Rongxing, L., Jiang, R.: Eapsg: efficient authentication protocol for secure group communications in maritime wideband communication networks. Peer-to-Peer Netw. Appl. **8**(2), 216–228 (2015)

15. 3rd Generation Partnership Project; Security Aspects of Machine-Type Communications (MTC) release 12. *3GPP TR 33.868 Ver.12.0.0* (2014)
16. Avispa automated validation of internet security protocols. http://www.avispa-project.org
17. Choi, D., Choi, H.-K., Lee, S.-Y.: A group-based security protocol for machine-type communications in lte-advanced. Wirel. Netw. **21**(2), 405–419 (2015)

Reduced Latency Square-Root Calculation for Signal Processing Using Radix-4 Hyperbolic CORDIC

Aishwarya Kumari and D. P. Acharya

Abstract In this paper, we are extending the usage of CORDIC algorithm from trigonometric mode, which has been the primary use of it from a very long time, to hyperbolic mode. Here, we have implemented Radix '4' hyperbolic CORDIC in vectoring mode for fast and efficient computing of Square-Root of a number. The simulation results are implemented on FPGA platform. Based on simulation results, comparison of Radix '4' CORDIC with Radix '2' CORDIC is presented too.

Keywords Hyperbolic · Trigonometric · CORDIC algorithm
Radix '4' · Pipelined architectures

1 Introduction

The ***CO****ordinate* ***R****otation* ***DI****gital* ***C****omputer* (CORDIC) has been a prevailing computational tool for a wide range of applications. This strategy has expanded in preponderance, especially in mathematical applications. Volder introduced CORDIC algorithm method for calculation of trigonometric functions as well as for complex binary techniques using CORDIC algorithm with its two different operation modes, which are vectoring and rotation mode. CORDIC-based architectures have been used for inversion of matrix, eigenvalue calculations, singular value decomposition (SVD) algorithms, logarithmic functions, multiplication of complex numbers, orthogonal transformations, etc.

Conventionally, CORDIC algorithm is implemented using Radix '2' microrotations. Later, various algorithms and architectures have been developed for increasing throughput of the algorithm through the pipelined implementation. The CORDIC algorithm has become very much popular majorly because of its efficiency for cost-efficient implementation of various applications. Its main attraction is its ability

A. Kumari (✉) · D. P. Acharya
Department of Electronics and Communication Engineering, National Institute of Technology, Rourkela, India
e-mail: aish.sumy1991@gmail.com

P. K. Sa et al. (eds.), *Recent Findings in Intelligent Computing Techniques*, Advances in Intelligent Systems and Computing 708,
https://doi.org/10.1007/978-981-10-8636-6_23

to provide simple architecture because it only uses shift, add, and subtract operations. It is also clear that it is really going to get much better shape in the future. In the present scenario, CORDIC is finding its great use in embedded system processors. The CORDIC algorithm is described suitable for the use in a special purpose computer where most of the computations involve basic trigonometric functions [1]. After completing 50 years of its invention, the basic evolutions made in the CORDIC algorithm and their architectures along with their effectiveness and applications in the coming times are illustrated in [2]. Various authors presented the restrictions of the numerical values of functional arguments, which are given to the CORDIC units with an emphasis on the binary as well as the fixed-point implementations [3]. Villalba discussed the Radix '4' CORDIC algorithm in circular vectoring mode [4]. In paper [4], authors presented that the proposed Radix '4' circular CORDIC algorithm in vectoring mode has a similar recurrence as the Radix '4' division algorithm and some dedicated studies are presented concerning the vectoring mode. In [5], the parallel Radix '4' architecture is implemented to check the latency and the improvements in hardware to reduce the area. Radix '4' architecture for rotation mode is designed in [6] where it can be seen that the total iterations' count in Radix '4' is half as compared to Radix '2.' Hence, we can see that most of the work in the area of CORDIC is limited to rotational mode only specially for calculating trigonometric functions. Therefore, Radix '4' CORDIC can be used for hyperbolic vectoring mode too. The time required for the computation of Square-Root can be reduced, which can be very useful for the works of signal processing.

2 CORDIC Algorithms

The Radix '2' and Radix '4' CORDIC algorithms are presented here in brief. The CORDIC algorithm can be altered to compute various hyperbolic functions. Hence, it is reformulated to a generalized form, good enough to execute rotations in linear, circular, and hyperbolic coordinate systems. For this, a variable 'p' is added extra, which takes distinct values for different coordinate systems. The value of 'p' can be p = 0.1 or −1 and

$$\beta_{\mathrm{m}} = tan^{-1}(2^{-\mathrm{m}}), (2^{-\mathrm{m}})\, or, tanh^{-1}(2^{-\mathrm{m}}),$$

where the generalized CORDIC algorithm is working, respectively, in linear, circular, or hyperbolic coordinate system. The formulae used for generalized CORDIC are as follows [2]:

$$\begin{aligned} x_{m+1} &= x_m - p\sigma_m 2^{-m} m \\ y_{m+1} &= y_m + \sigma_m 2^{-m} x_m \\ z_{m+1} &= z_m - \sigma_m \beta_{\mathrm{m}} \end{aligned} \tag{1}$$

where,

$$\sigma_m = \begin{cases} \text{sign}(z_m);\ \text{in rotation mode} \\ -\,\text{sign}(y_m);\ \text{in vectoring mode} \end{cases}$$

In Radix '2' CORDIC algorithm, to have n-bit output precision, n clock cycles are required. Hence, latency is more. The latency of computation is a major drawback of CORDIC algorithm. In various signal-processing applications, fast computation of Square-Root is required. So, attempts are made to reduce latency of computation for calculation of Square-Root.

In the following section, a vectoring mode Radix '4' hyperbolic CORDIC algorithm is developed, which is used in the calculation of Square-Root. To ensure the convergence, the values of w_m are taken as $w_m = 4^m y_m$.

The equations for Radix '4' CORDIC are as follows [4]:

$$\begin{aligned} x_{m+1} &= x_m + \sigma_m 4^{-2m} w_m, \\ w_{m+1} &= 4(w_m + \sigma_m x_m), \\ z_{m+1} &= z_m - \beta_m(\sigma_m). \end{aligned} \tag{2}$$

$$\text{Here,}\quad \beta_m(\sigma_m) = \tanh^{-1}(\sigma_m 4^{-m}).$$

Here, σ_i takes values $\{-2, -1, 0, 1, 2\}$. The scale factor here is

$$K = \prod_m \left(1 + \sigma_m^2 4^{-2m}\right)^{1/2}.$$

The scaling factor 'K' varies with iterations, as it varies with the σ_m values. Its value ranges from 1.0 to 2.62. The Radix '4' CORDIC algorithm has two problems, viz complexity of selection of σ_m and variable nature of scale factor.

By selecting four different comparison points, the value of σ_m is calculated. In case of circular coordinate CORDIC, the four different comparison points are taken as

$$P_m(\pm 1) = \begin{cases} \pm\frac{x_0}{2};\ \textit{if}\ m = 0 \\ \pm\frac{x_1}{2};\ \textit{if}\ m \geq 1 \end{cases} \tag{3}$$

$$P_m(\pm 2) = \begin{cases} \pm\frac{3x_i}{2};\ \textit{if}\ m \leq 0 \\ \pm\frac{3x_2}{2};\ \textit{if}\ m \geq 2 \end{cases} \tag{4}$$

Since we are using hyperbolic coordinate CORDIC, iteration $m = 0$ is invalid here. The iteration count will start from $m = 1$. The calculated values for σ_m are

$$\sigma_m = \begin{cases} +2; \; if \; w_m > \mathrm{P}_m(2) \\ +1; \; if \; \mathrm{P}_m(1) < w_m \leq \mathrm{P}_m(2) \\ 0; \; if \; \mathrm{P}_m(-1) < w_m \leq \mathrm{P}_m(1) \\ -1; \; if \; \mathrm{P}_m(-2) < w_m \leq \mathrm{P}_m(-1) \\ -2; \; if \; w_m \leq \mathrm{P}_m(-2) \end{cases} \quad (5)$$

Due to most popular area–delay–accuracy trade off [2], reducing latency in case of Radix '4' will increase the area by a smaller amount and will decrease the accuracy too.

3 Proposed Architecture for Radix '4' Hyperbolic CORDIC Algorithm

In the section explained below, we have shown the pipelined architecture of Radix '4' hyperbolic CORDIC algorithm (Fig. 1).

The proposed architecture consists of two parallel operations:

(i) Unscaled CORDIC architecture
(ii) Scale factor computation architecture.

This architecture uses 33-bit precision. The number of iterations is three, which is half of number of iterations required in Radix '2' CORDIC (Fig. 2).

To calculate the value for σ, combinational block consisting of comparators and multiplexers is used. The output of this block is used in both parallel operations. The computation for scale factor is carried out parallely with CORDIC iterations. After calculating scale factor for each operation, these values are stored in the LUT, which provides the final scale factor 'K.' The final output of unscaled CORDIC block is divided by this value of scale factor to get the Square-Root of given input.

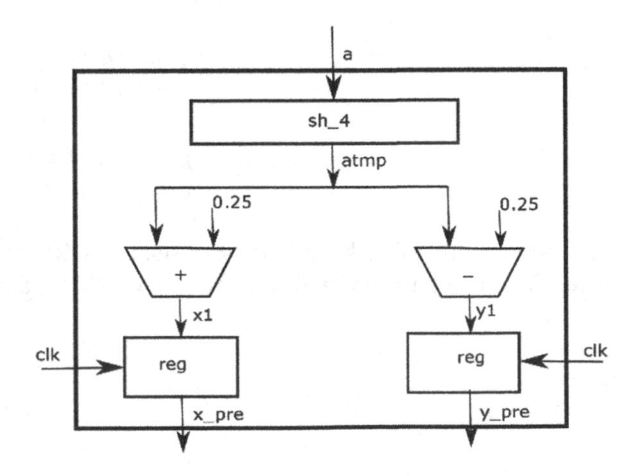

Fig. 1 Architectural representation of preprocessing block in unscaled CORDIC block

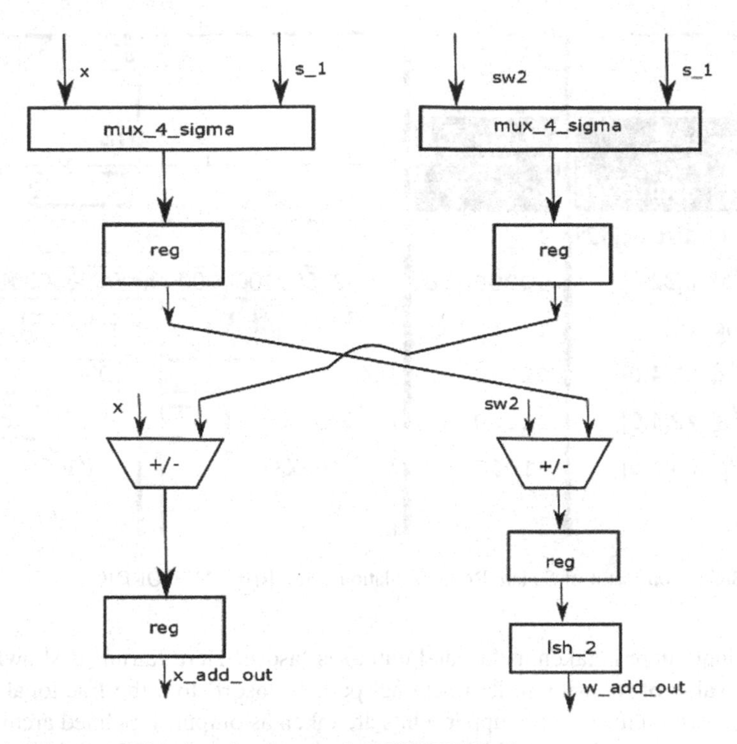

Fig. 2 Architecture of first iteration in unscaled CORDIC block

4 Results of Implementation in FPGA

The CORDIC architecture presented here consists of three stages and a word length of 33 bits. Out of these 33 bits, 9 bits are used as integer points and remaining as fractional points. The MSB is taken as sign bit. The FPGA used here has the following specifications (Table 1):

To calculate the Square-Root of a number 'a,' in CORDIC, initial values of x and y are taken as (Fig. 3).

$$x = a + 0.25 \text{ and } y = a - 0.25.$$

Table 1 FPGA device and simulation environment

FPGA	Xilinx Virtex7
Device	XC7VX690T
Package	FFG1157
Synthesis tool	XST (VHDL/Verilog)
Speed	−3
Preferred language	Verilog
Simulator	Isim

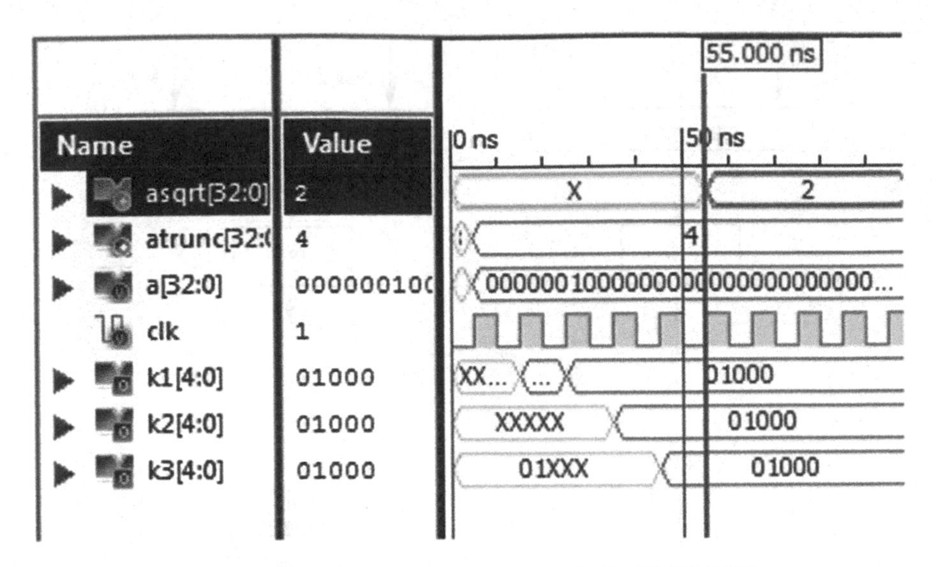

Fig. 3 Simulation result of Square-Root calculation using Radix '4' CORDIC

The input here is taken as 'a,' and output is 'asqrt.' Here 'atrun' is showing the integral value of 'a' without its fractional part. In 'asqrt' too, the fractional part is truncated, which means only upper 9 bits are taken as output. Pipelined architecture uses structure similar to that of parallel implementation of CORDIC. The only difference is pipelining registers are inserted after every iteration. The scale factor for each iteration is truncated to five bits in which first two bits represent integral part and remaining three represent fractional part (Fig. 4).

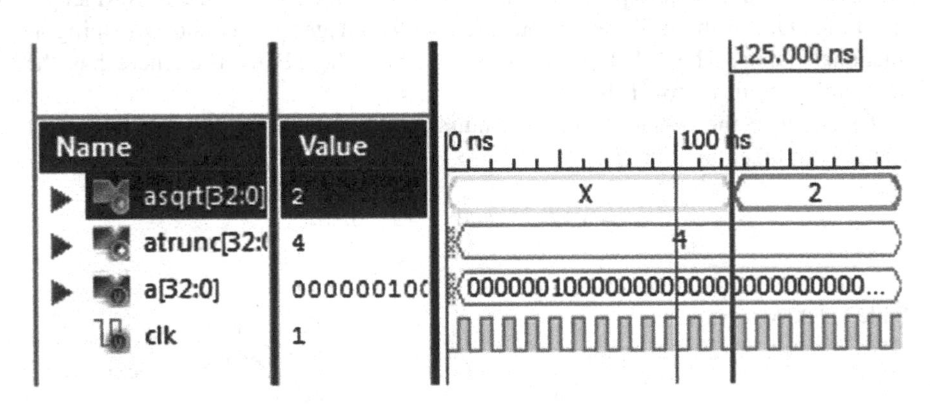

Fig. 4 Simulation result of Square-Root calculation using Radix '2' CORDIC

5 Conclusion

In the proposed work, a normally scaled Radix '4' hyperbolic CORDIC architecture is presented. The calculated latency of operation here is five clock cycles, which is approximately half of the Radix '2' CORDIC architecture. But, it involves comparatively more hardware than Radix '2' CORDIC because it parallelly computes the scale factor. This study reveals that the speed-optimized Radix '4' CORDIC architecture designed can be suitable for applications in real time.

References

1. Volder, J.E.: The CORDIC trigonometric computing technique. In: IRE Trans. Electron. Computers, vol. EC-8, pp. 330–334, Sept. 1959
2. Meher, P.K.: 50 years of CORDIC: algorithms, architectures, and applications. IEEE Trans. Circuits Syst. (2009)
3. Hu, X.: Expanding the range of convergence of the CORDIC algorithm. IEEE Trans. Comput. (1991)
4. Villalba, J.: Radix-4 vectoring CORDIC algorithm and architectures. IEEE Trans. Appl. Specif. Syst. Archit. Process. (1996)
5. Lakshmi, B.: VLSI architecture for parallel Radix-4 CORDIC. ELSEVIER Trans. Microprocess. Microsyst. **37**, 79–86 (2013)
6. Antelo, E.: High performance rotation architectures based on Radix-4 cordic algorithm. IEEE Trans. Comput. **46** (1997)

High-Gain A-Shaped EBG-Backed Single-Band Planar Monopole Antenna

Shridhar Desai, Umesh Madhind and Mahesh Kadam

Abstract This paper presents the use of a backed electromagnetic band gap (EBG) structure which gives the radiation efficiently as a director for our designed monopole antenna. For the wearable applications, proposed planar monopole antenna A single EBG structure exploited towards radiation efficiently. The gap between the ground and the EBG layer, and shape of the EBG structure are fixed to work the antenna at 2.45 GHz in wireless local area network. A proposed model with unique EBG-backed monopole antenna is presented. The proposed antenna is fabricated on a 64.23 × 38 × 1.57 mm^3 board of FR4 substrate with relative permittivity of 4.4. The reported efficient and robust radiation performance with the high gain and the compact size, which make the proposed antenna to work for wearable communication applications. The detailed theoretical explanations are supported by simulations and experiments.

Keywords Monopole antenna · Electromagnetic band gap (EBG) structure
Wearable · Antenna

1 Introduction

In communication system, recently, we are seeing fast advancement. Nowadays, wearable systems integrated with wireless communication technology innovation get to be distinctly essential for human. To monitor and record data like body

S. Desai (✉) · U. Madhind · M. Kadam
Electronics and Telecommunication Engineering, Terna Engineering College,
Navi Mumbai 400706, India
e-mail: shri.desai90@gmail.com

U. Madhind
e-mail: umahind10@gmail.com

M. Kadam
e-mail: maheshkadam@ternaengg.ac.in

P. K. Sa et al. (eds.), *Recent Findings in Intelligent Computing Techniques*,
Advances in Intelligent Systems and Computing 708,
https://doi.org/10.1007/978-981-10-8636-6_24

temperature, blood pressure, heart rate [1] need to design wearable device with smart clothes. That supports various applications such as rescue systems, health care systems, patient monitoring system, combat zone survival and wearable gaming consoles [2, 3]. Through wireless communication systems, the information observes and sends to receiving station. In such systems, antenna is an essential device to communicate between receiving station and sensor devices [4]. Such a commercial applications, frequency band is allotted globally. For these applications, industrial scientific, ultra-wideband and medical implantable communication system band are available.

In medical applications, antenna size must be as small as possible with features like lightweight, preferably comfortable. For narrowband applications, designing highly efficient antenna with characteristics like low profile, lightweight is a difficult task [5].

The microstrip antenna [6] and antenna with cavity backed [7, 8] are a good choice for wearable applications; be that as it may, they do not show conformal characteristics. A few textile-based conformal antennas such as microstrip antenna with full ground [9], wearable magnetoelectric dipole [10], SIW technology-based textile antenna [11] and textile fractal [12] are proposed to be flexible for communication in these applications; however, they have a generally substantial impression.

In this paper, we designed an exceptionally smaller planar monopole antenna with specific-structured EBG backed and displayed the experimental realization at 2.45 GHz operating frequency in ISM band. Its applications are most common in the UHF [13, 14], ultra-wideband (UWB) [15], etc. In this paper, we concentrate on designing off-body antenna for wireless body area network with efficient, low-profile and narrowband application. In Sect. 1, we gave the background for our proposed antenna. In Sect. 2, we describe the proposed antenna structure. In this study, we present a planar monopole antenna for single band, with and without EBG structure. Then in Sect. 3, we discuss the result for various parameters. Section 4 is comparison of performance of antenna without electromagnetic band gap (EBG) structure as well as with EBG structure. Then we conclude the topic in Sect. 5.

2 Antenna Design

From antenna theory, it is cleared that by using proper EBG structure, enhancement in performance of low-profile antenna demonstrated. The schematic diagram of proposed antenna is shown in Fig. 1 which comprises of a radiating monopole which is backed by single EBG structure. On either side of the substrate, monopole antenna and partial ground plane are imprinted. Along with partial ground, we imprinted EBG structure on same side. This reduces the size of the antenna.

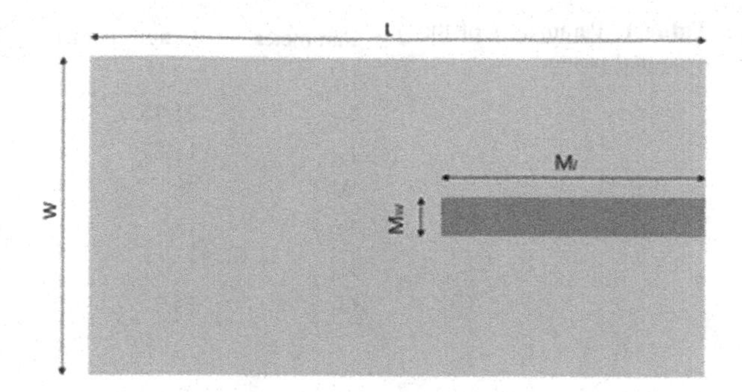

Fig. 1 Design of single-band monopole antenna

2.1 Design of Single-Band Monopole Antenna

The FR4 substrate material of 1.57 mm thickness is used to fabricate antenna. This material has permittivity of 4.4. As shown in Fig. 1, this structure is of size 64.2 × 38 mm^2. On upper side of the substrate, we imprinted monopole antenna of dimension 15.2 × 1 mm^2. This structure is designed for single band of frequency 2.4–2.5 GHz.

2.2 Design of EBG Structure

On the bottom side of the substrate, EBG structure is imprinted along with partial ground plane of 38 × 14.2 mm^2. Area cover by EBG structure is 38 × 48 mm^2, and the geometry of EBG structure is shown in Fig. 2. The EBG structure resembles to the English letter ‘A’.

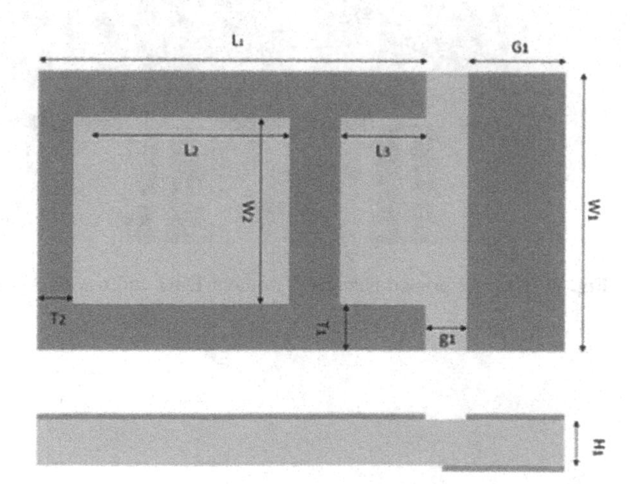

Fig. 2 Design of EBG structure

Table 1 Parameters of the designed printed

Parameter	Length (mm)	Parameter	Length (mm)
L_1	48	T_1	10
L_2	21.75	T_2	6.25
L_3	11.87	L	64.2
W_1	38	M_1	15.2
W_2	18	M_W	1
H_1	1.57	g_1	2
G_1	14.2		

For monopole antenna, 2.45 GHz is operating frequency for monopole antenna design. Hence, this must match with surface wave frequency band gap of EBG layer. Hence, at the initial stage of antenna design, for antenna, we combined operating frequency band. The EBG layer structure is with an elongated length L_1. To define the EBG surface wave frequency band gap, L_1, L_2, W_2, T_1, T_2 and g_1 optimized.

In the following Table 1, we have given all the dimensions of monopole antenna and EBG structure, respectively. For this, schematic is as shown in Fig. 1 and Fig. 2, respectively.

As shown in Fig. 3a, b and c, we actually designed the antenna structure, without EBG and with A-shaped EBG, respectively.

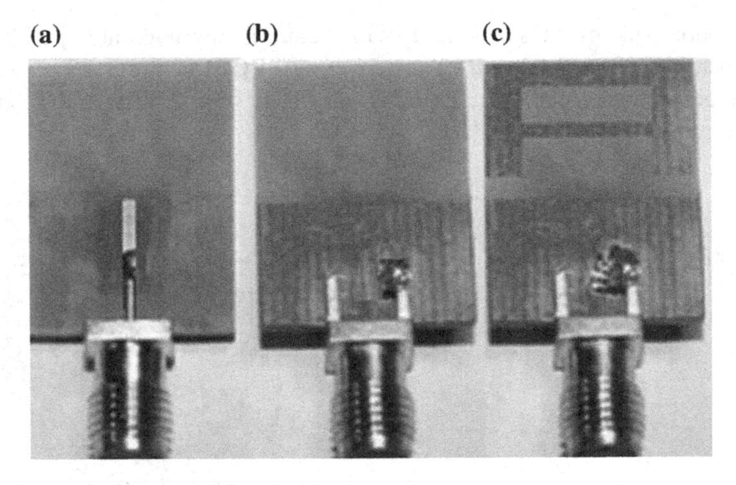

Fig. 3 **a** Actual printed antenna, **b** without EBG and **c** with A-shaped EBG

3 Result and Discussions

3.1 Return Loss

For simulation, we have used HFSS software. The simulated reflection coefficient monopole antenna without EBG and with EBG structure is shown in Fig. 4 and Fig. 5, respectively. By using HFSS, we investigated the radiation pattern and S-parameter. In Fig. 3, S-parameter of antenna without EBG structure is given. And in Fig. 4, S-parameter of proposed antenna with A-shaped EBG structure is given. Here, from simulation, it is clear that notch indicates that frequency band covers 2.38–2.52 GHz. Return loss of has been simulated as depicted in Fig. 4, which shows that the return loss is −12 and −21.45 dB in Figs. 3 and 4, respectively. Hence, −21 dB is sufficient for the wireless communication application as compared to other antennas. As shown in Fig. 2, the bandwidth of an antenna is about 435 MHz.

3.2 Radiation Performance

In Fig. 6, the E-plane radiation pattern of antenna without EBG structure is shown. In Fig. 7, the E-plane radiation pattern of antenna with EBG structure is shown. The operating frequency of the proposed antenna is 2.45 GHz. Figures 8 and 9 show the 3D radiation pattern of antenna without EBG and with EBG structure, respectively.

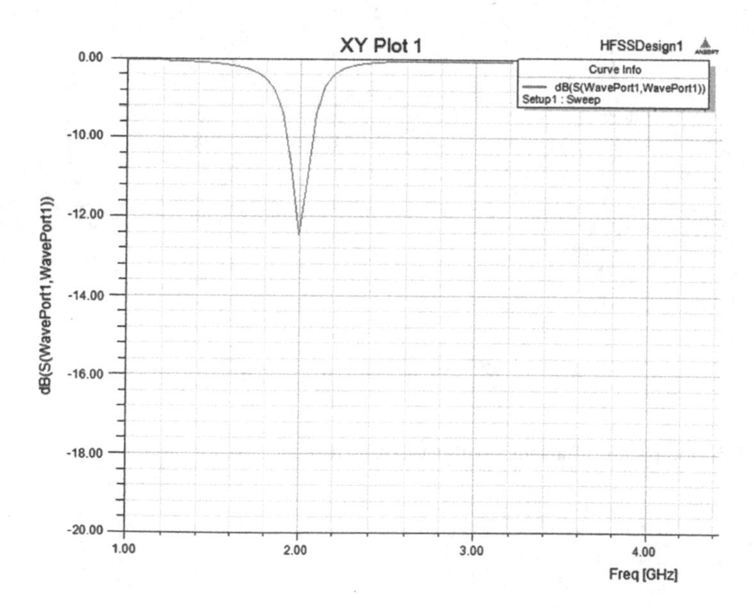

Fig. 4 Return loss of without EBG structure

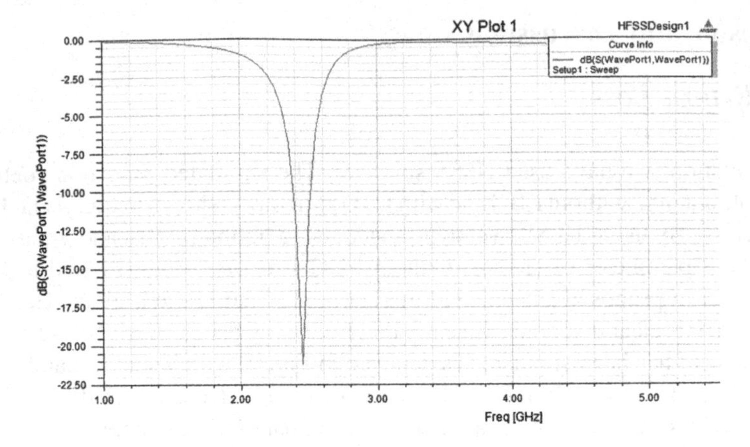

Fig. 5 Return loss of with EBG structure

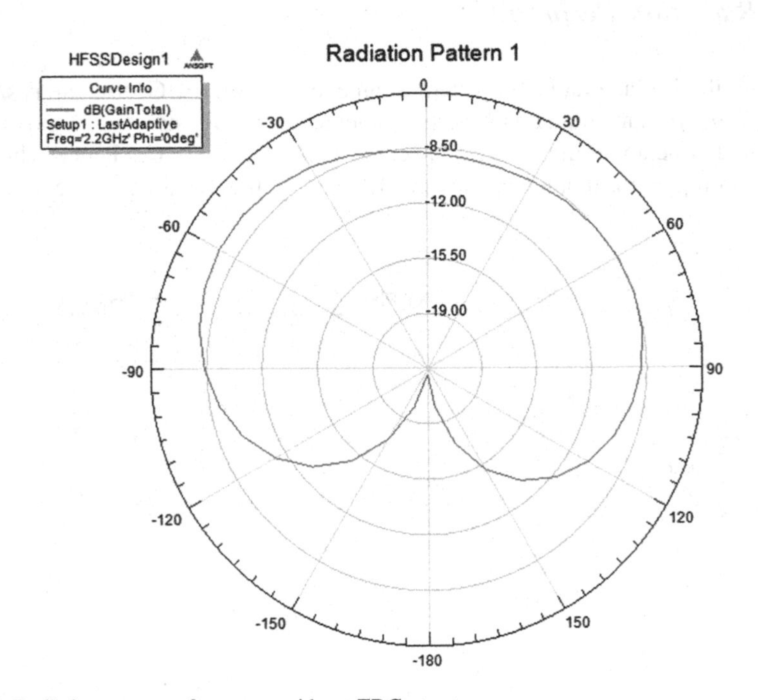

Fig. 6 Radiation pattern of antenna without EBG structure

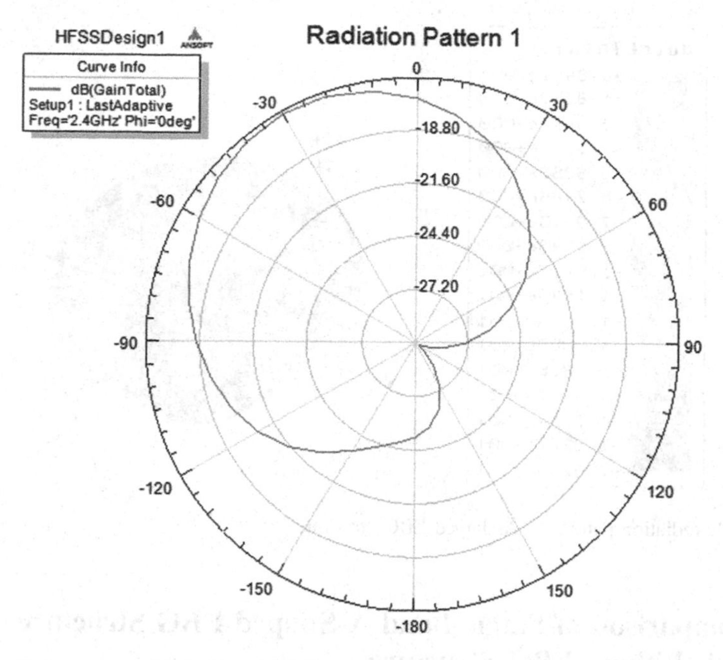

Fig. 7 Radiation pattern of A-shaped EBG structure

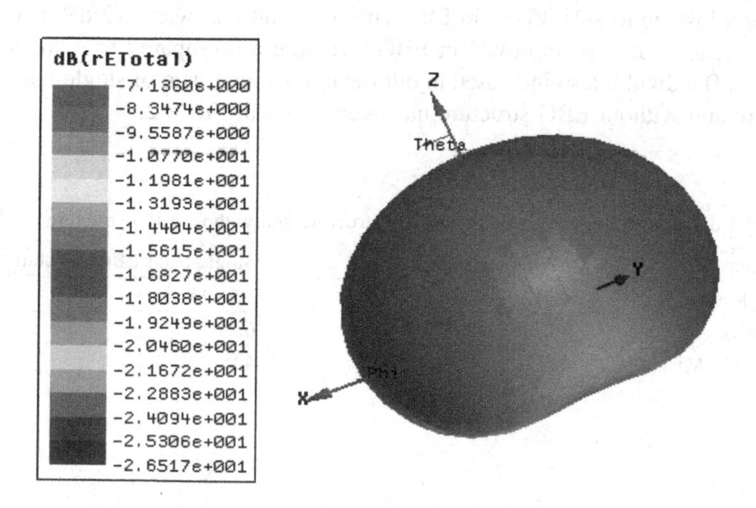

Fig. 8 3D radiation pattern of without EBG structure

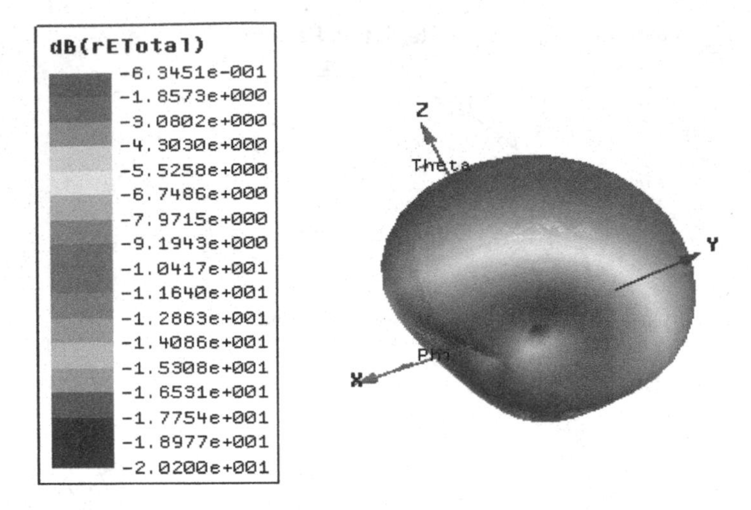

Fig. 9 3D radiation pattern of A-shaped EBG structure

4 Comparison of Single-Band A-Shaped EBG Structure and Without EBG Structure

In this section, we compare the result of our novel antenna design with single-band EBG structure and without EBG structure. Due to this design, we are able to reduce the return loss up to −21.45 dB in EBG structure and that was −12 dB in without EBG structure. Gain is improved in EBG structure as compared to without EBG structure. Bandwidth also increased in our design. Comparison of single-band EBG structure and without EBG structure has been given in Table 2.

Table 2 Comparison table of single-band EBG structure and without EBG structure

Parameter	Without EBG structure	Single-band EBG structure
Return loss (dB)	−12	−21.45
Gain (dB)	3.21	6.21
Bandwidth (MHz)	121	425

5 Conclusion

We have simulated and tested our proposed planar monopole antenna with A-shaped EBG-backed structure. It can be useful for wearable application because of its compact size, lightweight, efficiency ease in fabrication, etc. Our proposed antenna has high-gain characteristic, good bandwidth, very good return loss as listed in Table 2. Hence, such systems are very much useful for systems like medical monitoring or wearable biosensors.

References

1. Zhu, S., Langley, R.: Dual-band wearable textile antenna on an EBG substrate. IEEE Trans. Antenna Propag. **57**(4), 926–935 (2009)
2. Hall, P., Hao, Y.: Antenna and propagation for body-centric wireless communications, Norwood, MA, USA. Artech House (2012)
3. Roblin, C.: Antenna design and channel modeling in the BAN context—Part I: antennas. Ann. Telecommun. **66**(3), 139–155 (2011)
4. Raad, H., et al.: Flexible and compact AMC based antenna for telemedicine applications. IEEE Trans. Antenna Propag. **61**(2), 524–531 (2013)
5. Federal Communications Commission Office of Engineering, and Technology, Washington, DC, USA: Evaluating compliance with FCC guidelines for human exposure to radiofrequency electromagnetic fields, Supplement C (2001)
6. Sheikh, S., et. al.: Directive stacked patch antenna for UWB applications. Int. J. Antennas Propag. **2013**, 6 (2013), Article ID 389571
7. Agneessens, S., Rogier, H.: Compact half diamond dual-band textile HMSIW on-body antenna. IEEE Trans. Antennas Propag. **62**(5), 2374–2381 (2014)
8. Yun, S., et al.: Folded cavity-backed crossed-slot antenna. IEEE Antennas Wirel. Propag. Lett. **14**, 36–39 (2015)
9. Samal, P., et al.: UWB all-textile antenna with full ground plane for off-body WBAN communications. IEEE Trans. Antennas Propag. **62**(1), 102–108 (2014)
10. Yan, S., et al.: Wearable dual-band magneto-electric dipole antenna for WBAN/WLAN applications. IEEE Trans. Antennas Propag. **63**(9), 4165–4169 (2015)
11. Yan, S., et al.: Dual-band textile MIMO antenna based on substrate-integrated waveguide (SIW) technology. IEEE Trans. Antennas Propag. **63**(11), 4640–4647 (2015)
12. Karimyian-Mohammadabadi, M., et al.: Super-wideband textile fractal antenna for wireless body area networks. J. Electromagn. Waves Appl. **29**(13), 1728–1740 (2015)
13. Trajkovikj, J., Skrivervik, A.: Diminishing SAR for wearable UHF antennas. IEEE Antennas Wirel. Propag. Lett. **14**, 1530–1533 (2015)
14. Koohestani, M., et al.: A novel, low-profile, vertically-polarized UWB antenna for WBAN. IEEE Trans. Antennas Propag. **62**(4), 1888–1894 (2014)
15. Mohamed-Hicho, N., et. al.: Wideband high-impedance surface reflector for low-profile high-gain UHF antenna. In: 2015 9th European Conference on Antennas and Propagation (EuCAP), Lisbon, pp. 1–4

5 Conclusion

References

An Improvement to the Multicarrier Complementary Phase-Coded Radar Signal Using Signal Distortion Technique

C. G. Raghavendra, R. Sriranga, Siddharth R. Rao, Sanath M. Nadig, M. Vinay and N. N. S. S. R. K. Prasad

Abstract This paper aims to reduce the Peak-to-Mean Envelope Power Ratio (PMEPR) of a Multicarrier Complementary Phase-Coded (MCPC) signal. A MCPC signal consists of N subcarriers which are phase modulated by N distinct phase sequences. Each of these N subcarriers is spaced $1/t_b$ apart from each other, where t_b is the duration of each phase element, constituting an orthogonal frequency division multiplexing (OFDM) signal. A signal distortion technique, companding, is used to reduce the PMEPR of the generated MCPC signal. Further, the use of level shifting technique on the MCPC signal and consequent application of the companding technique reduces the abrupt variations in the MCPC signal. A comparison of the complex envelopes, autocorrelation, and ambiguity functions of the MCPC signal obtained by the above-mentioned methods is performed which exemplifies the advantage of application of level shifting and companding over the conventional method.

Keywords MCPC · PMEPR · OFDM · Companding · Level shifting Autocorrelation · Ambiguity function

---

C. G. Raghavendra (✉) · R. Sriranga · S. R. Rao · S. M. Nadig · M. Vinay
Department of Electronics and Communication Engineering, M.S. Ramaiah Institute of Technology, Bengaluru 560054, India
e-mail: cgraagu@msrit.edu

R. Sriranga
e-mail: srirangar2008@gmail.com

S. R. Rao
e-mail: siddharth19rao@gmail.com

S. M. Nadig
e-mail: sanathmnadig@gmail.com

M. Vinay
e-mail: vinaymadhugiri1@gmail.com

N. N. S. S. R. K. Prasad
Aeronautical Development Agency, Vimanapura Post, Bengaluru 560067, India
e-mail: nnssrkprasad2007@gmail.com

P. K. Sa et al. (eds.), *Recent Findings in Intelligent Computing Techniques*, Advances in Intelligent Systems and Computing 708,
https://doi.org/10.1007/978-981-10-8636-6_25

1 Introduction

Range and resolution [1] are the most important characteristics of a radar signal. As the pulse width of a radar signal increases, the range of the signal improves but the resolution reduces. Conversely, if the resolution has to be enhanced, the pulse width has to be reduced. This reduction of the pulse width limits the range of the radar signal. In order to balance these trade-offs, pulse compression technique is used. Pulse compression can be achieved by phase modulating the transmitted radar signal.

Usage of a single carrier system for transmission of the radar signal results in utilization of large bandwidth [2]. But, the usage of a multicarrier system (OFDM) for transmission of the radar signal helps reduce bandwidth. OFDM technology forms the foundation for a number of communication systems such as Digital Audio and Video Broadcasting, IEEE 802.11 g, Digital Subscriber Lines (xDSL). The latest applications include LTE and LTE Advanced. OFDM has also been applied to radar systems for object tracking and target detection. This application has been realized in different types of multipath and clutter environments.

Multicarrier signals inherently have a high value of Peak-to-Mean Envelope Power Ratio (PMEPR). The PMEPR of the signal indicates the variations present in the complex envelope. A high value of PMEPR implies that the signal has abrupt amplitude variations and the amplifier at the transmitter has to be extremely sensitive to these changes.

Further, phase coding of these multicarrier signals is performed to achieve pulse compression resulting in better range and resolution of the radar signal. The radar signals are phase coded using P3, P4 [3] phase sequences which form a complementary set. The signal thus generated is a MCPC signal as described by N. Levanon in [4]. The limitation of the MCPC signal is its high PMEPR value.

There are three domains [5] that can be used to reduce variations in the complex envelope for data transmission applications. They are: signal distortion techniques, coding schemes, and probabilistic methods. Envelope clipping and companding are well-known signal distortion techniques. Golay codes, turbo codes, and block codes are examples for coding schemes. Partial Transmit Sequences and Selective Mapping are probabilistic methods. The intention of this paper is to address the issue of high PMEPR of a MCPC radar signal using companding technique whose implementation until now has only been restricted to data transmission systems.

The structure of this paper is as follows. In Sect. 2, we describe the generation of the complex envelope of a MCPC signal. Section 3 elaborates on the various companding techniques. Section 4 implements the companding technique to the MCPC signal. The demerit of the companded signal is addressed in Sect. 5 through the use of level shifting technique. A comprehensive comparison between a generic MCPC signal and the signals obtained using the above-mentioned techniques is made, and conclusions are drawn in Sect. 6.

2 The Conventional MCPC Signal

A multicarrier phase-coded signal utilizes N sequences which are transmitted simultaneously on N carriers that are separated in frequency by the inverse of the bit duration t_b. This separation in frequency yields orthogonal frequency division multiplexing (OFDM). Each sequence consists of N phase-modulated bits, and the associated phases are generated based on a certain scheme.

The complex envelope of the MCPC signal is described by:

$$s(t) = \begin{cases} \sum_{k=1}^{N} A_k\, exp\left[j\left\{2\pi f_s t\left(\frac{N+1}{2} - k\right) + \theta_k\right\}\right] \\ \sum_{l=1}^{N} u_{k,l}[t-(l-1)t_b], \quad 0 \le t \le Nt_b \\ 0 \quad , \quad elsewhere \end{cases} \tag{1}$$

where

$$u_{k,l}(t) = \begin{cases} \exp(j\phi_{k,l}), & 0 \le t \le t_b \\ 0 \quad , & elsewhere \end{cases} \tag{2}$$

A_k is the amplitude weight applied to the subcarriers, and θ_k is the random phase shift introduced by the transmitter to each carrier. $\phi_{k,l}$ is the lth phase of the kth subcarrier. The P3 and P4 phase coding schemes produce signals that have ideal periodic thumbtack autocorrelation functions, i.e., zero periodic autocorrelation sidelobes. The MCPC signal is based on cyclic shifts of these phase sequences.

The P3 phase sequence is described by:

$$\phi_l = \frac{\pi}{N}(l-1)^2, \quad l = 1, 2, \ldots, N. \tag{3}$$

Phases of the set of five complementary phase-coded sequences are as shown in Table 1.

The P4 phase sequence is described by:

$$\phi_l = \frac{\pi}{N}(l-1)^2 - \pi(l-1), \quad l = 1, 2, \ldots, \mathrm{N}. \tag{4}$$

Table 1 P3 phase sequences

Seq 1	0°	36°	144°	324°	216°
Seq 2	36°	144°	324°	216°	0°
Seq 3	144°	324°	216°	0°	36°
Seq 4	324°	216°	0°	36°	144°
Seq 5	216°	0°	36°	144°	324°

Table 2 P4 phase sequences

Seq 1	0°	−144°	−216°	−216°	−144°
Seq 2	−144°	−216°	−216°	−144°	0°
Seq 3	−216°	−216°	−144°	0°	−144°
Seq 4	−216°	−144°	0°	−144°	−216°
Seq 5	−144°	0°	−144°	−216°	−216°

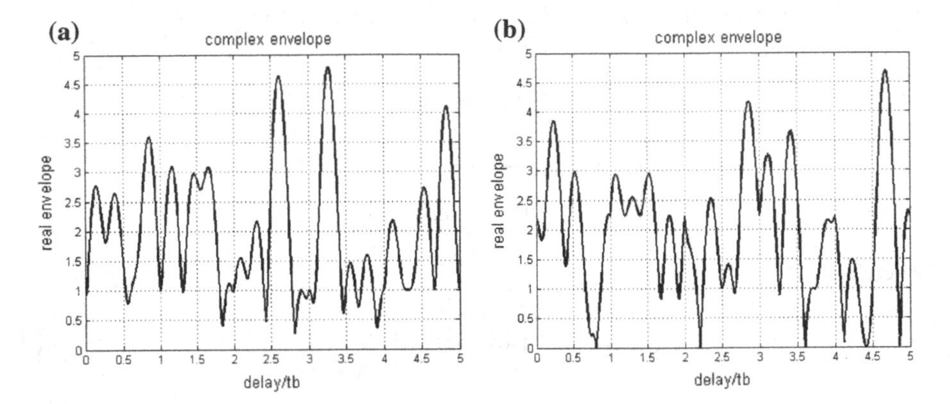

Fig. 1 **a** Complex envelope using P3. **b** Complex envelope using P4

Phases of the set of five complementary phase-coded sequences are as shown in Table 2.

The complex envelope, autocorrelation, and ambiguity function [6] of the MCPC signals based on cyclic shifts of P3 and P4 for the order 35214 are illustrated in Fig. 1, Fig. 2, and Fig. 3, respectively.

Autocorrelation is the correlation of a signal with a delayed copy of itself as a function of delay. Ambiguity function is a two-dimensional function of delay and Doppler frequency that measures the correlation between a waveform and its Doppler distorted version. Autocorrelation and the ambiguity function together help analyze the target detection capabilities of the radar signal. When we have multiple point targets, we have a superposition of ambiguity functions. A weak target located near the strong target can be masked by the sidelobes of the ambiguity function centered around the strong target. Hence, we have to minimize the minor lobes for the perfect detection of secondary targets [7].

From Fig. 2a, b, we can observe that the sidelobe power levels for the P4 plot are significantly lower than that obtained using P3 phases. Figure 3a, b depicts the ambiguity functions of P3 and P4, respectively. We can observe that the ridges for the P4 plot are lesser than that of P3 for different Doppler shifts.

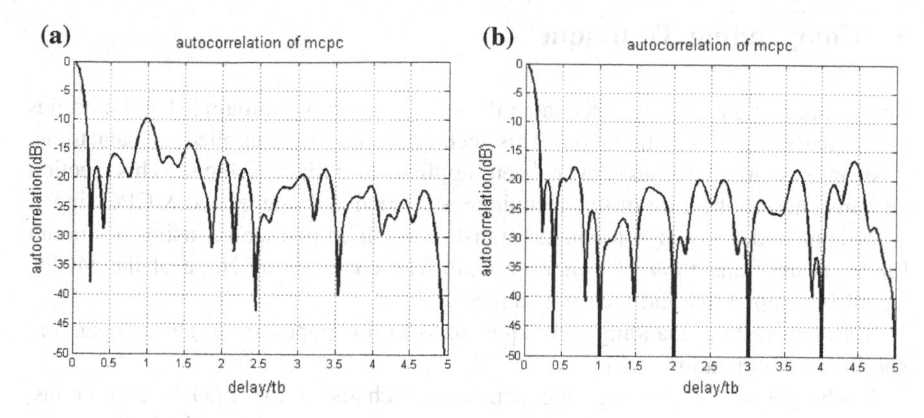

Fig. 2 **a** Autocorrelation for P3. **b** Autocorrelation for P4

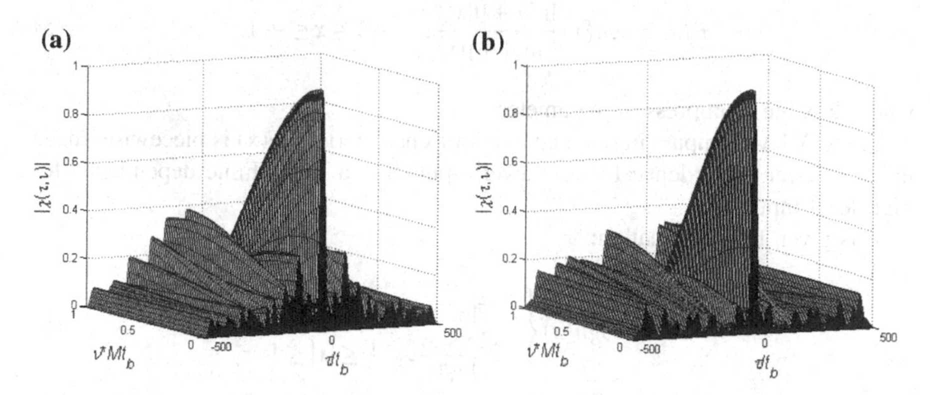

Fig. 3 **a** Ambiguity function of P3. **b** Ambiguity function of P4

Table 3 PMEPR comparison of P3 and P4

Sequence order	PMEPR using P3	PMEPR using P4
3 5 2 1 4	4.58	4.39
3 4 5 1 2	3.58	1.73
3 1 2 5 4	4.58	2.97

From Table 3, it is evident that P4 phase sequences produce a lower value of PMEPR when compared to P3 phase sequence. We also observe that the autocorrelation and ambiguity function of P4 is more favorable than P3. P4 also has a better bandwidth tolerance than P3.

Hence, we consider P4 phase sequences for generation of MCPC signals in the future sections.

3 Companding Technique

The companding technique falls under the signal distortion domain [8] and modifies the signal prior to amplification. This technique has been applied to data transmission systems but has never been applied to radar systems. This section demonstrates application of this technique to a previously generated MCPC signal.

A compander [9] is a nonlinear transformation implemented using a smooth limiter at the transmitter of a radar system. The complex envelope of the MCPC signal is compressed and then transmitted.

Two types of companding techniques used in this paper are μ-law companding and A-law companding [10].

In the μ-law companding, the compressor characteristic F(x) is continuous, approximating a linear dependence on x for low input levels and a logarithmic one for high input levels. μ-law companding equation is given by:

$$F(x) = sgn(x)\frac{\ln(1+\mu|x|)}{\ln(1+\mu)}, \quad -1 \leq x \leq +1. \tag{5}$$

where μ is the 'compression parameter.'

In the A-law companding, the compressor characteristic F(x) is piecewise, made up of a linear dependence for low-level inputs and a logarithmic dependence for high-level inputs.

It is given by the equation:

$$F(x) = sgn(x)\begin{cases} \frac{A|x|}{1+\ln(A)}, & |x| < \frac{1}{A} \\ \frac{1+\ln(A|x|)}{1+\ln(A)}, & \frac{1}{A} \leq |x| \leq 1 \end{cases}. \tag{6}$$

where A is the 'compression parameter.'

Using A-law or μ-law companders, the dynamic range of the signals is altered; i.e., weak signal amplitude is raised, and strong signal amplitude is lowered. The block diagram of the generation of MCPC using companding is shown in Fig. 4.

The N subcarriers are generated and are phase modulated by using P4 sequences. The subcarriers are then summed, and the resultant signal is fed as input to a compander which compresses the signal prior to transmission. In Sect. 4, μ-law and A-law companding techniques have been applied to the MCPC signals in order to reduce PMEPR.

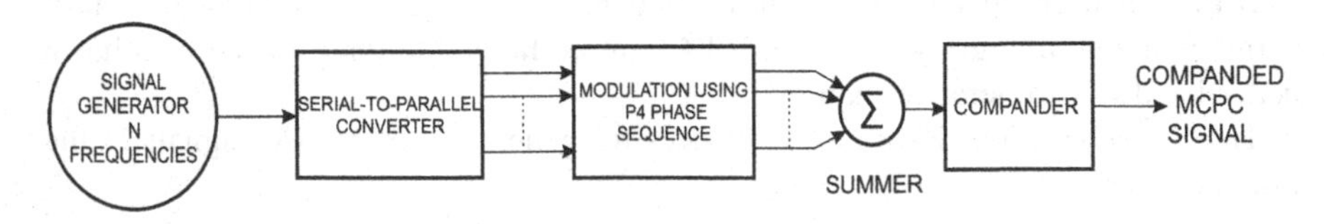

Fig. 4 Generation of MCPC using companding technique

4 Implementation of the Companding Technique

In this section, a comparison is made between the conventional MCPC signal and the signal subjected to the above-mentioned companding techniques.

Table 4 portrays the comparison of PMEPR values.

The companding technique has been applied to the MCPC signal that is based on the cyclic shifts of the P4 phase sequences for the order 3 5 2 1 4. The complex envelope, autocorrelation, and ambiguity function are plotted for 3 5 2 1 4 phase sequence using both μ-law and A-law companding techniques in Fig. 5, Fig. 6, and Fig. 7, respectively.

From Table 4, it is evident that the PMEPR of the companded signal using either μ-law or A-law is considerably lesser than that of the conventional MCPC signal. This quantifies the fact that the companded signal has lesser envelope variations. Figure 6a, b illustrates the autocorrelation plots of the companded signals using μ-law and A-law, respectively. Upon comparison with that of the conventional MCPC signal, it can be observed that the peak sidelobe power levels for both companded and conventional signals are around −17 dB. This ensures that the target detection capabilities of the radar signal are preserved after applying the technique. Figure 7a, b shows the ambiguity function of the companded signals using μ-law and A-law, respectively. A similar comparison as that for autocorrelation can be done to observe that the sidelobe power levels for delay and Doppler shifts have not considerably changed after application of companding.

5 Level Shifting and Companding

The merit of the companding technique as seen in Sect. 4 is that the PMEPR has been effectively reduced. The nature of the compander is such that zero points in the conventional MCPC signal are mapped to zero points after companding. Abrupt variations exist in the vicinity of the zero points in Fig. 5a, b which may prove to be difficult for the power amplifier to track. A novel approach to resolve this issue would be to remove the zero points before companding by employing the method of level shifting.

The block diagram shown in Fig. 8 illustrates the process of level shifting and companding.

The results obtained after the application of level shifting and companding for μ-law and A-law companding techniques are shown in Figs. 9, 10, and 11.

Table 4 PMEPR of companded signal

Sequence order	Conventional MCPC	μ-law	A-law
3 5 2 1 4	4.39	1.49	1.51
3 4 5 1 2	1.73	1.13	1.14
3 1 2 5 4	2.97	1.31	1.31

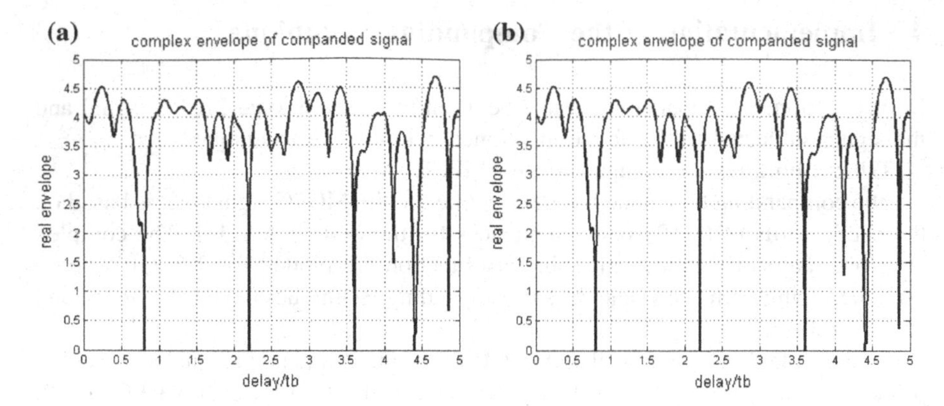

Fig. 5 **a** Complex envelope of μ-law companded signal. **b** Complex envelope of A-law companded signal

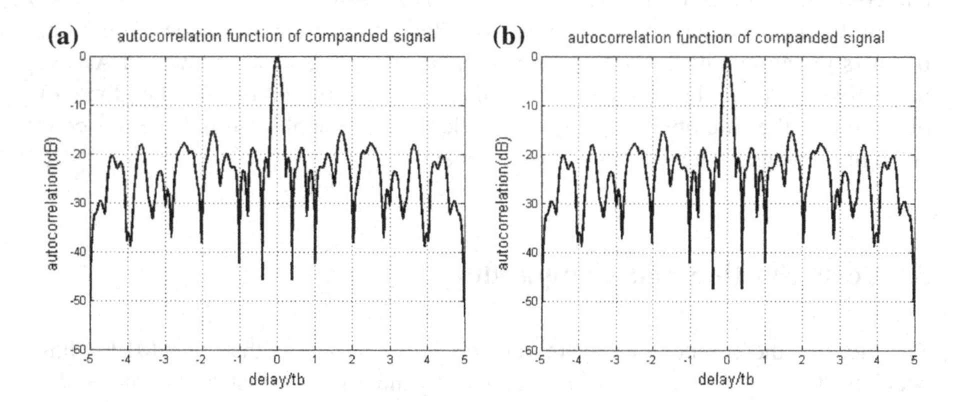

Fig. 6 **a** Autocorrelation of μ-law companded signal. **b** Autocorrelation of A-law companded signal

Figure 9a, b shows that the zero points in the complex envelope have been eliminated, thus removing the abrupt variations near the zero points that existed previously as shown in Fig. 5a, b.

The PMEPR for the shifted and companded signal is given in Table 5.

From Table 5, it can be observed that the value of PMEPR remains almost unchanged when compared to Table 3 with an improvement in the complex envelope. The autocorrelation and ambiguity functions shown after the application of level shifting and companding technique are very similar to those obtained in Sect. 4 and are acceptable.

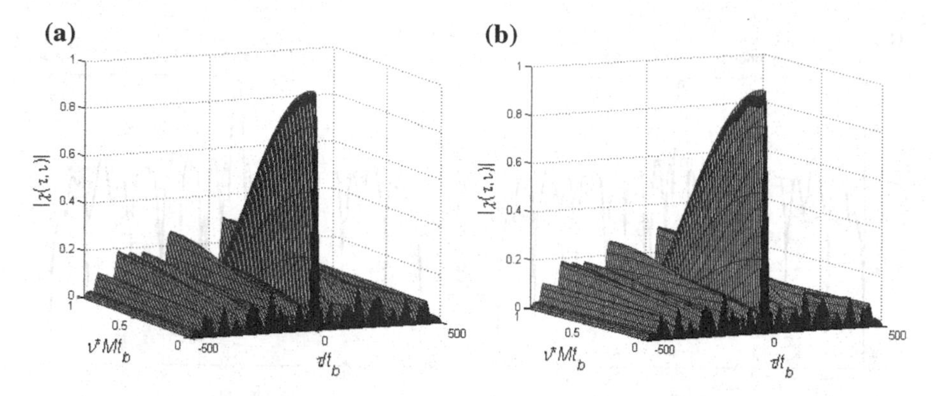

Fig. 7 **a** Ambiguity function of μ-law companded signal. **b** Ambiguity function of A-law companded signal

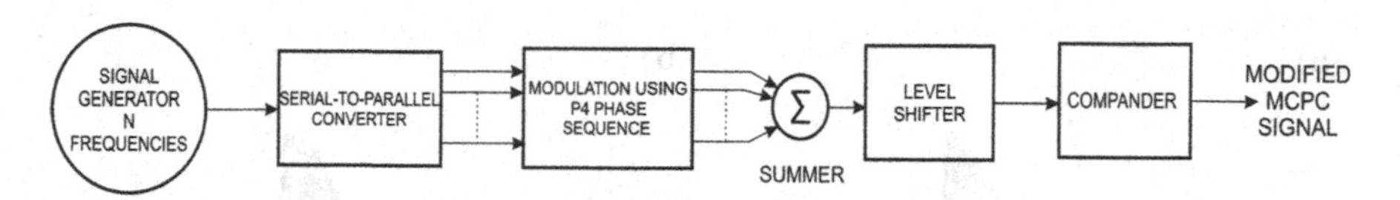

Fig. 8 MCPC generation by shifting and companding technique

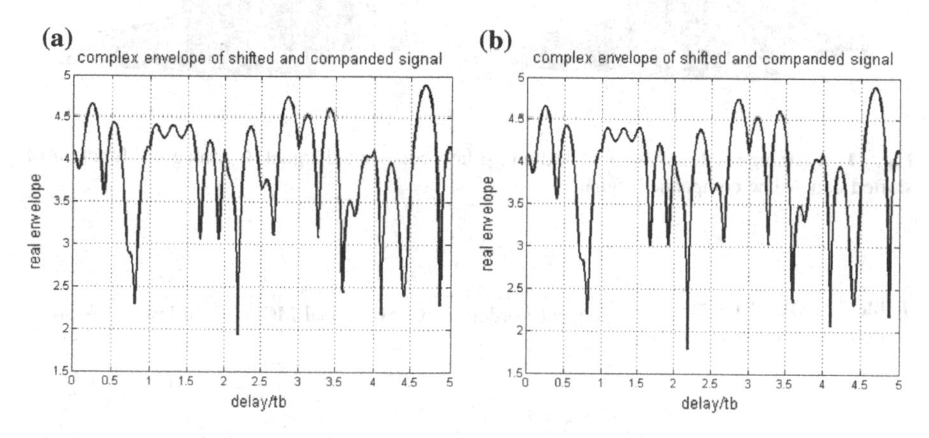

Fig. 9 **a** Complex envelope of shifted and μ-law companded signal **b**: complex envelope of shifted and μ-law companded signal shifted and A-law companded signal

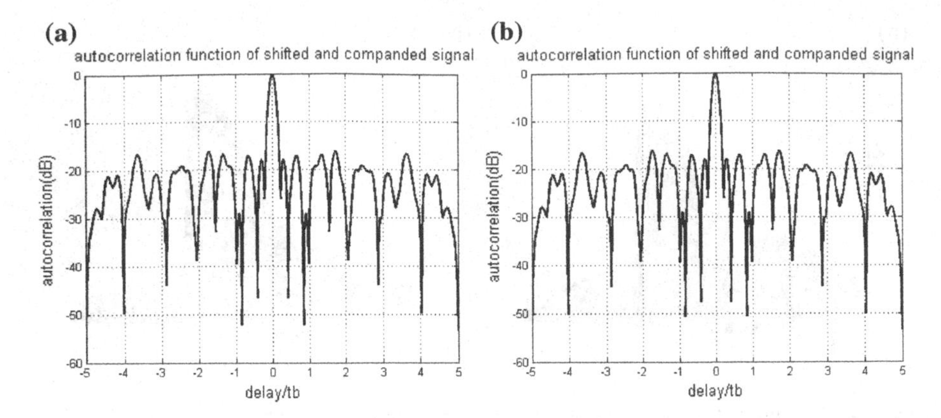

Fig. 10 **a** Autocorrelation of shifted and μ-law companded signal. **b** Autocorrelation of shifted and A-law companded signal

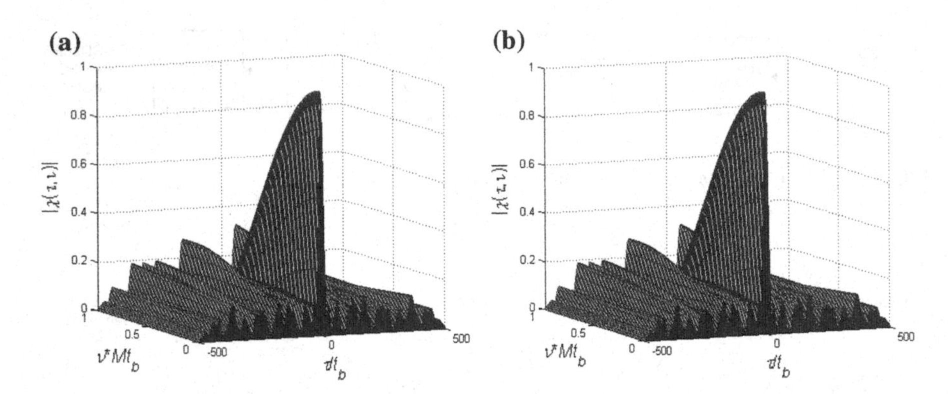

Fig. 11 **a** Ambiguity function of shifted and μ-law companded signal. **b** Ambiguity function of shifted and A-law companded signal

Table 5 PMEPR for the shifted and companded signal

Sequence order	Conventional MCPC	μ-law	A-law
3 5 2 1 4	4.39	1.51	1.53
3 4 5 1 2	1.73	1.16	1.16
3 1 2 5 4	2.97	1.34	1.35

6 Conclusion

The MCPC signal has many advantages in terms of bandwidth efficiency and pulse compression capability when compared to other radar signals which makes it more suitable for radar applications. P3 and P4 phase sequences are used to phase modulate the multicarrier system, and it is seen that P4 phase sequence is more suitable as it produces a much better autocorrelation function and complex envelope resulting in a lower value of PMEPR. P4 also has better bandwidth tolerance than P3.

From results obtained in Sect. 4, it is evident that companding techniques can be used to reduce the PMEPR of the MCPC signal effectively. The problem with employing just the companding technique is solved by the process of level shifting as shown in Sect. 5. The results thus obtained from level shifting are favorable as it still maintains a low value of PMEPR, an improved complex envelope, and low sidelobe power levels in the autocorrelation and ambiguity functions. Hence, we can infer that this a more suitable approach for the generation of an MCPC radar signal.

References

1. Levanon, N., Mozeson, E.: Radar Signals. Wiley (2004)
2. Jankiraman, M., Wessels, B.J., van Genderen, P.: System design and verification of the PANDORA multifrequency radar. In: Proceedings of International Conference on Radar Systems, Brest, France, Session 1.9, 17–21 May 1999
3. Levanon, N., Mozeson, E.: PhaseCoded Pulse, pp. 100–167, 1st edn. Wiley, IEEE Press (2004)
4. Levanon, N.: Multifrequency complementary phase-coded radar signal. In: IEEE Proceedings-Radar, Sonar Navigation, vol. 147, No. 6, Dec 2000
5. Rahmatallah, Y., Mohan, S.: Peak-to-average power ratio reduction in OFDM systems: a survey and taxonomy. IEEE Commun. Surv. Tutor IEEE (2013)
6. Mozeson, E., Levanon, N.: MATLAB code for plotting ambiguity functions. IEEE Trans. Aerosp. Electron. Syst. **38**(3), 1064–1068 (2002)
7. Yedla, K.N., Srinivasu, C.H.: Importance of using gold sequence in radar signal processing. J. Theor. Appl. Inf. Technol. **78**(3), 31st Aug 2015
8. Wang, L., Tellambura, C.: An overview of peak-to-average power ratio reduction techniques for OFDM systems. In: Proceedings of the IEEE International Symposium on Signal Processing and Information Technology, pp. 840–845 (2006)
9. Ermolova, N.Y.: New Companding Transform for Reduction of Peak-to-Average Ratio, Vehicular Technology Conference, pp. 1404–1407. IEEE (2002)
10. Haykin, S.: Digital Communications, 4th edn. Wiley (1988)

A Wavelet Transform-Based Filter Bank Architecture for ECG Signal Denoising

Ashish Kumar, Rama Komaragiri and Manjeet Kumar

Abstract In the present work, a wavelet transform-based filter bank architecture suitable for ECG signal denoising is proposed. Firstly, wavelet transform functions are used to filter the signals in Matlab R2013b, and then, the resulting signal is converted into 16-bit binary data. This data is used further as an input of QRS detection block. Modified architecture contains only three low-pass filters and a high-pass filter, which is less compared to previously designed architectures. One of the key advantages of the proposed architecture is that no multiplexer and multiplier circuits are required for the further processing. The proposed architecture consumes less area and is relatively fast compared to previously designed architectures.

Keywords Cardiovascular diseases (CVDs) · Electrocardiogram (ECG)
Wavelet filter bank (WFB)

1 Introduction

This cardiovascular diseases (CVDs) or diseases related to the heart are due to abnormalities or disorders of the heart and blood vessels. CVDs are classified further as coronary heart disease and rheumatic heart disease. Most important reasons for CVD are the usage of tobacco, an unhealthy diet, and excessive consumption of alcohol. CVDs are the leading causes of deaths globally. As per World Health Organization report, the global mortality rate due to CVDs is far greater than

---

A. Kumar (✉) · R. Komaragiri · M. Kumar
Electronics and Communication Engineering, Bennett University,
Greater Noida 201310, India
e-mail: ashish.kumar1@bennett.edu.in

R. Komaragiri
e-mail: rama.komaragiri@bennett.edu.in

M. Kumar
e-mail: manjeetchhillar@gmail.com

P. K. Sa et al. (eds.), *Recent Findings in Intelligent Computing Techniques*,
Advances in Intelligent Systems and Computing 708,
https://doi.org/10.1007/978-981-10-8636-6_26

any other reason. According to WHO estimates, in the year 2012, 17.5 million people who died from CVDs contribute to 31% of the overall global deaths [1].

The most often used clinical cardiac test is electrocardiogram analysis or most popular ECG analysis. Hence, there is a strong focus from researchers to technologically advance the cardiac function assessment to improve the conventional cardiovascular analysis technologies used in hospitals and clinics. Another thrust area is on replacing the Holter devices by devices that are small in size and require less power. Holter devices do not diagnose arrhythmias automatically in real time and do not provide real-time information to the concerned when a critical condition occurs. Thus, there is a huge scope for researchers to put more effort in developing efficient ECG denoising algorithms for wearable devices or portable devices.

In the present work, wavelet transform-based filter bank architecture for ECG signal denoising is proposed.

2 Literature Overview

An ECG signal corruption, generally called as noise, can be classified into following categories [2]: (i) muscle contraction, (ii) motion artifacts, (iii) baseline drift, (iv) powerline interface, (v) ECG amplitude modulation with respiration, (vi) electrode contract noise, (vii) electrosurgical noise, (viii) instrumentation noise, and (ix) interference (electromyographic). In this work, for the simulation purpose, only four different noise sources are selected [3].

(1) Baseline drifts due to respiration and its low-frequency properties.
(2) Powerline interference, which is ubiquitous
(3) Electromyographic (EMG) is random in nature and has a high-frequency content.

For the denoising of the noises mentioned above, many approaches have been proposed. As suggested by Pan and Tompkins [4], very first real-time ECG denoising algorithm based on the band-pass filter uses a band-pass filter which uses cascaded high- and low-pass filters. This method needs less time to denoise the signal, and hardware implementation is easy. The major drawback of this approach is that the frequency bands of the ECG signal and the noise overlap which degrades the performance. Different noise removal techniques are proposed to improve numerical efficiency and accuracy to improve the quality of ECG signal after denoising. They are: (i) first and second derivatives, (ii) digital filtering, (iii) mathematically morphology, (iv) filter banks, and (v) wavelet transform. Based on the hardware complexity and detection accuracy approaches, the wavelet transform is a more suitable technique for ECG signal denoising when compared to other methods. The wavelet transform-based filter bank consists of a biphasic and a monophasic filter function which approximates biphasic and monophasic morphologies. The basic architecture of wavelet transform-based filter bank was proposed by Mallet [5]. The parameters of ECG signal are obtained using wavelet decomposition tree. Initially, ECG signal is decomposed into smooth and detail constituents. The smooth component is then further decomposed into smooth and detail constituents. The process of decomposing a signal into the smooth and detailed process is repeated over the desired number of scales [6] (Fig. 1).

The architecture in [6] is further modified by Rodrigues [7] as shown in Fig. 2. They use un-decimator-based wavelet filter bank for ECG signal denoising. A large chip area due to the requirement of the constant clock and a large number of registers are the main drawback of the method in [7].

To minimize the area problem in [7], Min et al. [8] proposed a decimator-based wavelet filter bank architecture, which reduces the number of registers as shown in Fig. 3. However, even after significant modifications, this architecture still requires a large area. Some filter design techniques used in ECG denoising are presented in [9].

In this work, to minimize the problem of the area in the methods mentioned above, a modified wavelet transform-based filter bank architecture for ECG signal denoising is proposed.

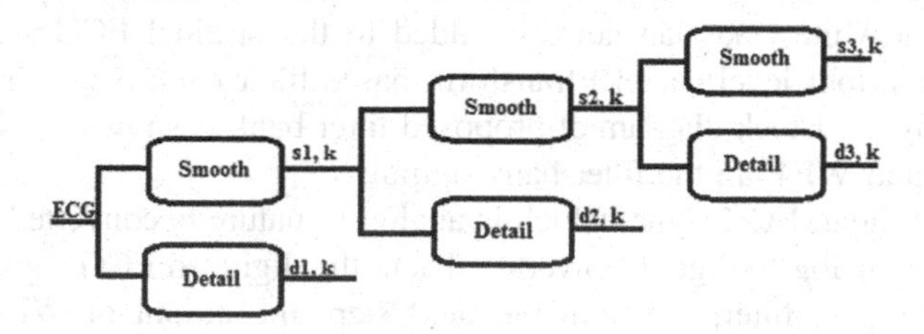

Fig. 1 Wavelet decomposition of ECG signal

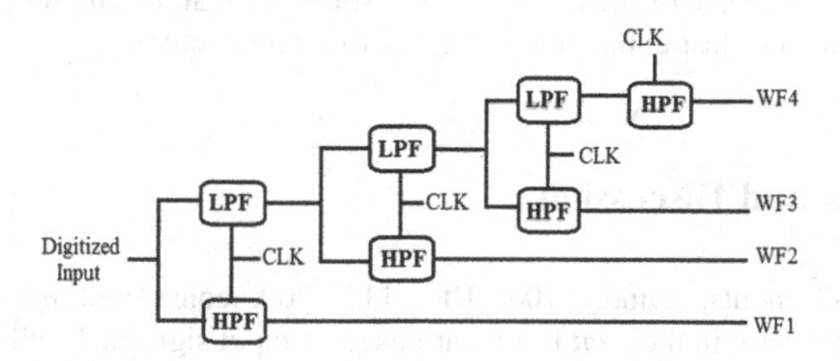

Fig. 2 Un-decimator-based wavelet filter bank

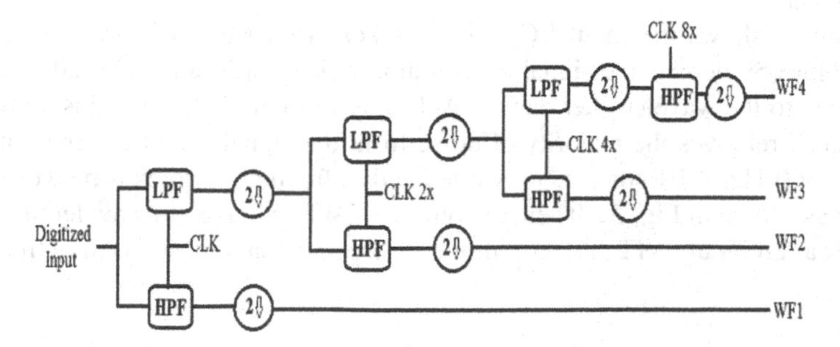

Fig. 3 Decimator-based wavelet filter bank

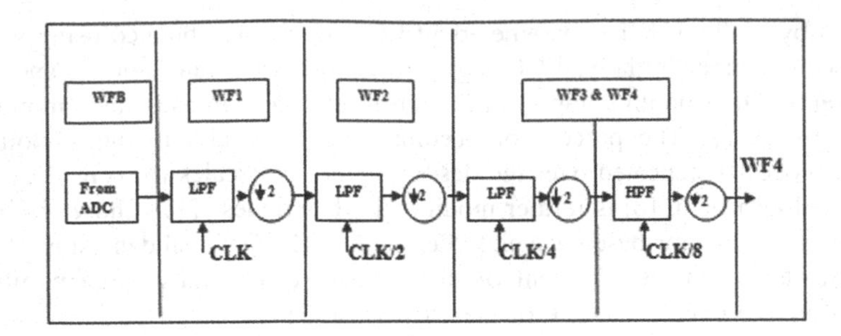

Fig. 4 Block diagram of the proposed filter bank

3 Proposed Work

ECG from MIT-BIH arrhythmia database in *.mat* format is considered as input. The frequency of the signal is 360 Hz. In reality, as various noises corrupt an acquired ECG signal, a white Gaussian noise is added to the original ECG signal. In the present work, a four-level wavelet transform-based filter bank is used to analyze a noisy ECG signal. Block diagram of proposed filter bank is shown in Fig. 4. WF1, WF2, WF3, and WF4 are the filter bank outputs.

First of all, input ECG signal which is analog in nature is converted into digital form using an analog-to-digital converter. Then, the digitized ECG signal is passed through a low-pass filter, WF1. In the next step, the output of WF1 is passed through a low-pass filter, WF2, and then through the low-pass filter, WF3. The output of WF3 is finally passed through a high-pass filter, WF4. As this architecture requires a less number of filters, thus, it consumes less area. The proposed architecture is relatively fast compared to the existing architectures.

4 Results and Discussion

Various ECG inputs, namely 100, 106, 117, 200, considered from MIT-BIH arrhythmia database in the *.mat* format are used as input signals. To all of the input signals, a white Gaussian noise is added. The ECG signal with noise is then passed through the filter bank. Figure 5 shows the wavelet filter bank output of 100*.mat* ECG signal.

Figure 5a shows the input ECG signal taken from MIT-BIH arrhythmia database. Figure 5b depicts the signal after adding noise. The signal after adding noise is then fed to the wavelet filter bank 1 (WF1) as an input. WF1 contains a low-pass filter which removes the majority of noise from the signal. The frequency of input signal is 360 Hz. WF1 contains only one band of frequency which is between 0 and 180 Hz as shown in Fig. 5c. Then, the output of WF1 is given to wavelet filter bank 2 (WF2) as an input. WF2 further removes noise and contains the band of frequency

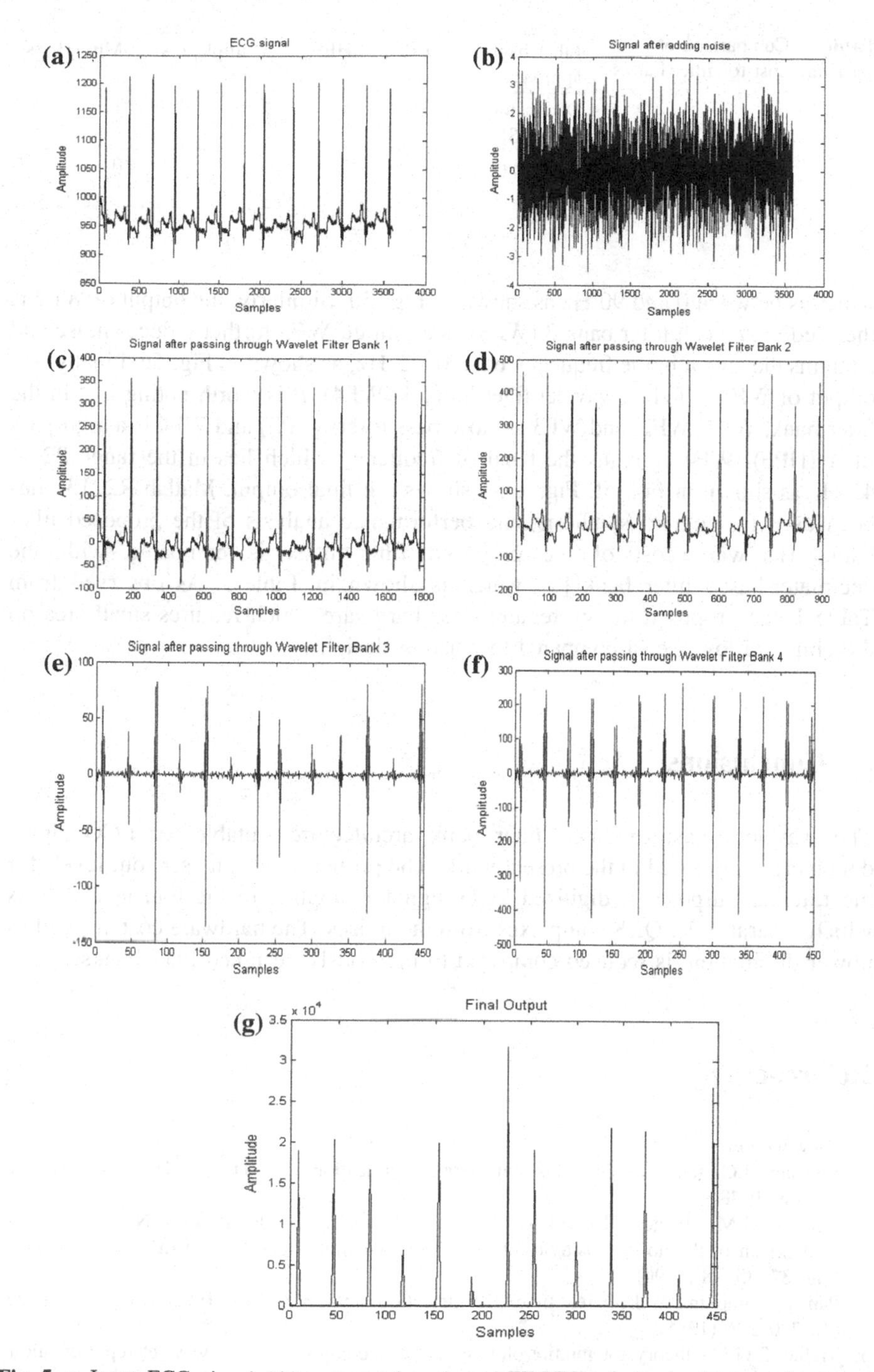

Fig. 5 **a** Input ECG signal 100.*mat* as taken from MIT-BIH arrhythmia database, **b** output waveform after adding noise to the signal, **c** output waveform after passing signal with noise through filter WF1, **d** output waveform after passing output from filter WF1 through filter WF2, **e** output waveform after passing output from filter WF2 through filter WF3, **f** output waveform of after passing output from filter WF3 through filter WF4, **g** final output after downsampling

Table 1 Comparison of hardware cost for filter banks

Filter bank approach	LPF	HPF	Multiplexers	Multipliers
Decimator based [8]	3	4	2	1
Proposed design	3	1	0	0

which is between 0 and 90 Hz as shown in Fig. 5d. Similarly, the output of WF2 is then fed to wavelet filter bank 3 (WF3) as an input. WF3 further reduces noise and contains the signal in the frequency band 0–45 Hz, as shown in Fig. 5e. Finally, the output of WF3 is fed to wavelet filter bank 4 (WF4). It is worth noting that in the filter bank, WF1, WF2, and WF3 are low-pass filters (LPF) and WF4 is a high-pass filter (HPF). WF4 contains the band of frequency which lies in the range 22.5–45 Hz as shown in Fig. 5f. Figure 5g shows the final output. Matlab R2013b has been used for coding, synthesis, and performance analysis of the proposed filter bank. Hardware cost of the proposed filter bank is compared with the decimator-based filter bank [8], which is shown in Table 1. As observed from Table 1, the proposed design requires less hardware which requires small area on the chip and lower cost compared to existing designs.

5 Conclusions

The wavelet transform-based filter bank architecture suitable for ECG signal denoising is proposed in the present work. The proposed design uses four levels for the filtering purpose. A digitized ECG signal is applied to the four-level WFBs which separates the QRS complexes from the noises. The hardware cost, as well as power dissipation, is reduced compared to previously designed filter banks.

References

1. www.who.int
2. Webster, J.G. (ed.): Medical Instrumentation-Application, and Design. Houghton Mifflin, Boston (1978)
3. Friesen, G.M., Jannett, T.C., Jadallah, M.A., Yates, S.L., Quint, S.R., Nagle, H.T.: A comparison of the noise sensitivity of nine QRS detection algorithms. IEEE Trans. Biomed. Eng. **37**, 85–98 (1990)
4. Pan, J., Tompkins, W.J.: A real-time QRS detection algorithm. IEEE Trans. Biomed. Eng. **32** (3), 230–236 (1985)
5. Mallat, S.G.: A theory for multiresolution signal decomposition: the wavelet representation. IEEE Trans. Pattern Anal. Mach. Intell. **11**(7), 674–693 (1989)

6. Sivannarayana, N., Reddy, D.C.: Biorthogonal wavelet transform for ECG parameters estimation. Med. Eng. Phy. **21**, 167–174 (1999)
7. Rodrigues, J.N., Olsson, T., Sornmo, L., Owall, V.: Digital implementation of a wavelet-based event detector for cardiac pacemakers. IEEE Trans. Circ. Syst. I Reg. Papers **52**(12), 2686–2698 (2005)
8. Min, Y.J., Kim, H.K., Kang, Y.R., Kim, G.S., Park, J., Kim, S.W.: Design of a wavelet-based ECG detector for implantable cardiac pacemaker. IEEE Trans. Biomed. Cir. Syst. **7**(4), 426–435 (2013)
9. Aggarwal, A., Rawat, T.K., Kumar, M., Upadhyay, D.K.: Design of Hilbert transformer using the L1-method. In: Proceedings of INDICON-2015, pp. 1–6, Dec 2015

Comparative Assessment for Power Quality Improvement Using Hybrid Power Filters

Soumya Ranjan Das, Prakash K. Ray and Asit Mohanty

Abstract Nowadays, power quality is one of the sensitive issues considered in power system. In order to develop this superiority of power, the disturbances present in the signal have to be eliminated. This disturbance is nothing but the harmonics which was traditionally filtered out by the use of passive filter, and later on it claims superiority by use of active power filter due to certain problems raised in passive filters like resonance and fixed compensation. Particularly, the hybrid active power filter is used to eliminate the harmonics present in the fundamental signal and also compensates reactive power. In this paper, methods like instantaneous power theory (p-q) and synchronous reference frame theory (id-iq) are used to calculate the compensating current. Both the above methods are simulated with help of PI and fuzzy controller for different voltage conditions. The obtained results show the active behavior of fuzzy logic controllers over PI controllers.

Keywords Active filters · Harmonics · Fuzzy logic controller
Hybrid filters · PI controller · Power quality · Total harmonic distortion (THD)

1 Introduction

Power electronic equipments are heavily used in the environment like industrial and domestic fields. Due to the use of power electronic equipments, produce unwanted signals called harmonics in the utility system and due to use of these devices,

S. R. Das (✉) · P. K. Ray
Department of Electrical and Electronics Engineering, IIIT Bhubaneswar,
Bhubaneswar, India
e-mail: srdas1984@gmail.com

P. K. Ray
e-mail: pkrayiiit@gmail.com

A. Mohanty
Department of Electrical Engineering, CET Bhubaneswar, Bhubaneswar, India
e-mail: asithimansu@gmail.com

P. K. Sa et al. (eds.), *Recent Findings in Intelligent Computing Techniques*,
Advances in Intelligent Systems and Computing 708,
https://doi.org/10.1007/978-981-10-8636-6_27

reactive powers [1] also exit in the utility system. The transmission losses in lines are dependent on excessive reactive power which would also enhance generating capacity of generating stations. Therefore, it is essential to supply reactive power at the receiving end. These nonlinear loads at the receiving end draw current which are not sinusoidal in nature and generate voltage drops in the conductors connected in the supply system. In order to mitigate these disturbances and for improving power factor, traditionally passive filters were used but the factors like resonance problems with fixed compensation make this device with less use. To overcome these issues, active filters [2] were used. Out of several control strategies used for reference current calculations, only two control techniques are used frequently called instantaneous power theory (p-q) [3–5] and synchronous reference frame theory (id-iq) [6, 7].

Present paper mostly focused on two control strategies basically (p-q and id-iq) with PI controller [7, 8] and fuzzy controller [7, 8] used in hybrid active power filter (HAPF) [9, 10]. To authenticate present observations, simulations are performed with PI and fuzzy controller for both p-q and id-iq methods and the results are compared.

2 System Configuration

2.1 Hybrid Filters

Hybrid filters topology overcomes the problem related with passive and APF by providing better quality performance. These topologies at the same time decrease the switching noise as well as reduce electromagnetic intervention. Main role of HAPF is to progress dynamically the performance of filtering of higher order harmonics by supplying low order with cost-effective harmonics mitigation technique. Figure 1 shows the power circuit arrangement of HAPF as arrangement of shunt passive connected in series with shunt active filters. Application of HAPF is based on the configurations of an active filter (shunt and series) along with a passive filter having harmonic frequencies of 5th harmonics and 7th harmonics and a high-pass filter (second order) with 11th harmonic frequency. Even if these HAPF are slightly unusual in circuit design, but still these filters are equivalent in principle of operation and filtering performance. The major job of an active filter is relatively to attain isolation of harmonics between source and load. Eventually, occurrence of harmonics is absent in the supply system.

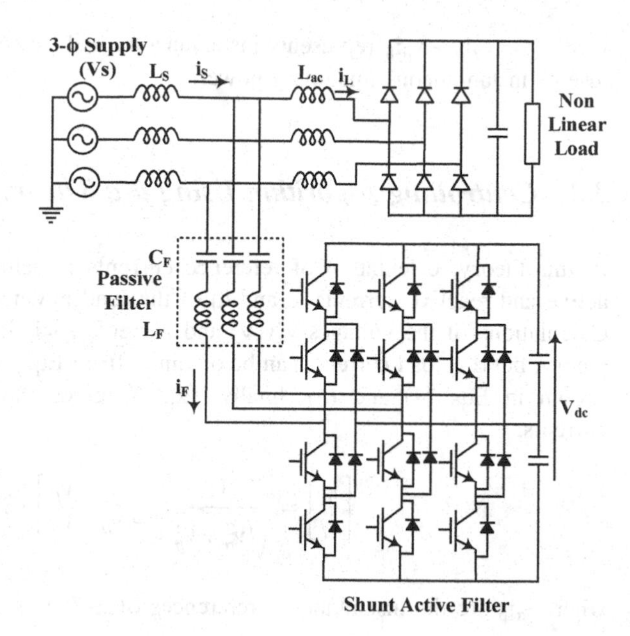

Fig. 1 Configuration of hybrid filters as series connected shunt passive and shunt active power filters

3 Control Algorithms

3.1 Controlling Algorithm Using p-q Theory

In this theory, the voltages $v_a v_b v_c$ and currents $i_a i_b i_c$ are calculated and transformed into $\alpha\beta0$ coordinates as described in Eq. (1) and the reference current controller is calculated back to the abc frame.

$$\begin{bmatrix} v_\alpha \\ v_\beta \\ v_0 \end{bmatrix} = \sqrt{\frac{2}{3}} \begin{bmatrix} 0 & \frac{\sqrt{3}}{2} & -\frac{\sqrt{3}}{2} \\ 1 & -\frac{1}{2} & -\frac{1}{2} \\ \frac{1}{\sqrt{2}} & \frac{1}{\sqrt{2}} & \frac{1}{\sqrt{2}} \end{bmatrix} \begin{bmatrix} v_a \\ v_b \\ v_c \end{bmatrix} \tag{1}$$

$$\begin{bmatrix} i_\alpha \\ i_\beta \\ i_0 \end{bmatrix} = \sqrt{\frac{2}{3}} \begin{bmatrix} 0 & \frac{\sqrt{3}}{2} & -\frac{\sqrt{3}}{2} \\ 1 & -\frac{1}{2} & -\frac{1}{2} \\ \frac{1}{\sqrt{2}} & \frac{1}{\sqrt{2}} & \frac{1}{\sqrt{2}} \end{bmatrix} \begin{bmatrix} v_a \\ v_b \\ v_c \end{bmatrix} \tag{2}$$

The active and reactive components of power are obtained from the $\alpha\beta$ coordinates of current and voltage.

$$\begin{bmatrix} p \\ q \end{bmatrix} = \begin{bmatrix} v_\alpha & v_\beta \\ -v_\beta & v_\alpha \end{bmatrix} \begin{bmatrix} i_\alpha \\ i_\beta \end{bmatrix} \tag{3}$$

where $p = v_\alpha i_\alpha + v_\beta i_\beta$ represents instantaneous real power and $q = v_\alpha i_\beta - v_\beta i_\alpha$ represents instantaneous imaginary power.

3.2 *Controlling Algorithm Using d-q Theory*

In this theory, calculation of reference currents is achieved using instantaneous active and reactive currents id and iq of the load having nonlinear characteristics. Calculations of the voltages $v_a v_b v_c$ and currents $i_a i_b i_c$ into $\alpha\beta 0$ are similar to p-q theory, but d-q load currents can be obtained from Eq. (4). Numerical relations are shown in Eqs. (4) and (5); finally, Eq. (6) gives the calculation of reference currents.

$$\begin{bmatrix} i_d \\ i_q \end{bmatrix} = \frac{1}{\sqrt{v_\alpha^2 + v_\beta^2}} \begin{bmatrix} v_\alpha & v_\beta \\ -v_\beta & v_\alpha \end{bmatrix} \begin{bmatrix} i_\alpha \\ i_\beta \end{bmatrix} \tag{4}$$

where $i_\alpha i_\beta$ are the instantaneous references of α–β axis current.

$$\begin{bmatrix} i_d \\ i_q \end{bmatrix} = \begin{bmatrix} \cos\theta & \sin\theta \\ -\sin\theta & \cos\theta \end{bmatrix} \begin{bmatrix} i_\alpha \\ i_\beta \end{bmatrix} \tag{5}$$

$$\begin{bmatrix} ic_\alpha \\ ic_\beta \end{bmatrix} = \frac{1}{\sqrt{v_\alpha^2 + v_\beta^2}} \begin{bmatrix} v_\alpha & -v_\beta \\ v_\beta & v_\alpha \end{bmatrix} \begin{bmatrix} ic_d \\ ic_q \end{bmatrix} \tag{6}$$

where $ic_d ic_q$ are the compensation currents.

4 Fuzzy Control

Fuzzy logic is considered one of the effective tools in soft computing techniques. Basically, fuzzy controller is very effective in dealing with problems like variation of parameter, uncertainty and complexity in system design. For a conventional control system, the design is based on mathematical design of the plant. To analyze any model, the system design or model with known parameters is needed. But no mathematical model is needed in fuzzy logic controller and can supply robust performance of the linear system and nonlinear control system with parameter variation. A FLC can be classified as fuzzification, knowledge base, and defuzzification. Fuzzification is the process of converting into linguistic variable from crisp value based on certain membership function. And knowledge base assembles the membership functions with different rules which are necessary like "IF-THEN", and the fuzzy inputs are processed to the rule set finally producing the output with

Table 1 Fuzzy logic controller linguistic parameters for fuzzy rule

e / ce	NB	NM	NS	Z	PS	PM	PB
NB	NVB	NVB	NVB	NB	NM	NS	Z
NM	NVB	NVB	NB	NM	NS	Z	PS
NS	NVB	NB	NM	NS	Z	PS	PM
Z	NB	NM	NS	Z	PS	PM	PB
PS	NM	NS	Z	PS	PM	PB	PVB
PM	NS	Z	PS	PM	PB	PVB	PVB
PB	Z	PS	PM	PB	PVB	PVB	PVB

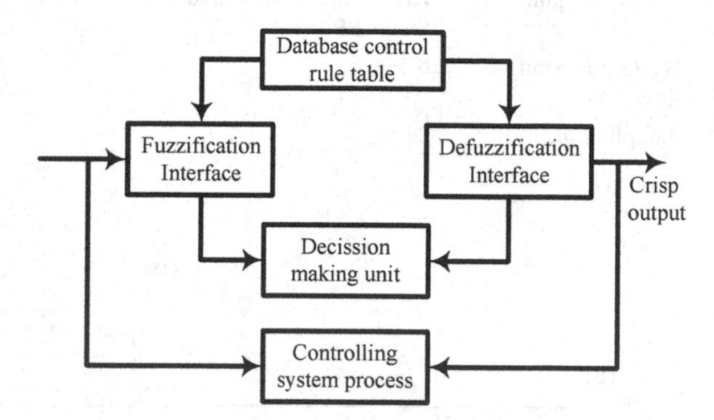

Fig. 2 Block diagram of fuzzy logic controller

fuzzy value. The rules for knowledge base are shown in Table 1. The process of transforming into numerical crisp value from the obtained fuzzy outputs is done through the process of defuzzification. Detailed process is shown in Fig. 2.

5 Simulation Results and Analysis

5.1 Harmonic Compensation Using Hybrid Filters Based on d-q Theory (with PI Controller)

The conventional p-q method used in hybrid filter is modified with a new technique based on d-q theory. The above two techniques are being implemented in a hybrid active power filter (HAPF) configuration. The power system under research is simulated in MATLAB/Simulink environments. The parameters of the system are given in Appendix section in Table 2. Simulations are done using the "Power System Blockset" simulator. The performance of the power system is tested under two operating conditions: (a) power system operating under the connection of

Table 2 Parameters of the system

Parameters	Value
Line voltage and frequency	415 V, 50 Hz
Line and load impedance	Ls = 0.15 mH, Lac = 1.5 mH, CDC = 1500 μF, RL = 60 Ω, L = 10 μF; (R1 = 2 Ω, R2 = 4 Ω, R3 = 6 Ω)
Tuned passive filter	Cf5 = 50 μF Lf5 = 8.10 mH; Cf7 = 20 μF Lf7 = 8.27 mH; Cf11 = 20 μF Lf11 = 8.270 mH
Ripple filter	CRF = 50 μF, LRF = 0.68 mH
Active filter parameters	CD = 1500 μF VD = 750 V
Filter coupling inductance	2.5 mH
Controller gain	KP = 0.032, KI = 0.00004, for Series PF; KP = 24, KI = 1.2, for Shunt PF
Hysteresis band limit	0.5 A
Sampling time	2e s

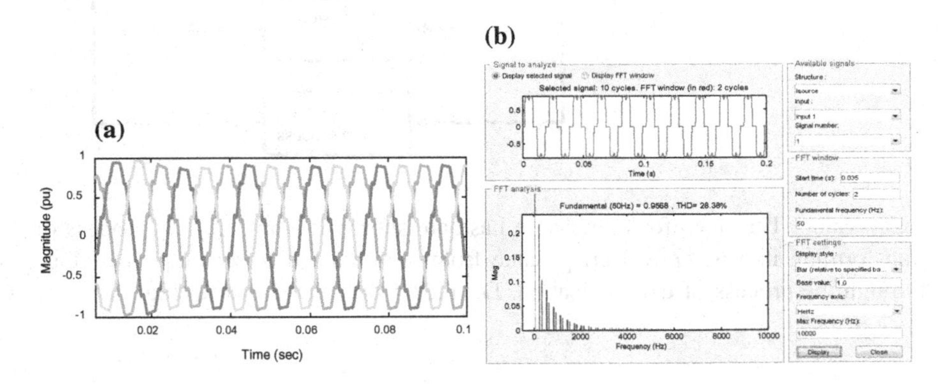

Fig. 3 **a** Load currents and **b** THD without filters

nonlinear load and (b) operation of the proposed hybrid filters in power system with nonlinear load. First, the power system is operated without the presence of filters connecting with non-sinusoidal loads and the simulated results are presented in Fig. 3a. This shows that currents in the load side, i.e., ia, ib, and ic, contain harmonics components with their waveform distorted because of the nonlinear load. The THD value is obtained to be 28.38% and shown in Fig. 3b.

5.2 *Harmonic Compensation Using Hybrid Filters Based p-q Theory (with Fuzzy PI Controller)*

This subsection represents the study of harmonics improvement using the proposed hybrid filter configurations based on p-q theory and using a fuzzy PI controller. Due to the presence of nonlinear load, it is observed that the distortion in load and source current is due to the imbalance in reactive power in the utility system. The performance of the hybrid active power filter is enhanced with the incorporation of fuzzy PI controller, and the simulated results are presented in Fig. 4a, b, and c, respectively. Figure 4a gives details of the three-phase load current with the DC link voltage which is required to be kept constant. The load current is improved because of a compensated current injection by the hybrid filter which indirectly provides better reactive power compensation at load side. It is analyzed that the THD value in the presence of fuzzy controller significantly improves to 4.76% by the action of shunt- and series-based hybrid active power filter. It is also being

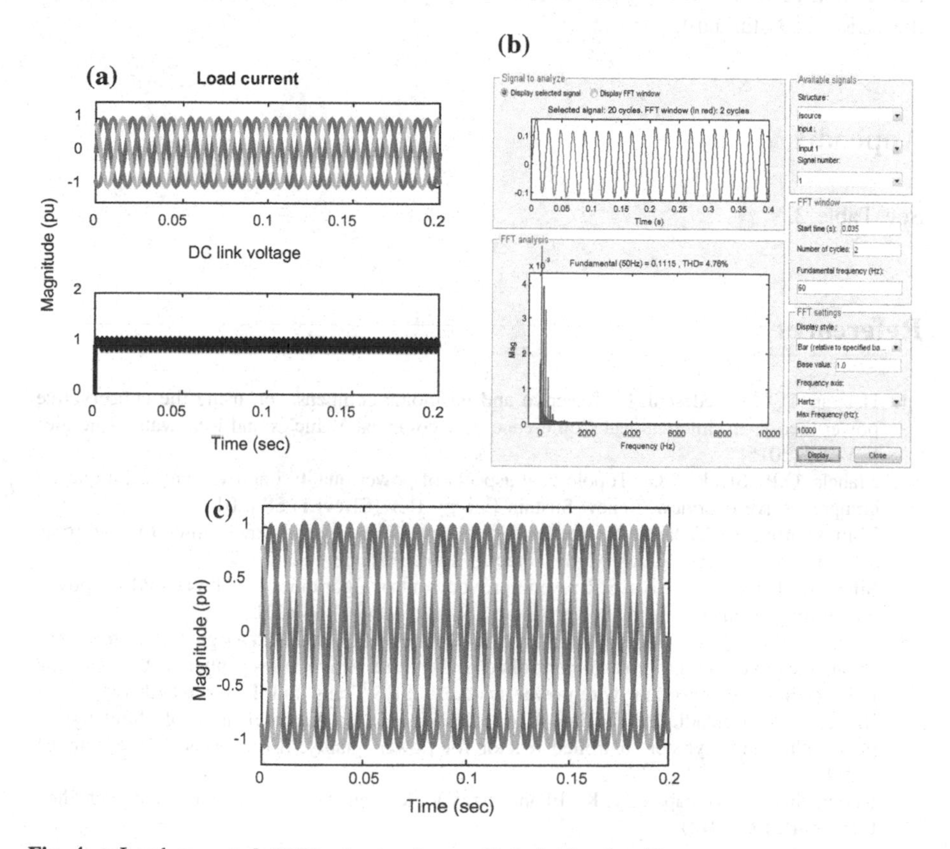

Fig. 4 **a** Load currents **b** THD **c** load voltage with hybrid active filter

observed that the load voltage is also becoming free from harmonics contents which will improve the power system performance in terms of operation and control.

6 Conclusion

In this paper, modeling and simulation of hybrid active power filter controlled by a PI and fuzzy PI are presented. The simulation was performed in MATLAB/ Simulink. Design of HAPF with FLC is used to eliminate the current and voltage harmonics of power system in load side. The performance of hybrid active power filter connected through fuzzy logic controller-based PI has been examined, and the results are compared with the classical PI controller. During the process, it is found that compared to that of PI controller the FLC system has shown improved performance in improving the current and voltage harmonics present in the system. The output results with its THD value are illustrated in Fig. 3b and Fig. 4b, respectively, for PI and fuzzy PI controllers. This shows that FLC is proving to be better than PI in eliminating harmonics in the presence of hybrid filter and follows the IEEE-519 standard.

Appendix

See Table 2.

References

1. Haugan, P.T.S., Tedeschi, E.: Reactive and harmonic compensation using the conservative power theory. In: International Conference on Ecological Vehicles and Renewable Energies (EVER) (2015)
2. Mahela, O.P., Shaik, A.G.: Topological aspects of power quality improvement techniques: a comprehensive overview. Renew Sustain Energy Rev (Elsevier) **58** (2016)
3. Thirumoorthi, P., Yadaiah, N.: Design of current source hybrid power filter for harmonic current compensation. Simul. Model. Pract. Theor. (Elsevier) **52** (2015)
4. Singh, B., Dube, S.K., Arya, S.R.: An improved control algorithm of DSTATCOM for power quality Improvement. Electr. Power Energy Syst. (Elsevier) **64** (2015)
5. Lam, S., Wong, M.C., Choi, W.H., Cui, X.X., Mei, H.M., Liu, Z.J.: Design and performance of an adaptive low-DC-voltage-controlled LC-hybrid active power filter with a neutral inductor in three-phase four-wire power systems. IEEE Trans. Ind. Electr. **61** (2014)
6. Rahmani, S., Hamadi, A., Haddad, K.A., Dessaint, L.A.: A combination of shunt hybrid power filter and thyristor-controlled reactor for power quality. IEEE Trans. Ind. Electr. **61** (2014)
7. Karuppanan, P., Mohapatra, K.K.: PI and fuzzy logic controllers for shunt active power filter. ISA Trans. **51** (2012)

8. Mikkili, S., Panda, A.: PI and fuzzy logic controller based 3 phase 4 wire shunt active filters for the mitigation of current harmonics with the I_d-I_q control strategy. J. Power Electr. **11** (2011)
9. Thirumoorthi, P., Yadaiha, N.: Design of current source hybrid power filter for harmonic current compensation. Simul. Model. Pract. Theor. **52** (2015)
10. Singh, B., Bhuvaneswari, G., Arya, S.R.: Review on power quality solution technology. Asian Power Electron. J. **6** (2012)

Part III
Big Data and Recommendation Systems

A Patient-Centric Healthcare Model Based on Health Recommender Systems

Ritika Bateja, Sanjay Kumar Dubey and Ashutosh Bhatt

Abstract Health recommender system (HRS) has been quite useful in providing effective recommendations not only to the patients but also to the physicians. However, this paper focusses on the patient-centric model and our major concern will be towards the recommendations that will serve as a boon to the patients by recommending suitable treatments to them based on analysis of other patient's profile with similar attributes. Patients have to provide their contextual information such as diagnosis and symptoms which will be maintained in the form of personal health records (PHR). HRS using patient-centric healthcare model as proposed in this paper will not only help the patients in the early prediction and treatment of disease, but also help in reducing the costs of treatments. Model will also be leveraged by patients in selection of healthcare provider based on successful treatment delivered by them to patients with similar profiles and symptoms.

Keywords Health recommender system · Electronic health records
Patient health record · Electronic medical record

1 Introduction

Patient-centric health care as the name suggests revolves around the patients only. There is vast amount of data available across different sources over Internet satisfying different healthcare needs. But this data is not well organized and does not provide contextual information based on patient profile. Due to the overload of such

R. Bateja (✉) · S. K. Dubey
Amity University, Sec-125, Noida, Uttar Pradesh, India
e-mail: ritika.fet@mriu.edu.in

S. K. Dubey
e-mail: skdubey1@amity.edu

A. Bhatt
Birla Institute of Applied Sciences, Bhimtal, Uttrakhand, India
e-mail: ashutoshbhatt123@gmail.com

P. K. Sa et al. (eds.), *Recent Findings in Intelligent Computing Techniques*,
Advances in Intelligent Systems and Computing 708,
https://doi.org/10.1007/978-981-10-8636-6_28

premature information and scattered data over different sites/channels, users are lost at times or are uncertain while looking for patient care information on their own. Heterogeneous vocabulary used by medical professionals further makes it difficult for patients who need information in laymen-friendly language [1]. Personalizing and targeting the contents in the context of the patients can support them in finding relevant information.

Key need for any patient-centric health care is to recommend and connect patient to other patients with similar profile (age, disease, etc.) to discuss and understand treatment for their own health problems. This helps them to either do self-diagnosis of the problem based on symptoms or choose right healthcare provider for them. A framework like RS is well suited for this purpose. RS is originated from e-commerce domain, and HRS is the evolution of it in healthcare domain. Earlier, recommending doctors to patients using 'Decision Tree' as a prediction algorithm to predict the success of treatment is considered as one of the best methods to be used in health care [2]. RS also uses various other algorithms like linear algebra algorithms [3], Fuzzy rule-based algorithms [4] for increasing the performance. In addition to this, various approaches of RS had been discussed in the past for making effective recommendations like collaborative, content-based, hybrid and knowledge-based approach [5]. In this paper, we are going to propose a framework on how HRS can be used in patient-centric health care to deliver personalized healthcare contents to the patients using appropriate filtering techniques for recommendations.

The remainder of the paper is organized as follows: Sect. 2 describes different types of health records used by the patients, healthcare professionals, hospitals, etc. Section 3 will elaborate the different components of the proposed framework. Section 4 comprises the challenges and their solutions followed by Sect. 5 which consists of the conclusion.

2 Different Types of Health Records

Electronic health records (EHRs) and electronic medical records (EMRs)—these are the medical records in electronic forms which also enable the sharing of electronic data between different authorized healthcare providers [6]. Basically, EMRs are the records managed by the physician at his/her clinic in digital form, and EHRs are the records managed by hospitals, healthcare providers, etc.

Patient health records (PHRs)—these are the records managed and maintained by the patients themselves. They have the right to control, access and manage their own data. According to the Markle's Foundation [7], PHR is described as an electronic application used by individuals for accessing and managing their health-related information and also sharing with the other stakeholders of health care in a private, secure and confidential environment.

The four types of categories of PHR's as defined by International Organization for Standardization (ISO) in [8] are as follows:

(a) a self-contained EHR that the patient controls and maintains.
(b) EHR maintained by a third party such as web service provider.
(c) EHR maintained by health professionals and partially controlled by patients.
(d) Integrated care EHR (ICEHR) completely controlled and maintained by the patients.

3 Proposed Framework

We have proposed a framework which utilizes electronic healthcare record (EHR) and patient healthcare record (PHR) and recommends solutions using HRS. EHR is electronic records from different hospitals, clinics and health professionals. PHR is healthcare records uploaded by patients themselves. HRS is used over with EHR and PHR to deliver personalized recommendations. Usage of HRS to deliver personalized healthcare content is detailed in further sections.

(a) **Patient Profile**

In order to find the similarity between two patients, it is important to have the profile setup for different patients. Patient's profile will not only include information such as their age, sex, location, height and weight but also include demographics information, as well as the information on related symptoms and their treatments. Based on the profile, each patient will be assigned a unique id title 'PID'. This unique ID will be used to extract patient's information. Patient's profile can be created either manually by the patients by entering their structured information on their own or there can be automated generation of it. Moreover, it is the combination of structured as well as unstructured data [9].

(b) **Patient Health Repository (PHRep) and Electronic Health Repository (EHRep)**:

In order to capture the patient health record, patient will maintain and control data related to their medical history such as any disease they would have had in the past and treatment received by them. Patient will enter this data using a web interface. Data will be maintained and persisted in a data store titled 'Patient Health Repository' (PHRep). Likewise, the data stored by the health professionals on the portal is persisted and stored in 'Electronic Health Repository' (EHRep). Data storage in EHRep and PHRep is shown in Fig. 1.

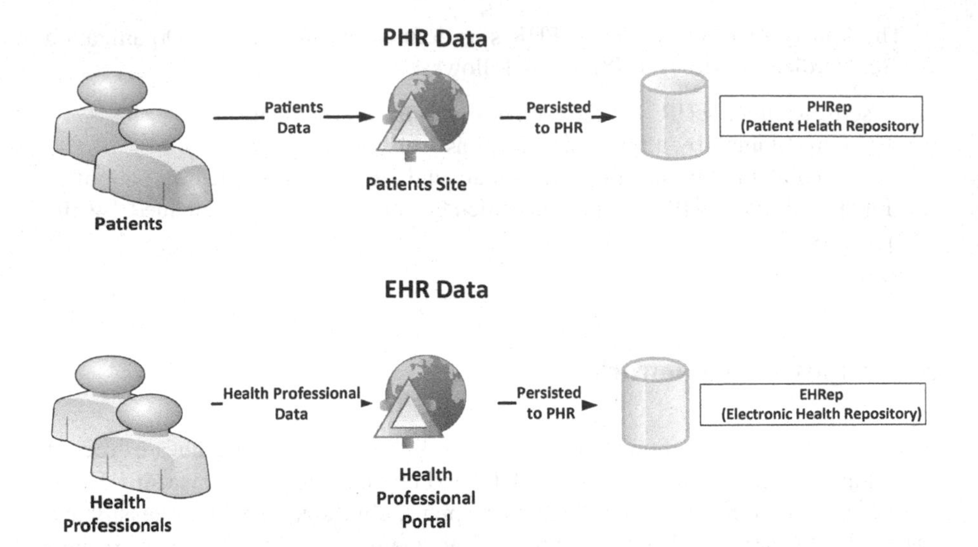

Fig. 1 Data storage in PHRep and EHRep

(c) HRS and Filtering Techniques

In order to find the similar patient profiles, relevant filtering technique/algorithm needs to be used. Considering the confidential nature of the patient's data, it is not recommended to use collaborative filtering as that generally provides recommendations by accessing the data across the profiles. Content-based filtering is better suited for this purpose as that provides recommendation by matching relevant content from a given profile with other profiles with similar attributes.

(d) Patient's Interface

User interface title 'Patients Self Care' will be used by the patients to find and get the relevant information. As explained in PHR section, each patient will be assigned with the patient ID. In order to find the relevant information related to any disease, patient will key in the following information as shown in Fig. 2, which includes as follows:

- Patient ID
- Disease for which they are looking for treatment/diagnosis
- Symptoms they are experiencing
- Search period as number of historical years for which information needs to be searched.

Inputs will be provided to HRS-based 'Recommendation Engine' to filter different profiles and deliver relevant information to the patients. Recommendation engine is detailed in coming sections.

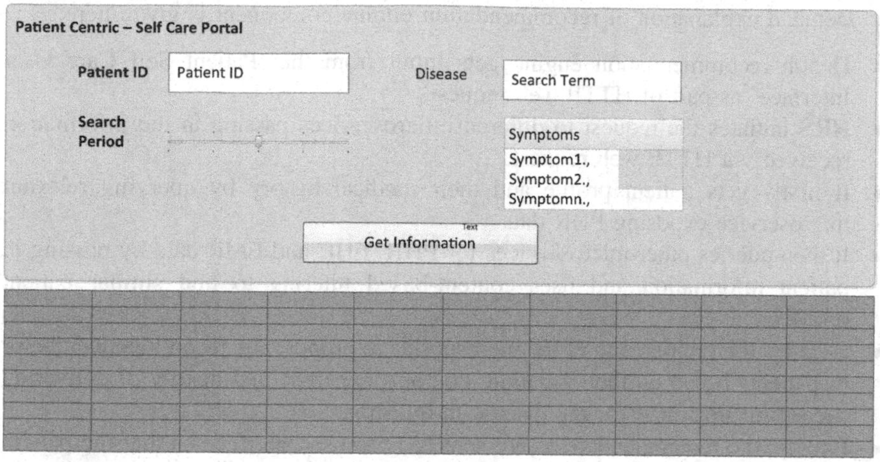

Patient Centric – Self Care Portal
Patient ID
Patient ID
Disease
Search Term
Search Period
Symptoms
Symptom1.,
Symptom2.,
Symptomn.,
Text
Get Information

Fig. 2 Patient-centric portal

(e) Patient Self-Service Data Flow

As depicted in Fig. 1, information from different data store is exposed using the microservice. There is separate underlying data source used for each microservice as per recommended architecture for developing them. In order to deliver information exposed by microservice to the patient user interface, some middleware process is required which can analyse/filter the different patient profiles to provide the recommendations. This is the responsibility of recommendation engine component which searches the recommended treatments based on patient satisfactions by viewing similar patient profiles and the corresponding outcome of treatments. Patient self-service Recommendation Engine as HRS is shown in Fig. 3.

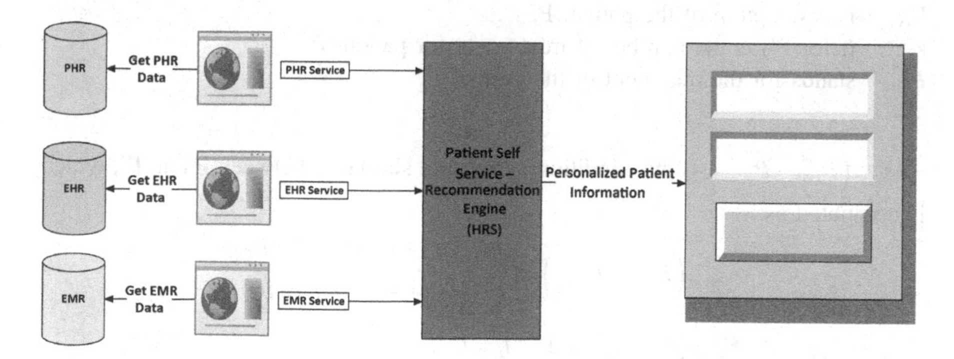

Fig. 3 Patient self-service data flow

Detailed explanation of recommendation engine component is given here:

- Health recommendation engine gets input from the 'Patient Self Care User Interface' as part of HTTP web request.
- HRS initiates the request to different microservices passing in the information received via HTTP web request.
- It firstly gets patient profile and their medical history by querying relevant microservice exposing PHR data.
- It then queries other microservices for PHR, EHR and EMR data by passing in patient information and uses content-based filtering to find similar patient profiles.
- A call is then initiated to EHR microservice to propose the recommended doctor to patients based on their location, doctor's expertise and history of delivering successful treatment for the disease in question.
- Finally, the recommendations are sent back to user interface so that the patient can view recommended doctor/healthcare provider and patients with similar attributes.

Following is high-level overview of computations used by recommendation engine to find the matching patient profile:

- It computes similarity between different patient profiles using the similarity measure proposed in [10], which computes similarity between patient's profiles based on symptoms and treatments.

$$s\left(P,P'\right) = \frac{\alpha \sum_{i=1}^{k} \sum_{j=1}^{n} s_{sym}\left(P_{sym_i} \cdot P'_{sym_j}\right)}{kn} + \frac{(1-\alpha) \sum_{i=1}^{z} \sum_{j=1}^{r} s_t\left(P_{t_i} \cdot P'_{t_j}\right)}{zr}$$

where

k (n for P′) represents the number of symptoms that patient P is having,
P_{sym} is a symptom of the patient P,
z (r for P′) is the number of treatments for patient P,
P_t stands for the treatment of the patient p

$s_{sym}\left(P_{sym_i} \cdot P'_{sym_j}\right)$ is the condition(symptom) similarity between P and P′, which is computed as:

$$s_{sym}\left(P_{sym_i} \cdot P'_{sym_j}\right) = \begin{cases} log \frac{P}{P_{sym_i}} \; if \; sym_i = sym_j \\ \frac{1}{sp\left(sym_i \cdot sym_j\right)}, otherwise \end{cases}$$

$$s_t\left(P_{t_i} \cdot P'_{t_j}\right) = \begin{cases} 1 \; if \; t_i = t_j \\ 0, otherwise \end{cases}$$

similarity score s_{sym} for patients having the same symptom sym_i is computed as logarithm of the ratio between P (total number of patients) and P_{sym_i} (number of patients affected by that symptom). If the symptoms are different, s_{sym} is computed as the reciprocal of the length of the shortest path (sp) that connects the two conditions sym_i and sym_j in the disease hierarchy. The treatment similarity, $s_t\left(P_{t_i} \cdot P'_{t_j}\right)$, is a binary score, that is 1, if the treatments are same, 0 otherwise. The effect of symptom similarity and treatment similarity on patient similarity can be controlled by the parameter α.

4 Challenges and Solutions

4.1 Challenges

- One of the major challenges faced by the healthcare industry is the lack of interoperability between different healthcare systems as they lack the unified terminology set [11].
- Privacy of patient's data is also one of the major concern [12].
- Data is coming from different sources available in different formats and is not consolidated.
- Problems related to accuracy and security of data, digital divide and literacy issues [13].

4.2 Possible Solutions

- Solution is to deliver a self-service portal which patients can use to input their current condition and know recommended treatments.
- Deliver a search tool to search data across patient's, doctor's and treatment's data store followed by consolidation of results to provide relevant information in the context of patients.

5 Conclusion

Right information to right audiences is the need of the day. In today's world where there is lot of information overload, there is major need of delivering contextual information to the people seeking information on their health. By maintaining profiles for the patient and using HRS over PHR and EHR, we can deliver personalized information to the health information seekers. This will help them to make informed decision on their health-related problems such as self-treatment, early diagnosis of diseases and reaching out to healthcare provider best suited for their needs. This will be a big milestone in personalized healthcare information delivery to people and lead to better lifestyle and prevention of diseases.

References

1. Hardey, M.: Doctor in the house: the internet as a source of lay health knowledge and the challenge to expertise. Sociol. Health Illn. **21**(6), 820–835 (1999). https://doi.org/10.1111/1467-9566.00185
2. Ahire, S.B., Khanuja, H.K.: HealthCare recommendation for personalized framework. Int. J. Comput. Appl. **110**(1), 0975–8887 (2015)
3. Sharifi, Z., Rezghi, M., Nasiri, M.: Alleviate sparsity problem using hybrid model based on spectral co-clustering and tensor factorization. In: 5th International eConference on Computer and Knowledge Engineering (ICCKE), Mashhad, pp. 89–91 (2015)
4. Cordn, O., Herrera, F., de la Montana, J., Sanchez, A., Villar, P.: A prediction system for cardio vascularity diseases using genetic fuzzy rule-based systems. In: Proceedings of the 8th Ibero–American Conference on AIm, pp. 381–391 (2002)
5. Aguilar, J., Valdiviezo-Diaz, P., Riofrio, G.: A general framework for intelligent recommender systems. Appl. Comput. Inform. (2016). http://dx.doi.org/10.1016/j.aci.2016.08.002
6. Wiesner, M., Pfeifer, D.: Health recommender systems: concepts, requirements, technical basics and challenges. Int. J. Environ. Res. Public Health **11**(3), 2580–2607 (2014)
7. Tang, P.C., Ash, J.S., Bates, D.W., Overhage, J.M., Sands, D.Z.: Personal health records: definitions, benefits, and strategies for overcoming barriers to adoption. J. Am. Med. Inform. Assoc. **13**(2), 121–126 (2006). https://doi.org/10.1197/jamia.M2025
8. Health Informatics–Electronic Health Record–Definition, Scope and Context; Standard ISO/TR 20514:2005; International Organization for Standardization: Geneva, Switzerland (2005)
9. Bissoyi, S., Mishra, B.K., Patra, M.R.: Recommender systems in a patient centric social network—a survey. In: International Conference on Signal Processing, Communication, Power and Embedded System (SCOPES) (2016)
10. Narducci, F., Musto, C., Polignano, M., Gemmis, M.D., Lops, P.G., Semeraro, G.: A recommender system for connecting patients to the right doctors in the health net social network. In: 24th International Conference on World Wide Web, WWW'15 Companion, Florence, Italy, pp. 81–82 (2015)
11. Senthilkumar, S., Tholkappia, G.: Implementation of context aware medical ontology and health recommender framework. J. Chem. Pharm. Sci. **9**(4) (2016). ISSN 0974-2115
12. Datta, A., Dave, N., Mitchell, J., Nissenbaum, H., Sharma, D.: Privacy challenges in patient-centric health information systems. In: 1st Usenix Workshop on Health Security and Privacy (2010)
13. Kyungsook, K., Nahm, E.: Benefits of and barriers to the use of personal health records (PHR) for health management among adults. Online J. Nurs. Inf. (OJNI) **16**(3) (2012). http://ojni.org/issues/?p=1995

SIoT Framework to Build Smart Garage Sensors Based Recommendation System

A. Soumya Mahalakshmi, G. S. Sharvani and S. P. Karthik

Abstract The Internet of Things landscape has garnered increasing attention in recent times, with its untapped potential and possibilities of its integration into next generation cellular systems. Social media and social networks, on the other hand, have formed a domain that has been inextricably linked to a variety of emerging technologies, giving rise to innovations that are crowd-sourced, intelligent, and evolutionary. Social Internet of Things (SIoT) is a paradigm that aims to converge the technologies of Social Networks and Internet of Things. An SIoT framework works to establish a social network for connected objects in the IoT window, so that heterogeneous devices can find, communicate, collaborate to achieve a common objective. Further, social network platforms can be integrated with existing IoT solutions to make them more dynamic. The aim of this paper is to understand the plethora of emerging techniques in the field of SIoT and use it to build a smart recommendation system using garage sensors. The objective of this paper is to propose a design module for this system that is autonomous with minimum interference from the user. The system consists of three subsystems—proximity sensors in the garage, car GPS, and social media which are heterogeneous and yet, are able to collaborate with each other to recommend places for dining and leisure to the user, based on time of arrival and history of places visited. The paper further elucidates on the logistics and market potential of such an implementation, thereby providing the possibilities of integrating SIoT solutions for mainstream applications.

Keywords Internet of Things · Social Internet of Things · Social networks
Smart recommendation systems · Proximity sensors · Global positioning system

---

A. Soumya Mahalakshmi (✉) · G. S. Sharvani · S. P. Karthik
Department of Computer Science and Engineering, R. V. College of Engineering, Bengaluru, India
e-mail: a.soumyamahalakshmi@gmail.com

P. K. Sa et al. (eds.), *Recent Findings in Intelligent Computing Techniques*, Advances in Intelligent Systems and Computing 708,
https://doi.org/10.1007/978-981-10-8636-6_29

1 Introduction

The growing popularity of Internet of Things is proving to be its own limiter for its continued growth. With the vast number of devices being integrated into the framework every minute, it becomes immensely important to devise new means by which objects can communicate with each other. The population of the world stands at 7.4 billion today, and when such a large population wishes to make friends, interact and collaborate with the rest of the world, they resort to social networks. There are several trillion objects in the world today. The idea of Social Internet of Things (SIoT) is to be able to provide a social network for connected heterogeneous objects in a manner such that they can provide collaborative, autonomous, crowd-sourced, and dynamic services to the user. SIoT strives to achieve separation between people and things, and yet provide a social network for objects, without hindering basic privacy. It is inevitable that many services in the future would require using groups of objects in their entirety by taking advantage of cohesiveness and collaboration within the group. The rise of swarm intelligence and swarm robotics prove this hypothesis. The paper titled "The Social Internet of Things and the RFID-based Robots" by Cristina Turcu et al. throws light on how robot–robot interactions were established to support unity of service amongst the robots by integrating the social network platform, Twitter. The robots send Twitter messages to each other through radio frequency identification technology, thereby introducing the concept of swarm robotics into the SIoT paradigm to produce better results [1].

Therefore, SIoT has high relevance in today's dynamic scenarios, and the path ahead will only keep growing. To align any application along with the specifications of SIoT, it becomes important to understand its structure. In the paper titled "SIoT: Giving a Social Structure to the Internet of Things," Luigi Atzori et al. have given a comprehensive framework for the SIoT architecture. The system architecture as proposed by the authors consists of three layers on the server side—the base layer, component layer, and application layer. While the base layer handles the database for storage and management of data and communications, the component layer hosts tools for component implementation. The application layer encompasses interfaces to objects, humans, and third-party services [2].

The architecture of the component layer is essentially the heart of the SIoT design. It helps to modularize data from the base layer into carefully designed components, each of which provides a unique dimension to the application's footprint in SIoT. Hence, the design of the components in the component layer will be borrowed to implement one of the layers in the design of the smart recommendation system that is proposed in this paper. With the components in place, a mechanism has to be devised for object interaction. In an SIoT design, objects interact on the basis of predefined relationships. In the paper titled "Socialite: A Flexible Framework for Social Internet of Things," Ji Eun Kim et al. have proposed a specialized framework for SIoT with a special emphasis on defining relationships between the objects, thereby classifying it into four broad classes of kinship, ownership, shared ownership, and friendship. Their model ensures seamless connection and cooperation [3]. Considering the large network of objects that may be integrated into an SIoT circuit, the paper titled "Network Navigability in the Social Internet of Things" by Michele Nitti et al. provides a solution for quicker navigability. The underlying idea of link selection for better navigability is used, so as to heuristically select a narrow set of links that can manage friendships more effectively [4]. In order to enable negotiations between objects during connections and transfers, the policy language of Ponder was considered most appropriate according to a study conducted by Lijuan He et al. and summarized in the paper titled "Design of policy language expression in SIoT" [5].

Solutions have also been devised to ensure trustworthiness management, which provides a standardization of rules to be followed by objects to process information being rendered by a variety of sources, friends, or otherwise. The system must be reliable, yet subjective as elucidated by Michele Nitti et al. in their paper titled "A Subjective Model for Trustworthiness Evaluation in the Social Internet of Things." Trust is developed based on own experience as well as the opinion of peers and fueled by a feedback system [6]. Any model being built on SIoT paradigm is based on the key idea of Social Governance, which necessitates a device to participate in setting up a collaborative protocol that outlines the policies that govern the space that it is using. Further, Muthucumaru Maheswaran et al. in their paper titled "Toward a Social Governance Framework for Internet of Things" advocate that incentives ensure the adherence of the devices to the policies formulated by the collaborative process [7].

Lianhong Ding et al. in their paper titled, "The Clustering of Internet, Internet of Things and Social Network," have elaborated on the necessity of clustering the three exclusive macro-elements of information, objects, and people, thereby clustering the Internet, the Internet of Things, and the social network. The paper sheds light on the growing importance of understanding the behaviors of objects and people as data, hence necessitating the requirement of a platform to carry out the clustering. This clustering, once achieved, can greatly simplify the SIoT framework [8]. A solution was proposed in the form of the paper titled "Taking the SIoT down from the Cloud: Integrating the Social Internet of Things in the INPUT Architecture" by I. Farris et al. which aimed at integrating SIoT over the platform of INPUT architecture, a cloud infrastructure platform. This technique greatly improved efficiency and network utilization [9]. Orfefs Voutyras et al. proposed integration of SIoT into another platform called COSMOS, in their paper titled "Social Monitoring and Social Analysis in Internet of Things Virtual Networks." The model proposed by the authors aims at enhancing services like discovery, recommendation, and sharing between things enriched with social properties. It places a special emphasis on understanding trustworthiness, reliability, monitoring, and analysis within the COSMOS platform by transforming objects into Virtual Entities [10]. Since, the COSMOS platform for SIoT provides enhanced recommendation services; it is a worthy candidate for integration into the proposed design of smart garage sensors based recommendation system. However, it has to be lightweight for the proposed application.

A large amount of effort and research has gone into making a framework for SIoT design feasible, as furnished by the above examples. The combination of the above principles makes the SIoT framework, a truly comprehensive and well-built solution that has taken every possibility into account. It remains to be seen how well the framework will be used to build revolutionary applications. Antonio J. Jara et al. in the paper titled “Social Internet of Things: The potential of the Internet of Things for defining human behaviors” have explained how the triangle formed by Big Data, Smart Cities, and Personal/Wearable Computing will be game changers in the society [11]. When integrated with SIoT, the emerging paradigm will add the ultimate dimension to human dynamics and will provide a platform for truly people-centric applications. Considering the example quoted by Tiago Marcos Alves et al. in their paper titled “Exploring the Social Internet of Things concept in a University Campus using NFC,” where SIoT is used to connect heterogeneous intelligent devices on a college campus to provide quick, customized and intelligently autonomous service to students, it becomes evident as to where the strength of SIoT lies [12]. The strength of SIoT to take autonomous decisions based on the behavior of the user rather than conscious commands is indeed incredible and is further validated by the application proposed by Asta Zelenkauskaite et al. in their paper titled “Interconnectedness of Complex Systems of Internet of Things through Social Network Analysis for Disaster Management,” where sensors worn by people alert the central server, if their vitals are low and if there geographical coordinates indicate them to be positioned at disaster hit area. It helps the rescue forces locate victims who are most in need. This system couples the strength of GPS, weather data analysis modules, blood pressure sensors, and heart rate sensors and allows them to talk to each other and collaborate to rescue the victim [13]. It has therefore been established that SIoT is a technology that will inevitably be infused in our daily lives in the not so distant future. Hence, it becomes important to embrace the change armed with the possibilities of utilizing its complete potential.

2 Design and Architecture

The objective of this paper is to delve into the details of designing and building a smart recommendation system using garage sensors. The proposed system is completely autonomous and takes decisions and makes recommendations based on observing human behavior rather than using a command-driven approach. It is built on the SIoT framework by taking advantage of the architectural nuances provided in this paradigm. At the outset, the system strives to create a collaborative and cooperative social network between three subsystems: The garage sensors, car GPS, and APIs from social media platforms. The proposed system implements the following functionality. Consider a user who arrives home from his workplace at a set time every day, offset by a few minutes. A proximity sensor fitted on his garage

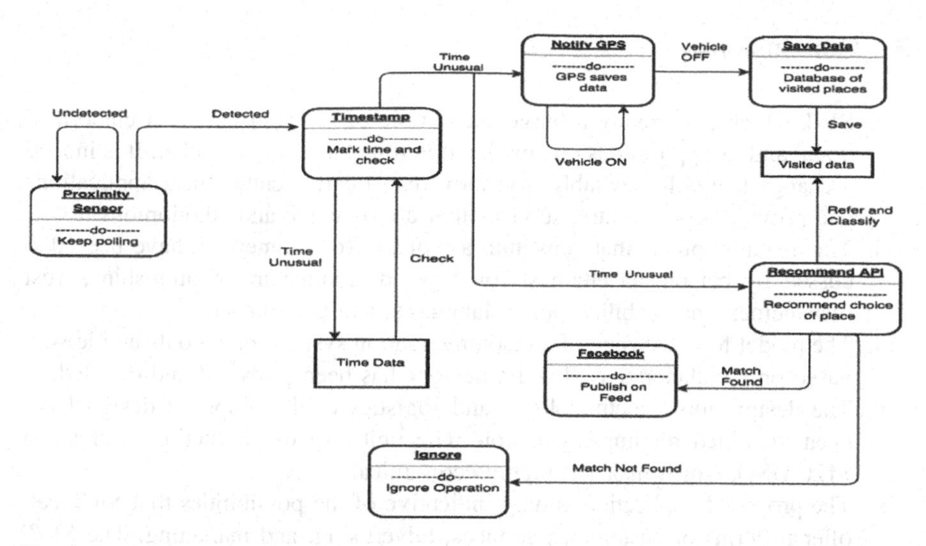

Fig. 1 Prototype implementation

doors, which will be referred to as a garage sensor, will timestamp his arrival every day and store it in a database. It is noticed that whenever his arrival time varies by a large margin than his usual arrival time in terms of coming early, he visits a place of leisure, such as dining, malls, clubs. Hence, whenever this large deviation in his arrival time is observed by the sensors from the database, the garage sensor requests the GPS module on his car to start tracking and save the user's whereabouts. Hence, the GPS provides the information about the restaurants, clubs, and malls he visits whenever he comes early and gives this data to recommendation Web sites, whose APIs can be used to determine similar places that would suit the taste of the user. The next time the garage sensors timestamp the early arrival of the user, he will get a recommendation on where to visit, on his phone, in the form of a notification through a social networking website, for example, his Facebook feed. The steps employed in the design process have been elaborated as follows.

A. System Design

The design follows a layered architecture divided into base layer, component layer, and application layer. It is to be noted that this layered architecture is aligned with the SIoT framework as suggested by Luigi Atzori et al. in their paper titled "SIoT: Giving a Social Structure to the Internet of Things" [2].

B. Prototype Implementation

The state transition model has been of the prototype has been shown in Fig. 1.

3 Conclusions

i. SIoT which is currently a burgeoning topic of research has been elaborated upon, and its applications and market potential have been detailed. It is indeed a change that will inevitably fuse into mainstream technological applications and provide people-centric services that are dynamic and autonomous.
ii. The essential pillars that constitute a typical SIoT framework have been discussed by placing an emphasis on layered architecture, relationships, trust management, navigability, policy language, and governance.
iii. The model for the design of a recommendation system for eat outs and leisure based on a collaborative SIoT framework has been proposed and detailed.
iv. The design, tools, methodology, and logistics of the proposed design have been identified for implementation. The unit cost of production is close to $120 USD, which makes it highly economical.
v. The proposed application is only indicative of the possibilities that SIoT can offer in terms of customized services, advertising, and marketing. The SIoT applications, including the proposed system, are highly specific. Yet, they have immense market potential, owing to the sheer personalization and customization that they offer.

References

1. Cristina Turcu, C.T.: The social internet of things and the RFID-based robots. In: IV International Congress on Ultra Modern Telecommunications and Control Systems, pp. 77–83. IEEE (2012)
2. Luigi Atzori, A.I.: SIoT: giving a social structure to the internet of things. IEEE Commun. Lett. **15**(11), 1193–1195 (2011)
3. Kim, J.E., Maron, A.: Socialite: a flexible framework for social internet of things. In: 16th IEEE International Conference on Mobile Data Management, pp. 94–103. IEEE (2015)
4. Michele Nitti, L.A.: Network navigability in the social internet of things. In: IEEE World Forum on Internet of Things (WF-IoT), pp. 405–410. IEEE (2014)
5. He, L., Qiu, X.: Design of policy language expression in SIoT, pp. 321–326. IEEE (2013)
6. Michele Nitti, R.G.: A subjective model for trustworthiness evaluation in the social internet of things. In: 23rd Annual IEEE International Symposium on Personal, Indoor and Mobile Radio Communications, pp. 18–23. IEEE (2012)
7. Muthucumaru Maheswaran, S.M.: Towards a social governance framework for internet of things (2015)
8. Ding, L., Shi, P.: The clustering of internet, internet of things and social network. In: 3rd International Symposium on Knowledge Acquisition and Modeling, pp. 417–420. IEEE (2010)
9. Farris, I., Girau, R.: Taking the SIoT down from the cloud: integrating the social internet of things in the INPUT architecture. IEEE (2015)
10. Orfefs Voutyras, P.B.: Social monitoring and social analysis in internet of things virtual network. In: 18th International Conference on Intelligence in Next Generation Networks, pp. 244–251. IEEE (2015)

11. Antonio, J., Jara, Y.B.: Social internet of things: the potential of the internet of things for defining human behaviours. International Conference on Intelligent Networking and Collaborative Systems, pp. 581–585. IEEE (2014)
12. Tiago Marcos Alves, C.A.: Exploring the social internet of things concept in a university campus using NFC. In: XLI Latin American Computing Conference (CLEI). IEEE (2015)
13. Asta Zelenkauskaite, N.B.: Interconnectedness of complex systems of internet of things through social network analysis for disaster management. In: Fourth International Conference on Intelligent Networking and Collaborative System, pp. 503–508. IEEE (2012)

KEA-Based Document Tagging for Project Recommendation and Analysis

M. G. Thushara, S. A. Sreeremya and S. Smitha

Abstract This paper proposes an innovative approach in managing project-related documents, project domain analysis, and recommendation of open areas from current project document pool. Using keyterm extraction technique, documents are tagged under appropriate categories and subcategories for better management of project documents. Hence, this tagged document serves as a reference for the students who are planning to take up new projects. The system generates various reports for statistical analysis of projects carried out in each research domain. These statistics benefit users to get an overview of the trends of project works done over the past few years. There are also reports illustrating the number of open areas over respective academic years. The open areas are identified and listed for the students. This novel approach would help the students who are seeking new project. Our system helps the students, faculty, and other academicians to get involved in ongoing projects and also to obtain ideas in their respective research domain. We have modified the stemming method in basic keyterm extraction algorithm (KEA) by adding Porter stemmer rather than Lovins stemming method, and our experimental results confirm that our modified keyterm extraction method outperforms the KEA method while tagging English documents.

Keywords Keyterm extraction · Open areas · Project document
Report · Stemming · Tagging · TF-IDF

M. G. Thushara (✉) · S. A. Sreeremya · S. Smitha
Department of Computer Science & Applications, Amrita School of Engineering,
Amrita Vishwa Vidyapeetham, Amrita University, Amritapuri, Kollam, India
e-mail: thusharamg@am.amrita.edu

S. A. Sreeremya
e-mail: sreeremyasa@gmail.com

S. Smitha
e-mail: smithasadasivan25193@gmail.com

P. K. Sa et al. (eds.), *Recent Findings in Intelligent Computing Techniques*,
Advances in Intelligent Systems and Computing 708,
https://doi.org/10.1007/978-981-10-8636-6_30

1 Introduction

Dissertation or project is an integral part of any higher education system, which involves preparation of various documentations and maintains these documents for further reference and analysis. Every student is required to prepare various project documents such as Software Requirement Specification, Software Test Documentation, Software Design Document, Survey Papers, Thesis Report. In each academic year, hundreds of such documents are pooled. This pooled document becomes the reference for future students and faculty guiding in various projects. For the proper use of this reference, categorization and clustering of the documents are very much relevant. In categorizing the documents, there are various algorithms. KEA algorithm [1] is the basic keyterm extraction method used for document clustering. In our proposed method, we have adopted KEA algorithm for tagging the project documents and we made a slight modification in KEA algorithm. Instead of using Lovins stemmer [2], we have used Porter stemmer. Finally, we compared the result of using these two stemming algorithms, and from our result, we arrived at a conclusion that Porter stemmer is best compared to Lovin stemmer while dealing with tagging of English documents. This document tagging helps our system to manage the academic project documents for future reference. In the paper, we also propose a new project recommendation, wherein open areas from various research domains are listed for getting new ideas unlike other recommendation systems which work based on user ratings. With the help of various statistical reports [3], an analysis study can be done on various domains.

We apply supervised learning methods for keyterm extraction from project documents. In [1], the authors suggest an algorithm, which extracts keyterms automatically from English documents by applying naive Bayesian algorithm [4]. In KEA, the authors used Lovins stemming [2] algorithm. Experimental results in [5] prove that Porter stemmer works well in English document clustering. English documents are used in [1] so it will be better using Porter stemmer rather than Lovins stemmer. Porter stemmer is used in [6] for preprocessing the text file which is used for text clustering. In [7], the most commonly used weighting scheme, TF-IDF, which can be used in fields such as natural language processing, machine learning, text mining, and search engines, is projected.

2 Methodology

We propose an ideal system helping in project coordination by tagging the project documents for easy management, generate reports for analysis purpose, and recommend new ideas through existing open areas mined from project documents. The implementation of this system starts with the extraction of keyterms from the project documents. We adopted some ideas from KEA algorithm [1] in our implementation regarding keyterm extraction technique.

2.1 Tagging the Project Documents

In most of the university, there will be certain research domains in which they concentrate. For example, consider the Department of Computer Science where most common domains of research are data mining, networking, language engineering, and bioinformatics. In each academic year, a new set of projects is carried out in these domains. After evaluation and verification, valid documents are uploaded to the project pool. In order to organize and make these documents available as a reference, document tagging is a significant procedure. This is made possible by applying a supervised learning method, that is, by identifying and comparing the key terms extracted from these documents with that of the trained set of keyterms related to each specific research domain.

2.2 Supervised Learning Method for Tagging

Project documents are tagged under the categories like UG or PG, research domain, faculty, and year. For this tagging purpose, we need associable valid keyterms for each research domain and hence the keyterm extraction technique is employed. In the supervised learning method for keyterm extraction, the two phases are used—training and testing phases.

Training Phase: A training set of almost 200 research papers is collected by Google Scholar for each research domain. These research papers are considered as training phase documents for obtaining valid keyterms. These keywords are associable keyterms for each research domain. From these documents, valid keyterms are generated. The following steps are followed in the extraction of keyterms:

- Tokenization, Stopword Removal, and Stemming: The documents obtained for training phase are tokenized. These tokens are the contents of the document that have been extracted as word by word. This list of tokens is then subjected to stopword removal process, using a standard list of stopwords obtained from KEA tool. Stopwords [1] are most common words in a language like 'the,' 'is,' 'at,' 'which,' 'on.' So after stopword removal, we obtain a new list of relevant tokens which are free from stopwords. Then, we use Porter stemming algorithm to derive the root word of these tokens. Finally, we obtain a set of possible keyterms for all research domains. The next step is to select the most valid keyterms from these set of possible keyterms for each research domain. TF-IDF weighting scheme is used for this purpose.
- TF-IDF Calculation [8]: In order to identify the most valid keyterms of a particular research domain, the list of keyterms obtained from the training set of documents is considered. To this list of keyterms, TF-IDF calculation is performed. Following is the mathematical formula for calculating TF-IDF weights for each extracted keyterm.

For [7] a key t and a document d in a document corpus D, then we get

$$TF(t) = f(t, d) \tag{1}$$

$$IDF(t) = \log\left(\frac{|D|}{f(t, D)}\right) \tag{2}$$

$$TF \times IDF = TF(t) \times IDF(t) \tag{3}$$

where d is an element of D and $f(t, d)$ is the frequency of that term compared to the total number of terms in the document. Term Frequency (TF) is used to find the highest weighed term that occurs in a document, and Inverse Document Frequency (IDF) gives a measure of the frequency of a term across all documents present in a global corpus. After calculating TF × IDF values for the list of keyterms related to a particular research domain, a threshold value is set. This threshold value is the average of all TF × IDF values obtained for the above list of keyterms under that particular research domain. We choose only those keyterms which have the TF IDF value less than this threshold value. This means, if the value is less than the threshold, then they are the frequently occurring and associable terms to a particular research domain. These chosen keyterms are then stored in a text file under corresponding research domain for easy and hassle-free access. These keyterms are used as training set keyterms for tagging purposes, and they are considered as 'valid' training set keyterms.

Table 1 shows the result after TF × IDF calculation using Porter stemming method. It shows the keyterms and their corresponding TF, IDF, and TF × IDF values. These keyterms are obtained after the training phase for data mining research domain, where some research papers from Google Scholar related to data mining are collected. Then, TF × IDF value is found by applying the keyterm extraction steps. After setting the threshold value by taking the average of TF IDF values of all the terms in these research papers, we eliminate those keyterms having TF × IDF value greater than the threshold value. These eliminated terms are rarely occurring terms which are not suitable for tagging purpose. We choose only those terms having TF × IDF value less than the threshold value as they are the most often occurring terms and are suitable for tagging purpose.

Using (1), (2), and (3), TF, IDF, and TF × IDF values are calculated, respectively. For instance, here, *mine* is a term occurring in all the selected research papers and hence it has TF-IDF value as 0.000. Then, we calculate the threshold value and here it is 0.02. TF × IDF value of *mine* is less than 0.02 so it is one of the most frequently occurring and valid keyterms under data mining research domain. Similarly, *classifi*, *regress*, *cluster*, *naive*, *predict*, *kdd*, *confid*, etc., are valid keyterms for data mining research domain. The above procedure is done individually for all other research domains in order to generate 'valid' training set keyterms associated with the domain. This helps in tagging the project documents under appropriate research domains.

Table 1 TF-IDF values of keyterms using Porter stemming method

Project documents	Keyword	TF	IDF	TF × IDF
Project doc 1	Mine	0.0170	0.0000	0.0000
	Classif	0.0120	0.0969	0.0009
	Regress	0.0022	0.3010	0.0006
	Cluster	0.0011	0.3010	0.0003
	Naïve	0.0011	1.0000	0.0011
	Factor	0.0056	0.6989	0.0039
Project doc 2	Mine	0.0170	0.0000	0.0000
	Classif	0.0102	0.0969	0.0009
	Associ	0.0022	0.3010	0.0006
	Summari	0.0004	1.0000	0.0004
Project doc 3	Extract	0.0201	0.3010	0.0060
	Predict	0.0057	0.2218	0.0012
	Mine	0.0129	0.0000	0.0000
	Kdd	0.0129	0.2218	0.0028
Project doc 4	Mine	0.0112	0.0000	0.0000
	Classif	0.0117	0.0969	0.0011
	Associ	0.0099	0.3010	0.0298
	C45	0.0067	0.5228	0.0035
	Confid	0.0063	0.3979	0.0025

Testing Phase: In this phase, project documents that need to be tagged are taken. The tagging is possible only after the keyterm extraction and comparison processes on these documents that are to be tagged. That is, in the training phase, we apply the steps like tokenization, stopword removal, stemming, and TF × IDF to these valid documents. After TF × IDF calculation for all the tokens obtained after stopword removal and stemming in this testing phase, we take the average of these TF × IDF values as the threshold value. In training phase, tokens having TF × IDF value less than the threshold value are chosen as 'valid' keyterms. These valid keyterms are suitable and decisive in tagging the project under appropriate research domain.

2.3 Comparison and Tagging

- Document Tagging: Finally, we compare the keyterms obtained for a project document in the testing phase with the trained set of keyterms under each research domain. The project document is tagged to that particular research domain where the largest number of keyterms matches occurs. Hence, the new project document is tagged under corresponding research domains. A data source is maintained for storing the tagged project documents.

- Faculty Tagging: After this tokenization and stopword removal steps, the name of the faculty is also extracted from project documents for further tagging of these documents. When a new project document arrives, the corresponding faculty who guided that project will be given a score under the respective research domain that the project belongs to. This could help in the later statistical analysis of faculty's expertness by this system.

This tagging of project documents occurs automatically when new project documents are uploaded by the students. Based on the tagging, a tree structure of the entire tagged document file system is generated.

2.4 Recommendation of New Project Proposals

- Project Recommendation: It deals with the extraction of the title of project document and the paragraph under the heading 'Conclusion and Future Scope' in it. The 'Conclusion and Future Scope' paragraph helps the user to get the overall idea conveyed in the project document as well as helps them to obtain the open areas contained in this particular document. The approved project documents at the university follow a specific template where there is a section named 'Conclusion and Future Scope.' When approved project documents get uploaded and tagged, the 'Conclusion and Future Scope' paragraph of these tagged documents is extracted. Furthermore, we maintain separate data stores for storing these extracted 'Conclusion and Future Scope' paragraphs with respect to its respective research domain.
- Listing the Open Areas: Our system provides options like 'year' and 'research domain' for students or academicians to obtain project open areas. Also once they select their interested research domain, there is another option asking to input the keyterm for more specific search of open areas. Based on their options and input, the open areas are listed based on the 'Conclusion and Future Scope' of the project document. If they are interested in any of the open areas listed, they are given accessibility to download the original project document where the particular open area exists. The document tagging based on keyterm extraction plays a pivotal role in implementing this 'project recommendation' for new project proposals.

2.5 Report Generation

For statistical analysis [9], reports are generated based on the projects that have been carried out in the past years and those that are ongoing during the current academic year. Also, there are reports demonstrating the contribution of the faculty under various research domains. These reports help the user to get the statistical trends of projects and faculty expertise under each research domain, for instance, over a period of time in the past. In addition, there is report demonstrating the number of open areas pertaining to each research domain in respective years. We used JFreeChart [10] and DynamicReports [11] for generating reports and charts. These reports and charts are generated dynamically based on a data source. This data source is updated automatically during the tagging phase. Thus, document tagging based on keyterm extraction is vital in report generation process.

3 Result and Analysis

The system we proposed would help the students and other faculties in efficient project proposal and management. It helps them by suggesting innovative ideas in the form of open areas being listed out. For better analysis, reports are generated providing statistical knowledge about the projects already carried out in past years, which are ongoing in the current year, projects faculty's expertness, and report on number of open areas in respective years under each research domain. Our experiment is based on focusing major research domains under Computer Science Department. They are data mining, networking, compilers, and bioinformatics (Table 2).

Table 3 shows the result of document tagging based on keyterm extraction using Lovins stemmer.

Data set consists of project documents which are in the form of template recommended by the university. For our experiment purpose, we have taken the documents of past three years. The result of our project document tagging based on keyterm extraction technique is given in Table 2. From a total of 180 documents, 164 documents are correctly tagged under their respective categories, whereas the remaining documents are falsely categorized.

For calculating the accuracy [12] of our document tagging technique, we consider all four research areas:

$$Accuracy\ Rate = \frac{(TP + TN)}{(P + N)} \tag{4}$$

Table 2 Result of document tagging under various research areas based on keyterm extraction using Porter stemmer

Research area	Number of correctly tagged documents	Total number of documents
Data mining	61	65
Networking	56	60
Compilers	30	35
Bioinformatics	17	20

Table 3 Result of document tagging under various research areas based on keyterm extraction using Lovins stemmer

Research area	Number of correctly tagged documents	Total number of documents
Data mining	59	65
Networking	55	60
Compilers	27	35
Bioinformatics	14	20

In (4), let True Positive (TP) be the number of documents correctly categorized under a particular research area and True Negative (TN) be the number of documents correctly categorized under other research areas other than TP. (P + N) is the total number of documents. Hence, while calculating the accuracy rate of our Porter stemmer-based document tagging technique, it scores 91.11%, whereas Lovins stemmer-based document tagging scores only 86.11%.

Figure 1 illustrates columnar report and pie chart, which demonstrates the number as well as the percentage of project works done under each research domain for the last three consecutive years by PG students in the university. By analyzing this report, the user can infer the trends in project works carried out under various research domains. For instance, here, this report shows that in a period of past three

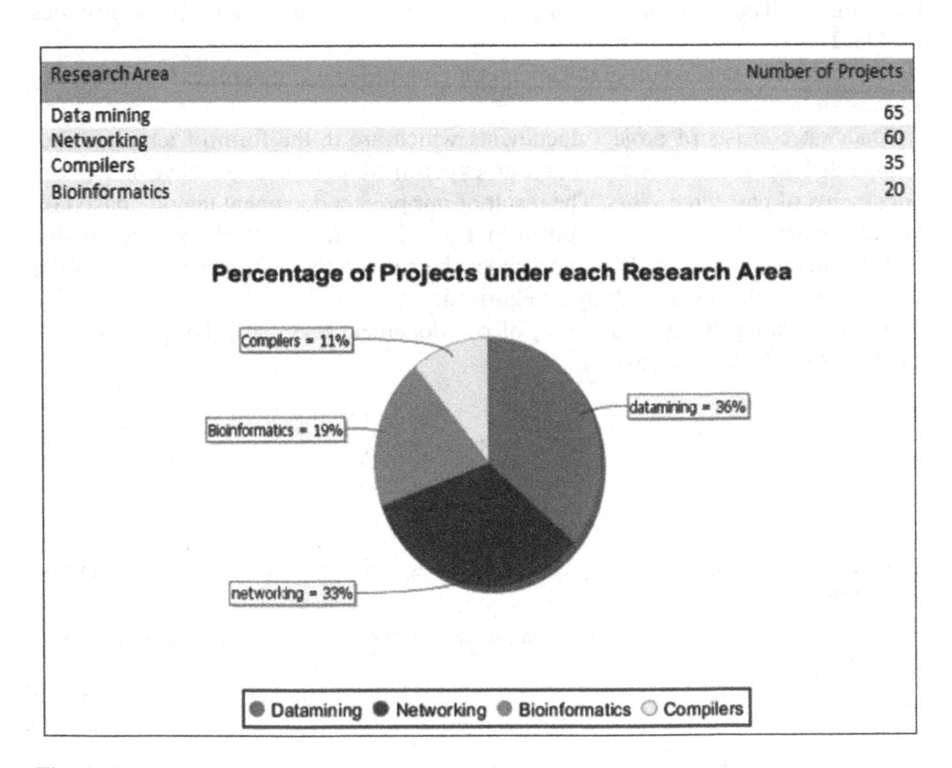

Fig. 1 Report of projects done in three consecutive years

consecutive years, maximum numbers of projects are done under data mining (65 projects).

Figure 2 illustrates the ongoing project works in the current academic year. It represents the inclination of students under various research domains in the current year.

Figure 3 illustrates the bar chart showing faculty involved in projects under various research domains such as data mining for the last three consecutive years. This helps students to choose the right faculty for guidance, according to the 'faculty expertise' for the specific research domain.

Figure 4 illustrates the display of the open areas extracted from project document pool for a selected year, based on their opted research domain and keyword input. This would help a student with their decision making while choosing a new project. Moreover, for further elaborate reference, they can download the source paper that finds their interest.

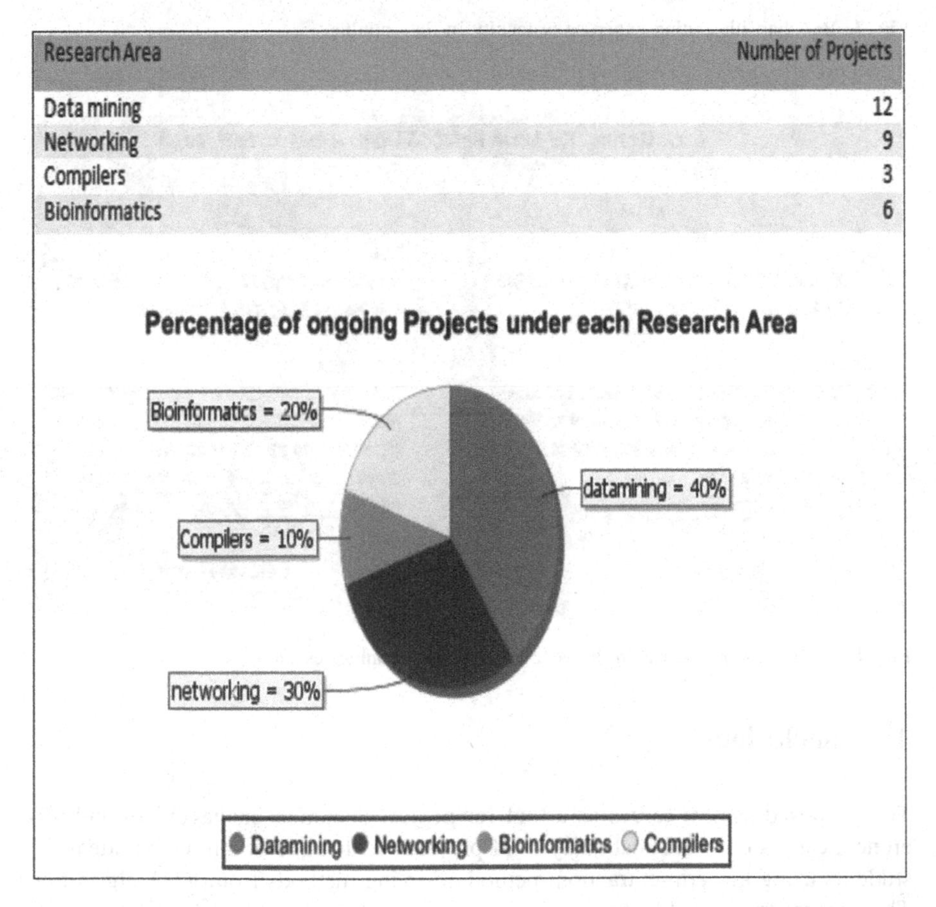

Research Area	Number of Projects
Data mining	12
Networking	9
Compilers	3
Bioinformatics	6

Fig. 2 Report of ongoing projects in the current academic year

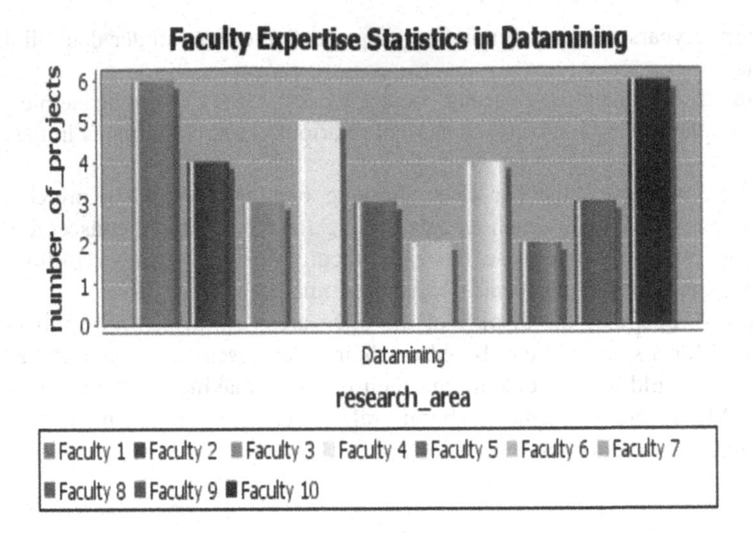

Fig. 3 Bar chart illustrating expertise of faculty in data mining

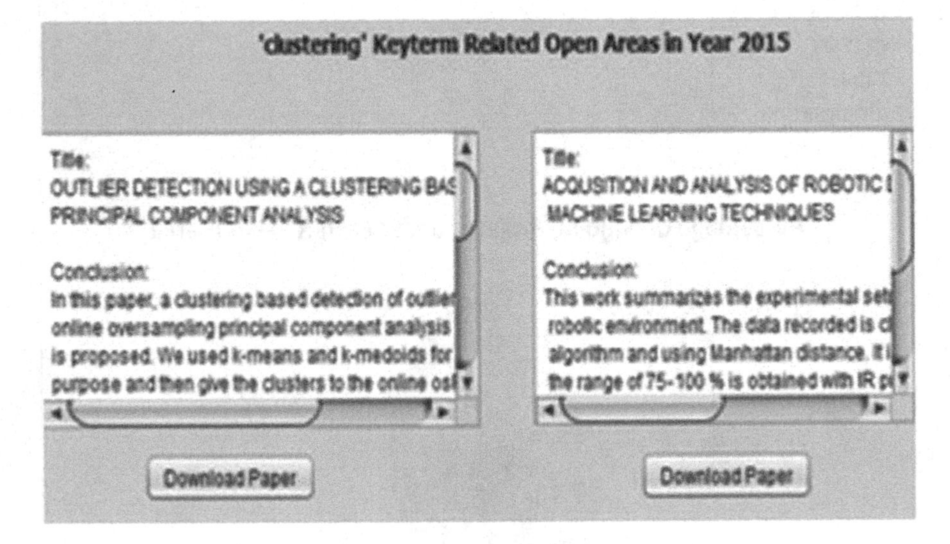

Fig. 4 Listing the open areas in the year 2015 of data mining research

4 Conclusion

The proposed system serves as a tool for project document management and reference, carries out statistical analysis of projects, and suggests innovative ideas for students using keyterm extraction method adopting methods from KEA algorithm. This paper provides different statistical reports which demonstrate the trends of projects done under each research domain over the years in the past. Also, the report illustrates the ongoing projects in the current academic year. Moreover, charts represent the expertness of the faculty under each research domain which helps the students to choose a proper guide in their domain of interest. It provides suggestion of new project topics and ideas for the students and academicians who are seeking for one.

We compared document tagging with Porter stemming keyterm extraction technique and Lovins stemmer keyterm extraction. The accuracy results prove that Porter stemmer is better than Lovins stemmer while dealing with English document tagging. Thus, our system proves to be an effective aid to all those who seek to learn the trends of projects so far done and to get novel ideas for new project proposals.

Acknowledgements We would like to express our sincere gratitude to the faculty of Department of Computer Science and Applications of Amrita Vishwa Vidyapeetham, Amritapuri, for providing help and guidance.

Our sincere thanks to Dr. M. R. Kaimal, Chairman, Computer Science Department, Amrita Vishwa Vidyapeetham, Amritapuri, for his prompt support.

References

1. Witten, I.H., Paynter, G.W., Frank, E., Gutwin, C., Nevill-Manning, C.G.: KEA: practical automatic keyphrase extraction. In Proceedings of the Fourth ACM Conference on Digital Libraries, pp. 254–255. ACM (1999)
2. Han, P., Shen, S., Wang, D., Liu, Y.: The influence of word normalization in English document clustering. In 2012 IEEE International Conference on Computer Science and Automation Engineering (CSAE), vol. 2, pp. 116–120. IEEE (2012)
3. Menon, R.R.K., Kini, N.V.: Harnessing the discriminatory strength of dictionaries. IEEE (2016)
4. Sathyadevan, S., Athira, U.: Improved document classification through enhanced naive Bayes algorithm. IEEE (2014)
5. Jivani, A.G.: A comparative study of stemming algorithms. Int. J. Comp. Tech. Appl. **2**(6), 1930–1938 (2011)
6. Rose, J.D.: An efficient association rule based hierarchical algorithm for text clustering. Int. J. Adv. Eng. Tech/Vol. VII/Issue I/Jan.-March **751**, 753 (2016)
7. Patil, L.H., Atique, M.: A novel approach for feature selection method TF-IDF in document clustering. In: 2013 IEEE 3rd International Advance Computing Conference (IACC). IEEE (2013)
8. Pillai, P.G., Narayanan, J.: Question categorization using SVM based on different term weighting methods. Int. J. Comput. Sci. Eng. (IJCSE) **4**(05) (2012)
9. Mii, M., Lazi, M., Proti, J.: A software tool that helps teachers in handling, processing and understanding the results of massive exams. In: Proceedings of the Fifth Balkan Conference in Informatics. ACM (2012)
10. JFreeChart project (2012). http://www.jfree.org/
11. Dynamics Reports. http://www.dynamicreports.org/
12. Han, J., Kamber, M., Pei, J.: Data Mining Concepts and Techniques, 3rd edn (2011)

Toward Scalable Anonymization for Privacy-Preserving Big Data Publishing

Brijesh B. Mehta and Udai Pratap Rao

Abstract Big data is collected and processed using different sources and tools, which leads to privacy issues. Privacy-preserving data publishing techniques such as *k*-anonymity, *l*-diversity, *t*-closeness are used to de-identify data, but chances of re-identification are there as data is collected from multiple sources. Due to a large amount of data, less generalization or suppression is required to achieve same level of privacy, which is also known as "large crowd effect," but to handle such a large data for anonymization is also a challenging task. MapReduce handles a large amount of data, but it distributes data into small chunks, so the advantage of large data cannot be achieved. Therefore, scalability of privacy-preserving techniques has become a challenging area of research, and we are trying to explore it by proposing an algorithm for scalable *k*-anonymity for MapReduce. Based on comparison with existing algorithm, our approach shows significant improvement in running time.

Keywords Big data · Big data privacy · *k*-anonymity

1 Introduction

Scalability can be defined as the ability of a system to handle growing amount of work without degrading its performance. We are also trying to develop such a scalable *k*-anonymity approach for MapReduce. To understand scalable *k*-anonymity for MapReduce, first, we need to understand big data, and MapReduce (as it is generally used to process the big data). After that, the pros and cons of MapReduce with respect to existing privacy preserving techniques. At last, *k*-anonymity, the most popular data anonymization technique, is explored.

B. B. Mehta (✉) · U. P. Rao
Sardar Vallabhbhai National Institute of Technology, Surat, India
e-mail: brijeshbmehta@acm.org

U. P. Rao
e-mail: upr@coed.svnit.ac.in

P. K. Sa et al. (eds.), *Recent Findings in Intelligent Computing Techniques*,
Advances in Intelligent Systems and Computing 708,
https://doi.org/10.1007/978-981-10-8636-6_31

Big data is a collection of data sets, large and complex, that becomes difficult to process using traditional data processing applications. The term big data analytics is used for the processing of big data. Big data is having characteristics of 3Vs, Volume (a large amount of data), Variety (structured, unstructured, or semi-structured form of data), and Velocity (fast generation and processing of data, viz. real-time or stream data).

Big data analytics collects and processes a large amount of data from different sources, which leads to privacy issues. Therefore, privacy-preserving big data publishing is required to address such issues. In general, privacy is defined as the ability of an individual or group to seclude themselves, or information about themselves, and thereby express themselves selectively. User is going to decide that which data he wants to share with others and which he wants to keep secret. Mehta and Rao [15] have discussed some of the privacy issues with respect to big data such as privacy issues in big mobile data, privacy issues in social media data, privacy issues in healthcare data.

MapReduce [3, 10], a distributed programming framework, is used to process big data. MapReduce takes an advantage of distributed computing in processing a large amount of data, and at the same time, it creates challenges for anonymization. As data is distributed among small chunks, it is difficult to take advantage of whole data in anonymization process. Therefore, we are proposing an approach which can take advantage of both large size of data and distributed computing. Mainly, there are three steps in map reduce: mapping, shuffling, and reducing.

A hospital wants to publish patient's records. They will remove the column having patient name and believe that patient's privacy will be preserved. But Samarati and Sweeney [20, 21] have shown that if anyone is having a background knowledge, such as voter list details, about any individual present in dataset, then that person can be identified.

Samarati and Sweeney have proposed a privacy model in year 2002 to prevent such attacks, which is known as k-anonymity. They have divided all the attributes of a dataset into four categories: personal identity information (PII), quasi identifier (QID), sensitive attributes (SA), and non-sensitive attributes (NSA). k-Anonymity checks that if one record in the dataset has some value of QID, then at least $k - 1$ other records also have same QID values. In other words, at least k records in the dataset must have the same QID value. Such a table is known as k-anonymous. Therefore, probability of identifying an individual in such dataset becomes 1/k. A large amount of data requires less generalization or suppression to achieve same level of privacy, which is also known as "large crowd effect." User get more cover to hide information; hence, less data perturbation is required. But whole data should be considered for the anonymization process which is a challenging task in distributed programming framework such as MapReduce.

Discussion of existing privacy-preserving big data publishing approaches for MapReduce, such as two-phase top-down specialization for data anonymization using MapReduce, Hybrid approach for scalable sub-tree anonymization, and MapReduce-based algorithm, has been given in Sect. 2. Section 3 discusses about our proposed approach. Implementation details and results are discussed in Sect. 4. At last, conclusion and future scope have been given.

2 Related Work

The privacy-preserving data publishing has started with the work of Samarati and Sweeney on k-anonymization [20, 21]. After that, many researchers had came up with different anonymization techniques for privacy-preserving data publishing such as l-diversity [13, 14], t-closeness [12], (c, t)-isolation [1, 2] differential privacy [4–6], multi-relational k-anonymity [17–19]. A detailed survey of these techniques has been given in a paper from Fung et al. [7]. Each technique is having an assumption with respect to attacker's knowledge and work against specific types of attacks. The implicit assumption which is not mentioned in any of the privacy-preserving technique is with respect to size of data, which should be small or moderate to achieve desirable results. Mehta et al. [16] have given a comparison of existing privacy-preserving techniques with 3Vs of big data and found that k-anonymity can be a good candidate to address all 3Vs of big data.

There is relatively very less work has carried out in the direction of scalability of privacy-preserving big data publishing techniques:

Zhang et al. [25] have extended the privacy preserving technique, "Top-Down Specialization (TDS)" proposed by Fung et al. [8] for MapReduce framework as two-phase top-down specialization (TPTDS). Zhang et al. further found that TDS approach is having high running time for smaller value of k, whereas bottom-up generalization (BUG) approach proposed by Wang et al. [22] is having a high running time for higher value of k. Hence, Zhang et al. proposed a hybrid approach for scalable sub-tree anonymization [24] in which based on value of k, anonymization technique is selected. Main drawback of these approaches is the distribution of data. Here, we cannot take advantage of large crowd effect of large data instead it becomes distributed data anonymization. We further found that sub-tree anonymization approaches cannot be used for velocity of data as it requires all domain value of attribute apriori to generate taxonomy tree.

Zakerzadeh et al. [23] have proposed a novel approach for MapReduce-based anonymization by implementing Mondrian [11] algorithm for MapReduce framework. Mondrian is a multidimensional k-anonymization algorithm. Therefore, in Mondrian algorithm, equivalence classes are generated using all the attributes of dataset.

The major drawbacks of this technique are multiple iterations and file management. As the number of iteration increases, the performance of the system decreases and file management using MapReduce is difficult task. Therefore, we propose a new approach to overcome these drawbacks.

3 Proposed Approach

As discussed in previous chapter, existing privacy-preserving techniques for MapReduce have some drawbacks. To overcome these drawbacks, we are proposing a novel approach for scalable k-anonymity for MapReduce.

We are taking the weighted sum of all the attribute of a record to generate initial equivalence class. In next step, we are checking for end of all class by comparing *class–id* with total number of class n. Then, we are checking whether the number of records in each equivalence class is fulfilling the condition of k-anonymity or not. If the number of records in any equivalence class is less than k, then we are merging equivalence classes to generate a bigger class which fulfills k-anonymity criteria. These steps are repeated until all the classes are processed, and then, information loss is calculated in terms of normalized cardinality penalty (NCP) [9].

Now we are going to discuss each algorithm in detail. Starting with Algorithm 1, mapper for scalable k-anonymity:

- In step 2, we are obtaining a data file from distributed file system and it will be automatically distributed among different mappers in small chunks. So our mapper is going to process those chunks individually.
- In step 3, we have initialized an array variable name *output* to store the output value.
- In step 4–8, for each *tuple*, we are calculating the average value and storing that value in new attribute named *avg*; then we are merging the value of *avg* and *tuple* and give it to the *output* which will update the input dataset for next phase. Now we are emitting the value of *output* and 1 so that combiner can read this value as (*key*, *value*) pair.

Algorithm 1 Mapper for Scalable k-anonymity

```
1: procedure SKMAPPER(k, v)
2:     //obtain data file from distributed file system
3:     output = empty //an array of size data
4:     for each tuple in data file do
5:         find average value and store in avg
6:         append pair(avg, tuple) to output
7:         emit(output, 1)
8:     end for
9: end procedure
```

In Algorithm 2, combiner for scalable k-anonymity, we are generating initial equivalence class by merging the similar tuples based on *key* received from mapper and adding the *value* to get the number of tuples in each equivalence class. At last, we are emitting the pair of *key* and updated *value* in $emit(k, v)$.

Algorithm 2 Combiner for Scalable k-anonymity

1: **procedure** SKCOMBINER(k, v)
2: **for** i in [1, ..., (length of $v - 1$)] **do**
3: **if** $k[i] == k[i+1]$ **then**
4: *add*($v[i], v[i+1]$)
5: **end if**
6: **end for**
7: *emit*(k, v)
8: **end procedure**

In Algorithm 3, reducer for k-anonymity:

- In step 2, we are obtaining the anonymization value K from user to check k-anonymity condition for each equivalence class.
- In step 3, we are initializing an array to store intermediate equivalence class data.
- In step 4, we are taking a variable *sum* as the counter of number *tuples* in each equivalence class. It is also used to check whether the number of *tuples* in each class is fulfilling the k-anonymity condition or not.
- In step 5, we are initializing an integer variable named *eqId* to assign identifier to each generated equivalence class.
- In step 6–16, we are processing each equivalence class generated by combiner and checking for the number of *tuples* in each class. If a number of *tuples* are less than K value, we are merging next class in current equivalence class and again check for number of *tuples* for merged class. If number of *tuples* are greater than or equal to K, then we are considering it as desired equivalence class and merging *eqId* to it and finding NCP value for that class. At the same time, we are resetting the value of *sum* and *eqClassData* and also updating the value of *eqId*.

4 Implementation Details and Result Discussion

We have used a Hadoop cluster of seven nodes, each having 4 GB of RAM and 1 TB of hard disk. We have implemented our approach using two programming languages Apache Pig and Python. As Pig is a data flow scripting language for Hadoop, it is very easy to write data flow statement and it generates mapper and reducers automatically for corresponding data flow statements. We found difficult to calculate NCP using Pig, so we used Python UDF to calculate NCP value from equivalence classes generated from Pig. Therefore, we are processing data to generate equivalence class in Pig and then calculating the value of NCP using python UDF.

We have used NCP to measure information loss because it is the widely used standard and gives result in percentage of information loss after applying anonymization. The formula to calculate NCP of class and global NCP has been given by Ghinita et al. [9].

Algorithm 3 Reducer for k-anonymity

1: **procedure** SKREDUCER(k, v)
2: //Obtain anonymization value K
3: $eqClassData = empty$ //an array of size data
4: $sum = 0$
5: $eqId = 1$
6: **for** i in [1, .., (length of v)] **do**
7: append $pair(k[i], v[i])$ to $eqClassData$
8: $sum = sum + v[i]$
9: **if** $sum >= K$ **then**
10: append $pair(eqId, eqClassData)$
11: $NCPofClass = findNCPofClass(eqClassData)$
12: $emit(NCPofClass)$
13: $eqClassData = empty$
14: $sum = 0$
15: $eqId = eqId + 1$
16: **end if**
17: **end for**
18: **end procedure**

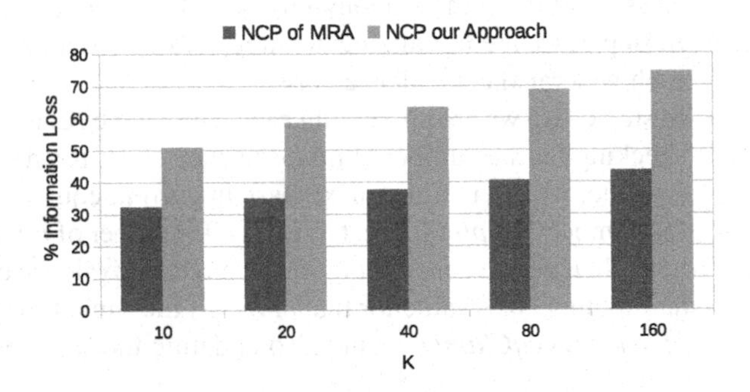

Fig. 1 NCP comparison of MapReduce anonymization [23] and our proposed approach

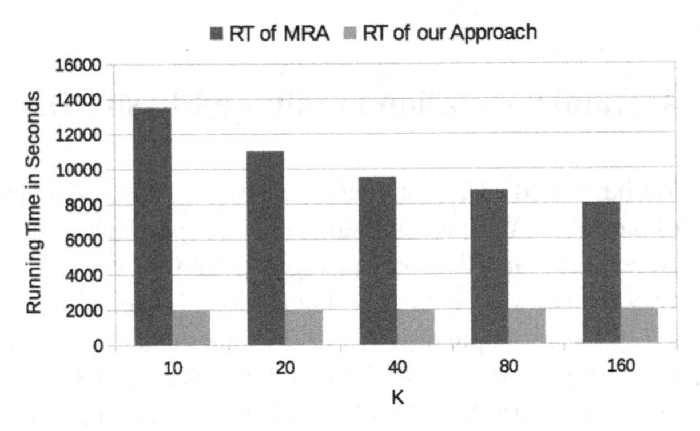

Fig. 2 Running time comparison of MapReduce anonymization [23] and our proposed approach

As shown in Figs. 1 and 2, we have compared results of our proposed algorithm with MapReduce anonymization algorithm. From the results, it is clear that our proposed algorithm is having a relatively more information loss as compared to MapReduce anonymization, but we are getting significant improvement in running time.

As in privacy-preserving techniques, we have to maintain a trade-off between privacy and data utility, and our approach is having an issues with data utility. Based on these experiments, we have found that if we can decrease the number of records in initial equivalence class generation, then it is possible to get low information loss. We have to find some classification mechanism by which we can generate small initial equivalence class.

5 Conclusion and Future Scope

We have proposed a novel approach for scalable k-anonymity for MapReduce and compared it with existing MapReduce-based anonymization approach. Though we have not got desirable results in terms of data utility, we have got significant improvement in running time and a way to explore the area of scalable anonymization.

In future, we are planning to find out a classification method to achieve small initial equivalence class so that information loss in our approach can be reduced. We further like to extend our approach for velocity as well as variety of data.

References

1. Chawla, S., Dwork, C., McSherry, F., Smith, A., Wee, H.: Toward privacy in public databases. In: Proceedings of the Second International Conference on Theory of Cryptography, TCC'05, pp. 363–385. Springer Berlin Heidelberg, Cambridge, MA, USA, Feb 2005
2. Chawla, S., Dwork, C., McSherry, F., Talwar, K.: On privacy-preserving histograms. In: Proceedings of the Twenty-First Conference on Uncertainty in Artificial Intelligence (UAI2005), pp. 120–127. Association for Uncertainty in Artificial Intelligence Press, Edinburgh, Scotland, July 2005. http://research.microsoft.com/apps/pubs/default.aspx?id=64359
3. Dean, J., Ghemawat, S.: MapReduce: simplified data processing on large clusters. Commun. ACM **51**(1), 107–113 (2008)
4. Dwork, C.: Differential privacy. In: Bugliesi, M., Preneel, B., Sassone, V., Wegener, I. (eds.) Automata, Languages and Programming. Lecture Notes in Computer Science, vol. 4052, pp. 1–12. Springer, Berlin Heidelberg (2006)
5. Dwork, C.: Ask a better question, get a better answer a new approach to private data analysis. In: Schwentick, T., Suciu, D. (eds.) Database Theory ICDT 2007. Lecture Notes in Computer Science, vol. 4353, pp. 18–27. Springer, Berlin Heidelberg (2007)
6. Dwork, C.: Differential privacy: a survey of results. In: Agrawal, M., Du, D., Duan, Z., Li, A. (eds.) Theory and Applications of Models of Computation. Lecture Notes in Computer Science, vol. 4978, pp. 1–19. Springer, Berlin Heidelberg (2008)
7. Fung, B.C.M., Wang, K., Chen, R., Yu, P.S.: Privacy preserving data publishing: a survey of recent developments. ACM Comput. Surv. **42**(4), 14:1–14:53 (2010)
8. Fung, B.C.M., Wang, K., Yu, P.S.: Anonymizing classification data for privacy preservation. IEEE Trans. Knowl. Data Eng. **19**(5), 711–725 (2007)
9. Ghinita, G., Karras, P., Kalnis, P., Mamoulis, N.: Fast data anonymization with low information loss. In: Proceedings of the 33rd International Conference on Very Large Data Bases, VLDB '07, pp. 758–769. VLDB Endowment, Vienna, Austria Sep 2007. http://dl.acm.org/citation.cfm?id=1325851.1325938

10. Lämmel, R.: Google's mapreduce programming model–revisited. Sci. Comput. Program. **70**(1), 1–30 (2008)
11. LeFevre, K., DeWitt, D.J., Ramakrishnan, R.: Mondrian multidimensional k-anonymity. In: Proceedings of the 22nd International Conference on Data Engineering, ICDE '06, pp. 1–11. IEEE Computer Society, Washington, DC, USA, Apr 2006
12. Li, N., Li, T., Venkatasubramanian, S.: t-closeness: privacy beyond k-anonymity and l-diversity. In: Proceedings of the 23rd International Conference on Data Engineering, pp. 106–115. IEEE, Istanbul, Turkey, Apr 2007
13. Machanavajjhala, A., Gehrke, J., Kifer, D., Venkitasubramaniam, M.: L-diversity: privacy beyond k-anonymity. In: Proceedings of the 22nd International Conference on Data Engineering, pp. 13–24, Atlanta, GA, USA, Apr 2006
14. Machanavajjhala, A., Kifer, D., Gehrke, J., Venkitasubramaniam, M.: L-diversity: privacy beyond k-anonymity. ACM Trans. Knowl. Discov. Data **1**(1), 1–52 (2007)
15. Mehta, B.B., Rao, U.P.: Privacy preserving unstructured big data analytics: issues and challenges. Procedia Comput. Sci. **78**, 120–124 (2016). In: 1st International Conference on Information Security and Privacy 2015
16. Mehta, B.B., Rao, U.P., Kumar, N., Gadekula, S.K.: Towards privacy preserving big data analytics. In: Proceedings of the 2016 Sixth International Conference on Advanced Computing and Communication Technologies, ACCT-2016, pp. 28–35. Research Publishing, Rohtak, India, Sept 2016
17. Nergiz, M.E., Clifton, C., Nergiz, A.E.: Multirelational k-anonymity. IEEE Trans. Knowl. Data Eng. **21**(8), 1104–1117 (2009)
18. Nergiz, M., Clifton, C., Nergiz, A.: Multirelational k-anonymity. In: Proceedings of IEEE 23rd International Conference on Data Engineering, pp. 1417–1421, Istanbul, Turkey, Apr 2007
19. Nergiz, M.E., Atzori, M., Clifton, C.: Hiding the presence of individuals from shared databases. In: Proceedings of the 2007 ACM SIGMOD International Conference on Management of Data, SIGMOD '07, pp. 665–676. ACM, New York, NY, USA, Jun 2007
20. Samarati, P., Sweeney, L.: Generalizing data to provide anonymity when disclosing information. In: Proceedings of the Seventeenth ACM SIGACT-SIGMOD-SIGART Symposium on Principles of Database Systems, PODS '98, pp. 188–188. ACM, New York, NY, USA, June 1998
21. Samarati, P., Sweeney, L.: Protecting privacy when disclosing information: k-anonymity and its enforcement through generalization and suppression. Technical report, SRI International (1998). http://www.csl.sri.com/papers/sritr-98-04/. Accessed 18 Feb 2015
22. Wang, K., Yu, P.S., Chakraborty, S.: Bottom-up generalization: a data mining solution to privacy protection. In: Proceedings of the Fourth IEEE International Conference on Data Mining, ICDM'04, pp. 249–256. IEEE, Brighton, UK, Nov 2004
23. Zakerzadeh, H., Aggarwal, C.C., Barker, K.: Privacy-preserving big data publishing. In: Proceedings of the 27th International Conference on Scientific and Statistical Database Management, SSDBM '15, pp. 26:1–26:11. ACM, New York, NY, USA, June 2015
24. Zhang, X., Liu, C., Nepal, S., Yang, C., Dou, W., Chen, J.: A hybrid approach for scalable sub-tree anonymization over big data using mapreduce on cloud. J. Comput. Syst. Sci. **80**(5), 1008–1020 (2014)
25. Zhang, X., Yang, L.T., Liu, C., Chen, J.: A scalable two-phase top-down specialization approach for data anonymization using mapreduce on cloud. IEEE Trans. Parallel Distrib. Syst. **25**(2), 363–373 (2014)

VDBSCAN Clustering with Map-Reduce Technique

Ashish Sharma and Dhara Upadhyay

Abstract Clustering techniques are used for the partition of the data points in clusters. In DBSCAN clustering algorithm, it deals with dense data points, but DBSCAN algorithm does not deal with varied density data. So, for variable density, VDBSCAN algorithm is suitable, since the existing VDBSCAN algorithm is unable to find the exact radius. The existing algorithm VDBSCAN is based on distance. Due to more distance and large data sets, some data points cannot become the part of any cluster. To overcome this problem, the map-reduce technique is used. Using map reduce, the values of k can be identified correctly. It provides a proper value of k on the basis of frequency. This new approach is relatively more effective than VDBSCAN.

Keywords DBSCAN · VDBSCAN · Map-reduce

1 Introduction

Today every business acquires a huge amount of volumes of data from different sources. The data is used for getting the right information and analysis. This analysis supports for effective decision-making. The main requirement of data analytics is scalability, simply due to the large volume of data that need to be extracted, processed, and analyzed in a time-bound manner. To do the right analysis on this data for the scalability, fault-tolerance, ease of programming, and flexibility point of view, the Map Reduce technique is more suitable because it can process massive amounts of unstructured data in parallel across a distributed environment [1]. The distribution of the data is done prior to analysis. It will also improve the

A. Sharma (✉)
Computer Engineering and Applications, GLA University, Mathura, Uttar Pradesh, India
e-mail: ashishs.sharma@gla.ac.in

D. Upadhyay
Arya Institute of Engineering & Technology, Jaipur, India
e-mail: dhara.upadhyay017@gmail.com

P. K. Sa et al. (eds.), *Recent Findings in Intelligent Computing Techniques*,
Advances in Intelligent Systems and Computing 708,
https://doi.org/10.1007/978-981-10-8636-6_32

performance of the Map Reduce technique. Here, clustering techniques can be used to perform such tasks.

Clustering techniques are used for the partition of the input data sets in clusters. In clustering, data set are classified into groups. In clustering, several types of techniques are used like—partitioning, hierarchical, and density based. Density-based clustering methods are very useful in dense data. For the dense data, we use density-based algorithm such as Density-Based Spatial Clustering of Application with Noise [2] algorithm (DBSCAN). In DBSCAN algorithm, density is estimated on the basis of two parameters, i.e., radius (Eps) and threshold value (Minpts). This particular approach classifies the points into core point, border point, and noisy point. Points within the radius are considered as a core point and the points on the border of the cluster are termed as border points.

However, DBSCAN algorithm does not deal with the varied density. So, for varied density, we use Varied Density-Based Spatial Clustering of Application with Noise (VDBSCAN) [3] algorithm. VDBSCAN is used basically for varied density data sets' analysis. The basic approach of VDBSCAN is based on DBSCAN algorithm. VDBSCAN algorithm is used to find the value of k and according to that value of k clusters are formed. In VDBSCAN algorithm, two parameters—Epsilon (Eps) and Minimum Points (minpts)—are used. For Eps, we are supposed to find the k-distance of each point. That value of k-dist is regarded as an Eps and in accordance with those values of Eps and Minpts clusters are synthesized. Complication arises with VDBSCAN algorithm when it is applied in case of extremely large distance data points, since the value of k is determined on the basis of distance among data points, on account of which data points are separated to each other with an extremely large distance are misinterpreted as noisy points. Measurement of distance plays a significant role because if we have identical objects, which are quite distant to each other, will fall in different clusters. This particular limitation of VDBSCAN algorithm can be resolved by employing map-reduce technique.

Map-reduce technique involves two parameters—key and value. In map-reduce technique, key and value are used for frequency calculations. So, the frequency of those specific data points or keys can be obtained by employing map-reduce technique. In proposed approach, map-reduce technique with VDBSCAN is implemented. The above-mentioned technique is used for finding the frequency and then the value of k is calculated with the help of frequency and distance.

This paper contains several sections: Sect. 2 embodies related work of DBSCAN and VDBSCAN; Sect. 3 reveals proposed approach; Sect. 4 encompasses experiment; Sect. 5 contains the analysis, and Sect. 6 comprises conclusion.

2 Related Work

In the paper, authors Yugandhar and Vinod Babu [4] discuss the new algorithm EVDBSCAN algorithm. In this paper, author introduces a new approach Extended Varied Density-Based Spatial Clustering of Applications with Noise (EVDBSCAN) to find the value of k automatically according to the data sets. In this, the value of k is dependent on the behavior of data set. Problem with this approach is that, in this method according to data set the value of k is change and if new points entered in data sets, it should have again found the value of k and whole process is repeated, which is time consuming. In this paper, authors Wang et al. [5] proposed a new method Optimize VDBSCAN (OVDBSCAN). In OVDBSCAN, the value of K is calculated by examining the k-distance of all objects in data sets. In this method, the same problem arising as previous method, also the value of k is dependent on the data set. So, if the data set is changed, whole process is repeated which consumes time. Author has implemented map-reduce technique on large size of data sets. In this, map-reduce technique is implemented with k-means for massive data and for parallel processing. The efficiency of k-means algorithm is improved, but the limitations of k-mean are not resolved by this technique. In this paper [6], Dai, B., and Lin, I., have given a new algorithm, which is a mixture of Map/Reduce and DBSCAN algorithm called DBSCAN-MR to solve the scalability problem. In DBSCAN-MR, the input dataset is partitioned into smaller parts and then on the Hadoop platform these smaller parts of data are parallel processed. In this method, partition is done with reduce boundary points (PRBP), to select partition boundaries based on the distribution of data points. Using this method the data is divided into number of clusters which makes the approach more efficient and scalable. This method is not suitable for data which is varied in density. He and Tan [7] shared that the data clustering is important for growing data on daily basis. However, it is challenging due to the size of datasets. Every day the size of data set grows, this rapidly growth of data makes an extra-large scale in the real world. To analyze better data the Map Reduce is a desirable parallel programming platform. In this paper, author proposed a four stages Map Reduce algorithm with parallel density-based clustering algorithm. Author have given a quick partitioning strategy for large scale non-indexed data. They have merged the bordering partitions and make optimizations on it. In this paper author Weizhong et al. [8] also proposed K-means algorithm with map-reduce technique for parallel processing. This process increased speedup, scale-up and size-up of the process. This shows that the proposed approach can process large amount of data size. But still K-means algorithm contains some drawback which is not removed by this technique. It is difficult to predict the value of k. In this paper author extended the K-means algorithm with map-reduce technique for parallel processing. The value of k is justified using the technique.

Borah, B., Bhattacharyya, D.K., have contributed in [9] and [10] to make DBSCAN more robust. The DBSCAN requires large volume of memory support because it operates on the entire database. Author has improved sampling-based DBSCAN which can cluster large-scale spatial databases effectively.

Rakshit and Sikka [11] have given a new approach of DBSCAN to provide it a facility to identify to detect the correct clusters, if there is density variation within the clusters. Their approach is based on oscillation of clusters which is obtained by applying basic DBSCAN algorithm to conflation it in a new cluster. Ram et al. [12] has contributed DBSCAN in terms of density variation. As DBSCAN is not able to handle the local density variation that exists within the cluster. Author has proposed an Enhanced DBSCAN algorithm which keeps track of local density variation within the cluster. Beckmann et al. [13] have given a new data structure the R-tree, it is based on the heuristic optimization of the area. It efficiently supports points and spatial data.

In the paper [14], Sander, J., Ester, M., Kriegel H.P., Xu X., have generalized the concept of DBSCAN [2] in two important directions. Their algorithm can cluster point objects and can separate the objects for spatial and their non-spatial attributes. This generalization makes it GDBSCAN. The real life application area of DBSCAN is given by Sharma et al. [15]. Their approach is evolved using density in the work presented in this paper wherein two stage solutions is proposed for Uncapacitated Facility Location Problem.

Density-Based Spatial Clustering of Application with Noise (DBSCAN) algorithm which is a form of clustering technique used for dense data. So, for varied density another approach Varied Density-Based Spatial Clustering of Application with Noise (VDBSCAN) algorithm is applied.

The objective of our proposed work is to create the cluster for any type of data points. It covers all types of data point whether they are closer points or varied density data points. The proposed approach gives far more accurate result as compared to the existing algorithm.

3 Proposed Work

Clustering techniques are utilized for the partition of the data points in clusters. Density-based clustering methods are very useful in dense data. DBSCAN is one of the algorithms which are applied to density-based data, but DBSCAN algorithm does not operate with varied density data. However, most of the data set is present in the form of the very dense area. DBSCAN algorithm works only with the points that are close to each other, but it is not effective for varied density data points.

The VDBSCAN algorithm approach works well in case of varied density data points, and subsequently, the clusters are formed on the grounds of distance between each data point. So, if the data point is available at a very large distance, then it rates that point as a noisy point. VDBSCAN algorithm is implemented to formulate the value of k and according to that value of k clusters are created. Since the value of k is determined on the basis of distance among data points on account of which data points are separated to each other with an extremely large distance which are misinterpreted as noisy points, so when it is applied in case of extremely large distance data points, it fails to deliver the expected results. Map-reduce technique has been innovated to get rid of this type of predicament. The concept of frequency forms the foundation of map-reduce technique. It evaluates the most frequent values of k which are far more accurate and reliable when compared to the simple values of k.

In new improvised approach, noisy data points are also taken into account and then the values of k are obtained on the basis of two parameters—distance and frequency. The new determined value of k is more optimized and accurate. Map-reduce technique provides us the value of k and that particular value is entertained as Eps. When we implement map-reduce with VDBSCAN algorithm, it finds the value of k, on the basis of distance frequency between each data points.

The distance function is used for finding the distance between points. Generally, the Euclidean distance function is applied in determining the distance between the data points. After determining the distance between each data point, the clusters of the data points are created with the help of Eps and those core point, border point, and noisy point are determined.

3.1 Map-Reduce

Map-reduce technique involves two parameters—key and value. In map-reduce technique, key and value are used and using those parameters the frequency is evaluated. So, the frequency of those specific data points or keys can be obtained by employing map-reduce technique.

$$\text{Map: } (\text{k1, v1}) \rightarrow (\text{K2, v2}) \tag{1}$$

$$\text{Reduce: } (\text{K2, v2}) \rightarrow (\text{K3, v3}) \tag{2}$$

Map-reduce technique conducts parallel processing on the large data sets. However, DBSCAN with map-reduce technique carries out fast and parallel processing on the large amount of dense data sets.

In the proposed approach, map-reduce technique with VDBSCAN is implemented. VDBSCAN with map-reduce technique is used for finding the frequency and then the value of k is calculated using frequency and distance (Fig. 1).

	Input: D: dataset
Step 1	Finding the value of k-dist plot; 1) Initialize all objects in D as unprocessed: Compute k: Creating cluster with core point and noisy points; 2) Finding the k on the basis of two parameters for the noisy points- distance and frequency; Compute: Finding the average distance between each point; $Pi = Points(1,2,....n)$ $d(Pi) = \sum_{i=1}^{n} \frac{distance(P_i, X_i)}{n-1}$ d(Pi)=Average distance between each point Find the value of d(Pi) for each points; Finding the frequency of each distance value from each core point; Count (key, value)
Step 2	Consider frequent value as a value of k; k=Eps; Minpts=Threshold value; Adopt VDBSCAN algorithm for clustering;
	Display data point in cluster and noisy point as an outlier

Fig. 1 Algorithm

4 Implementation and Experimental Result

4.1 Experimental Data

In this section, the performance of the proposed approach is analyzed. Two-dimensional data set has been selected in order to get experimental results. First, the system assimilates the data from different sources to conduct VDBSCAN with map-reduce technique. The proposed approach tested on different sizes of data sets. In our proposed method, ASP.NET (2012) is used and coordinates with our data set. We are working on the data set of spaeth. Spaeth is a data set directory, which contains data for clusters. The data files described and made available on Web page are distributed under the GNULGPL license. In this spaeth, a FORTRAN90 library contains the sorts of data. In this collection, data set of 2D points is available [16]. In the proposed approach, we are using four types of data sets with different size. In the spaeth, dataset is the collection of hundred pieces of information about each case of US Supreme Court decided by the Court between 1946 and 2013 terms the proposed approach is analysis. Two-dimensional data set has been selected in order to get experimental results. In this approach, various size of the data set is used.

5 Analytical Analysis

Clusters evaluated with "internal" as well as "external" measures. External measures consider whether the current cluster contains correct data points or not. For the external measures, purity, entropy, F-measures are evaluated, and internal measures consider the inter- and intra-cluster distance of data points.

External measure: In external measure, we find the purity, entropy, and F-measure of cluster. Purity is one of the basic methods to validate the quality of the cluster. Greater value of purity indicates good clustering. In addition, the disorder in quality is found by entropy method. Therefore, for the quality of each cluster, the entropy should be lower. F-measure, combine the precision and recall idea. The higher value of F-measure indicates better clustering.

Internal Measure: When a cluster result, evaluated, based on the data points were cluster itself, this is called internal measure. So, in inter-cluster similarity, we find the similarity between the clusters, and in intra-cluster similarity, we find the similarity within a cluster.

Experiment 1: In experiment 1, measure the purity of clusters, which is known purity measure (Tables 1, 2, 3, 4).

Table 1 Purity measure of 110 size data points of spaeth data file

Number of clusters	Purity of VDBSCAN	Purity of VDBSCAN with map-reduce
Cluster 1	0.66	0.75
Cluster 2	0.53	0.62
Cluster 3	0.50	0.83

Table 2 Purity measure of 200-size data points of spaeth file

Number of clusters	Purity of VDBSCAN	Purity of VDBSCAN with map-reduce
Cluster 1	0.69	0.71
Cluster 2	0.53	0.68
Cluster 3	0.50	0.88

Table 3 Purity measure of 280-size data points of spaeth file

Number of cluster	Purity of VDBSCAN	Purity of VDBSCAN with map-reduce
Cluster 1	0.46	0.64
Cluster 2	0.78	0.84
Cluster 3	0.50	0.60
Cluster 4	0.62	0.80

Table 4 Purity measure of 370-size data points

Number of cluster	Purity of VDBSCAN	Purity of VDBSCAN with map-reduce
Cluster 1	0.56	0.64
Cluster 2	0.78	0.88
Cluster 3	0.40	0.60
Cluster 4	0.57	0.80

$$Purity(\cap, C) = \frac{1}{N} \sum_{k} {}_{J}^{Max} \|W_k \cap C_j\|$$

Experiment 2: It shows the entropy of each size of data sets (Table 5).

$$H(X) = -\sum_{i=0}^{n-1} P(x_i) * \log_2(P(x_i))$$

The entropy of VDBSCAN with map-reduce is low than VDBSCAN algorithm (Tables 6, 7, 8).

Experiment 3: In this experiment, Table 9 presents the various sizes of data sets and represents precision and recall of each algorithm.

Table 5 Entropy measure of 110 size data points of spaeth file

Number of clusters	Entropy of VDBSCAN	Entropy of VDBSCAN with map-reduce
Cluster 1	1.16	0.74
Cluster 2	1.27	0.95
Cluster 3	1.50	0.64

Table 6 Entropy measure of 200 size data points of spaeth file

Number of cluster	Entropy of VDBSCAN	Entropy of VDBSCAN with map-reduce
Cluster 1	0.94	0.85
Cluster 1	1.22	0.89
Cluster 3	1.50	0.58

Table 7 Entropy measure of 280 size data points of spaeth file

Number of cluster	Entropy of VDBSCAN	Entropy of VDBSCAN with map-reduce
Cluster 1	1.42	0.92
Cluster 2	0.83	0.69
Cluster 3	1.50	0.95
Cluster 4	0.95	0.71

Table 8 Purity measure of 370 size data points of spaeth file

Number of cluster	Entropy of VDBSCAN	Entropy of VDBSCAN with map-reduce
Cluster 1	1.29	0.92
Cluster 2	0.87	0.51
Cluster 3	1.50	0.95
Cluster 4	1.12	0.71

Table 9 Comparison between VDBSCAN with map-reduce versus VDBSCAN based on precision percentage

Data points	VDBSCAN with map-reduce (%)	VDBSCAN (%)
110	89.0	77.2
200	88.5	78.5
250	87.5	76.5
370	88.1	77.0

6 Conclusion

The experiment authenticates that VDBSCAN with Map-Reduce technique is highly effective in obtaining the clusters for variable density. VDBSCAN with Map-Reduce technique algorithm enables us to successfully overcome one of the primary shortcomings of the traditional VDBSCAN algorithm. The traditional VDBSCAN algorithm selects the value of Eps on the basis of value of k, which is determined with the help of distance. VDBSCAN algorithm is used to find the value of k and according to that value of k clusters are formed. Complication arises with VDBSCAN algorithm when it is applied in case of extremely large distance data. So, if the data point is available, at a very large distance, then it rates that point as a noisy point, which is not appropriate. The experiments were done on new approach and proved that new approach overcomes the limitation of VDBSCAN in very efficient manner.

The experiments were done using two-dimensional data sets. In future work, this algorithm may be evaluated on high-dimensional data sets.

References

1. Doulkeridis, C., Nørvåg, V.: A survey of large-scale analytical query processing in map reduce. VLDB J. **23**(3), 355–380 (2014)
2. Ester, M., Kriegel, H.P., Sander, J., Xu, X.: A density-based algorithm for discovering clusters in large spatial database with noise. In: Proceedings of the 2nd International Conference on Knowledge Discovery and Data Mining, Portland, OR, pp. 226–231 (1996)
3. Peng, L., Dong, Z., Naijun, W.: VDBSCAN: varied density based spatial clustering of applications with noise. In: Proceedings of the International Conference on Service Systems and Service Management (ICSSSM'07), Chengdu, China, pp. 528–531, June 2007
4. Yugandhar, B., Vinod Babu, P.: Extended varied DBSCAN on multiple data sources. Int. J. Eng. Res. Technol. **1**(6) (2012). ESRSA Publications
5. Wang, W., Shuang, Z., Qingqing, X.: Optimum VDBSCAN (O-VDBSCAN) for identifying downtown areas. Int. J. Digit. Inf. Wirel. Commun. (IJDIWC) **3**(3), 271–276 (2013)
6. Dai, B., Lin, I.: Efficient map/reduce based DBSCAN algorithm with optimized data partition. In: International Conference on Cloud Computing. IEEE (2012)
7. He, Y., Tan, H.: MR-Dbscan: an efficient parallel density-based clustering algorithm using map-reduce. In: 17th International Conference on Parallel and Distributed Systems. IEEE (2011)
8. Weizhong, Z., Ma, H., He, Q.: Parallel k-means clustering based on Map Reduce. In: IEEE International Conference on Cloud Computing. Springer, Berlin, Heidelberg (2009)
9. Borah, B., Bhattacharyya, D.K.: An improved sampling-based DBSCAN for large spatial database. In: Proceedings of International Conference on Intelligent Sensing and Information Processing. IEEE (2004)
10. Borah, B., Bhattacharyya, D.K.: A clustering technique using density difference. In: International Conference on Signal Processing, Communications and Networking (ICSCN). IEEE (2007)
11. Rakshit, A., Sikka, G.: An optimized approach for density based spatial clustering application with noise. In: ICT and Critical Infrastructure: Proceedings of the 48th Annual Convention of Computer Society of India. Springer International Publishing (2014)
12. Ram, A., Sharma, A., Jalal, A. S., Agrawal, A., Singh, R.: An enhanced density based spatial clustering of applications with noise. In: International Advance Computing Conference (IACC). IEEE International, pp. 1475–1478 (2009)

13. Beckmann, N., Kriegel, H.P., Schneider, R., Seeger, B.: The R*-tree: an efficient and robust access method for points and rectangles. In: Proceedings of the ACM SIGMOD, International Conference on Management of Data, Atlantic City, NJ, pp. 322–331 (1990)
14. Sander, J., Ester, M., Kriegel, H.P., Xu, X.: Density-based clustering in spatial database: the algorithm GDBSCAN and its applications. Data Min. Knowl. Discov. Int. J. **2**(2), 169–194 (1998). Kluwer Academic Publishers
15. Sharma, A., Jalal, A. S. Krishnakant.: A Density based model for facility location problem. In: Annual IEEE India Conference (INDICON), Pune (2014)
16. https://people.sc.fsu.edu/~jburkardt/datasets/spaeth/spaeth.html

SprIntMap: A System for Visualizing Interpolating Surface Using Apache Spark

Chandan Misra , Sourangshu Bhattacharya and Soumya K. Ghosh

Abstract High-resolution gridded meteorological dataset of various environment variables, e.g., max–min land surface temperature, precipitation, humidity etc., are not easily available for most of the Earth's surface. This imposes a bottleneck on the research ideas that require the gridded dataset for experimentation and testing. Such data are often expensive and thus cannot be used for academia and individual researchers. In this paper, we present SprIntMap, a Web service which allows users to fetch and process climatic datasets with different spatial interpolation methods and visualize and extract the resulting interpolated gridded data on-the-fly. The system facilitates access to NOAA's a century-old archive which is one of the comprehensive sensor data archives in the world. Yet, it is challenging to obtain these large and noisy data and apply interpolation methods on it. SprIntMap consists of Google Map-based front end for easy visualization of sensor data as well as gridded surface and employs an interpolation package written in Apache Spark, providing fast parallel implementations of the interpolation algorithms for efficient and easy extraction of the gridded data in standard format.

Keywords Spatial interpolation · Apache Spark · Visualization

1 Introduction

Numerical weather prediction is the branch of science which requires numerical techniques like interpolation and aggregation. They act on climatic dataset containing thousands of points with different attributes (e.g., maximum and minimum land

---

C. Misra (✉) · S. Bhattacharya · S. K. Ghosh
Department of Computer Science and Engineering, IIT Kharagpur, Kharagpur 721302, West Bengal, India
e-mail: chandan.misra1@gmail.com

S. Bhattacharya
e-mail: sourangshu@cse.iitkgp.ernet.in

S. K. Ghosh
e-mail: skg@iitkgp.ac.in

P. K. Sa et al. (eds.), *Recent Findings in Intelligent Computing Techniques*, Advances in Intelligent Systems and Computing 708,
https://doi.org/10.1007/978-981-10-8636-6_33

surface temperatures, precipitation, humidity) at very high spatial and temporal resolution [1] to analyze the trend in climate changes. However, the climatic model creation is hindered as the number of weather stations and the observations from them are not enough. As a way out, very high-resolution gridded meteorological surface is developed to fulfill the need of various ecological and climatic applications. Gridded meteorological datasets are aggregated using irregularly placed sensors or weather stations readings and with the help of various interpolation methods like Inverse Distance Weighting (IDW) [14], Kriging [4], Regression [8], and Thin Plate Spline [7]. These gridded datasets are then archived online [9, 10] and with the help of data processing and delivery system applied scientists can have on-demand access to these datasets.

However, it is not very easy to obtain these archive data for two reasons—(1) the data are expensive and (2) the data are created mostly on country level. Therefore, one has to buy datasets from different sources to make climatic experiments in entire earth scale. Additionally, researchers often try to compare results of different interpolation methods on large dataset to come up with a best gridded surface. But, creating such huge gridded dataset from available large scattered sensor data poses several problems like—(1) there is no single service by which the user can experiment with various interpolation algorithms and visualize it with minimal effort, (2) the size of input data is very large, i.e., data intensive, (3) the interpolation algorithms that act on the dataset are computationally intensive, (4) the output data, i.e., the interpolated surface is very large when the region of interest is very large (sometimes the entire Earth surface), (5) the degree of resolution is very fine, and (6) it is not very easy to get the data of any particular area from the whole dataset and also eliminate the noisy ones to make it ready for any further processing.

A recently developed system, Shahed [6] allows users to query and visualize satellite gridded data using a user-friendly interface. It employs SpatialHadoop [5] to index and run spatiotemporal queries on the dataset. Also, it employs two-dimensional interpolation technique to recover missing data and uses a novel spatiotemporal indexing algorithm. Shahed displays gridded surface while SprIntMap acts on scattered data and provides different surface visualization for different interpolation methods. FetchClimate [15] is another Web service which is developed by Microsoft to access mean values of various climatic attributes for selected geographic regions of the Earth surface and for a selected time duration. Unlike these systems, SprIntMap reveals the interpolation methods to the user providing them to access and compare different interpolation algorithms.

Li et al. in [11] proposed a scientific workflow framework for big geoscience data analytics using cloud computing, MapReduce, and service-oriented architecture. Caradonna et al. in [3] proposed an approach to customize and integrate two open-source platforms: WebGIS and MapServer for publishing and querying on spatial data. Li et. al. in [12] presented an online data visual analytics system which overcome the data format, management and storage-related issues, client and server response time. The data preloaded into the system and analytics are done for several temporal resolution. However, these systems do not include the processing of spatial datasets using any interpolation algorithm.

In this paper, we present SprIntMap, a Web service that addresses the above mentioned shortcomings in an efficient way by storing and visualizing the scattered sensor data from the NOAA archive [16]. SprIntMap allows users to create interpolated surfaces data on-the-fly using distributed interpolation algorithms based on Apache Spark [2] and visualizes the surface on Google Map. It consists of three major components: Model-View-Controller (MVC) module, that takes user requests and contacts Hadoop cluster for storage and processing of data, *SprInt* package built on Apache Spark where all the parallel interpolation methods are implemented, and Hadoop cluster for storage and processing of parallel algorithms.

2 System Architecture

Figure 1 depicts the overall architecture of SprIntMap. It is built on the MVC architecture in order to make it portable to any other desktop or mobile-based systems. The system is divided into four modules. They are: (1) user interface (UI) or the front end, (2) controller or the HTTP request handler, (3) model or utility classes that opens, establishes, and closes the SSH and FTP connections, and (4) SprInt library or package containing Java classes of the implementations of the interpolation algorithms.

2.1 NOAA Data and Preprocessing

We have downloaded the large dataset from the global hourly integrated surface database (ISD). The data contain surface reading of approximately 20,000 stations

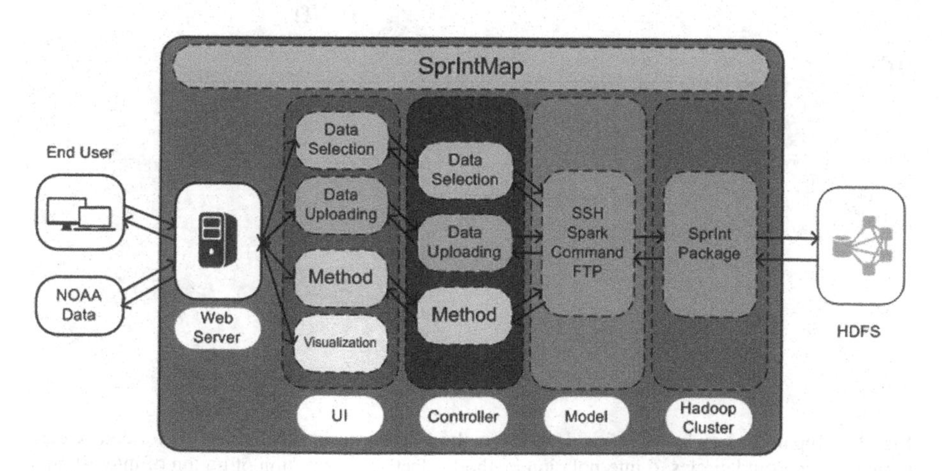

Fig. 1 SprIntMap architecture

around the globe from as early as 1901 and are updated in hourly basis with the latest data, and have gone through numerous quality control phases [13]. The data are arranged in its FTP directory from 1901 to last day of 2017. Each such link provides *gzipped* files indexed in STATION ID-YEAR manner. From the files, we have extracted all the hourly data for each station, and then made an average over all the readings for each date. We have also eliminated the erroneous data from the file. There are two types of erroneous data—(1) missing values and (2) quality code of the data. If the temperature reading is 9999, then the data are missing. The ISD data consist of 23 types of quality code among which codes 0, 1, 4, 5, and 9 are safe to collect. After cleaning, the data have been stored inside the HDFS.

2.2 User Interface

Figure 2 shows the primary SprIntMap interface, which is a Google Map of the whole world. We have designed an easy-to-use interface, eliminating the hassles which come with various points of interaction. Thus, the system consists of only a text box, which is used as the primary source of user input and interaction as shown in Fig. 2a. Users can select either data or the interpolation method to use, at a particular instant.

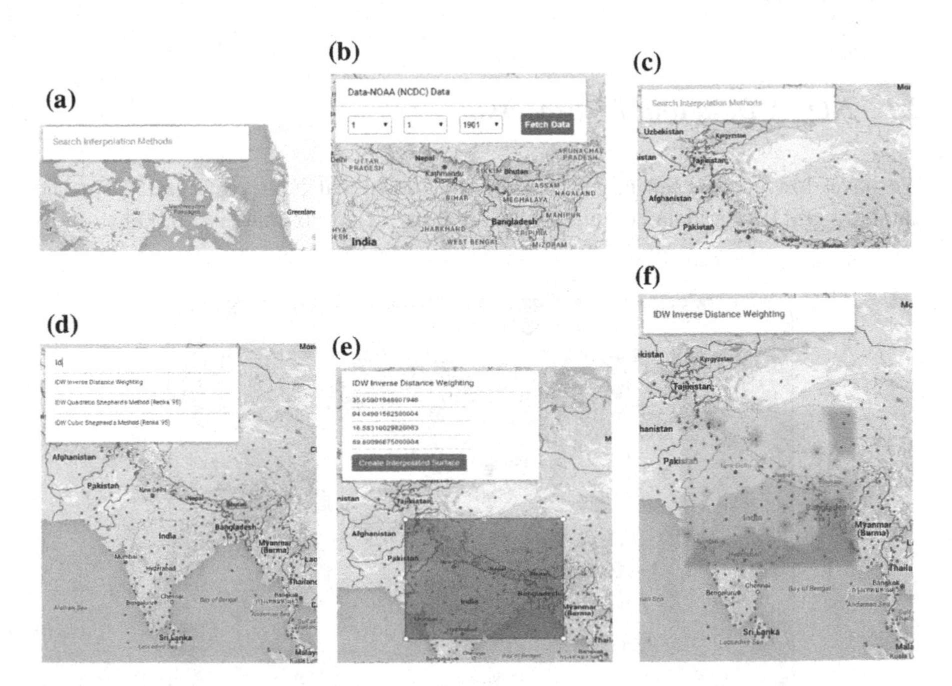

Fig. 2 **a** Input text box for selecting data and method, **b** drop-down date selection for data, **c** data points shown as red circles, **d** interpolation method selection, **e** selection of region of interest, and **f** final interpolated surface

Users cannot select any interpolation method prior to the data selection. The text box senses the keystrokes and opens a drop-down for each of the purpose stated above. Users can select the data by typing *data* in the text box. It automatically opens a drop-down that contains day–month–year selector as shown Fig. 2b. Users can select any date from the selector and click on the *Fetch Data* button. As soon as the user clicks on the *Fetch Data* button, the entire dataset is shown on the map with red circles on the locations with centers at the latitude–longitude of the data points. The red circles are clickable and show the location along with the selected attribute value of the data points.

In order to select the interpolation method, the user needs to type the name of the method. If that method matches with any name that is implemented in the system, a list of match methods is shown as drop-down list. For example, if the user types letter 'i', then IDW-related methods are shown in the list as shown in Fig. 2d. Depending on the interpolation method selected, another appropriate drop-down appears containing all the blank parameter text boxes. For example, if user selects IDW as an interpolation method, the drop-down will contain five text boxes: four text boxes for northeast and southwest lat-longs of the bounding box of the target interpolated surface, one text box for the resolution of the interpolation surface. Other methods contain parameter inputs for respective requirements. Selecting method simultaneously opens a rectangular selection tool on the map to allow users to select the bounding box of the region where the surface will be created. The select area tool is draggable and resizable to fulfill the needs of the user.

As soon as the user clicks on the *Create Interpolated Surface* button, the back-end processes start. The entire Web service is built on the MVC architecture. The request to calculate the interpolated surface is carried out by the servlet, and it calls appropriate models for different tasks. For example, visualizing the training data points and the resulting interpolated surface have different servlet requests respectively. For showing the points, it reads the file from the server and renders it on the map using the JavaScript library. To calculate the interpolated surface, the servlet calls a model which establishes a SSH connection to the cluster master node where Spark is running. It then calls the utility classes that accomplish the following five tasks: (1) execute the appropriate interpolation class, setting the parameters coming from the front end, (2) delete any existing results from the HDFS and create a fresh data for currently running program, (3) copy the result through FTP from HDFS to the user home directory, (4) delete result from HDFS, and (5) return the result as a list to the servlet. The servlet then returns the data to the JavaScript function, and it converts the list to JSON string and shows the surface on the map.

2.3 Controller and Model

As shown in Fig. 1, the controller consists of three sub-modules. They are: (1) data selection, (2) data uploading, and (3) the interpolation. Each sub-module gets activated from different user interactions and transfers control to the model. The model is a collection of utility classes that contain methods to establish and close connection to a remote server using SSH and transfer files using FTP. For uploading NOAA data to the HDFS, the controller transfers control to the classes of the module that make a FTP connection to the NOAA server and copy the latest file to the Web server. Then it makes an SSH connection to the Hadoop cluster and issues a command to upload the file in the HDFS. For creating the interpolated surface, the model is used to issue the spark-submit command for running appropriate interpolation method selected by the user in the user interface.

2.4 *SprInt Package*

SprInt is an Apache Spark package which implements interpolation algorithms in a distributed manner. It is a collection of Java classes, each of which is a parallel implementation of an interpolation algorithm in Spark. The first step in tackling the interpolation problem is to preprocess the input, which is then fed to the individual algorithms mentioned above for interpolation. The preprocessing step takes the bounding box of the target region and enumerates all the test data points available within the bounding box, considering the resolution parameter as well. Following this, a cartesian product between the test and training data (which is stored in the HDFS) is computed, and the result is stored as an RDD object. This RDD is now consumed as an input for the other interpolation methods.

3 Performance Evaluation

To evaluate the performance of the system, we have performed two kinds of tests. The first test case provides insight on the system's response time with a particular job. Here job refers to the entire process of reading the input data, creating the output grid and interpolation surface creation on those grid points. The second test case deals with how the system scales with increasing number of nodes. We have selected Inverse Distance Weighting (IDW) as the interpolating method.

For both the test cases, we have taken the LST for a particular date of 10,000 sensors around the world as an input to the system. The ROI of the interpolated surface is taken as a rectangular bounding box from point (Lat: 37° N, Lon: 67° E) to point (Lat: 7° N, Lon: 97° E). For the first test case, we measure the response time of the system by comparing the response time of serial and Spark version of the IDW algorithm. For that, we have performed two experiments. In the first case, we keep the grid resolution constant and increase the input points from 1000 to 6000 selected randomly from 10,000 input data points. In the second case, keeping the input data points constant, we increase the grid resolution from $1° \times 1°$ to $0.01° \times 0.01°$.

The test cases are carried out on two separate systems of nodes. The distributed version of the IDW algorithm has been performed on a physical cluster of 10 nodes.

Each node has two Intel Xeon 2.60 GHz processors, each consists of six cores, 132 GB of memory, and 15 TB of hard disks. All the nodes run on CentOS 6.5 operating system and having JDK 1.7 and Spark 1.3.0 version installed. For measuring the performance for the serial IDW algorithm on the same dataset, we use a single server having 132 GB of memory, two Intel Xeon 2.60 GHz processors, each consists of six cores.

In the first experiment, as the input data size increases, the time required for serial method increases dramatically, whereas for the Spark distributed approach increases gradually. Due to the framework overhead, the first few tests are dominated by the serial algorithm. Same response scenario is achieved from the second experiment. Both the experiment results are shown in Fig. 3.

For investigating the scalability of distributed IDW implementation, we generate six test cases, each containing a different set of input data containing 1000–6000 points respectively. For each input set of points, we select a constant resolution (0.05° × 0.05°) and report the running time for increasing number of Spark

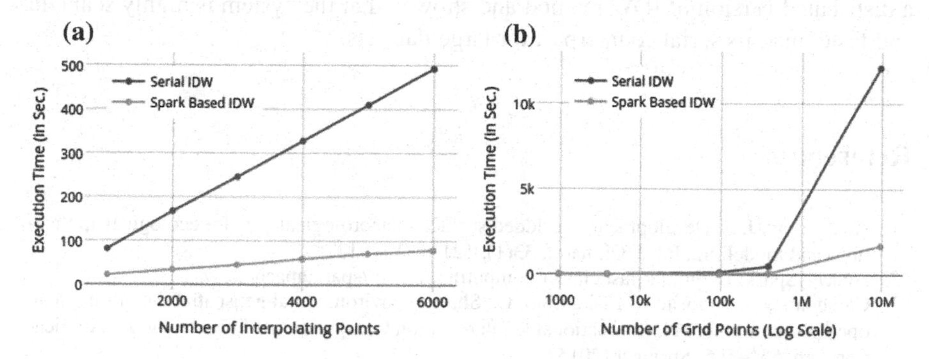

Fig. 3 Time consumed for interpolating **a** with increasing interpolating points and **b** with increasing output grid points using IDW method

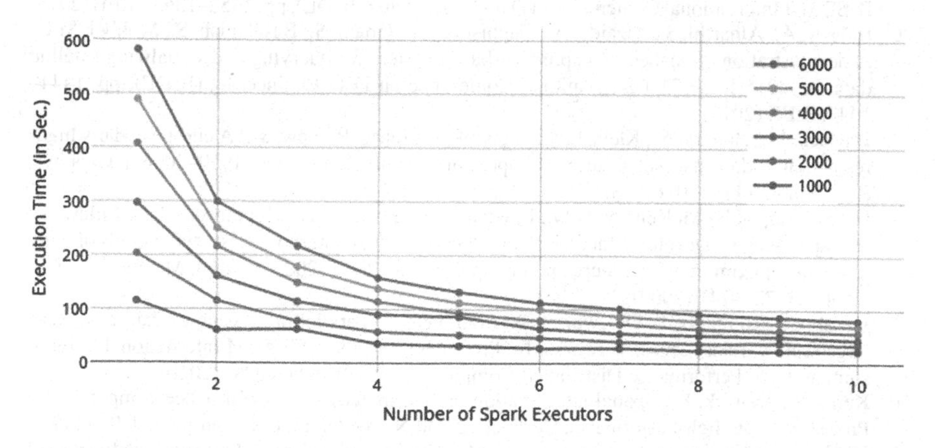

Fig. 4 Scalability of IDW algorithm in Spark

executors. It can be seen from Fig. 4 that our algorithm has a strong scalability and as the input size is increased, the algorithm becomes more scalable.

4 Conclusion

In this paper, we present a Web-based system which helps in accessing, visualizing, and predicting large climatic datasets using Apache Spark framework. The system allows users to access and visualize large scattered vector datasets on a map-based interface, enable them to predict environmental variables for unknown points using different interpolation methods and download the interpolated surface for future use. The system consists of a Google Map-based front end to visualize the data and surface and an Apache Spark-based parallel interpolation package, SprInt, which uses Spark's in-memory data processing to perform interpolation algorithms on large datasets in reasonable time. We investigated the performance of the system with a distributed version of IDW method and showed that the system is highly scalable and faster than its serial counterpart for large datasets.

References

1. Abatzoglou, J.T.: Development of gridded surface meteorological data for ecological applications and modelling. Int. J. Climatol. **33**(1), 121–131 (2013)
2. Apache Spark: Lightning-fast cluster computing. https://spark.apache.org/
3. Caradonna, G., Figorito, B., Tarantino, E.: Sharing environmental geospatial data through an open source webgis. In: International Conference on Computational Science and Its Applications, pp. 556–565. Springer (2015)
4. Cressie, N.: Statistics for Spatial Data. Wiley (2015)
5. Eldawy, A., Mokbel, M.F.: Spatialhadoop: A mapreduce framework for spatial data. In: 2015 IEEE 31st International Conference on Data Engineering (ICDE), pp. 1352–1363. IEEE (2015)
6. Eldawy, A., Alharthi, S., Alzaidy, A., Daghistani, A., Ghani, S., Basalamah, S., Mokbel, M.F.: A demonstration of shahed: a mapreduce-based system for querying and visualizing satellite data. In: 2015 IEEE 31st International Conference on Data Engineering (ICDE), pp. 1444–1447. IEEE (2015)
7. Haylock, M., Hofstra, N., Klein Tank, A., Klok, E., Jones, P., New, M.: A european daily high-resolution gridded data set of surface temperature and precipitation for 1950–2006. J. Geophys. Res. Atmos. **113**(D20) (2008)
8. Hutchinson, M.F., McKenney, D.W., Lawrence, K., Pedlar, J.H., Hopkinson, R.F., Milewska, E., Papadopol, P.: Development and testing of canada-wide interpolated spatial models of daily minimum-maximum temperature and precipitation for 1961–2003. J. Appl. Meteorol. Climatol. **48**(4), 725–741 (2009)
9. Jardak, C., Riihijärvi, J., Oldewurtel, F., Mähönen, P.: Parallel processing of data from very large-scale wireless sensor networks. In: Proceedings of the 19th ACM International Symposium on High Performance Distributed Computing, pp. 787–794. ACM (2010)
10. Kerry, K., Hawick, K.: Spatial interpolation on distributed, high-performance computers. In: Proceedings of High-Performance Computing and Networks (HPCN) Europe, vol. 98 (1997)
11. Li, Z., Yang, C., Huang, Q., Liu, K., Sun, M., Xia, J.: Building model as a service to support geosciences. Comput. Environ. Urban Syst. **61**, 141–152 (2017)

12. Li, Z., Yang, C., Sun, M., Li, J., Xu, C., Huang, Q., Liu, K.: A high performance web-based system for analyzing and visualizing spatiotemporal data for climate studies. In: International Symposium on Web and Wireless Geographical Information Systems, pp. 190–198. Springer (2013)
13. Lott, J.N.: 7.8 the quality control of the integrated surface hourly database (2004)
14. Shepard, D.: A two-dimensional interpolation function for irregularly-spaced data. In: Proceedings of the 1968 23rd ACM National Conference, pp. 517–524. ACM (1968)
15. Fetchclimate2. http://fetchclimate2.cloudapp.net/
16. National centers for environmental information. https://www.ncdc.noaa.gov/

Development of an ARIMA Model for Monthly Rainfall Forecasting over Khordha District, Odisha, India

S. Swain, S. Nandi and P. Patel

Abstract The assessment of climate change, especially in terms of rainfall variability, is of giant concern all over the world at present. Contemplating the high spatiotemporal variation in rainfall distribution, the prior estimation of precipitation is necessary at finer scales too. This study aims to develop an ARIMA model for prediction of monthly rainfall over Khordha district, Odisha, India. Due to the unavailability of recent rainfall data, monthly rainfall records were collected for 1901–2002. The rainfall during 1901–82 was used to train the model and that of 1983–2002 was used for testing and validation purposes. The model selection was made using Akaike information criterion (AIC) and Bayesian information criterion (BIC), and ARIMA (1, 2, 1) $(1, 0, 1)_{12}$ was found to be the best fit model. The efficiency was evaluated by Nash–Sutcliffe efficiency (NSE) and coefficient of determination (R^2). The model forecasts produced an excellent match with observed monthly rainfall data. The outstanding accuracy of the model for predicting monthly rainfall for such a long duration of 20 years justifies its future application over the study region, thereby aiding to a better planning and management.

Keywords Rainfall forecast · ARIMA · Khordha district · AIC
BIC

S. Swain (✉) · P. Patel
IDP in Climate Studies, IIT Bombay, Mumbai 400076, Maharashtra, India
e-mail: sabyasachiswain16@gmail.com

P. Patel
e-mail: pratimaniitb@gmail.com

S. Nandi
Department of Civil Engineering, IIT Bombay, Mumbai 400076, Maharashtra, India
e-mail: ce.saswata@gmail.com

P. K. Sa et al. (eds.), *Recent Findings in Intelligent Computing Techniques*,
Advances in Intelligent Systems and Computing 708,
https://doi.org/10.1007/978-981-10-8636-6_34

1 Introduction

Climate change has become one of the alarming issues for the current generation. The cumulative losses incurred due to climate change are very high compared to any other issues [1]. As evident from numerous studies, the impacts are going to be severe for developing countries due to poor adaptation and lack of timely preparedness [2]. Since climate change refers to the long-term changes in the normal meteorological conditions, it is essential to assess the meteorological variables among which rainfall is the prominent one. The abundance or deficit of rainfall leads to flood or drought-like situations, thereby causing havoc. Moreover, it varies to a wide range over time and space. Thus, assessment of rainfall variability is of giant concern, which requires prior quantification. Not surprisingly, the analysis and forecasting of rainfall time series has been pivotal around the globe [3, 4]. In India, mostly rainfall forecasting is performed at a very high spatial scale. Precipitation has been forecasted all over India or in various states [5, 6]. Some prior works have also been done on river basin scale [7]. But the inferences drawn for a large area may not hold good for all its segments, as rainfall possesses remarkably high spatial variation even within a small region. So it is required to carry out rainfall forecasting at finer spatial resolution too.

The methods used for rainfall prediction are broadly categorized into two approaches, viz. dynamic approach and empirical approach. The dynamic approaches mostly deal with models that are physics-based, e.g., numerical weather prediction models. The empirical approaches, e.g., regression, fuzzy logic, artificial neural network (ANN), ARIMA, deal with assessment of past rainfall data and their link with other meteorological variables [8]. An ARIMA model is a combination of three operators, i.e., autoregressive (AR) function processed on the historical values of the parameter (rainfall), moving average (MA) function processed on entirely random values, and an integration (I) part to reduce the difference between them, thereby being regarded as the differencing operator [9, 10].

In the present study, ARIMA modeling approach has been applied at a finer scale, i.e., over Khordha district, Odisha, India. Finally, the model is also employed to forecast rainfall for 20 years. The details of the study area, ARIMA modeling, and results are presented in the subsequent sections.

2 Study Area and Data Used

The area selected for this study is Khordha (also called Khurda) district, Odisha, India. Area of Khordha district is 2888 km^2 (1115 square miles). The capital of the state Bhubaneswar lies within this district. The geographical location of the district is 20.13°N latitude and 85.48°E longitude. The location of Khordha (Khurda) district is shown in Fig. 1.

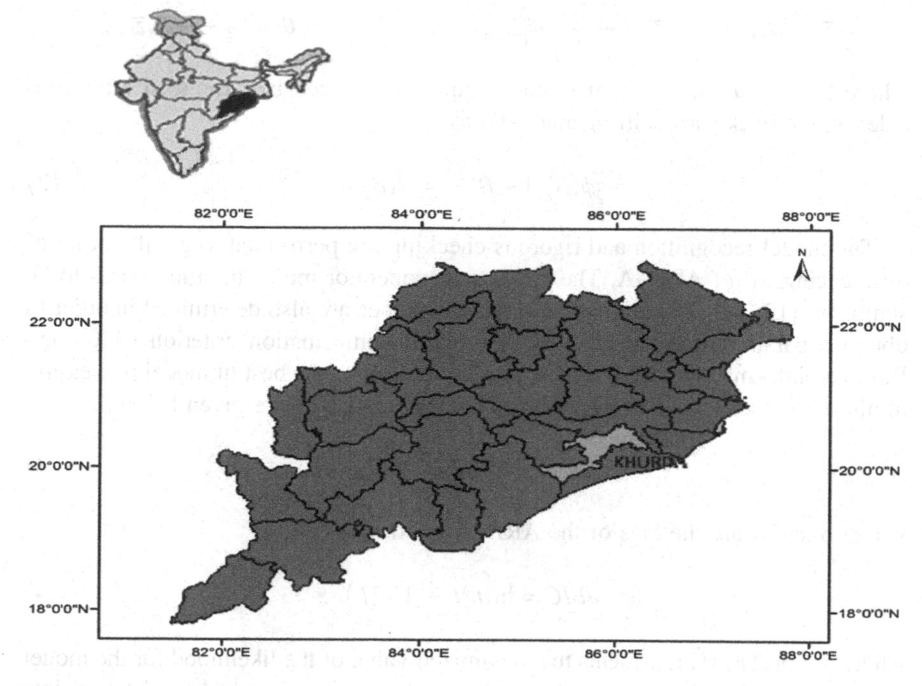

Fig. 1 Location of Khurda District, Odisha

From climatological point of view, the average annual precipitation over the district is about 1200–1400 mm. Moderate temperature prevails over the area throughout the year barring the summer season (March–June), where the maximum temperature even exceeds 45 °C. The average minimum temperature over the district is 9.6 °C [4].

The monthly rainfall data over Khordha district is collected for a period of 102 years, i.e., 1901–2002 from India Water Portal. The study limitation is the unavailability of recent rainfall data that hinders exploration of current rainfall patterns. But the availability of century-long records is advantageous as that would boost the robustness of the model. The period of 1901–82 is used for training the ARIMA model, whereas the later 20 years, i.e., 1983–2002, was used for testing and validation purposes.

3 Methodology

In this study, rainfall prediction is done for 20 years, which is considered as a very long duration from meteorological point of view. ARIMA model predicts future rainfall on the basis of historical records (training period). The expression of ARIMA (p, d, q) is as follows [11, 12]:

$$\overline{z_t} = \phi_1 \overline{z_t}_{-1} + \phi_2 \overline{z_t}_{-2} + \cdots + \phi_p \overline{z_t}_{-p} + a_t - \theta_1 \overline{z_t}_{-1} - \theta_2 \overline{z_t}_{-2} - \cdots \theta_q \overline{z_t}_{-q} \quad (1)$$

where $\overline{z_t} = z_t - \mu$, and a_t is the shock. Equation (1) can be expressed after considering the backward shift operator (B) as:

$$\phi(B)(1-B)^d z_t = \theta(B) a_t \quad (2)$$

The model recognition and rigorous checking are performed to get the order of differencing 'd' of ARIMA. The differencing operator molds the time series to be stationary [13, 14]. The optimal AR and MA values are also determined in order to obtain a parsimonious model [15]. The Akaike information criterion (AIC) and Bayesian information criterion (BIC) are used to select the best fit model possessing minimum values [16]. The expressions for AIC and BIC are given below.

$$AIC(p,q) = N \ln(\sigma_\varepsilon^2) + 2(p+q) \quad (3)$$

where p and q are the lags of the ARIMA (p, d, q).

$$BIC = \ln(n)k - 2\ln(\hat{L}) \quad (4)$$

where $\hat{L} = p(x|\psi, M)$ represents the maximized value of the likelihood for the model M using ψ as the parameters, x is the observed data, n is the number of data points in x (sample size), and k represents number of free parameters to be estimated.

4 Results and Discussion

The model training is carried out for the period of 1901–82, and the best fit model is estimated based on the minimum values of both AIC and BIC. The ARIMA (1, 2, 1) $(1, 0, 1)_{12}$ fitted the best with the observed data. The $(1, 0, 1)_{12}$ here represents the seasonal ARIMA (also called SARIMA), which is generally used for modeling the variables with significant seasonal component. Typically, the ARIMA (1, 2, 1) and SARIMA $(1, 0, 1)_{12}$ are no different from each other for this case. As it is well known that for Indian context of climatology, rainfall is extremely seasonal, being maximum during summer monsoon months (June–September) and scanty in other months. Thus, the seasonal ARIMA will be more reliable in rainfall prediction over the study area. However, the performance of any empirical model should be tested prior to application. Here, the model is tested for its ability to forecast rainfall for 1983–2002. A comparative plot of the forecasted monthly rainfall values with observed values is presented in Fig. 2. The scatter plot of both forecasted and observed monthly rainfall is also presented below in Fig. 3.

The ARIMA model produced excellent agreement with observed rainfall values as evident from Fig. 2. The mean values of the forecasted rainfall are parsimonious in

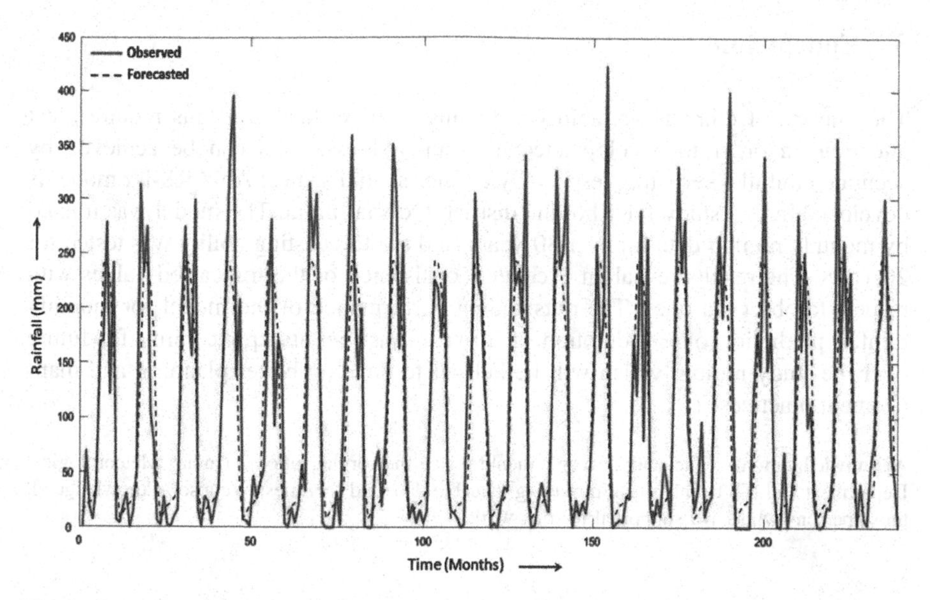

Fig. 2 Comparison of observed and model forecasted rainfall over Khordha for 1983–2002

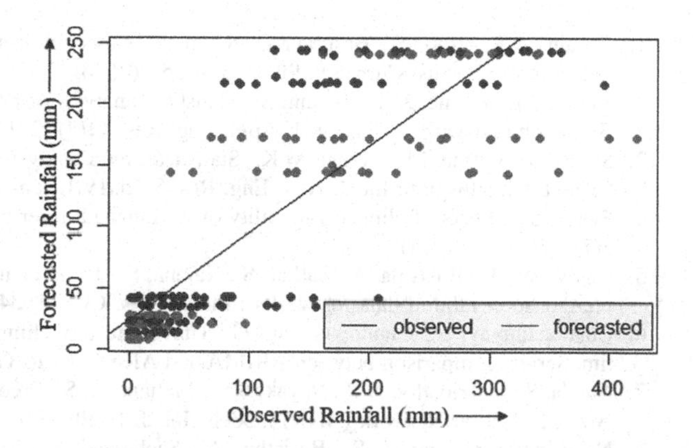

Fig. 3 Scatter plot of observed and forecasted rainfall values

matching the actual rainfall. From Fig. 3 too, it can be noticed that the forecasted and observed values possess a very good correlation. The established efficacy measures, i.e., Nash–Sutcliffe efficiency (NSE) and coefficient of determination (R^2), are used to validate the model outputs. The NSE and R^2 values are determined to be 0.72 and 0.86, respectively, which is considered excellent for any model. However, one interesting thing to notice is that the model is able to capture the lower quantile values exquisitely, but it is not very effective for the higher quantiles. This might be due to changes in trend of extreme rainfall events over the study area [4]. The extreme rainfall events and floods have been noticed to increase in Odisha after 1980. Nevertheless, the outstanding performance of the model, that too for forecasting monthly rainfall over 20 years, substantiates its ability for future application.

5 Conclusion

The impacts of climatic variability are going to be critical and thus require great attention, in order to develop adequate preparedness. This can be achieved by accurate rainfall forecasting, especially at finer spatial scales. An ARIMA model is developed in this study for Khordha district, Odisha, India. The model was trained by monthly rainfall data for over 80 years, and the forecasting ability was tested for 20 years. The results reveal an excellent consistency of the forecasted values with respect to observed ones. The outstanding performance of the model for monthly rainfall prediction over a duration of 20 years justifies its applicability for future over the study region, which will be helpful to develop better planning and management practices.

Acknowledgements The authors are thankful to the organizations (India Meteorological Department and IIT Bombay) for providing valuable data and facilities. We also acknowledge all those persons who have supported for this work.

References

1. Gosain, A.K., Rao, S., Basuray, D.: Climate change impact assessment on hydrology of Indian River Basins. Curr. Sci. **90**(3), 346–353 (2006)
2. Verma, M., Verma, M.K., Swain, S.: Statistical analysis of precipitation over Seonath River Basin, Chhattisgarh, India. Int. J. Appl. Eng. Res. **11**(4), 2417–2423 (2016)
3. Swain, S., Verma, M., Verma, M.K.: Statistical trend analysis of monthly rainfall for Raipur District, Chhattisgarh. Int. J. Adv. Eng. Res. Stud./IV/II/Jan-March, 87–89 (2015)
4. Swain, S.: Impact of climate variability over Mahanadi river basin. Int. J. Eng. Res. Technol. **3**(7), 938–943 (2014)
5. Narayanan, P., Basistha, A., Sarkar, S., Kamna, S.: Trend analysis and ARIMA modelling of pre-monsoon rainfall data for western India. C. R. Geosci. **345**(1), 22–27 (2013)
6. Chattopadhyay, S., Chattopadhyay, G.: Univariate modelling of summer-monsoon rainfall time series: comparison between ARIMA and ARNN. C. R. Geosci. **342**(2), 100–107 (2010)
7. Nanda, S.K., Tripathy, D.P., Nayak, S.K., Mohapatra, S.: Prediction of rainfall in India using Artificial Neural Network (ANN) models. Int. J. Intell. Syst. Appl. **5**(12), 1–22 (2013)
8. Narayanan, P., Sarkar, S., Basistha, A., Sachdeva, K.: Trend analysis and forecast of pre-monsoon rainfall over India. Weather **71**(4), 94–99 (2016)
9. Kaushik, I., Singh, S.M.: Seasonal ARIMA model for forecasting of monthly rainfall and temperature. J. Environ. Res. Dev. **3**(2), 506–514 (2008)
10. Valipour, M., Banihabib, M.E., Behbahani, S.M.R.: Comparison of the ARMA, ARIMA, and the autoregressive artificial neural network models in forecasting the monthly inflow of Dez dam reservoir. J. Hydrol. **476**, 433–441 (2013)
11. Khashei, M., Bijari, M.: A novel hybridization of artificial neural networks and ARIMA models for time series forecasting. Appl. Soft Comput. **11**(2), 2664–2675 (2011)
12. Valipour, M.: How much meteorological information is necessary to achieve reliable accuracy for rainfall estimations? Agriculture **6**(4), 53 (2016)
13. Salahi, B., Nohegar, A., Behrouzi, M.: The modeling of precipitation and future droughts of Mashhad plain using stochastic time series and Standardized Precipitation Index (SPI). Int. J. Environ. Res. **10**(4), 625–636 (2016)

14. Dastorani, M., Mirzavand, M., Dastorani, M.T., Sadatinejad, S.J.: Comparative study among different time series models applied to monthly rainfall forecasting in semi-arid climate condition. Nat. Hazards **81**(3), 1811–1827 (2016)
15. Rahman, M.A., Yunsheng, L., Sultana, N.: Analysis and prediction of rainfall trends over Bangladesh using Mann–Kendall, Spearman's rho tests and ARIMA model. Meteorol. Atmos. Phys. 1–16 (2016)
16. Kumar, U., Jain, V.K.: ARIMA forecasting of ambient air pollutants (O3, NO, NO2 and CO). Stoch. Environ. Res. Risk Assess. **24**(5), 751–760 (2010)

Developing an Improvised E-Menu Recommendation System for Customer

Rajendra Pawar, Shashikant Ghumbre and Ratnadeep Deshmukh

Abstract Various operations performed by waiters like starting from taking orders till delivery of food/menu to the customer, also billing by cashier made manually. Due to manual process and paperwork may cause time delay, ignorance of customer, errors in billing leads to dissatisfaction of customers. As in today's digital era, customers expect high quality, smart services from restaurant. So to improve quality of service and to achieve customer satisfaction, we proposed improvised E-Menu Recommendation System. This system can build e-reputation of restaurant and customer community in live. All orders and expenses are stored in database and give statistics for expenses and profit. The proposed recommender system uses wireless technology and menu recommender to build improvised E-Menu Recommendation System for customer-centric service. Professional feels and environment are provided to the customers/delegates with additional information about food/menu by using interactive graphics. Outcomes of experimental are obtained by comparing results of two data mining algorithms Apriori and FP-growth which have practical potential in providing customer-centric service.

Keywords Improvised E-Menu Recommendation System · Apriori and FP-growth · Frequency

---

R. Pawar (✉) · R. Deshmukh
Department of Computer Science and Information Technology,
Dr. Babasaheb Ambedkar Marathwada University, Aurangabad, India
e-mail: rajendra.pawar@mitcoe.edu.in

R. Deshmukh
e-mail: rrdeshmukh.csit@bamu.ac.in

S. Ghumbre
Department of Computer Engineering, Government College of Engineering and Research, Avasari, Pune, India
e-mail: shashi.ghumbre@gmail.com

P. K. Sa et al. (eds.), *Recent Findings in Intelligent Computing Techniques*,
Advances in Intelligent Systems and Computing 708,
https://doi.org/10.1007/978-981-10-8636-6_35

1 Introduction

Recommendation system is a very powerful tool which provides quality recommendations of any kind of service and product based on customer's choice and preferences [1]. This system enables waiters to immediately identify customers based on e-mail id and then actively recommend the most appropriate menus through E-menu recommender for customers according to their consumption records. For those customers who visiting restaurant first time, the members of restaurant give recommendations based on feedback given by previous customers and popularity of menu which are stored in the back-end database. Therefore, this system provides customer satisfaction by pruning the workload. This system provides menu ordering, as well as it gives special personal recommendation of food to the customer [2]. Identification of potential customers and providing customer-centric services through digital medium are the main goal behind E-Menu Recommendation System [3, 4].

Emerging countries like Taiwan where most of food restaurant build mobile application with touch screen for processing order [5], where keypad or mouse interface is used by cashier to understand needs of customers as well as optical scanner is used for reading 2-dimensional barcode to understand the order. This motivates to build improvised E-Menu Recommendation System using advanced technology.

2 Architecture of Improvised E-Menu Recommendation System

Figure 1 shows complete architectural overview of the proposed improvised E-Menu Recommendation System. As this system provides various functions with respect to administrator and customer which include online menu order, reservation, and personalized menu recommendation to the customer. At the beginning of the system, the user will login into the system and system will authenticate him/her by his e-mail id. After successful login, the user can see the menu of the restaurant from which he/she can order the food. The customer sees the categorized menu card in the digital form on the tablet. He will be having the tablet on the table via which he can place order. After placing the order, it goes to the kitchen desktop from where kitchen admin can see the customers' orders. Then the chef will prepare the food according to the customers' orders, and the waiter will serve the food to the respective customers.

Manager is responsible for checking the order status whether the order is delivered to the customer or not. After successful delivery of food to the customer, the customer will eat and make the payment to the manager. System is also responsible for giving the branch suggestions.

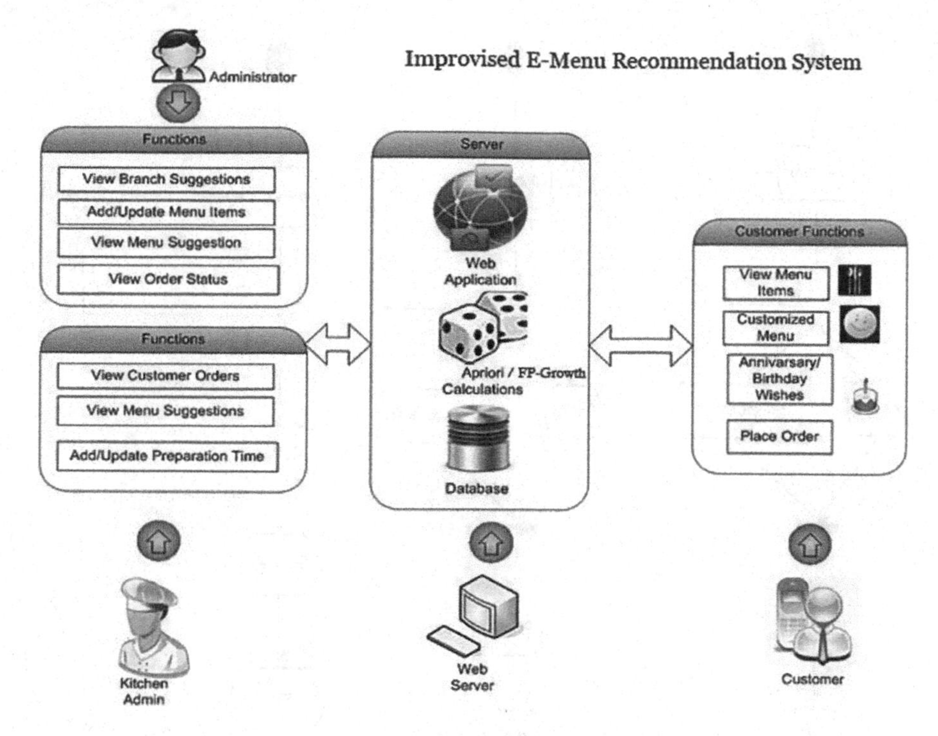

Fig. 1 Architecture of improvised E-Menu Recommendation System

Advantages:

- Allows staff to be more efficient.
- Reduces errors in ordering.
- Helps to keep track of customers and their history.
- Maintains customer relationship.
- Suggests menu to customer.

This system provides pleasure to customers for making orders, and management can meliorate their management (Fig. 2).

Activity diagrams for admin, user, and kitchen admin which show decisions that can be made by data mining are listed below [6, 7].

1. **Customer Profiling**—Attributes: Ordering Patterns, Birth Date, Bill Amount, no. of Visits. We will classify customers on the basis of their birth date along with the no. of visits so that they can be given discounts or any other complementary offers.

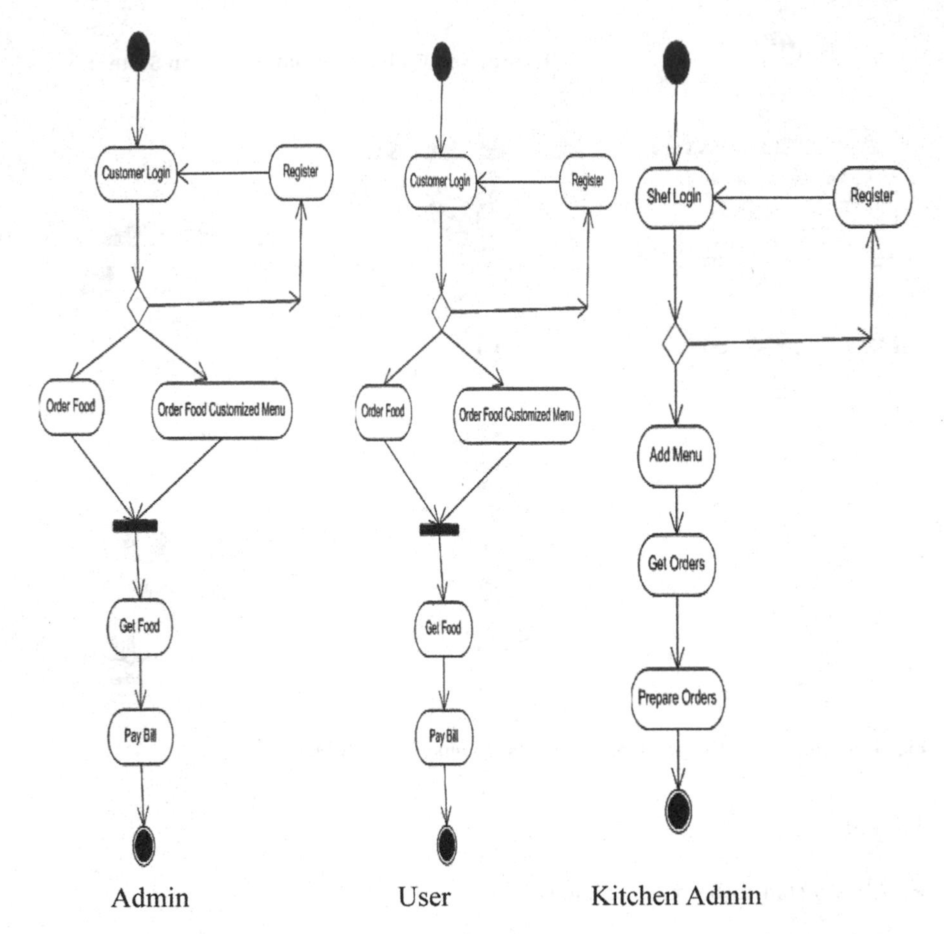

Fig. 2 Activity diagram

2. **Site Selection**—Attributes: Address of Customers, Occupation. It is important to know where the customer originated from. This will establish a trading area for the unit.
3. **Productivity Indexing**—Attributes: Total time between ordering and serving the customer. This data provides insight into average service time with respect to turnover of customer as well as waiting line statistics.
4. **Menu Engineering**—Attributes: Item ordered. By analyzing the pair of items preferred by the customers, we can put them together into meals.
5. **Forecasting**—Attributes: Customer transactional data, Amount of products sold. This will help estimating in advance how much and when menu items will need to be prepared.

3 Algorithms

Association rules are generated on different menu items by using Apriori and FP-growth algorithms. Both algorithms are compared on the basis of technique used, memory utilization, execution time, and number of scan.

3.1 Apriori Algorithm

This algorithm is used to find out the frequency of data into the database. In our application, we use to find out which menu items are frequently ordered. Most frequently ordered combinations will be displayed as a combo pack, so user can able to understand which menu item everyone is preferring on the basis of that he can order his menu item.

Testing Apriori algorithm with following example.

First customer is ordering menu item burger, fries, pizza, coke.

Second customer has ordered menu item burger, fries, coke.

Third customer has ordered menu item burger, fries.

Fourth customer has ordered menu item fries, pizza, coke.

Fifth customer has ordered menu item fries, pizza.

Sixth and seventh customer have ordered menu items pizza, coke and fries, coke, respectively.

So the combinations of menu items are obtained by using the Apriori algorithm. Most frequent menu items ordered by customers are recommended by considering frequency and support value of each menu items.

Support value for each menu item after first scan is calculated which is shown in Table 1.

The first step of Apriori is to count up the number of occurrences, measured in terms of the support for of each menu item separately, by scanning the given database first time. We obtain the following result. All the menu item sets of size 1 have a support of at least 3; it means menu item frequently occurs.

The second step is to generate a list of concerned all pairs of the frequent items and find support for the same which is shown in Table 2.

Table 1 Frequency and support of menu item

Menu item	Frequency	Support (%)
Burger	3	3 = 42
Fries	6	6 = 86
Pizza	4	4 = 57
Coke	5	5 = 71

Table 2 Frequency and support of menu item pairs

Menu item	Frequency	Support (%)
Burger, Fries	3	3 = 42
Burger, Pizza	1	1 = 14
Burger, Coke	2	2 = 29
Fries, Pizza	3	3 = 42
Fries, Coke	4	4 = 57
Pizza, Coke	3	3 = 42

The pairs {Burger, Fries}, {Fries, Pizza}, {Fries, Coke}, and {Pizza, Coke} all meet or exceed the minimum support of 3, so they are frequent. The pairs {Burger, Pizza} and {Burger, Coke} are not. Now, because {Burger, Pizza} and {Burger, Coke} are not frequent, any larger set which contains {Burger, Pizza} or {Burger, Coke} cannot be frequent. In this way, we can *prune* sets: we will see all triples in the database that are frequently occurring, but excluding all triples that contain one of these two pairs.

$$\text{Menu Items (Fries, Pizza, Coke) Frequency (2)}$$

So, {Fries, Pizza, Coke} is the best and first combo; and second, third, and fourth combos we will take as {Fries, Coke}, {Fries, Pizza}, and {Pizza, Coke}.

3.2 FP-Growth Algorithm

To reduce number of iterations than and as well as save memory, FP-growth algorithm is used which is based on FP tree structure. Every node of FP tree represents a menu item and its current count, and each branch represents a different association.

Example: Customer = [Burger, Fries, Pizza, milk], [Burger, milk, Pizza], [Burger, milk, cheese], [Burger, Pizza, Cokes, cheese], [Burger, cheese, Fries].

We count all the items in all the transactions. Apply the threshold we had set previously. Suppose we have a threshold of 25% so each item has to appear at least twice. As per count of each item, list is sorted in ascending order and then builds FP tree. So as per each transaction, add all items in sorted list (Table 3).

In order to get the associations, we visited every branch of the tree and that only include in the association all the nodes whose count passed the threshold which is shown in figures (Figs. 3, 4, 5, 6, and 7).

Table 3 Frequency of menu item using FP-growth

Menu item	Frequency
Burger	5
Fries	2
Pizza	3
Coke	1

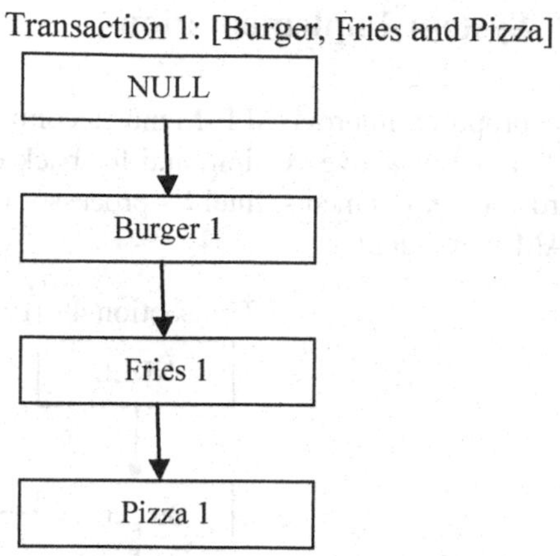

Transaction 1: [Burger, Fries and Pizza]

NULL

Burger 1

Fries 1

Pizza 1

Fig. 3 Transaction 1

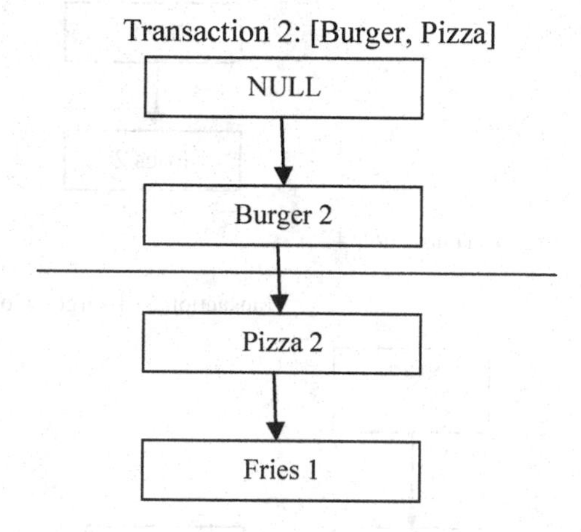

Fig. 4 Transaction 2

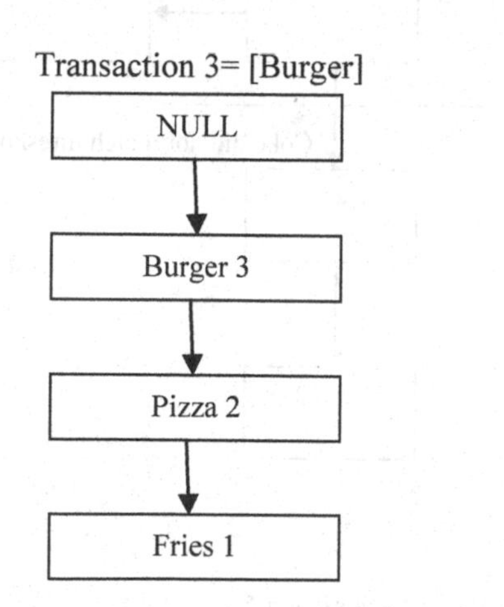

Fig. 5 Transaction 3

4 System Implementation

The proposed improvised E-Menu Recommendation System uses tools like Eclipse Indigo 3.3 or above version, and for back-end, MySQL GUI Browser is used. For hardware requirements, Intel P4 processor or above with at least minimum 256 MB RAM is required.

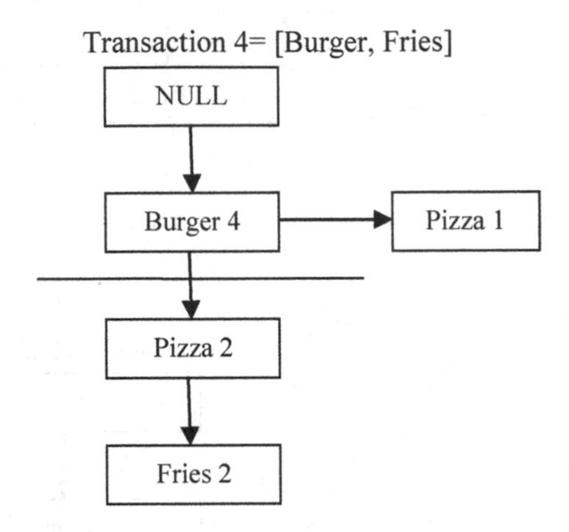

Fig. 6 Transaction 4

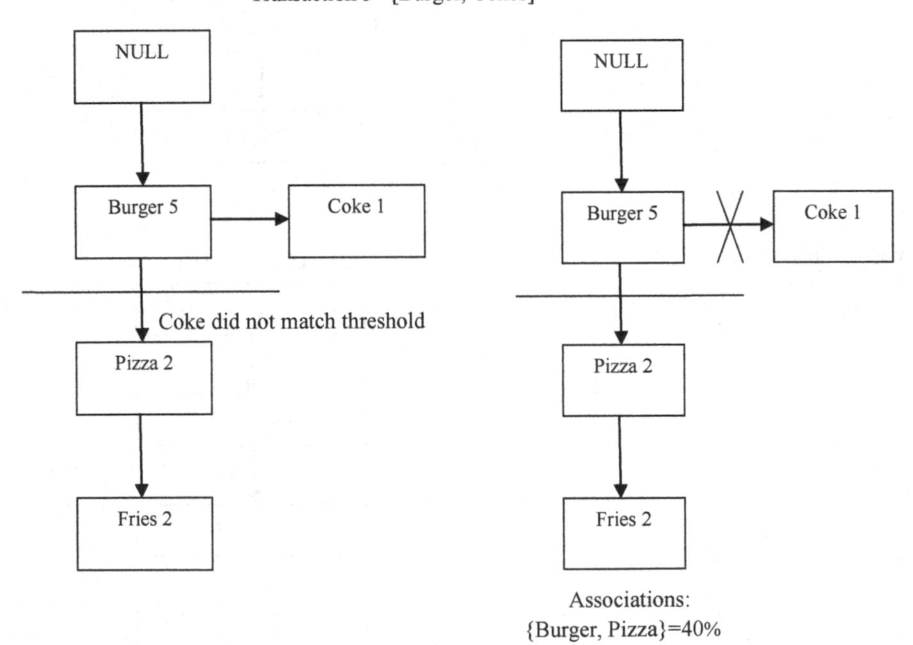

Fig. 7 Transaction 5

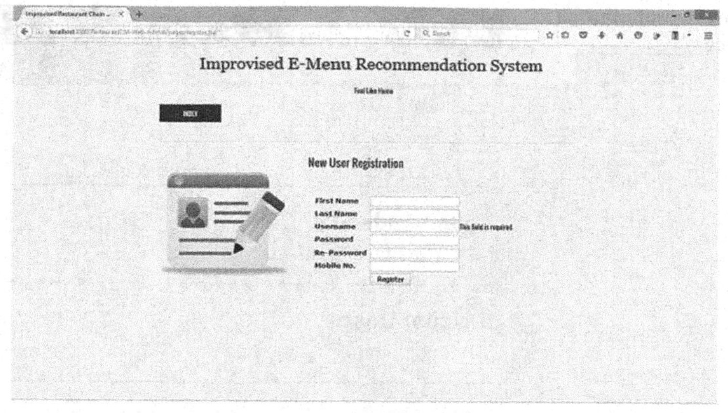

(a) User Registration

(b) Category menu addition

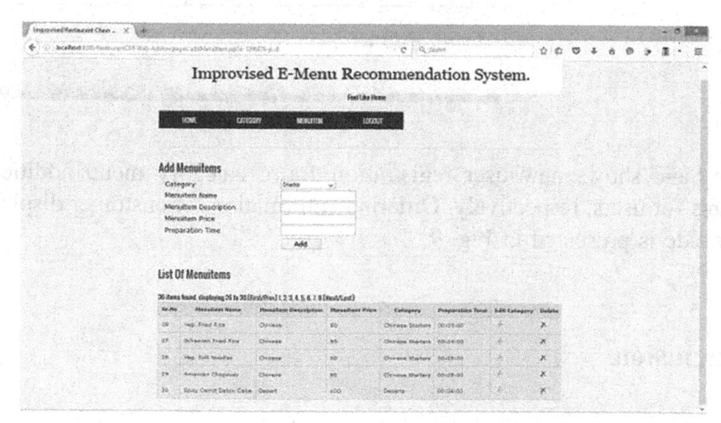

(c) Menu items for users

Fig. 8 **a** User registration **b** category menu addition **c** menu items for users

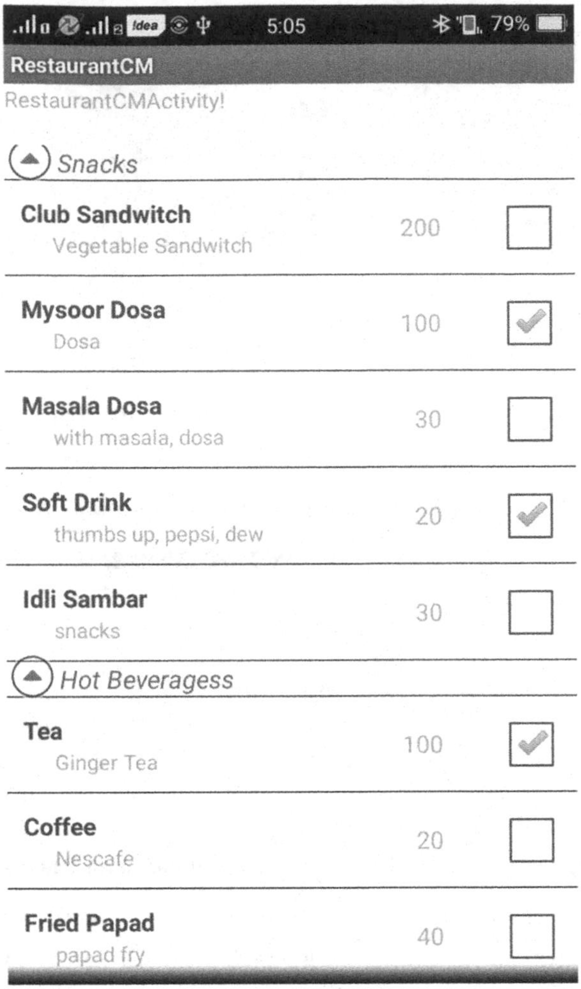

Fig. 9 Ordering information for customer

Figure 8a–c shows new user registration form, category menu addition, and menu items for users, respectively. Ordering information for customer displayed on customer side is presented in Fig. 9.

5 Conclusion

This work on improvised E-menu recommender can be helpful in minimizing waiting time and human errors such as ordering error. This system can help a lot during rush period. It also helps in reducing workload on waiters and hence makes system more efficient. It helps to improve profits to the restaurant owner.

This system can enhance working efficiency and quality of services by providing real-time ordering, proper menu choices and with good checkout services. User who experienced this kind of customer-centric services will recommend such kind of restaurants to other people, friends, etc. All orders given and expenses are in digitized format which permits owner of the restaurant for promotion purpose as well as to know statistics of expenses. As users will appreciate high quality provided by the restaurant, this will in turn build good reputation and increase profit ratio of restaurant.

References

1. Blocker, C.P., Flint, D.J.: Customer segments as moving targets: integrating customer value dynamism into segment instability logic. Ind. Mark. Manag. **36**, 810–822 (2007)
2. Kumar, A., Tanwar, P., Nigam, S.: Survey and evaluation of food recommendation systems and techniques. In: 3rd IEEE International Conference on Computing for Sustainable Global Development, Oct 2016
3. Golsefid, S.M., Ghazanfari, M., Alizadeh, S.: Customer segmentation in foreign trade based on clustering algorithms. World Acad. Sci. Eng. Technol. **28**, 405–411 (2007)
4. Hosseini, M., Anahita, M., Mohammad, R.G.: Cluster analysis using data mining approach to develop CRM methodology to assess the customer loyalty. Expert Syst. Appl. **37**, 5259–5264 (2010)
5. Tan, T.H., Chang, C.S., Chen, Y.F.: Developing an intelligent e-restaurant with a menu recommender for customer-centric service. IEEE Trans. Syst. **42**(5) (2012)
6. Kamber, M., Han, J.: Data Mining: Concepts and Techniques. Morgan Kaufmann (2008)
7. Huang, S.C., Chang, E.C., Wu, H.H.: A case study of applying data mining techniques in an outfitter's customer value analysis. Expert Syst. Appl. **36**(6), 5909–5915 (2009)

[illegible]

References

[illegible]

Real-Time Bigdata Analytics: A Stream Data Mining Approach

Bharat Tidke, Rupa G. Mehta and Jenish Dhanani

Abstract The outburst of Bigdata has driven a great deal of research to build and extend systems for in-memory data analytics in real-time environment. Stream data mining makes allocation of tasks efficient among various distributed computational resources. Managing chunk of unbounded stream data is challenging task as data ranges from structured to unstructured. Beyond size, it is heterogeneous and dynamic in nature. Scalability and low-latency outputs are vital while dealing with big stream data. The potentiality of traditional approach like data stream management systems (DSMSs) is inadequate to ingest and process huge volume and unbounded stream data for knowledge extraction. A novel approach develops architectures, algorithms, and tools for uninterrupted querying over big stream data in real-time environment. This paper overviews various challenges and approaches related to big stream data mining. In addition, this paper surveys various platforms and proposed framework which can be applied to near-real- or real-time applications.

Keywords Bigdata · Stream data mining · Real-time analytics
Real-time platforms

1 Introduction

Advancements in technology make data to grow exponentially from distinctive areas of society [1]. Bigdata is a term used to describe such data. Discovering knowledge from unstructured real-world data gains lot of attention from researchers and practitioners in traditional data mining environment [2–4]. However, a detailed study of data mining techniques is needed while dealing with Bigdata. The conventional data mining techniques which are based on the loading whole dataset may not be capable to handle challenges put by 3 Vs of Bigdata. Stream data mining

B. Tidke (✉) · R. G. Mehta · J. Dhanani
Department of Computer Engineering, SVNIT, Surat, India
e-mail: batidke@gmail.com

P. K. Sa et al. (eds.), *Recent Findings in Intelligent Computing Techniques*,
Advances in Intelligent Systems and Computing 708,
https://doi.org/10.1007/978-981-10-8636-6_36

(SDM) algorithms can efficiently induce a categorization or forecasting models in Bigdata environment [5]. These algorithms iteratively update the model automatically with upcoming data to avoid loading stored data every time. SDM algorithms have capability to handle unbounded data streams, to process data in memory, and to extract knowledge on the fly. SDM techniques having low latency are crucial in giving information to users and businesses to take correct and relevant decisions in real time. Framework-based technologies are the core in Bigdata environment. Some existing examples are Hadoop and Spark which are based on various components creating ecosystem to handle Bigdata challenges [6, 7]. But, there is still lack of capable and flexible real-time data analysis framework to manage big datasets, to create opportunities, and to overcome challenges put by 3 Vs in timely manner.

1.1 Motivation

Large volumes, variety in formats, and infinite instances of data arriving continuously have challenged the traditional data mining algorithms. The systems involved in data storage and processing pose many challenges due to intrinsically dynamic nature of data and ill-structured multidisciplinary data sources with nonlinear domain. The data can be multidimensional and multiscale. Such data is hard to think, impossible to visualize, and, due to exponential growth of number of possible value attached to it, difficult to enumerate. This motivates us to survey various available platforms, algorithms, and frameworks which can be integrated to handle various challenges for stream Bigdata analytics.

2 Stream Data Mining

The continuous input data is abbreviated as streaming data or data stream. Mining information from such data is called as stream data mining (SDM) [8]. SDM refers to information structure extraction as models and patterns from continuous data streams. Traditional method apply complex algorithms using several passes over stored data involves disk I/O overhead. Two major approaches, incremental approach and ensemble learning, for handling streaming data have been found in the literature to enhance the functionality of existing data mining algorithm. In incremental approach, the new incoming data or streams are incrementally added in existing model and the model will be updated based on new stream data. In ensemble learning, the infinite stream is stored as subset periodically, and new model is prepared from it which is most often outperform single model and is most suitable for well-known notion of stream mining "concept drift".

2.1 Challenges in Traditional Stream Mining Techniques

The traditional data mining algorithm is suitable for static data. Existing stream data and its processing have produced various challenges for real-time analytics [9, 10],

- *Data must be processed continuously*

Low latency can be achieved by processing input data without using time-intensive storage operation. The system need not be passive by incorporating continuous polling in the application which is a major cause of the reduced latency.

- *Removing stream data impurities*

The real-world data is prone to encompass the data quality issues including delayed, missing, and out-of-order data. Inbuilt mechanism is expected to deal with a stream.

- *Guarantee for repeatable outcome*

In case of failure, recovery of system or part of it, when replaying and reprocessing the same input stream should yield the same outcome, regardless of the time of execution.

- *Availability of application and integration of data with the application*

The application is expected to be up all the time and properly integrated with the data. In case of failure of the system, restarting the operating system and recovering the application will cause the lots of unacceptable overhead. To satisfy the need of the real-time system, preferably secondary system is managed and synchronized with the system.

2.2 Bigdata Stream Mining

Bigdata mining is a process to extract knowledge from dataset which involves huge volume, high-speed, scattered, and unstructured data. These datasets are beyond the capacity of conventional computational resources for ingestion, storage, processing, and visualization. Loads of sufficient and immensely valuable information are hidden in these big datasets.

To utilize the advantages of both processing engines, i.e., batch and stream, an architectural framework “Lambda Architecture” [11] has been proposed. Lambda architecture contains three layers: batch, speed, and serving.

2.2.1 Sources of Stream Bigdata

Concepts like digital India show transformation of physical world into virtually digitized world. Technologies like Web search engines (Google, Bing, etc.), Internet of Things (sensors, radio-frequency identification (RFID), mobile phones), social networking (Facebook, Twitter) generates a huge stream data with high velocity, and are main sources of stream Bigdata [3].

2.2.2 Data Ingestion

There are several key reasons to ingest data as a stream. In real-time systems, decisions are normally based on the current state or evolution of data. The relevant data with respect to current scenario should be ingested as a stream.

2.2.3 Data Storage

Traditional databases like RDBMs even with many partitioning and parallelizing capability fall short to simple and cost-effectively scale to ever-rowing data storage requirements. These systems suppose to work on structured data. Bigdata requires storage capable of storing raw unstructured data (Fig. 1).

Hadoop is an open-source framework to store and compute unstructured Bigdata with scalability. Features of Hadoop include distributed environment, fault-tolerant, reliable, highly scalable, cost-effective solution and can handle petabytes of data. Drawback of Hadoop is that it processes data in batch mode which is not suitable for real-time data analytics. Another open-source framework from apache is Spark, real-time framework for Bigdata analytics.

2.2.4 Data Preprocessing

Preprocessing techniques like data summarization and cleaning play a vital role in Bigdata analytics, as most of real-world data is unstructured [2]. Noisy data such as human error, sensor data due to failure and irrelevant data from unknown sources makes data scientist work more tedious.

2.2.5 Data Processing

Streaming processing paradigm is generally based on hidden information that comes with incoming data. This information can be used to obtain useful results to do further analysis. In this process, since data is arriving continuously and in huge amount, only a fraction of stream data is stored in limited memory databases and processed using stream processing system such as Spark Streaming [5].

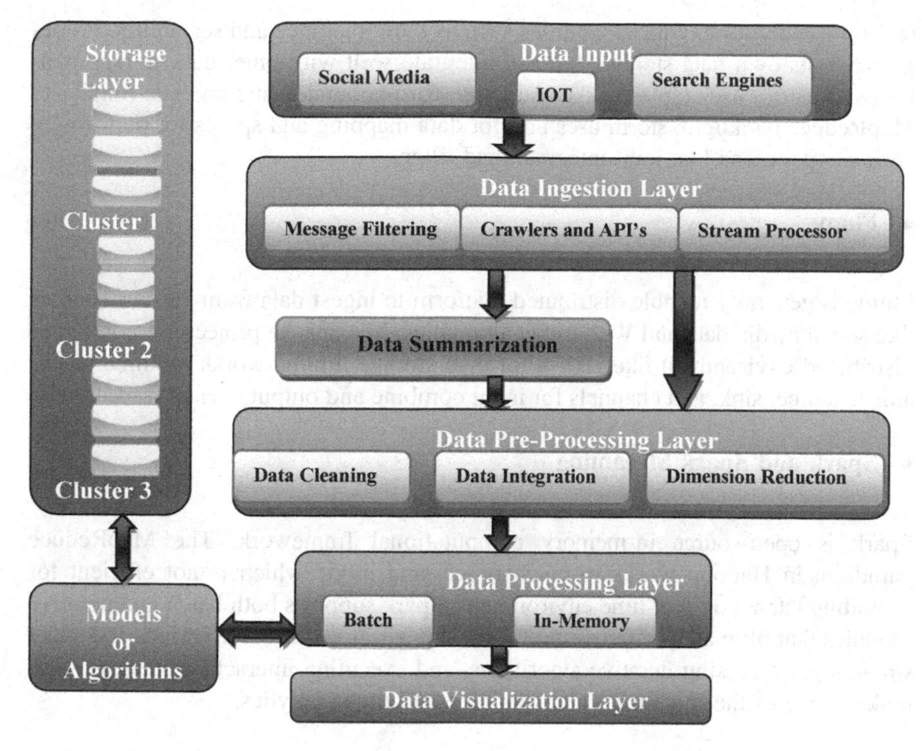

Fig. 1 Real-time architecture for Bigdata analytics

2.2.6 Bigdata Visualization

Data visualization has been used by academicians and stakeholders from industries for analyzing and interpreting their result using techniques such as plot, tables, charts, and graphs According to SAS report [11], data visualization is one of key components in Bigdata analytics, and they also highlighted some of problems while visualizing such large amount of data.

3 Real-Time Platforms

Existing data-warehousing environments are mainly works on batch data with extremely high latencies which seeks new technologies to overcome such challenges [5, 7].

- **Apache Storm**

Storm is a distributed open-source platform for big stream data ingestion and processing. It performs in-memory data processing, making it well suited for real-time analytics having capabilities such as fault tolerance and scalability. It does not have its own data storage, but can integrate well with other data storage platforms. A storm uses tuples as data structure for unbounded stream data. Similar to MapReduce paradigm, storm uses bolt for data mapping and spouts for performing various operations like join, integrate, and filter.

- **Flume**

Flume is generally reliable distributed platform to ingest data from various sources like social media data and Web traffic anomalies. It is apache project used on top of distributed environment like HDFS for data storage. Flume works on three mechanism source, sink, and channels for input combine and output events, respectively.

- **Spark and Spark Streaming**

Spark is open-source in-memory computational framework. The MapReduce paradigm in Hadoop works in batch processing mode which is not efficient for providing latency in real-time environment. Spark supports both batch and iterative modules that ultimately improve processing speed. It can execute various tasks like streaming, processing iterative algorithms, and executing interactive queries which make it one of the most used platforms for real-time analytics.

- **Kafka**

Kafka is real-time system ingest sequence of messages having unique id attached to it. These messages spread over many partitions across various nodes in a cluster maintained by Kafka. These messages are published for specified amount of time in Kafka to be utilized and subscribed keeping track of message.

4 Conclusion

Bigdata systems move from batch-oriented systems toward in-memory computations. Modern sources of data are dynamic in nature and normally require more real-time analysis. This growth calls for novel system frameworks for large-scale data collection, transmission, storage, and processing methods. Concrete example includes Spark which emerged as future-generation Bigdata processing engine, surpassing Hadoop MapReduce. In this paper, our objective is to study and understand various characteristic of Bigdata, and discover knowledge from it to take critical decisions for real-time applications. At the same time, this paper proposed a high-level framework for real-time Bigdata analysis based on stream unstructured data. In future, our plan is to implement various suitable real-time platforms and novel algorithms and to compare our results with existing approach.

References

1. Gantz, J., Reinsel, D.: The Digital Universe in 2020–Big Data, Bigger Digital Shadows, and Biggest Growth in the Far East, IDC IView (2016)
2. Romero, C., Ventura, S.: Educational data mining: a survey from 1995 to 2005. Expert Syst. Appl. **33**(1), 135–146 (2007)
3. Mitra, S., Pal, S.K., Mitra, P.: Data mining in soft computing framework: a survey. IEEE Trans. Neural Netw. **13**(1), 3–14 (2002)
4. Mukhopadhyay, A., Maulik, U., Bandyopadhyay, S., Coello, C.A.C.: A survey of multiobjective evolutionary algorithms for data mining: part I. IEEE Trans. Evol. Comput. **18**(1), 4–19 (2014)
5. Marz, N., Warren, J.: Big Data: Principles and Best Practices of Scalable Realtime Data Systems. Manning Publications Co. (2015)
6. Bello-Orgaz, G., Jung, J.J., Camacho, D.: Social big data: recent achievements and new challenges. Inf. Fusion **28**, 45–59 (2016)
7. Bolón-Canedo, V., Sánchez-Maroño, N., Alonso-Betanzos, A.: Recent advances and emerging challenges of feature selection in the context of big data. Knowl.-Based Syst. **86**, 33–45 (2015)
8. Gaber, M.M., Zaslavsky, A., Krishnaswamy, S.: Mining data streams: a review. ACM Sigmod Rec. **34**(2), 18–26 (2005)
9. Babcock, B., Babu, S., Datar, M., Motwani, R., Widom, J.: Models and issues in data stream systems. In: Proceedings of the Twenty-First ACM SIGMOD-SIGACT-SIGART symposium on Principles of Database Systems, pp. 1–16. ACM (2002)
10. Jiang, N., Gruenwald, L.: Research issues in data stream association rule mining. ACM Sigmod Rec. **35**(1), 14–19 (2006)
11. SAS Institute Inc.: Five big data challenges and how to overcome them with visual analytics, Report, pp. 1–2 (2013)

References

Part IV
Computational Intelligence: Algorithms, Applications and Future Directions

Obesity Prediction Using Ensemble Machine Learning Approaches

Kapil Jindal, Niyati Baliyan and Prashant Singh Rana

Abstract At the present time, obesity is a serious health problem which causes many diseases such as diabetes, cancer, and heart ailments. Obesity, in turn, is caused by the accumulation of excess fat. There are many determinants of obesity, namely age, weight, height, and body mass index. The value of obesity can be computed in numerous ways; however, they are not generic enough to be applied in every context (such as a pregnant lady or an old man) and yet provide accurate results. To this end, we employ the R ensemble prediction model and Python interface. It is observed that on an average, the predicted values of obesity are 89.68% accurate. The ensemble machine learning prediction approach leverages generalized linear model, random forest, and partial least squares. The current work can further be improvised to predict other health parameters and recommend corrective measures based on obesity values.

Keywords Obesity · Prediction · Machine learning · Ensemble
Accuracy

1 Introduction

Many people suffer from the problem of being overweight, i.e., the problem of obesity, without even being aware of how to check obesity, and body mass index (BMI). Obesity has multiple levels, i.e., levels 1, 2, and 3. These levels are determined by the BMI, which in turn depends on weight and height alone. However, obesity additionally depends on age and gender. For instance, if the ages

K. Jindal (✉) · N. Baliyan · P. S. Rana
Computer Science and Engineering Department, Thapar University, Patiala, India
e-mail: kapil.jindal786@gmail.com

N. Baliyan
e-mail: niyati.baliyan@thapar.edu

P. S. Rana
e-mail: prashant.singh@thapar.edu

P. K. Sa et al. (eds.), *Recent Findings in Intelligent Computing Techniques*,
Advances in Intelligent Systems and Computing 708,
https://doi.org/10.1007/978-981-10-8636-6_37

of two persons are 88 and 22 years, whereas the weight is the same for both ages, then the BMI is the same for both persons. However, the obesity level is different for both persons. In our knowledge, there is no mathematical formulation that explains and/or calculates this gap in obesity levels. Therefore, we are motivated to employ machine learning techniques [1] in order to achieve accurate results of obesity values in a wide variety of situations.

2 Background

In the following subsections, we summarize the terminology used in our work.

2.1 Body Mass Index

Body mass index is denoted by BMI and depends on the weight and height only. If weight is in kilogram (kg) and height in meter (m), then the equation is [2].

$$\mathrm{BMI} = \mathrm{Weight}/(\mathrm{Height} \times \mathrm{Height})$$

If weight is in pounds (lb) and height in inches (in), then the equation changes to:

$$\mathrm{BMI} = (\mathrm{Weight}/(\mathrm{Height} \times \mathrm{Height})) \times 703.0704$$

2.2 Basal Metabolic Rate

Basal metabolic rate (BMR) depends on age, weight, height, gender. It defines the rate of energy consumed by the human body. According to Harris and Benedict [3], if the gender is male, then

$$\mathrm{BMR} = (13.7 \times \mathrm{weight}) + (5 \times \mathrm{height}) - (6.8 \times \mathrm{age}) + 66$$

and if the gender is female, then

$$\mathrm{BMR} = (9.6 \times \mathrm{weight}) + (1.8 \times \mathrm{cm}) - (4.7 \times \mathrm{age}) + 655$$

where the unit of weight is kg, unit of height is cm, unit of age is year.

2.3 Resting Metabolic Rate

Resting metabolic rate (RMR) depends on age, weight, height, gender. According to [4], the equation of RMR has two variants: one for male and the other for female.

If the gender is male, then

$$\text{RMR} = (10 \times \text{weight}) + (6.25 \times \text{cm}) - (5 \times \text{age}) + 5$$

and if the gender is female, then

$$\text{RMR} = (10 \times \text{weight}) + (6.25 \times \text{cm}) - (5 \times \text{age}) - 161$$

where the unit of weight is kg, unit of height is cm, unit of age is year.

2.4 Body Fat Percentage

Body fat percentage (BFP) is researched by Deurenberg [5] and is calculated by using BMI, age, and gender. BFP has two variants: one for child and the other for adult.

If the person is a child, then

$$\text{Fat}\,\% = (1.51 \times \text{BMI}) - (0.70 \times \text{age}) - (3.6 \times \text{gender}) + 1.4$$

and if the person is an adult, then

$$\text{Fat}\,\% = (1.20 \times \text{BMI}) + (0.23 \times \text{age}) - (10.8 \times \text{gender}) - 5.4$$

where the unit of age is year, the gender value is kept at 1 for males and 0 for females.

2.5 Protein Recommended Dietary Allowance

Protein recommended dietary allowance is denoted by protein RDA and is used to calculate daily need of protein in grams. It depends on the body weight and work done by the body. The equation for calculating protein RDA has two variants.

If the person is a non-athlete, then Eq. [6] is

$$\text{Protein RDA} = \text{weight} \times 0.8$$

If the person is an athlete, protein RDA also depends on the amount of work done, which can range from 1.4 to 1.8.

$$\text{Protein RDA} = \text{weight} \times \text{range}$$

where the unit of weight is kg, unit of protein RDA is grams per day.

3 Proposed Work

Figure 1 outlines our ensemble approach for obesity prediction using machine learning.

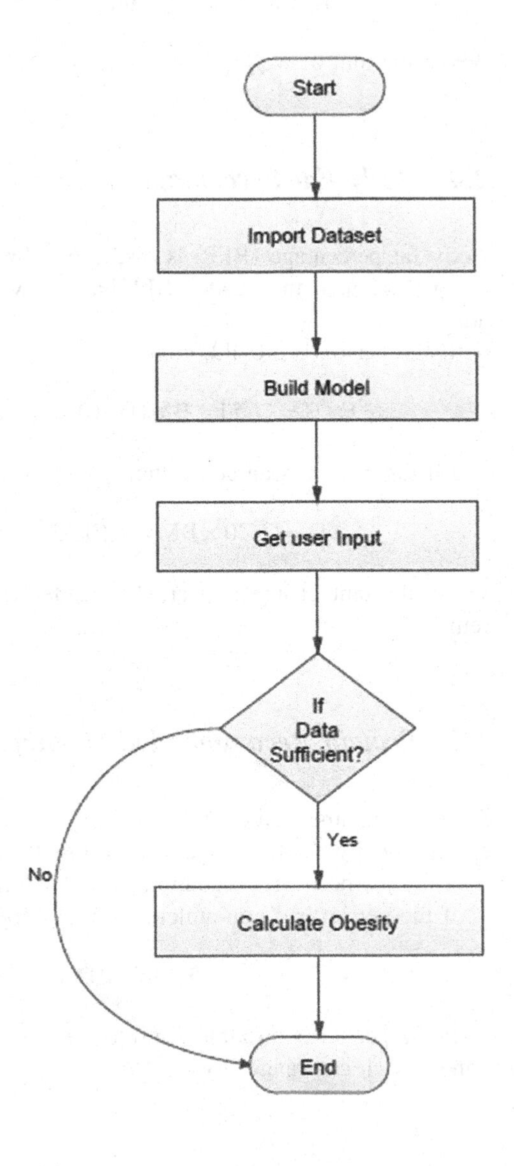

Fig. 1 Flowchart for calculating obesity value

Table 1 Sample dataset

Obesity	Age	Weight	Height	Gender	BMI
0.42	20	65	157.32	1	26.26
1.2	23	79	157.32	1	31.92
3.01	25	108	162.56	2	40.87
3.48	32	69	126	1	43.46
2.47	57	128	182.88	2	37.37
1.29	83	79	158.49	1	31.45

The accuracy of models depends on the dataset, so firstly we clean the dataset for better results. If the dataset is very large, then apply feature selection. The dataset contains the following parameters with their units mentioned as age in years, weight in kilograms, height in centimeters, gender as a 1 or 0 value, BMI in weight per meter square obesity levels 1, 2, or 3.

Table 1 presents sample dataset for the given problem.

The user inputs only age, weight, height, gender, and athlete/non-athlete attributes and gets as output the obesity level, BMI, BMR, RMR, BFP, and protein RDA. By using this information, they can prevent many diseases that are caused by obesity. Moreover, for a pregnant lady, BMI will use the same equation, but result of obesity is different because pregnant lady will gain the average weight 25–35 lb or 11–15 kg [7]. For normal person, obesity depends on the BMI range. They have three classes of obesity—class 1, class 2, and class 3. These classes further depend on the BMI [2].

Table 2 highlights the relationship between obesity classes and BMI range.

Obesity class 1 starts from BMI of 30. If we have BMI between 31 and 34.8, then the output of obesity class is 1, but by using machine learning model, we predict the value of obesity in float variable, which is more precise.

Table 3 describes the division of dataset across various machine learning models in our ensemble approach. Next, we take the arithmetic mean of the output of every machine learning model and then test with the fourth part of data. The code is executed 50 times for the verification of results. We choose the model which has the best accuracy.

Every model has method argument value, type, packages, and tuning parameters. Argument value is responsible for calling the function. Type defines the type of

Table 2 Obesity related to BMI

BMI (kg/m × m)	Weight classification	Obesity class	Disease risk
<18.5	Underweight	–	–
18.5–24.9	Normal	–	–
25.0–29.9	Overweight	–	Increased
30.0–34.9	Obese	Obesity class 1	High
35.0–39.9	Obese	Obesity class 2	Very high
≥40	Extremely obese	Obesity class 3	Extremely high

Table 3 Description of dataset

S. no.	Data count	Model name
1	200 rows	Generalized linear model
2	200 rows	Partial least squares
3	200 rows	Random forest

Table 4 Description of machine learning model

Model	Argument	Type	Package	Tuning parameters
Generalized linear model	Glm	Dual use	None	None
Partial least squares	kernelpls	Dual use	pls	Ncomp
Random forest	Rf	Dual use	Random Forest	Mtry

model, namely classification, regression, or dual use. Dual use defines the use of both classification and regression in their model. There are a number of packages needed to execute model [1], as given in Table 4.

Every machine learning model needs packages; however, some models have inbuilt packages and some models require to install that packages. Then, firstly set path of the environment variable of your system to bin folder of R. Then, package will be installed via command prompt or R scripts.

The dataset has two partitions: one for training data and the other for testing. The training data is used to build model, and if we use more than one model, then the training data divides the partitions. After building the model, we check the accuracy of model by using testing data. If the accuracy is high, then use this model; otherwise, execute the code again because the model has been trained for random sample of data.

Figure 2 depicts entire workflow of our model. Here, 0.5 refers to the control of entire model and user inputs are stored in .csv file. Model output is stored in .csv file. Python is used for the interface of R.

After building the model, create interface of the model using Python and take input from user and store it into file with extension csv. If R script executes the model and imports .csv file as input, then the model returns output via another .csv file. By using Python, we read this file. Python is used as the interface of R. It requires some library files for different functions.

For using csv file: Import csv

For opening csv file: Open ('file_name.csv', 'wb') Open ('file_name.csv', 'rb')

- W is used for writing mode, b is writing in binary mode,
- Rb is used for reading mode.

For executing R script:

```
Import os
os.system ("Rscript File_name.R")
```

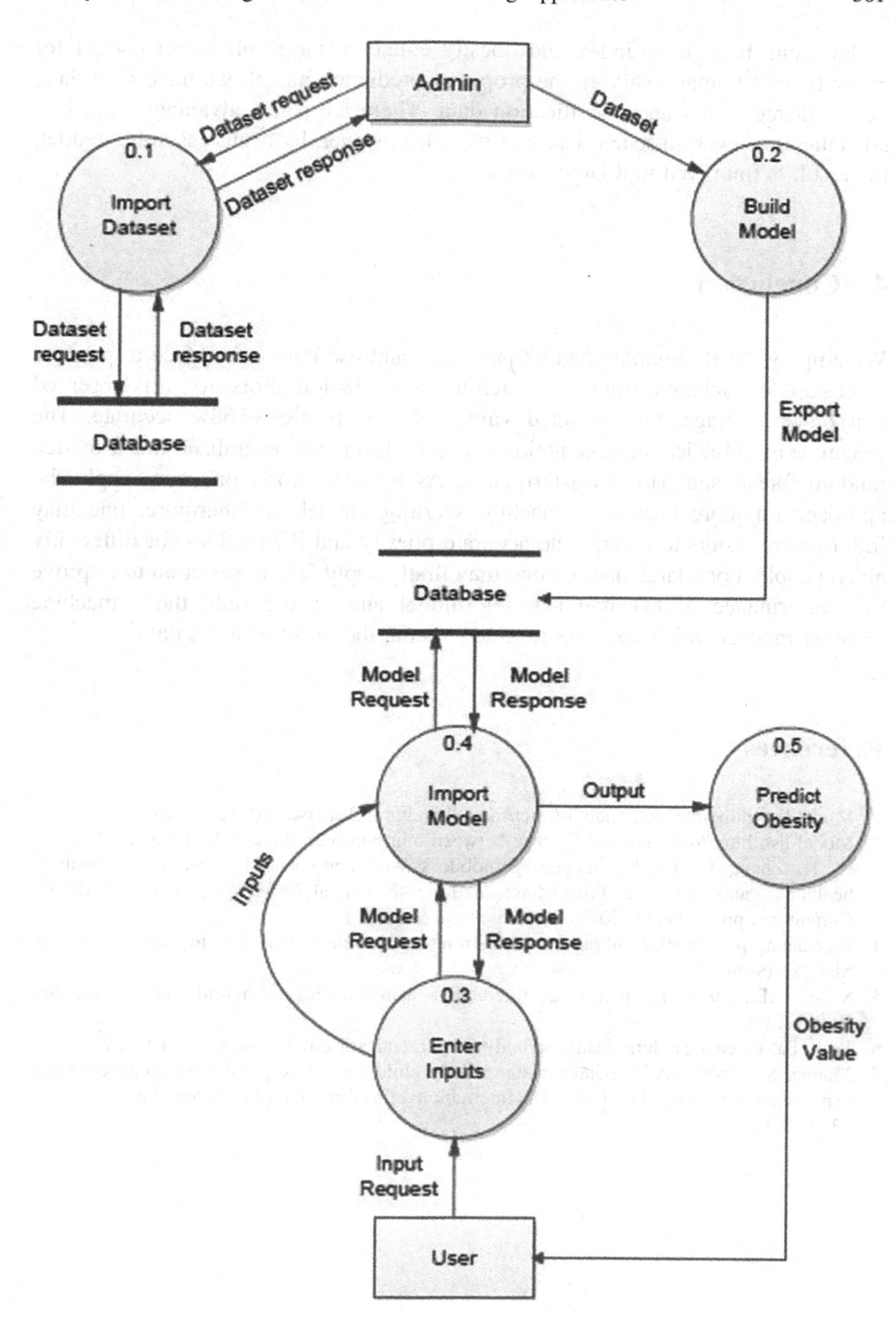

Fig. 2 Description of model workflow

By using body mass index and obesity equation, the result is not correct for every type of human, while in the proposed prediction model, we have dual data, i.e., both regression and classification data. There are some advantages and disadvantages of every machine learning model; however, by using ensemble model, the result is improved to a large extent.

4 Conclusion

We employ the R ensemble prediction model and use Python interface to propose an ensemble machine learning approach to the prediction of obesity. It is observed that on an average, the predicted values of obesity are 89.68% accurate. The ensemble machine learning prediction approach leverages generalized linear model, random forest, and partial least squares. As a future work, one may apply the approach on more than three machine learning models. Furthermore, one may improve the results to incorporate accurate obesity and BMI values for differently abled people. For a large dataset, one may firstly apply feature selection to improve the performance of machine learning model and choose only those machine learning models which take less time to execute the result of input data.

References

1. Model list. http://topepo.github.io/caret/modelList.html. Accessed 10 Aug 2016
2. Model list. http://pediaa.com/difference-between-bmr-and-rmr/. Accessed 12 Sep 2016
3. Li, H., Zhang, Q., Lu, K.: Integrating mobile sensing and social network for personalized health-care application. In: Proceedings of the 30th Annual ACM Symposium on Applied Computing, pp. 527–534 (2015)
4. Calculating protein RDA. http://www.livestrong.com/article/343966-how-to-calculate-protein-rda/. Accessed 15 June 2016
5. Shao, Y.E.: Body fat percentage prediction using intelligent hybrid approaches. Sci. World J. (2014)
6. Body Fat Percentage. http://halls.md/body-fat-percentage-formula/. Accessed 1 Oct 2016
7. Manna, S., Jewkes, A.M.: Understanding early childhood obesity risks: an empirical study using fuzzy signatures. In: 2014 IEEE International Conference on Fuzzy Systems, pp. 1333–1339 (2014)

Performance Analysis for NFBN—A New Fuzzy Bayesian Network Learning Approach

Monidipa Das and Soumya K. Ghosh

Abstract The fuzzy Bayesian networks (FBNs) have gained substantial research interest in recent years. This paper presents a probabilistic performance analysis for NFBN, a new fuzzy Bayesian network learning approach, proposed in our earlier work. Previously, it has been shown empirically that NFBN produces encouraging results with high accuracy for multivariate time series prediction. The present paper focuses on the *theoretical analysis* of the learning performance of NFBN, based on *four evaluation criteria*, namely (1) *consistency*, (2) *preciseness*, (3) *learning sensitivity to number of parents* of any node in the network, and (4) *learning sensitivity to domain size*. Finally, using *two case studies*, it has been shown that the theoretical assessment conforms to the empirical results.

Keywords Bayesian network · Fuzzy Bayesian learning
Causal dependency graph · Performance analysis

1 Introduction

Fuzzy Bayesian networks (FBNs) are the generalization of classical Bayesian networks where the networks contain variables having fuzzy states [5]. Though Bayesian networks (BNs) are considered to be effective tool for representing and reasoning with uncertain knowledge [3], sometimes it becomes hard to express knowledge in BNs because of ambiguity due to the lack of information and expert knowledge [2]. In that case, fuzzy Bayesian networks, which combine fuzzy methods and BNs, can be used to tackle the situation effectively.

Recently fuzzy Bayesian networks have gained growing interest for application in various fields including fault diagnosis [6], prediction [1], industrial management

M. Das (✉) · S. K. Ghosh
Department of Computer Science and Engineering, Indian Institute of Technology, Kharagpur, Kharagpur 721302, India
e-mail: monidipadas@hotmail.com

S. K. Ghosh
e-mail: skg@iitkgp.ac.in

P. K. Sa et al. (eds.), *Recent Findings in Intelligent Computing Techniques*, Advances in Intelligent Systems and Computing 708,
https://doi.org/10.1007/978-981-10-8636-6_38

[4]. This paper aims at analyzing the performance of a new fuzzy Bayesian network, termed as NFBN, proposed in our previous work [1]. NFBN is a variant of FBN [6] which is treated to be one of the most popular approaches for fuzzy Bayesian networks (FBNs). In our earlier work, NFBN has been proved to show better performance compared to FBN [6] in an empirical study on weather prediction with multiple climatological variables. However, no theoretical performance analysis for NFBN was carried out in [1]. The present paper analyzes the learning performance of NFBN from both theoretical and empirical perspectives. The assessment has been made with respect to *four criteria*, which have been formally defined in the subsequent subsection.

1.1 Proposed Performance Criteria

The various performance criteria used in our present study are defined below:

Definition 1 *Consistency:* A fuzzy Bayesian network approach is said to be consistent if the summation of the hypothesis probabilities related to any set of fuzzy evidences is unitary.

For example, if H_i be any hypothesis to be tested and $\tilde{\epsilon_k}$ be any fuzzy evidence, then as per the Definition 1, a corresponding FBN approach is consistent if,

$$\sum_{i=1}^{m} P(H_i|\tilde{\epsilon_1}, \tilde{\epsilon_2}, \ldots, \tilde{\epsilon_k}) = 1 \tag{1}$$

where m is the total number of values that can be achieved by the hypothesis.

Definition 2 *Preciseness*: If $\{\tilde{\epsilon_1}, \tilde{\epsilon_2}, \ldots, \tilde{\epsilon_k}\}$ be the set of fuzzy variables involved in either marginal or conditional probability estimation in a FBN approach, then preciseness in probability estimation can be defined as follows:

$$Preciseness(\tilde{\epsilon_1}, \tilde{\epsilon_2}, \ldots, \tilde{\epsilon_k}) = \frac{1 + |nonzero_{actual}|}{1 + |nonzero_{actual} \cup nonzero_{considered}|} \tag{2}$$

where $nonzero_{actual}$ = A set of observed values, actually having nonzero membership in $\tilde{\epsilon_1}, \tilde{\epsilon_2}, \ldots, \tilde{\epsilon_k}$, and $nonzero_{considered}$ = A set of observed values, considered to have nonzero membership in $\tilde{\epsilon_1}, \tilde{\epsilon_2}, \ldots, \tilde{\epsilon_k}$.

The more is the preciseness, the less is the uncertainty *introduced* by the FBN.

Definition 3 *Parent Sensitivity*: Let TC be the time complexity of a fuzzy Bayesian network learning approach, and K be the maximum number of parents of any node present in the network (causal dependency graph or CDG), then the network learning sensitivity on maximum number of parents K, denoted by SoP, can be defined as the rate of change of TC with respect to K, i.e.,

$$SoP = \frac{d(TC)}{dK} \quad (3)$$

where TC is a function of K.

Definition 4 *Domain size Sensitivity*: Let TC be the time complexity of a fuzzy Bayesian network learning approach, and D be the maximum size of domain for any node/variable present in the network (causal dependency graph or CDG), then the network learning sensitivity on maximum domain size D, denoted by SoD, can be defined as the rate of change of TC with respect to D, i.e.,

$$SoD = \frac{d(TC)}{dD} \quad (4)$$

where TC is a function of D.

The above-described performance criteria are not only applicable for NFBN, but for any other fuzzy Bayesian network learning approach as well.

1.2 Contributions

The major contributions in this work can be summarized as follows:

- *Defining* consistency, preciseness, and learning sensitivity as three important performance criteria for fuzzy Bayesian networks (FBNs);
- *Analyzing* the consistency, preciseness, and learning sensitivity of NFBN [1], in comparison with FBN [6];
- *Verification* of the theoretical analyses through empirical study, considering various structures of fuzzy Bayesian network.

The rest of the paper is organized as follows: Sect. 2 discusses various fuzzy Bayesian network learning techniques, especially FBN [6] and NFBN [1]. The theoretical performance analyses of NFBN have been presented in Sect. 3. The empirical study for verification of NFBN performance has been presented in Sect. 4. Finally, we conclude in Sect. 5.

2 FBN and NFBN—An Overview

Several models of BN with incorporated fuzzy logic [4, 6] have been proposed till date. Among the various approaches of FBN, the most popular is the one proposed by Tang and Liu [6]. The working principle of FBN [6] is as follows: Let

$\{A_1, A_2, \ldots, A_p\}$ and $\{B_1, B_2, \ldots, B_q\}$ be two sets of events. Also, let $\tilde{A}$ and $\tilde{B}$ be any two corresponding fuzzy events. Then, according FBN [6],

$$P(\tilde{B}|\tilde{A}) = \frac{\sum_{j=1}^{q} \sum_{i=1}^{p} \mu_{\tilde{B}}(B_j) \cdot \mu_{\tilde{A}}(A_i) \cdot P(A_i|B_j) \cdot P(B_j)}{P(\tilde{A})} \quad (5)$$

where μ denotes fuzzy membership value, and $P(\tilde{A})$ is fuzzy marginal probability, defined as:

$$P(\tilde{A}) = \sum_{i=1}^{p} \mu_{\tilde{A}}(A_i) \cdot P(A_i) \quad (6)$$

However, the two major limitations in FBN [6] are its *less preciseness* and *high computation power requirement* during parameter learning. Attempt has been made in [1] to overcome these limitations by devising a new fuzzy Bayesian network learning approach, termed as NFBN.

2.1 New Fuzzy Bayesian Network (NFBN)

The NFBN [1] is a variant of FBN [6] and produces more precise parameter estimates, considering the fuzzy membership of each individual observed values into the other ranges. Moreover, NFBN replaces the exhaustive computation involving each and every pair of ranges in FBN [6], with more simplistic computation involving only the observed values having nonzero membership in the considered range, and thereby reduces the time requirement.

The working principles of NFBN can be described as follows: Let $\{A_1, \ldots, A_p\}$ and $\{B_1, \ldots, B_q\}$ be two sets of events corresponding to the variables x and y, respectively—where $A_1, \ldots, A_p$ and $B_1, \ldots, B_q$ are in the form of range of values achieved by x and y; p and q are any positive integer $\in I$. Also, let $\tilde{A}$ and $\tilde{B}$ be any two corresponding fuzzy events. Then, according to *NFBN* [1],

$$P(\tilde{B}/\tilde{A}) = \frac{|\{m_i | \mu_{\tilde{B}}(y_{m_i}) > 0, \quad \mu_{\tilde{A}}(x_{m_i}) > 0\}|}{N \cdot P(\tilde{A})} = \frac{\left[P(A,B) + \frac{\gamma}{N}\right]}{\left[P(A) + \frac{\delta}{N}\right]} = P(B/A) \cdot \frac{\left[1 + \frac{\gamma}{N \cdot P(A,B)}\right]}{\left[1 + \frac{\delta}{N \cdot P(A)}\right]} \quad (7)$$

where A and B are corresponding crisp sets; $\{m_1, m_2, \ldots, m_N\}$ is a set of all the observations for the variable x and y; N is the total number of such observations; x_{m_i}= value of the variable x in the ith observation (m_i); y_{m_i}= value of the variable y in the ith observation (m_i); $\mu_{\tilde{A}}(x_{m_i})$ = membership of the value x_{m_i} in the fuzzy set $\tilde{A}$; $\mu_{\tilde{B}}(y_{m_i})$ = membership of the value y_{m_i} in the fuzzy set $\tilde{B}$; γ is the number (or count) of observations m_i such that $x_{m_i} \notin A$ and $y_{m_i} \notin B$, but $\mu_{\tilde{A}}(x_{m_i}) > 0$ and $\mu_{\tilde{B}}(y_{m_i}) > 0$; and δ is the count of observations m_i such that $x_{m_i} \notin A$, but $\mu_{\tilde{A}}(x_{m_i}) > 0$.

Here, in *NFBN*, the fuzzy marginal probability $P\left(\tilde{A}\right)$ is defined as:

$$P\left(\tilde{A}\right) = \frac{\mid \left\{m_i \mid \mu_{\tilde{A}}(x_{m_i}) > 0, m_i \in \left\{m_1, \dots, m_N\right\}\right\} \mid}{N} = P(A) + \frac{\delta}{N} = P(A) \cdot \left[1 + \frac{\delta}{N \cdot P(A)}\right] \quad (8)$$

where A is the corresponding crisp set; $\left\{m_1, \dots, m_N\right\}$ is a set of all observations for the variable x; N is the total number of observations for x; and δ is the number (or count) of observations m_i such that $x_{m_i} \notin A$, but $\mu_{\tilde{A}}(x_{m_i}) > 0$.

3 Performance Analysis of NFBN

This section presents the theoretical analysis of the performance of NFBN [1] (new fuzzy Bayesian network) from a probabilistic point of view. The analysis has been carried out with respect to three major aspects in any fuzzy Bayesian network performance, namely *consistency, preciseness,* and *learning sensitivity*.

3.1 Consistency Analysis

Theorem 1 *The new fuzzy Bayesian network (NFBN) approach becomes consistent iff* $\sum_i \gamma_i = \delta$, *while measuring the conditional probability.*

Proof Let H_i be any hypothesis to be tested and $\tilde{\epsilon_k}$ be any fuzzy evidence, then as per the principles of NFBN,

$$\sum_{i=1}^{m} P(H_i|\tilde{\epsilon_1}, \tilde{\epsilon_2}, \dots, \tilde{\epsilon_k}) = \sum_{i=1}^{m} \frac{P(H_i, \tilde{\epsilon_1}, \tilde{\epsilon_2}, \dots, \tilde{\epsilon_k})}{P(\tilde{\epsilon_1}, \tilde{\epsilon_2}, \dots, \tilde{\epsilon_k})}$$

$$= \sum_{i=1}^{m} \frac{P(H_i, \epsilon_1, \epsilon_2, \dots, \epsilon_k) \times \left(1 + \frac{\gamma_i}{N \cdot P(H_i, \epsilon_1, \epsilon_2, \dots, \epsilon_k)}\right)}{P(\epsilon_1, \epsilon_2, \dots, \epsilon_k) \times \left(1 + \frac{\delta}{N \cdot P(\epsilon_1, \epsilon_2, \dots, \epsilon_k)}\right)} \quad [N \text{ is the total no. of observations}]$$

$$= \frac{1}{\left(1 + \frac{\delta}{N \cdot P(\epsilon_1, \epsilon_2, \dots, \epsilon_k)}\right)} \times \left(\sum_{i=1}^{m} \left(P(H_i|\epsilon_1, \epsilon_2, \dots, \epsilon_k) + \frac{\gamma_i \cdot P(H_i|\epsilon_1, \epsilon_2, \dots, \epsilon_k)}{N \cdot P(H_i, \epsilon_1, \epsilon_2, \dots, \epsilon_k)}\right)\right)$$

Now, as per the definition in classical Bayesian network analysis, $\sum_{i=1}^{m} P(H_i|\epsilon_1, \epsilon_2, \dots, \epsilon_k) = 1$. Therefore, we get—

$$\sum_{i=1}^{m} P(H_i|\tilde{\epsilon_1}, \tilde{\epsilon_2}, \dots, \tilde{\epsilon_k}) = \frac{1}{\left(1 + \frac{\delta}{N \cdot P(\epsilon_1, \epsilon_2, \dots, \epsilon_k)}\right)} \times \left(1 + \sum_{i=1}^{m} \frac{\gamma_i \cdot P(H_i|\epsilon_1, \epsilon_2, \dots, \epsilon_k)}{N \cdot P(H_i, \epsilon_1, \epsilon_2, \dots, \epsilon_k)}\right)$$

$$= \frac{1}{\left(1 + \frac{\delta}{N \cdot P(\epsilon_1, \epsilon_2, \ldots, \epsilon_k)}\right)} \times \left(1 + \frac{\sum_{i=1}^{m} \gamma_i}{N \cdot P(\epsilon_1, \epsilon_2, \ldots, \epsilon_k)}\right)$$

$$= \frac{1}{\left(1 + \frac{\delta}{N \cdot P(\epsilon_1, \epsilon_2, \ldots, \epsilon_k)}\right)} \times \left(1 + \frac{\delta}{N \cdot P(\epsilon_1, \epsilon_2, \ldots, \epsilon_k)}\right) = 1 \quad [\because Given, \sum_{i=1}^{m} \gamma_i = \delta]$$

Hence, proved (since the Definition 1 is satisfied).

3.2 Preciseness Analysis

Preciseness can be treated as the measure of uncertainty involved in the fuzzy Bayesian procedure. Lesser is the preciseness, more is the uncertainty introduced in the probabilistic estimation.

Lemma 1 *NFBN is more precise than FBN [6]*

Proof Let $\{\tilde{\varepsilon_1}, \tilde{\varepsilon_1}, \ldots, \tilde{\varepsilon_k}\}$ be a set of fuzzy variables involved in marginal or conditional probability calculation. Also, let the domain of any such variable $\tilde{\varepsilon_i}$ be represented as the set: $\{\tilde{\varepsilon_i}^1, \tilde{\varepsilon_i}^2, \ldots, \tilde{\varepsilon_i}^K\}$. Then, according to Eqs. 7 and 8, NFBN always considers those observations, for which the fuzzy membership in each $\tilde{\varepsilon_i}$ is nonzero. Therefore, in case of NFBN, $nonzero_{actual} = nonzero_{considered}$ in Eq. 2. Hence, for NFBN, $Preciseness\_NFBN(\tilde{\varepsilon_1}, \tilde{\varepsilon_1}, \ldots, \tilde{\varepsilon_k}) = 1$ always.

On the other hand, while estimating probabilities (refer Eqs. 5 and 6), FBN [6] does not check whether the observed values actually have nonzero membership in the considered domain value or not. In any case, it considers the respective value of the fuzzy variables and the influence from other values in its domain.

For example, consider the case shown in Fig. 1. The figure shows a fuzzy membership function corresponding to various domain values of a fuzzy variable $\tilde{\varepsilon_i}$. The shaded portion indicates the range of values corresponding to various observations. Now, as shown in the figure, the observed values have nonzero membership in $\tilde{\varepsilon_i}^2$ only, not in other fuzzy values, i.e., $\tilde{\varepsilon_i}^1$ and $\tilde{\varepsilon_i}^3$. However, while estimating probability corresponding to $\tilde{\varepsilon_i}^1$ or $\tilde{\varepsilon_i}^3$, FBN [6] considers the influence from $\tilde{\varepsilon_i}^2$ and thus leads to a non-empty set corresponding to $nonzero_{actual} \cup nonzero_{considered}$, although $nonzero_{actual} = \phi$. Therefore, *Preciseness* becomes < 1. Hence, for FBN [6], *Preciseness_FBN* can sometimes become < 1 that means NFBN is more precise than FBN [6]. This concludes the proof.

3.3 Learning Sensitivity Analysis

Suppose $G(V, E)$ be a directed acyclic graph (DAG) of a fuzzy Bayesian network having the set of nodes V and the set of edges E. Also, let the network or graph G

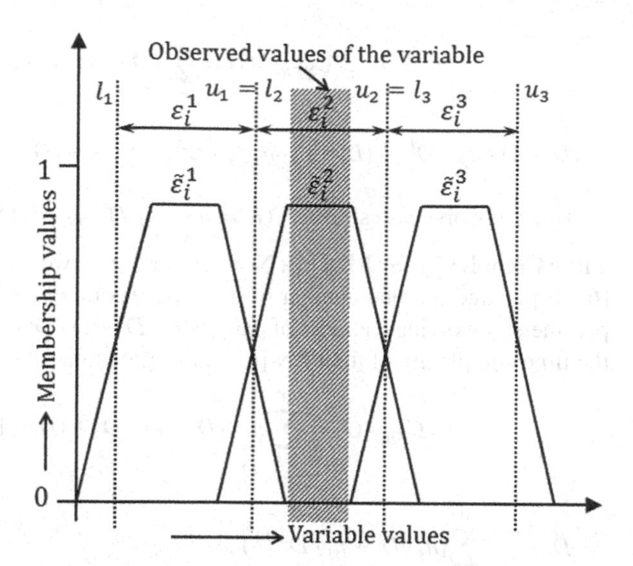

Fig. 1 Variable $\tilde{\varepsilon}_i$ along with the corresponding membership functions

contain n number of nodes, i.e., $|V| = n$ and the maximum number of parents for any node is K. Now, in case of a fuzzy Bayesian network, the learning time complexity (TC) for updating the belief (or learning the parameters) with respect to a single observation is dependent not only on K, but also on the domain size D for each variable or nodes involved in the network, and the total number of nodes n in the network as well. Therefore, $TC_{appr}(G) = f(n, K, D)$, where *appr* is the fuzzy Bayesian network learning approach.

This section provides a detailed analysis of learning time sensitivity of NFBN [1] and FBN [6], in terms of *total number of variables/nodes n, maximum domain size D*, and *maximum number of parents K* for any node in the network.

3.3.1 Time Complexity

Lemma 2 *NFBN performs better than FBN [6] with respect to time complexity*

Proof Suppose, in the DAG (G) considered, the number of nodes having K number of parents is n_K, number of nodes having $K-1$ number of parents is n_{K-1}, and so on, such that $n_K + n_{K-1} + \cdots + n_0 = n$.

Time Complexity for NFBN: As per the network learning equations for NFBN (Eqs. 8 and 7), a total number of iterations required for learning/updating a single parameter, having i number of parents, are $(D-1) \cdot D^i$, considering domain size for each variable to be D. Therefore, the time complexity of the NFBN learning phase becomes:

$$TC_{NFBN}(G) = \sum_{i=K}^{0} (D-1) \cdot n_i \cdot D^i \tag{9}$$

$$= (D-1) \cdot n_K \cdot D^K + (D-1) \cdot n_{K-1} \cdot D^{K-1} + \cdots + (D-1) \cdot n_0 \cdot D^0 = O(n \cdot D^{K+1}) \tag{10}$$

Thus, in worst case situation ($K = n - 1$), $TC_{NFBN}(G)$ becomes: $O(n \cdot D^n)$.

Time Complexity for FBN [6]: Now, as per the network learning equations for FBN [6] (Eqs. 5 and 6), total number of iterations required for learning/updating a single parameter, involving *i* number of parents, is $(D-1) \cdot D^i + (D-1) \cdot D^{2i+1}$. Therefore the time complexity of the FBN [6] learning becomes:

$$TC_{FBN}(G) = \sum_{i=K}^{0} n_i \left\{(D-1) \cdot D^i + (D-1) \cdot D^{2i+1}\right\} \tag{11}$$

$$= (D-1) \cdot \sum_{i=K}^{0} (n_i \cdot D^i + n_i \cdot D^{2i+1})$$

$$= (D-1) \cdot (n_K \cdot D^K + n_K \cdot D^{2K+1} + n_{K-1} \cdot D^{K-1} + n_{K-1} \cdot D^{2K-1} + \cdots + n_0 + n_0 \cdot D) \tag{12}$$

$$= O(n \cdot D^{2(K+1)}) \tag{13}$$

Therefore, in worst case (i.e., $K = n - 1$), $TC_{FBN}(G)$ becomes: $O(n \cdot D^{2n})$.

Hence, comparing Eqs. 10 and 13, it can be found that the NFBN is more efficient than FBN [6] in terms of run time complexity. This concludes the proof.

Lemma 3 *The NFBN learning time is less sensitive to the maximum number of parents (K) of any node in the causal dependency graph or DAG than FBN [6]*

Proof Suppose, in the DAG (G) considered, the number of nodes having K number of parents is n_K, number of nodes having $K - 1$ number of parents is n_{K-1}, and so on, such that $n_K + n_{K-1} + \cdots + n_0 = n$.

Now, from Eqs. 3 and 10,

$$SoP_{NFBN} = \frac{d(TC_{NFBN}(G))}{dK} \approx n \cdot D^{K+1} \cdot \log D \tag{14}$$

Similarly, from Eqs. 3 and 12,

$$SoP_{FBN} = \frac{d(TC_{FBN}(G))}{dK} \approx 2n \cdot D^{2(K+1)} \cdot \log D \tag{15}$$

From Eqs. 14 and 15,

$$SoP_{FBN} = 2 \cdot D^{K+1} \times SoP_{NFBN} > SoP_{NFBN} \qquad [\because D >= 1]$$

Hence, proved.

Lemma 4 *The NFBN learning time is less sensitive to the maximum domain size (D) of any variable in the network (or DAG) than FBN [6]*

Proof Suppose, in the DAG (G) considered, the number of nodes having K number of parents is n_K, number of nodes having $K-1$ number of parents is n_{K-1}, and so on, such that $n_K + n_{K-1} + \cdots + n_0 = n$.

Now, from Eqs. 4 and 10,

$$SoD_{NFBN} = \frac{d(TC_{NFBN}(G))}{dD} \approx n \cdot (K+1).D^K \quad (16)$$

Similarly, from Eqs. 4 and 12,

$$SoD_{FBN} = \frac{d(TC_{FBN}(G))}{dD} \approx 2n \cdot (K+1) \cdot D^{2K+1} \quad (17)$$

From Eqs. 16 and 17,

$$SoD_{FBN} = 2 \cdot D^{K+1} \times SoD_{NFBN} > SoD_{NFBN} \quad [\because D >= 1]$$

Hence, proved.

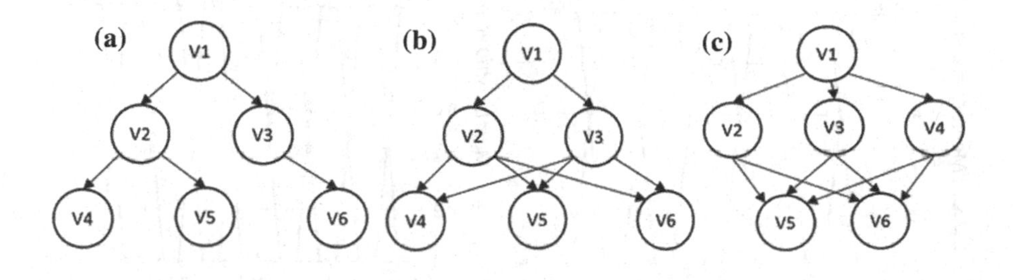

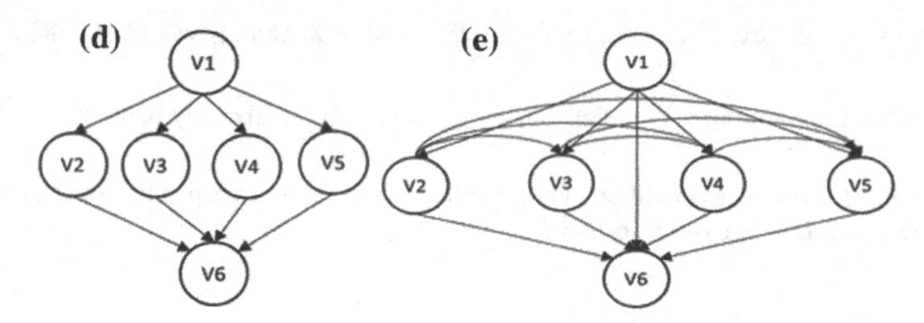

Fig. 2 CDG or DAGs: **a** DAG-1, **b** DAG-2, **c** DAG-3, **d** DAG-4, **e** DAG-5

4 Empirical Study for Measuring Performance of NFBN

In addition to the theoretical analysis as described in Sect. 3, the run time performance of NFBN learning technique has been validated using some empirical studies as well. Two main cases, regarding the sensitivity of learning time on Bayesian network structure, have been considered for this purpose. Fuzzy Bayesian networks with different network/graph structures have been used for experimental purpose. In each case, the networks have been trained with 1000 randomly chosen records. The details of the case studies have been thoroughly discussed in the subsequent part of this section.

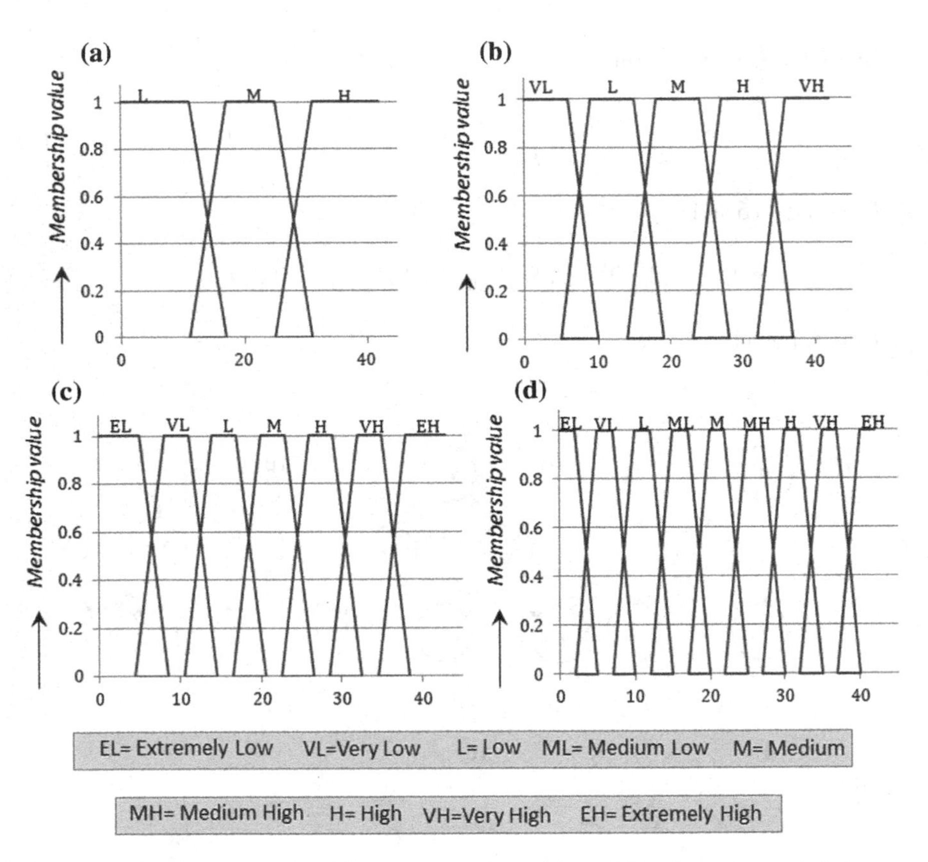

Fig. 3 Membership functions for different range values of the variables depending on their domain size (D): **a** $D = 3$, **b** $D = 5$, **c** $D = 7$, **d** $D = 9$

4.1 *Case-I: Study of Learning Time Sensitivity on the Maximum Number of Parents of Any Variable in the Network/DAG*

The empirical study over learning time sensitivity on maximum number of parents has been performed considering five different structures of network, shown in Fig. 2, and different domain size of fuzzy variables, shown in Fig. 3. As shown in Fig. 2, each of these network structures contains different numbers of maximum parents for a node within it. For example, the DAG-1 has maximum parent value = 1, whereas that for DAG-2, DAG-3, DAG-4, and DAG-5 are 2, 3, 4, and 5, respectively. On the other side, Fig. 3 shows the membership function of fuzzy variables having different domain size. For example, the fuzzy variables corresponding to Fig. 3a have a domain size = 3. Similarly, the fuzzy variables associated with Figs. 3b, 3c, and 3d have domain size of 5, 7, and 9, respectively.

In order to analyze the learning time sensitivity on maximum number of parents in the network, fuzzy variables with different domain size have been applied to each of the network structure separately. For each considered domain size of the variables, the change in learning time requirement with respect to maximum parent

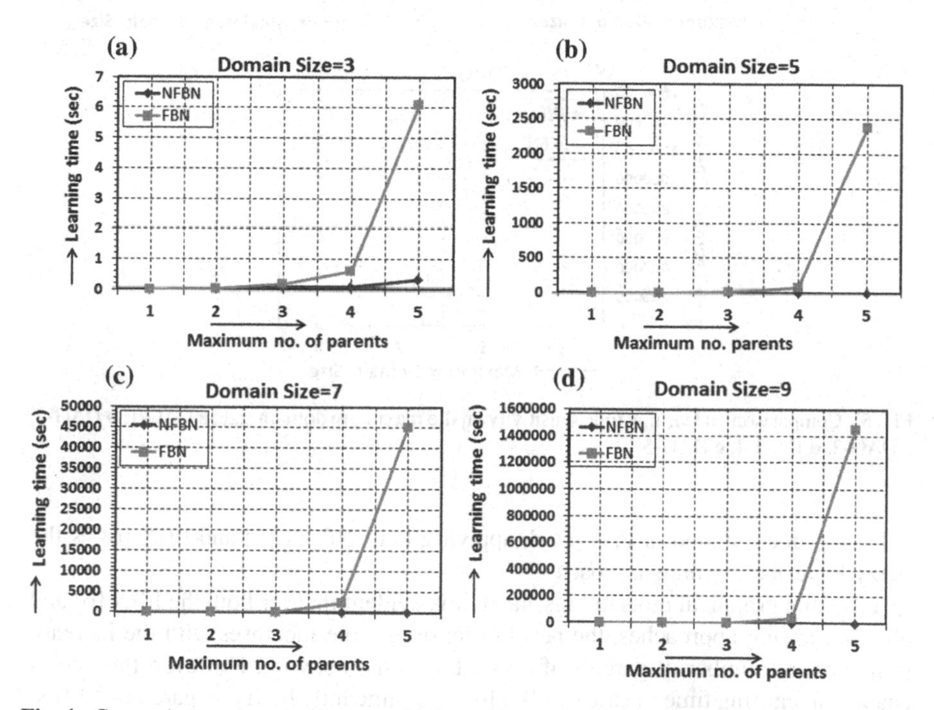

Fig. 4 Comparison on learning time sensitivity on the maximum no. of parents: **a** Domain size = 3, **b** domain size = 5, **c** domain size = 7, **d** domain size = 9

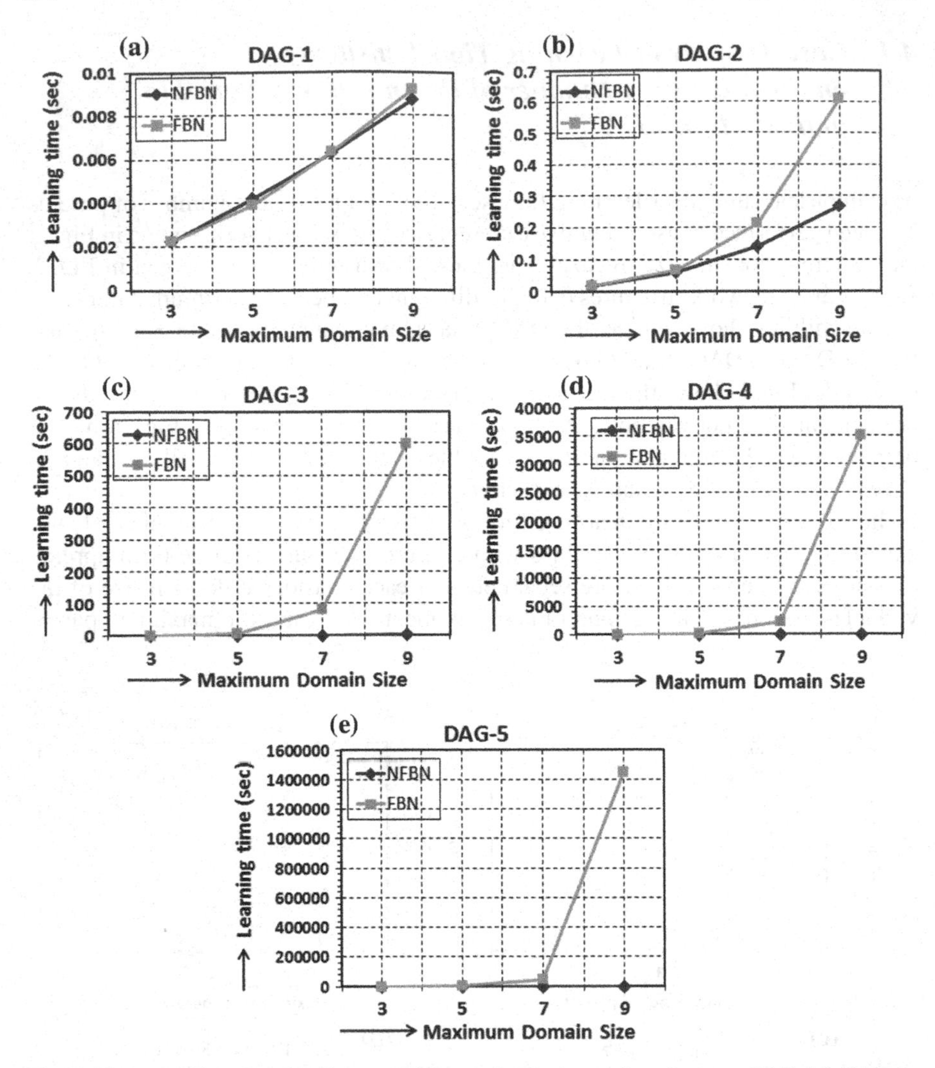

Fig. 5 Comparison on learning time sensitivity on the maximum domain size: **a** DAG-1, **b** DAG-2, **c** DAG-3, **d** DAG-4, **e** DAG-5

value has been plotted in Figs. 4a–d, applying both NFBN [1] and FBN [6] as the fuzzy Bayesian learning approach.

From the graphical plots in Figs. 4a–d, it is evident that for both the FBN [6] and NFBN learning approaches, the network learning time increases with the increase in maximum number of parents of a variable in the network. However, the rate of change in learning time in case of FBN [6] is significantly high compared to NFBN. This proves that the learning approach in FBN [6] is more sensitive to the maximum parents in the network structure than that of NFBN.

4.2 *Case-II: Study of Learning Time Sensitivity on the Maximum Domain Size of Any Variable in the Network/DAG*

In order to analyze the learning time sensitivity on maximum domain size of any variable in the network, the same structure of fuzzy Bayesian networks (Figs. 2a–e) and the same set of domain size of fuzzy variables (Figs. 3a–d) have been used in the empirical study. Each of the network structure has been learnt separately considering different domain size of the fuzzy variables. For each network structure, the change in learning time requirement with respect to the considered domain size of the variables has been plotted in Figs. 5a–e, applying both NFBN [1] and FBN [6] as the fuzzy Bayesian learning technique.

From the graphical plots in Figs. 5a–e, it is evident that for both the FBN [6] and NFBN learning approaches, the network learning time increases with the increase in maximum domain size of a variable in the network. Moreover, the rate of change in learning time in case of FBN [6] is exceptionally high compared to NFBN. This proves that the learning approach in FBN [6] is more sensitive to the maximum domain size of any variable than that of NFBN.

5 Conclusion

In the present work, we have analyzed the performance of NFBN, a new variant of fuzzy Bayesian network proposed in [1]. The performance analysis has been performed from both theoretical and empirical perspectives, considering various network structures. From theoretical analysis, it is found that the parameter learning in NFBN is more efficient than well-known FBN [6], in terms of preciseness and run time complexity. Further, the results of empirical study validate that the learning time of NFBN is significantly less sensitive to both maximum number of parents and maximum domain size of any variable in the network. In future, the work can be extended to more extensive performance study, in comparison with other variants of FBN and considering real-life applications.

References

1. Das, M., Ghosh, S.K.: A probabilistic approach for weather forecast using spatio-temporal interrelationships among climate variables. In: 2014 9th International Conference on Industrial and Information Systems (ICIIS), pp. 1–6. IEEE (2014)
2. Das, M., Ghosh, S.K.: semBnet: a semantic Bayesian network for multivariate prediction of meteorological time series data. Pattern Recognit. Lett. **93**, 192–201 (2017). https://doi.org/10.1016/j.patrec.2017.01.002

3. Das, M., Ghosh, S.K., Chowdary, V., Saikrishnaveni, A., Sharma, R.: A probabilistic nonlinear model for forecasting daily water level in reservoir. Water Resour. Manag. **30**(9), 3107–3122 (2016)
4. Ferreira, L., Borenstein, D.: A fuzzy-Bayesian model for supplier selection. Expert Syst. Appl. **39**(9), 7834–7844 (2012)
5. Fogelberg, C., Palade, V., Assheton, P.: Belief propagation in fuzzy Bayesian networks. In: 1st International Workshop on Combinations of Intelligent Methods and Applications (CIMA) at ECAI08, pp. 19–24 (2008)
6. Tang, H., Liu, S.: Basic theory of fuzzy Bayesian networks and its application in machinery fault diagnosis. In: Fourth International Conference on Fuzzy Systems and Knowledge Discovery (FSKD), vol. 4, pp. 132–137. IEEE (2007)

Implementing and Analyzing Different Feature Extraction Techniques Using EEG-Based BCI

H. S. Anupama, Raj V. Jain, Revanur Venkatesh, Rupa Mahadevan, N. K. Cauvery and G. M. Lingaraju

Abstract Brain–computer interface (BCI) is a method of communication between the brain and computer or machines, which use the neural activity of the brain. This neural activity communication does not occur using the peripheral nervous system and muscles, as is the usual case in human beings, but through any other mechanism. This paper focuses on different types of feature extraction techniques to explore a new kind of BCI paradigm and validate whether it can give a better ITR as compared to the existing paradigms.

Keywords Brain–computer interface • Spectral F Test • Fourier transform
Canonical correlation analysis • Continuous wavelet transform

1 Introduction

Artificial intelligence (AI) and machine learning (ML) are the two areas which help machines to think and perform the actions how human being does. They also help in processing the data and identifying the patterns given by the user. Brain–computer interface is one such area where AI and ML are used. One of the ways the performance of the BCI system is measured is using information transfer rate (ITR). The main aim of the proposed system is to achieve higher ITR by combining two different approaches, namely steady-state visually evoked potential and steady-state

H. S. Anupama (✉) · R. V. Jain · R. Venkatesh · R. Mahadevan
Department of Computer Science and Engineering,
R. V. College of Engineering, Bengaluru, India
e-mail: anupama_hs@rediffmail.com

N. K. Cauvery
Department of Information Science and Engineering,
R. V. College of Engineering, Bengaluru, India

G. M. Lingaraju
Department of Information Science,
M. S. Ramaiah Institute of Technology, Bengaluru, India

P. K. Sa et al. (eds.), *Recent Findings in Intelligent Computing Techniques*,
Advances in Intelligent Systems and Computing 708,
https://doi.org/10.1007/978-981-10-8636-6_39

auditory evoked potential. The system developed aims at checking if such a system can exist and if so, if it is providing accuracy that is high enough to be put to use in real-world applications.

BCI systems can be implemented in different ways [1, 2]. Evoked potentials from the brain have been used in this study. Potentials can be evoked from the brain using different types of stimuli. The ones that have been considered are as follows:

- Steady-State Visually Evoked Potential (SSVEP): This is the potential generated in the brain when a person is subjected to a visual stimulus at a particular frequency. This evokes signals in the brain which show a spike at the frequency of the visual stimulus.
- Steady-State Auditory Evoked Potential (SSAEP): This is the potential generated in the brain when a person is subjected to an auditory stimulus which is at a particular frequency. The signals evoked in the brain due to this stimulus show a spike at that frequency which is same as that of the auditory stimulus.

As can be seen from above, evoked potentials can be used in decision making. When two stimuli with different frequencies are presented to the user, the one that the user chooses to observe evokes signals in the brain with higher amplitude at that particular frequency. This observed frequency helps in identifying what the user chose. The different frequencies can be used to represent different choices to the user. This is particularly useful for people with severely impaired motor capabilities who have to resort to other methods of communication apart from the ones that are normally used.

The system proposed aims to enhance the decision-making capability of the person by providing two stimuli, which evoke two responses in different regions of the brain. This allows the user to make two decisions simultaneously. Since such a system has not been developed before, it should first be validated if the system provides results with sufficient accuracy to be able to be put to use for end users or patients.

Hence, a validation system has been developed which employs several feature extraction and classification algorithms. A comparison of different algorithms for this system has also been provided. The result of the experiment whether such a system can provide accurate results or not has also been discussed.

2 Methodology

The methods steady-state visually evoked potential (SSVEP) and steady-state auditory evoked potential (SSAEP) are two paradigms used widely till now to elicit decision from the user by using his/her attention visually or auditory, respectively. In SSVEP paradigm, two images flickering at different frequencies which are 7.5 Hz and 10 Hz, respectively. The frequencies are selected such that they are not harmonics of each other and well within the range of 6–24 Hz. SSAEP signals are generated using auditory input having two pure tones of different frequencies (amplitude modulated). The frequencies here are selected such that they are around 40 Hz so as to maximize the SNR. The characteristics of the frequencies define an SSAEP system. For example, in this system, it has been decided to give 37 and 43 Hz frequency audio input. Both the visual and auditory inputs are given concurrently to the subject.

Each of the choices in the visual and auditory stimuli stands for a particular value of a particular decision. For example, the visual stimuli can represent food selection and the auditory stimuli can represent drink selection. After making the subject understand the meaning of each of the choices, the training trials are started. Each training trial will begin by a rest period of approximately 2 s. Here the subject is asked to concentrate on one of the visual and auditory inputs as determined by the training program. This gives the subject to form his decision and also helps to get the baseline EEG data. Then, the visual and auditory inputs are given and EEG data is taken and used to get features.

- **Approaches and Background on Visual and Audio Stimuli**

SSVEP is seen when the flashing stimuli is above 6 Hz frequency. When the frequency is less than 6 Hz, it is termed as transient visual evoked potential (T-VEP). The reason for differentiating is that, when the frequency is above 6 Hz, the signal gets "embedded" into the person's brain. The effect is seen for some time even after the stimulus is removed. Whereas, in transient, the effect dies very quickly when the stimulus is removed, as compared to SSVEP.

Since, it is eyes which receive the signal, the effect is seen at the occipital region. Occipital region is a large region, and based on the 10–20 system, the response is extracted from O1, Oz, and O2 electrodes [3, 4].

Since, the brain can register objects of different types of shapes, sizes, colors, and many other visual parameters, it becomes difficult to know which would be the best combination of the parameters to provide the stimulus. It is generally seen that stimuli with simple shapes, like straight lines, square boxes, have a good effect. The reason falls down to the Hubel and Wiesel theory. In short, a lot of brain cells can register simple shapes, but to register complex shapes, a lot of cells need to work together. Thus, when simple shapes are used, the numbers of cells which can register is more; thus, the effect is pronounced. When a complex shape is used, a lot of cells need to work together (hyper-complex cells), like grouping. Thus, such groups will be less in number, and thus, the effect will not be as pronounced as in the previous case.

The usefulness of SSVEP as a BCI paradigm is there only when we can present more than one stimuli figure, so that decision can be made and extracted [5]. The effect at play is the Stiles–Crawford effect, which says the object at the centre of the visual field will get more registered as compared to those at the ends of the visual field [6–8].

Audio stimuli generate some of the least registrable responses. The reason is that it is very easy to get distracted to a visual or a tactile stimulus. And SSAEP responses require attention to the audio stimulus. Different ways of giving the audio stimulus are using click trains, pure sine tones, speech, localization-based stimuli, etc. SSAEP, also known as auditory steady-state response (ASSR), is most observed when the stimulation frequency is around 40 Hz [9–12].

Auditory stimuli register in the temporal region of the brain. According to the 10–20 system, the response can be recorded from T3, T4. According to these authors [9], Cz and Oz can be also used to discriminate responses when the stimuli are given at both the ears.

- **Approaches and Background on Feature Extraction**

The approaches can be broadly classified into frequency domain approaches, e.g., spectral F Test, Fourier transform, and time domain approaches, e.g., canonical correlation analysis, continuous wavelet transform.

1. ***Spectral F Test***

A hypothesis is a proposed explanation for a phenomenon [13]. For a hypothesis to be a scientific hypothesis, the scientific method requires that one can test it. In statistical hypothesis testing, a hypothesis is proposed for the statistical relationship between the two datasets, and this is compared as an alternative to an idealized null hypothesis that proposes no relationship between the two datasets. There are different tests available, and based on the statistical assumptions in the experiment, a suitable test must be chosen. Based on the test, a test statistic is stated, which represents the whole data. The test statistic's distribution under null hypothesis is derived, and a significance level, α (critical value), is set, generally 5 or 1%. From α, derive the possible value of test statistic for which the null hypothesis can be rejected. This partitions the distribution into two regions, one in which the null hypothesis can be rejected, called the critical region, and the other in which it cannot. Then, the observed data are used to calculate the observed test statistic. If the observed statistic falls in the critical region, the null hypothesis s rejected. If not, the test is said to be inconclusive. These kinds of studies come under a field called analysis of variance (ANOVA).

2. ***Fourier Transform (Power Spectrum)***

Fourier transform assumes that every signal is composed of sines and cosine waves with varying amplitudes and phases. The problem then boils down to finding those parameters for the given signal, which is solved with the below formula.

$$X_k = \sum_{n=0}^{N-1} x_n e^{-2\pi ikn/N}, \quad k \in Z$$

The only problem in Fourier transform is that it operates properly only on infinite signals, i.e., signals which are for infinite time. When used in its digital equivalent form known as the discrete Fourier transform (DFT) [14], this property causes problems. This fact was discovered when a digital sine wave's discrete

Fourier transform was not a single peak in the frequency domain, but rather a spread out wave. That was due to the reason that the function value outside the time limit defined is assumed to be zero. This leads to the invention of "sine" wave. Also, DFT was improvised with respect to its speed, and the algorithm formed is called fast Fourier transform (FFT). So when applied to digital signals which cannot be infinite, it causes a problem known as spectral leakage, which is exactly as mentioned previously. The amplitude or power of the signal gets spilled or leaked into other spectral components which might not be the case in actuality, but is being shown by the Fourier transform because it is been implemented in its digital form. The workaround this problem as talked by some of the researchers is to use window functions to reduce the spectral leakage. These functions have special properties which make the data have lesser spectral leakage [15]. Another problem which arises is that Fourier transform assumes that the particular sine or cosine wave is present throughout the signal. This generally is not the case. This problem is solved by dividing the data into segments of equal length (generally) and applying the Fourier transform on each segment separately. The segment length is also called epoch length. In this way, in which segment the wave of interested frequency has pronounced effect can be known. And since each segment is represented by a time, we can know at what time that wave was present. So we divide the data into segments causing a time-resolution phenomenon to occur, and this is seen in the frequency domain too. So based on the ratio of the length of the segment and the total data length, the time resolution and the frequency resolution will be obtained. So this brings in the question of what is the optimal ratio. This generally depends on the experiment paradigm or should be found with rigorous analysis codes.

3. *Canonical Correlation Analysis*

Canonical correlation analysis (CCA) was invented by Harold Hotelling [16] in 1936. CCA is a multivariable statistical method used when there are two sets of data, which may have some underlying correlation. CCA extends ordinary correlation to two sets of variables [17]. First, CCA finds a pair of linear combinations, called canonical variables, for two sets, such that the correlation between the two canonical variables is maximized. Then, it finds a second pair, which is uncorrelated with the first pair of canonical variables but has the next highest correlation. The process of constructing canonical variables continues until the number of pairs of canonical variables equals the number of variables in the smaller set.

4. *Canonical Correlation Analysis*

Wavelet transform theory, originally proposed in the field of mathematics, was introduced into the field of signal analysis by Goupillaud, Grossmann, and Morlet in 1984 [18]. Wavelet transform has its flexibility in the choice of analyzing functions, which overcomes the shortage of traditional Fourier transform method.

3 System Architecture

Systems architecture is a response to the conceptual and practical difficulties of the description and the design of complex systems. A system architecture or systems architecture is the conceptual model that defines the structure, behavior, and more views of a system [17]. The architectural design process is concerned with establishing a basic structural framework for a system. System architecture comprises of system components, the externally visible properties of those components, the relationships (e.g., the behavior) between them is shown in Fig. 1.

- **Data Acquisition Module**: Axxonett's BESS [19] has inbuilt stimulus control protocol, in which the visual and audio stimulus are loaded and kept. At run time, the signals are provided to the display monitor and the earphones, respectively. The subject is made to wear the headset, and the BESS software acquires all the signals from all channels and stores it as .EEG file. The .EEG file is converted to .EDF file using a module of BESS and stored in the database of all EDF files.
- **Preprocessing Module**: The EDF files are made to undergo two stages of preprocessing. The data are filtered using a notch filter (50 Hz) and bandpass filter (frequency range is based on the response being processed). The filtered data are made to undergo a range of feature extraction algorithms which extract information related to frequency, time, spatial domains. All the features are stored in a features database with labels added to them.
- **Training and Testing Module**: Training and testing is done as part of cross-validation. The training program uses part of the features as training features to train a machine learning model. That model is tested on rest of the features to get an accuracy measure. This procedure happens several times as

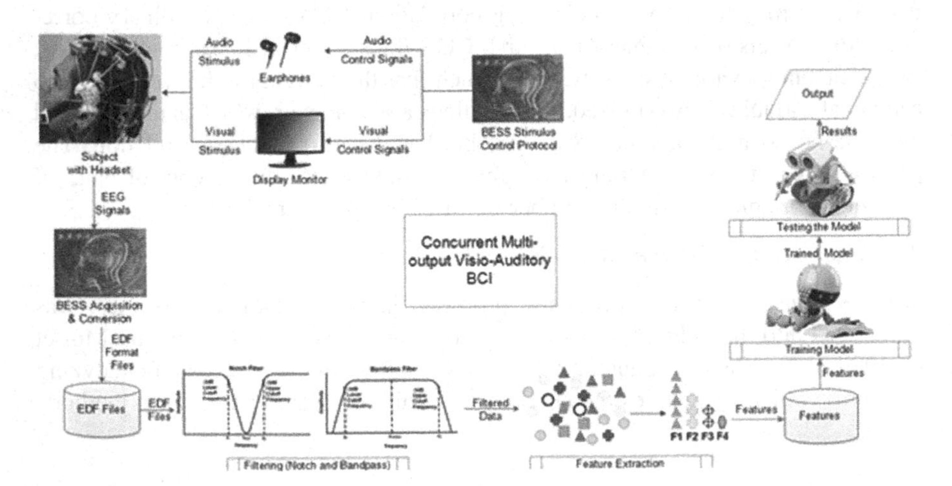

Fig. 1 System architecture

deemed by the cross-validation program. In this way, the accuracy measure is got for various machine learning models. Since each of them tackles the same problem differently, different accuracy measures are obtained and that is the analysis which is done and presented as output of the system.

4 Results and Analysis

Any machine learning application is data driven. Hence, the dataset considered for the application is of utmost importance. The details of dataset are as given below.

Data were collected from a total of 10 healthy subjects consisting of seven males and three females. All of them were aged between 18 and 23. Nobody had had any neurological disorder.

The experiment was conducted in an isolated room with no electronic devices running except for the data acquisition system. The participants were also instructed not to wear any metallic items and to keep away all electronic gadgets, to minimize any interference with the acquisition. The acquisition system consisted of an EEG cap that could be adjusted according to the participant's head dimensions, 16 Ag-AgCl electrodes, an amplifier, a Bluetooth device to record signals, a display on which the visual stimulus was presented, earphones through which the audio stimulus was presented, and a system with BESS software running on it, to collect the EEG data. The electrodes were placed according to the 10–20 system. The EEG data were sampled at the rate of 256 Hz as shown in Fig. 2. Also, a Notch filter of 45 Hz is applied to the data. Time-stamped data are then recorded using the BESS software. The BESS software saves data as a.EEG file. However, the MATLAB readable format is .EDF file. Hence, this conversion has to be done manually.

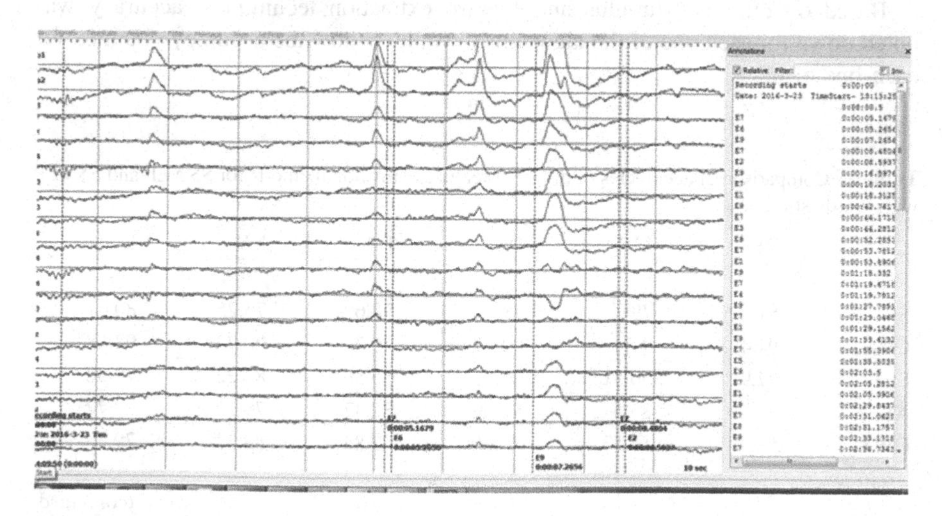

Fig. 2 EEG recordings of one of the subjects

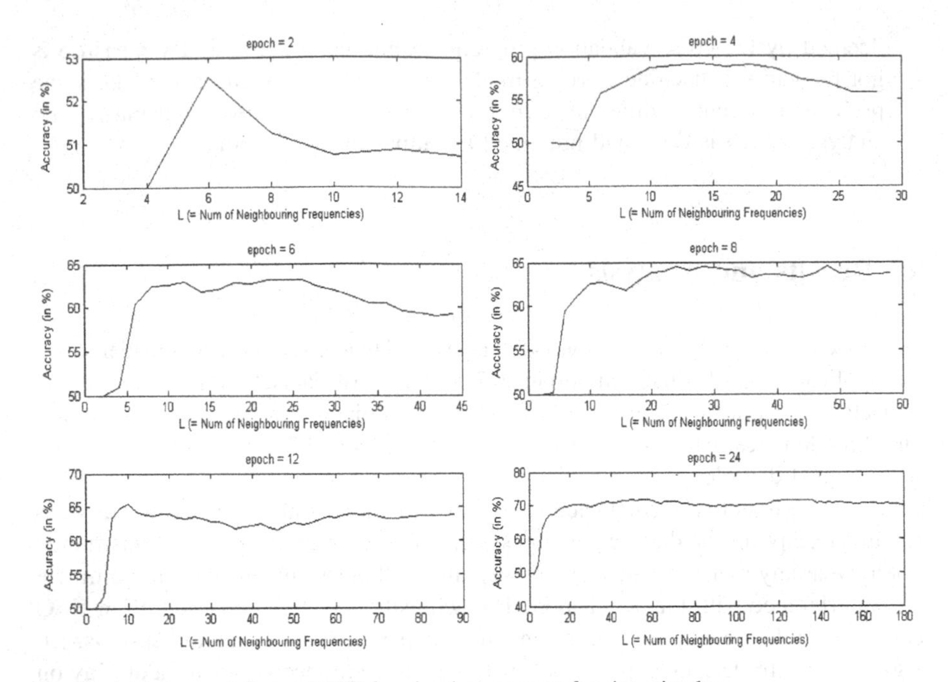

Fig. 3 Accuracy variations of SFT for visual component for sine stimulus

The output obtained from a system is evaluated against the different metrics to measure the performance of the system. Several algorithms have been used to validate the paradigm.

Figure 3 shows the accuracy variations of SFT for visual component for sine stimulus with respect to epoch and number of neighboring frequency variations.

Based on different stimulus and feature extraction techniques, accuracy was noted down for each one of the algorithms. Table 1 show the accuracy obtained for each one of it.

Table 1 Comparison of accuracies of different feature extraction methods for SSAEP and SSVEP with words stimulus

	RF	SVM	DT	DA	NB	KNN
Visual						
SFT	81	79	78	78.69	75.25	80.41
CCA	91.4	91.4	90.37	90.72	90.72	92.78
FFT	87.97	50.51	84.87	85.56	85.22	79.38
ASFT	81	83.5	81.09	82.47	79.72	79.72
CWT	86.9	84.53	86.25	83.84	83.33	79.38
Accuracy						

(continued)

Table 1 (continued)

	RF	SVM	DT	DA	NB	KNN
SFT	58.7	54.83	58.7	58.7	52.25	53.54
CCA	60.6	49.03	58.06	45.16	53.54	54.19
FFT	56.12	58.7	56.12	49.67	55.14	50.32
ASFT	59.3	55.48	52.9	54.83	53.06	54.83
CWT	56.12	52.9	61.93	46.45	48.95	54.19

5 Conclusion and Future Enhancement

BCI systems are gaining more attention over time because of their importance in neuro-prosthetics. BCIs are aimed at restoring some capability of a person. The focus has been shifted a lot toward helping paralyzed people.

With the results and analysis, the following conclusions can be made:

- Feature extraction techniques used in this work has got a good efficiency and also obtained good results.
- The implementation of a concurrent SSVEP and SSAEP system has achieved good results.
- The system implemented was only to check if the above paradigms can be combined to obtain acceptable levels of differentiation in the EEG data based on the choice made by the subjects.

Any system's limitations are areas which call for improvements. Additionally, other enhancements can also be made to make sure that the system can be finally converted into a product with market value. Some of the enhancements that can be done are stated below.

- Further classification algorithms can be implemented for the same extraction techniques and the results can be compared for each one of them.
- The system has to be transformed from a validation system to an end product that can be used by paralyzed people. The system so developed should have real-time processing capabilities to provide results as soon as possible to the end users.

6 Declaration

Authors have obtained all ethical approvals from appropriate ethical committee and approval from the subjects involved in this study.

Acknowledgements We would like to thank Axxonet solutions pvt limited for their immense support in getting the data for our work. They helped in providing the BESS software and the supporting hardware for data extraction from the human brain.

Reference

1. Vallabhaneni, A., et. al.: Brain-computer interface. Neural Eng. 85–121 (2005)
2. Lance, B.J., et al.: Brain-computer interface technologies in the coming decades. Proc. IEEE **100**, 1585–1599 (2012)
3. Zhu, D., et al.: A survey of stimulation methods used in SSVEP-based BCIs. Comput. Intell. Neurosci. (2010). https://doi.org/10.1155/2010/702357
4. Zhang, Y., et. al.: LASSO based stimulus frequency recognition model for SSVEP BCIs. J. Biomed. Signal Process. Control, 104–111 (2012). https://doi.org/10.1016/j.bspc.2011.02.002
5. Muller, S.M.T., et. al.: Incremental SSVEP analysis for BCI implementation. In: Annual International IEEE EMBS Conference, Buenos Aires, Argentina, no. 32, pp. 3333–3336 (2010)
6. Ng, KB, et. al.: Effect of competing stimuli on SSVEP-based BCI. In: Proceedings of the Annual International IEEE Engineering in Medicine and Biology Society Conference, pp. 6307–6310, (2011). https://doi.org/10.1109/iembs.2011.6091556
7. Tello, R.M., et. al.: Evaluation of different stimuli color for an SSVEP-based BCI. In: Congresso Brasileiro de Engenharia Biomedica, pp. 25–28 (2014)
8. Stiles, W.S., Crawford, B.H.: Luminous efficiency of rays entering the eye pupil at different points. Nature **139**(3510), 246–246 (1937)
9. Kim, D.-W., et. al.: Classification of selective attention to auditory stimuli: toward vision-free brain-computer interfacing. J. Neurosci. Methods, 180–185 (2011). https://doi.org/10.1016/j.jneumeth.2011.02.007
10. Nakamura, T., et. al.: Classification of auditory steady-state responses to speech data. In: Annual International IEEE EMBS Neural Engineering Conference, San Diego, California, pp. 1025–1028 (2013)
11. Power, A.J., et. al.: Extracting separate responses to simultaneously presented continuous auditory stimuli: an auditory attention study. In: International IEEE EMBS Neural Engineering Conference, Antalya, Turkey, pp. 502–505 (2009)
12. Higashi, H., et. al.: EEG auditory steady state responses classification for the novel BCI. In: Annual International IEEE EMBS Conference, Boston, Massachussets, pp. 4576–4579 (2011)
13. Mood, A.M., Graybill, F.A., Boes, D.C.: Introduction to the Theory of Statistics. McGraw-Hill (1974)
14. Oppenheim, A.V., Willsky, A.S., Young, I.T.: Fourier analysis for discrete-time signals and systems. In: Signals and Systems. Prentice-Hall, Chap. 5, pp. 291–396 (1983)
15. Heinzel, G., Rudiger, A., Schilling, R.: Spectrum and spectral density estimation by the DFT, including a comprehensive list of window functions and some new flat-top windows. Technical Report, Max-Planck-Institut fur Gravitationsphysik, Hannover, Germany (2002)
16. Hotelling, H.: Relations between two sets of variates. Biometrika **28**(3/4), 321–377 (1936)
17. Petrov, P., et. al.: A systematic methodology for software architecture analysis and design. In: International IEEE Eighth Information Technology: New Generations Conference, pp. 196–200 (2011)
18. Lin, Z., et al.: Frequency recognition based on canonical correlation analysis for SSVEP-based BCI. IEEE Trans. Biomed. Eng. **53**(12), 2610–2614 (2006)
19. BESS Manual

Domain Classification of Research Papers Using Hybrid Keyphrase Extraction Method

M. G. Thushara, M. S. Krishnapriya and Sangeetha S. Nair

Abstract Extracting thematic information from scientific papers has wide applications in information retrieval systems. Keywords give compact representation of a document. This paper proposes a document-centered approach for automatic keyword extraction and domain classification of research articles. Here, we induce a hybrid approach by adopting different methods in various phases of the system. Domain classification is important for researchers to identify the articles within their interest. The proposed system uses Rapid Automatic Keyword Extraction (RAKE) algorithm for automatic keyphrase extraction which gives best score of keywords. The classification process concerns semantic analysis which includes keyword score-matrix and cosine similarity. A comparative study of performance of RAKE algorithm which uses score-matrix against KEA based on term frequencies to extract relevant keyword was also performed.

Keywords RAKE algorithm · Cosine similarity · Automatic keyword extraction · Domain classification · Document-centered approach

1 Introduction

In recent years, there is an immense growth in the paper publications in various fields of research. The Internet provides a way to share data, especially in research fields. Categorizing the research articles into corresponding domains is

M. G. Thushara (✉) · M. S. Krishnapriya · S. S. Nair
Department of Computer Science & Applications, Amrita School of Engineering, Amrita Vishwa Vidyapeetham, Amrita University, Amritapuri, Kollam, India
e-mail: thusharamg@am.amrita.edu

M. S. Krishnapriya
e-mail: krishnapriya.ms009@gmail.com

S. S. Nair
e-mail: sangeethasnair01@gmail.com

P. K. Sa et al. (eds.), *Recent Findings in Intelligent Computing Techniques*, Advances in Intelligent Systems and Computing 708,
https://doi.org/10.1007/978-981-10-8636-6_40

time-consuming and error-prone. This paper proposes a Naive strategy for classifying the research articles into respective disciplines using hybrid keyphrase extraction method. The articles are classified into corresponding domains by cosine similarity computing.

The concept of keyword has wide range of possibilities in the area of research fields. Research has focused on automatic extraction of keyword from documents which can be used for tagging the research articles which better facilitates the search or information retrieval process. The proposed system suggests a framework of supervised [1] hybrid approach for automatic keyword extraction and subject classification of scientific articles into respective domain. The hybrid approach involves Rapid Automatic Keyword Extraction (RAKE) algorithm for automatic keyword extraction and cosine similarity for classifying document into relevant domain.

Early approaches such as Keyword Extraction Algorithm (KEA) use term weight or TF-IDF weight matrix for extracting the keyword rather than score of the terms in documents. TF-IDF describes the measure of term frequency in documents. Though this information is quite useful, some keywords that are relevant might be ignored. Instead of using TF-IDF as the feature values, RAKE calculates score-weight for each keyword which is used for extracting the relevant keyword. The extracted keywords can be used for further classification of research articles.

As the numbers of research articles related to various domains of research are being added every day, a lot of newborn phrases that are not described in the man-made corpus could not be identified. In contrast to this, RAKE extracts the keywords without the requirement of any thesaurus, while KEA uses man-made thesaurus for extracting the relevant keywords.

2 Related Work

Most of the work related to automatic keyword extraction algorithms depends on man-made thesaurus which sometimes ignores the relevant newborn phrases that might not occur in thesaurus.

A model based on perceptron training rule [2] which relies on weight of sentences is used for extracting keywords only from abstracts and titles. Rather than focusing on term frequencies, they have focused on importance of sentences based on the assumption that most of the relevant keywords are present in abstract and titles of research papers. Building word similarity [3] thesaurus using *Jenson–Shannon divergence* helps to extract semantic information from documents. TF-IDF, words' first occurrence, and node degree are used in word similarity-based keyword extraction which adopts Naive Bayes technique to build the prediction model.

Topic modeling [4] is one of the techniques for analyzing large corpus of documents. Latent Dirichlet Allocation (LDA) with collapsed Bayes and Gibbs sampling is used to extract semantic data from the document in this model. This further can be used for tagging the article into relevant domain. MAUI [5] algorithm extracts the keyphrases from the project documents, and these extracted terms are further used for automatic tagging of the documents.

Documents can be classified using a dictionary-based approach [6], where each dictionary is a collection of documents and each document in the dictionary is represented as a collection of vectors. Linear Discriminant Analysis (LDA) is a simple classification model which is used for learning the dictionary. Knowledge Cloud (Kloud) [7] combines together the information gathered from different areas, and Naive Bayes classifies the data that are pushed to the Kloud. Keyword Extraction Algorithm (KEA) [8] mainly focused on automatic keyphrase extraction by calculating feature values such as TF-IDF and first occurrences, and uses Naïve Bayes learning technique to determine whether extracted keyphrases are relevant or not.

Similarity ranking called HybridRank [9] using different weighting scores is used for improving research paper searching. TF-IDF is used as a weight to determine how important a word is to a document, and it is used for creating indices. Named entity extraction and citation [10] are used to extract all relevant and competing algorithms in a research paper. This is based on the assumption that all the algorithm names are syntactically and semantically similar to traditional names.

3 Methodology

We propose a novel methodology on automatic keyword extraction and domain classification of research articles. Automatic keyword extraction focuses on extracting relevant keywords from research articles using RAKE algorithm. These extracted keywords are used for article classification into relevant domain using cosine similarity. Figure 1 depicts the general architecture of proposed system. The system undergoes both training and testing phases. Figure 1 depicts the general architecture of proposed system.

A repository consists of many research articles, and training datasets with some manually annotated keywords are required in order to build the prediction model. During the training phase, training documents with known keyphrases undergo preprocessing and extraction, in which every relevant keyword is added to the dataset dynamically. This helps to expand the dataset with new keyphrases. Both training and extraction phrase use RAKE algorithm for extracting the relevant keyword from the documents. During the testing phase, each unclassified article undergoes preprocessing in which score-matrix is generated in order to extract the

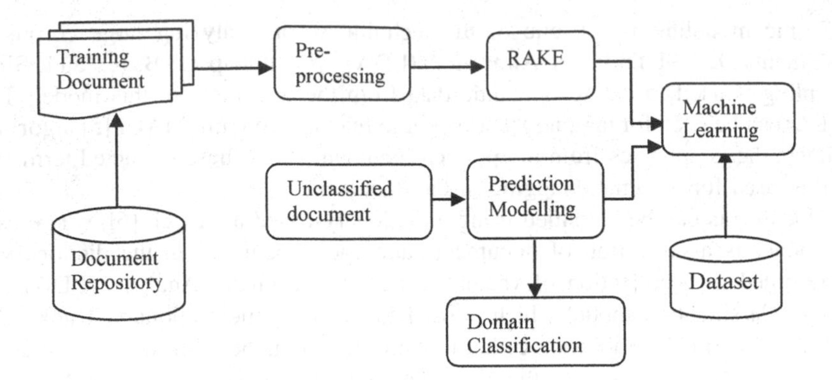

Fig. 1 System architecture of proposed system

relevant keywords. Preprocessing is the process of preparing the data before applying extraction algorithm. We use cosine similarity for classifying data into relevant domain. The main advantage of the proposed system is that newborn relevant keyword that is not previously existed in dataset can be added dynamically to the training dataset which can be used for classification in future.

3.1 Preprocessing

The preprocessing task focuses on processing the unstructured textual data using semantic features that serve to identify relevant keywords. It first cleans the input and extracts all possible words, phrases that can potentially be keywords. The input data is split into tokens, and all stopwords, punctuation marks, brackets, numbers, apostrophes are removed in this phase. It then extracts the candidate keyword which might not be the actual keywords.

3.2 Keyword Extraction

We use Rapid Automatic Keyword Extraction (RAKE) algorithm for extracting the keywords relevant to the research paper in the absence of manual annotations. RAKE extracts the keyword based on the score of each keyword. RAKE is an unsupervised and language-independent method for extracting keywords from the documents.

Algorithm1: RAKE Algorithm

```
Input: Research paper in text format
Output: extracted keywords

Begin
for every word not in stopwords
        phraselist [] =word
        calculate word scores for phraselist
end for
for every word in phraselist
        compute wordlist_degree=length (word)-1
        then ,  compute word_frequency[word]+=1
endfor
for every word in word_frequency
        computecandidate_word_degree+= word_frequency[word]
endfor
then sort the keywords
return keywords
```

The RAKE algorithm involves the following steps:

- Candidate Selection: It extracts all possible keyphrases after removing all stopwords and phrase delimiters.
- Feature Calculation: Instead of TF-IDF measure values which are used in many existing system, RAKE uses keyword score-matrix as the feature value.
- Keyword Selection: From the extracted candidate phrases, top-scored candidates are considered as the keywords.

3.3 Score Calculation

After extraction, a score-weight is calculated for every candidate phrase. Here, we try to model score-weight matrix using degree and frequency of candidate phrases. The score-matrix thus obtained can be used for domain classification. The different measures used for building the score-weight matrix are:

- Word Frequency: It indicates the number of occurrences of the frequent terms in the document.
- Word Degree: It is the degree of co-occurrence of each word in the document.
- Ratio of Degree to Frequency: The score-weight matrix thus created is used in the classification process.

Table 1 Discretization ranges

Feature	Discretization range				
Class	1	2	3	4	5
Score	<3.5	3.6–7.5	7.6–11.5	11.6–15.5	>15.5

3.4 Discretization

The individual score of each keyword in the score-matrix is real numbers which we convert to nominal data. Table 1 shows the discretization. Score values of the keywords are discretized into fixed levels. The discretization boundaries are given at the top of the table, which depicts the range into which the value falls. This discretized value is used for computing similarity.

3.5 Domain Classification

The research article is classified into relevant domain by cosine similarity. Cosine similarity finds the normalized dot product of two vectors. The two vectors have been represented as the discretized score value of extracted terms and the score of terms in training dataset. It is computed by:

$$\text{Similarity}\,(\text{A}, \text{D}) = \frac{\sum_{i=1}^{n} w_{A_i} w_{D_i}}{\sqrt{\sum_{i=1}^{n} w_{A_i}^2 \times \sum_{i=1}^{n} w_{D_i}^2}} \tag{1}$$

where A is the article to be classified into domain category and D is the dataset of particular domain. w_{A_i} is the score of the extracted terms of article, and w_{D_i} is the score of the corresponding terms in dataset. Similarity of each article is computed using the score of terms in the training dataset, so that article is classified into the corresponding domain which is having higher similarity.

4 Experimental Result

4.1 Dataset

The dataset contains 200 research papers published in Computer Science domains. The two topics covered in dataset are natural language processing and computer networks. The first 100 documents are collected from Journal of Machine NLP research in which 50 articles are taken as the training dataset and remaining 50 articles are taken as testing set. Research documents for network domain are taken from Journal of Networks. Training dataset consists of relevant keywords and its scores which are obtained during the training phase.

4.2 Comparison of KEA and RAKE

Early approaches such as KEA automatically extract keywords by evaluating corpus-oriented statistics of individual words. KEA identifies key phrases using TF-IDF measure and machine learning algorithm such as Naïve Bayes. Corpus-oriented methods typically operate only on single words based on the state of corpus. This will limit the possibility of determining phrases of multiple words as keywords. Therefore, it might be difficult to determine whether these words are relevant to documents. RAKE generates a score for each candidate phrase based on their:

i. Frequency
ii. Degree
iii. Ratio of degree and frequency (Figs. 2 and 3).

Table 2 shows the calculated score of each candidate keyword in the sample document using the metric *deg(w)/freq(w).* Candidate keywords that mainly occur in context which is more significant are determined by *deg(w)/freq(w);* the score of *deg(classification)/freq(classification) is* higher when compared to *deg(compact)/freq(compact).* From the score-weight matrix of relevant keywords, top *T* candidates with highest score can be considered as relevant keywords.

From example, the score of *automatic keyphrase extraction* is 9, that is, the sum of score of *automatic*, *keyphrase,* and *extraction*. We consider one-fourth of the total content words in the document.

Figure 4 represents analysis of keyword generation of KEA and RAKE algorithm. We have taken ten documents with known relevant phrases from natural language processing dataset. We have applied both algorithms on these ten research documents. In the graph, the *x*-axis represents documents in sample set of natural language processing domain and *y*-axis represents number of relevant keywords each algorithm generates. When compared to KEA, RAKE can extract relevant phrases of an individual even in the absence of any learning algorithm. From the analysis graph, it is clear that RAKE extracts more relevant keywords than KEA.

Keywords give a compact representation of a document. This paper proposes a document centered approach for automatic keyword extraction and domain classification of research articles. Extracting thematic information from scientific papers has wide applications in information retrieval systems. Domain classification is important for researchers to identify the articles within their interest. We use Rapid Automatic Keyword Extraction (RAKE) algorithm for automatic keyphrase extraction which gives best score of keywords.

Fig. 2 A sample abstract extracted from a research document

Keywords - compact representation - document centred approach - automatic keyword extraction -domain classification - research article - extracting thematic information - information retrieval system

Fig. 3 Most relevant keywords from the abstract in Fig. 2

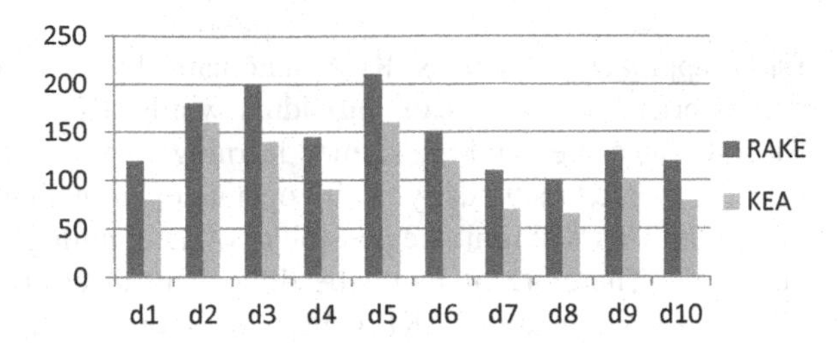

Fig. 4 Result analysis graph of KEA and RAKE

Table 3 lists out the score values of extracted keywords from sample set of three documents of natural language processing domain. The value clearly shows that the keywords with high score are more relevant to a domain. These keywords are used for domain identification using the mathematical model called cosine similarity. The similarity of keyword scores of test document and training documents is compared to determine the relevant domain. The document with higher similarity or low cosine angle is more similar to the respective domain.

5 Result and Analysis

In order to examine the accuracy of system, we need to evaluate the experimental results. Some measures such as precision and recall are used for evaluating the performance of the system. The F-measure is a harmonic mean of precision and recall.

$$\text{Recall} = \frac{\text{No. of relevant items retreived}}{\text{No. of relevant items in document}}$$

The recall or sensitivity is the ratio of relevant keywords retrieved to total number of relevant keywords in the document.

$$\text{Precision} = \frac{\text{No.of relevant items retreived}}{\text{Total No.of items in document}}$$

The precision is computed as the ratio of the number of relevant keywords retrieved to total number of keywords in the document.

Table 2 Candidate keywords and their calculated scores

	deg(w)	*freq(w)*	*deg(w)/freq(w)*
Automatic	9	3	3
Keyphrase	9	3	3
Classification	9	3	3
Information	2	2	1
Extraction	9	3	3
Compact	1	1	1
Centered	1	1	1
System	9	3	3
Retrieval	9	3	3

Table 3 Score calculated on sample set of documents

Document no.	Extracted keywords	Score
Doc1	Keyword extraction	9.92
	Perceptron training rule	9.0
	Automatic tagging	5.82
	Time-consuming	3.5
	Frequency	1.5
Doc2	Machine learning	9.0
	Automatic keyword extraction	8.81
	Candidate keyword	4.04
	Total frequency	3.5
	Stopwords	3.8
Doc3	Natural language processing	13.5
	Coherence document simulation	12.8
	Recurrent neural network	8.40
	Learning capability	3.8
	Processed keywords	2.98

Table 4 shows the performance of KEA and RAKE algorithm applied on ten documents. The table shows ability of each algorithm to extract the relevant and non-relevant keywords (Table 5).

Figure 5 shows the performance of KEA and RAKE, and it is clear that RAKE has higher precision and F-measure. A system with high precision and low recall returns more relevant keywords. Therefore, we can conclude that RAKE extracts more relevant keywords than KEA.

Table 4 Performance measures of algorithms

	Extracted keywords	Not extracted keywords	Relevant keywords	Not relevant keywords
KEA	7234	1086	4702	1446
RAKE	2818	113	366	2339

Table 5 Precision, recall, and F-measure of KEA and RAKE

	Precision	Recall	F-measure
KEA	0.654	0.961	0.663
RAKE	0.833	0.869	0.868

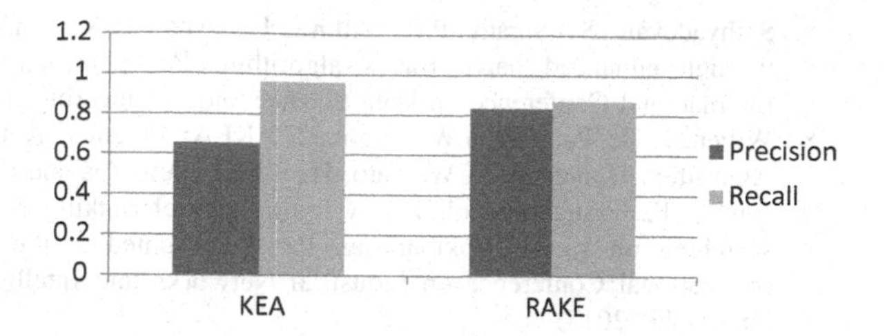

Fig. 5 Performance evaluation based on recall and precision

6 Conclusion and Future Work

In this paper, we proposed a hybrid approach for automatic keyword extraction and domain classification of research articles. The automatic keyword extraction consists of preprocessing and extracting relevant keywords. We used RAKE algorithm for automatic keyword extraction and cosine similarity for classifying articles into respective domains. We also evaluated the performance of RAKE algorithm which uses score-matrix to retrieve relevant keywords against KEA that relies on term frequencies for extraction. The results from the experiment show that algorithm using score-matrix is more efficient than that of algorithm using term frequencies.

Our future work will focus on novel methodology for extracting detailed information such as main contribution, techniques used from research articles which help to study the dynamics and evolution of research community.

Acknowledgements We would like to express our sincere gratitude to the Faculty of Department of Computer Science and Applications of Amrita Vishwa Vidyapeetham, Amritapuri, for the help and guidance. Our sincere thanks go to Dr. M. R. Kaimal, Chairman, Computer Science Department, Amrita Vishwa Vidyapeetham, Amritapuri, for his prompt support.

References

1. John, A.K., Di Caro, L., Boella, G.: A supervised keyphrase extraction system. Paper presented at the ACM International Conference Proceeding Series, 13–14 Sept 2016, pp. 57–62 (2016)
2. Bhowmik, R.: Keyword Extraction from Abstract and Title. IEEE (2012)
3. Meng, W., Liu, L., Dai, T.: A modified approach to keyword extraction based on word-similarity. Paper presented at the Proceedings—2009 IEEE International Conference on Intelligent Computing and Intelligent Systems (ICIS) (2009)
4. Anupriya, P., Karpagavalli, S.: LDA based topic modeling of journal abstracts. Paper presented at the ICACCS 2015—Proceedings of the 2nd International Conference on Advanced Computing and Communication Systems (2015)
5. Thushara, M.G., Dominic, N.: A template based checking and automate tagging algorithm for project documents. IJCTA (2016)
6. Menon, R.R.K., Kini, N.V., Krishnan, G.A.: Harnessing the discriminatory strength of dictionaries. Paper presented at the 2016 International Conference on Advances in Computing, Communications and Informatics (ICACCI) (2016)

7. Sathyadevan, S., Sarath, P.R., Athira, U., Anjana, V.: Improved document classification through enhanced naive Bayes algorithm. Paper presented at the Proceedings—2014 International Conference on Data Science and Engineering (ICDSE), pp. 100–104 (2014)
8. Witten, L.H., Painter, G.W., Frank, E.: KEA: Practical Automatic Key phrase Extraction Algorithm. University of Waikato, Hamilton, New Zealand (2014)
9. Jomsri, P., Prangchumpol, D.: A hybrid model ranking search result for research paper searching on social bookmarking. Paper presented at the Proceedings of the 2015 1st International Conference on Industrial Networks and Intelligent Systems, INISCom 2015, pp. 38–43 (2015)
10. Ganguly, S., Pudi, V.: Competing algorithm detection from research papers. Paper presented at the Proceedings of the 3rd ACM IKDD Conference on Data Sciences (CODS) (2016)

GLRLM-Based Feature Extraction for Acute Lymphoblastic Leukemia (ALL) Detection

Sonali Mishra, Banshidhar Majhi and Pankaj Kumar Sa

Abstract This work presents gray-level run length (GLRL) matrix as feature extraction technique for the classification of Acute Lymphoblastic Leukemia (ALL). ALL detection in an early stage is helpful in avoiding fatal hematopoietic ailment which might cause death. The GLRL matrix extracts textural features from the nucleus of the lymphocyte image which is used with Support Vector Machine (SVM) for classification. The public dataset ALL-IDB1 is used for the experiment, and we have obtained 96.97% accuracy for GLRL feature with SVM classifier.

Keywords Acute Lymphoblastic Leukemia · Gray-level run length
Marker-based watershed segmentation · CAD system

1 Introduction

The process of counting and grouping of blood cells from peripheral blood smear permits estimation and detection of a huge number of diseases. Illness related to hematopoietic cells influence the blood and bone marrow and also are major concerns for death [1]. By analyzing white blood cells (WBCs) or leukocytes, leukemia is usually detected. Based on the rate of progression of the disease, leukemia is categorized into two types, i.e., acute and chronic. Acute lymphoblastic leukemia (ALL) is a subtype of acute leukemia which primarily affects the lymphocyte (a type of WBC). One of the demonstrative strategies incorporates the microscopic examination of the white blood cells with abnormalities. From decades, this operation is

S. Mishra (✉) · B. Majhi · P. K. Sa
Pattern Recognition Research Lab, Department of Computer Science and Engineering,
National Institute of Technology, Rourkela 769008, India
e-mail: smishra.nitrkl@gmail.com

B. Majhi
e-mail: bmajhi@nitrkl.ac.in

P. K. Sa
e-mail: pankajksa@nitrkl.ac.in

P. K. Sa et al. (eds.), *Recent Findings in Intelligent Computing Techniques*,
Advances in Intelligent Systems and Computing 708,
https://doi.org/10.1007/978-981-10-8636-6_41

performed by the experts and skilled operators that suffers from several disadvantages like slowness and non-standard accuracy. Image processing techniques can be a way which provides information on the morphology of the cells. The primary objective of the research is to contribute a fully automated way to support medical activity by analyzing microscopic images.

The remainder of the paper is arranged as follows. Section 2 shows some of the valuable works for the discovery of the disease followed by the proposed system model in Sect. 3. Section 4 shows the experiments performed along with the comparison made with some standard classifiers. Finally, Sect. 5 presents the conclusion.

2 Related Work

According to the survey, some of the existing systems can analyze and classify the leukocytes form the peripheral blood smear. However, these systems are partially automatic. Especially, the work has been performed so as to count the number of WBCs through segmentation. Madhloom et al. [2] have proposed an automatic system based on arithmetic and threshold operations for segmenting the lymphoblast cells. In [3], the authors have extracted a single leukocyte by applying bounding box around the nucleus. The authors in [4] have used a low-pass filter and a thresholding technique to segment the white blood cell by removing the background. Halim et al. [5] have suggested a technique for detection of acute leukemia in blood that detects leukocytes by examining the S component of the HSV color space. We can conclude from literature survey that automation of the process depends entirely on the right segmentation and extraction of specific features. In this paper, a segmentation method proposed by [7] along with a new scheme is used for computerized investigation of the blood sample.

3 Proposed Work

The overall block diagram for the detection of ALL is given in Fig. 1. The step-wise description is given in the subsequent sections.

3.1 Nucleus Identification

The images made employing a digital microscope is in RGB color space. Color images are very difficult to segment. Therefore, RGB images are converted to CIELAB color space which helps in reducing the color dimension. Here, we have considered Otsu method [8] of thresholding technique for identification of leukocytes from the microscopic image. Figure 2a–c represents the steps for detection of leukocytes.

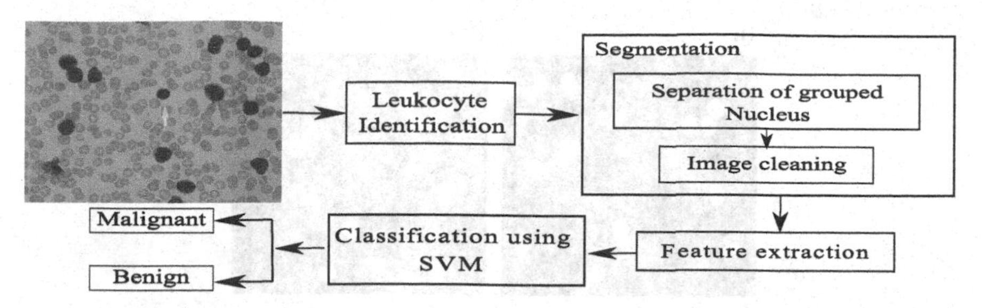

Fig. 1 Proposed block diagram for the detection of ALL

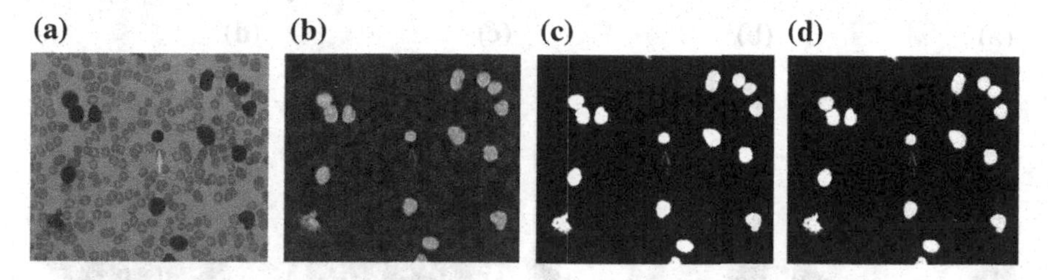

Fig. 2 Nucleus Identification: **a** original image, **b** a^* component of the CIELab color space, **c** threshold image, **d** separation of adjacent leukocytes using Marker-based watershed segmentation

3.2 Identification and Separation of Grouped Nucleus

The resulting image from the previous step contains only leukocytes. In this step, we have dealt with the separation of grouped leukocytes. This work analyses the presence of grouped leukocytes by taking the roundness value. Roundness value of a cell can be defined by,

$$roundness = \frac{4 \times \pi \times area}{convex\_perimeter^2} \tag{1}$$

In this work, we have considered a roundness value of 0.8 to distinguish between a single leukocyte and a grouped leukocytes. A Marker-controlled watershed [7] is used to refine the line of leukocytes having an irregular shape. The result for the separation of grouped leukocytes is given in Fig. 2d.

3.3 Image Cleaning

The next step after separation of grouped leukocytes is to clean the image. Image cleaning considers the elimination of all the components discovered at the edge of the smear. Cleaning operation can be performed by calculating the solidity value. Solidity of an object is defined as,

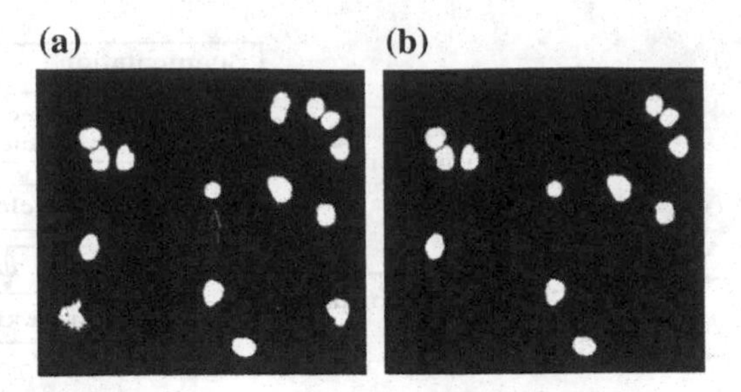

Fig. 3 **a** Image after edge cleaning, **b** Image after removing abnormal component

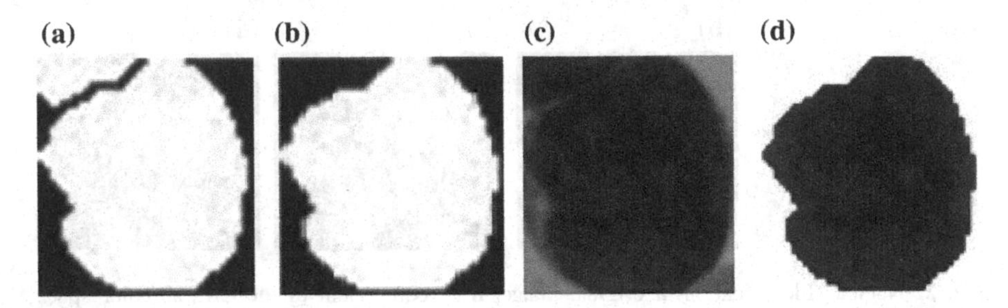

Fig. 4 Sub-imaging process: **a** nucleus sub-image, **b** image after border cleaning, **c**, **d** corresponding color image of **a** and **b**

$$solidity = \frac{area}{convex\_area} \tag{2}$$

Here, we have taken the solidity value as 0.85 which is the threshold and removed all the objects having intensity value less than 0.85. Figure 3 gives the final results after performing the cleaning operation. The individual nucleus can be found out using bounding box technique. Figure 4 shows the details of the sub-imaging process and the corresponding nucleus sub-image.

3.4 GLRLM-Based Feature Extraction

Many methods of texture analysis have been developed over the past decades [6]. Xiaoou Tang [9] has introduced 11 textural features calculated from the gray-level run length matrices which are used to characterize the nucleus of leukocytes. For a sub-image of size $M \times N$, the number of gray levels and the longest run (a string of continuous pixels having the same gray-level intensity in a specific linear direction) is represented as g and r, respectively. Z is denoted as the total number of run. The GLRLM is a two-dimensional matrix of $(g \times r)$ components in which each component $q(m, n)$ gives the times of occurrences of the run having length n of gray level m in a given direction θ. A run length matrix $q(m, n)$ is defined as,

$$q_{mn} = |m^n| \quad (3)$$

where m^n means m exhibits exactly n times, and $1 \leq m \leq g$ and $1 \leq n \leq r$. The feature matrix is calculated using GLRLM and the steps are is described in Algorithm 1.

Algorithm 1 GLRLM-based Feature Extraction

Require: N: Leukocyte samples
  GLRLM: Gray-level run length matrix
  θ: direction parameter
  P: number of directions (4: 0°, 45°, 90°, and 135°)
  S and M denotes the feature descriptor and number of features respectively.
Ensure: $X[N : M]$: Feature matrix,
1: Initialize S
2: $M \leftarrow P \times S$
3: **for** $i \leftarrow 1$ to N **do**
4: Determine the GLRL matrix using *grayrlmatrix*() for the input image (*ip*)
5: **for** $q \leftarrow 1$ to P **do**
6: $GLRLM_{\theta_q} \leftarrow grayrlmatrix(ip, \theta_q)$
7: **for** $x \leftarrow 1$ to S **do**
compute the $GLRLM_{\theta_q}$ and append it to X
8: **end for**
9: **end for**
10: **end for**

3.5 SVM-Based Classification

SVM is a binary classifier which generates a hyperplane by employing a subset of training vectors which are known as support vectors. This paper uses Support Vector Machine [10] for the classification process. To evaluate the effectiveness of SVM model, the proposed method is being compared to many standard models, namely *k*-NN (*k*-Nearest Neighbor), Naive Bayes, Back Propagation Neural Network (BPNN). Along with this, the proposed method is being tested with the most common kernel used in SVM.

4 Experimental Evaluation

The experiment is being carried out using MATLAB R2015b on Microsoft Windows 8.1 having a 4 GB RAM as internal memory. The proposed method is trained and tested with the public database ALL-IDB1 [11]. A fivefold cross-validation scheme is used to generalize the performance of the classifier. The lymphoblasts (malignant) and leukocytes (normal) are termed as positive and negative class, respectively. This

Table 1 Calculated GLRL features

Features	0°		45°		90°		135°	
	Healthy	Unhealthy	Healthy	Unhealthy	Healthy	Unhealthy	Healthy	Unhealthy
SRE	0.32	0.0186	0.33	0.01	0.32	0.018	0.33	0.018
LRE	29.59	60.50	30.510	60.78	0.32	60.46	30.47	60.78
LGRE	50.82	25.81	47.67	7.71	29.60	26.11	49.91	7.77
HGRE	180.95	34.09	275.003	62.46	49.81	34.29	276.77	62.35
SRLGE	0.564	0.20	0.89	0.36	177.005	0.19	0.894	0.36
SRHGE	0.099	0.04	0.132	0.11	0.53	0.04	0.12	0.11
LRLGE	3107.31	762.38	1127.42	252.64	0.097	743.68	1029.46	253.11
LRHGE	0.0127	0.0012	0.020	0.002	3294.28	0.001	0.018	0.002
GLNU	2439.16	12.87	880.90	4.40	0.011	12.62	782.4	4.40
RLNU	2.36	1.89	3.81	6.42	2613.5	1.86	3.79	6.45
RPC	32350.99	47631.85	10424.5	15644.35	2.44	0.018	10242.68	15663.13

Table 2 Evaluation of different classifiers with fivefold cross-validation

Classifier	True positive rate (TPR)	True negative rate (TNR)	Accuracy(%)
NB	0.99	0.99	96.27
k-NN	0.73	0.85	83.49
BPNN	0.81	0.99	95.36
SVM-L	**0.87**	**1.00**	**96.97**
SVM-Q	0.88	0.99	96.60
SVM-P	0.95	0.65	73.21
SVM-R	0.86	0.96	96.51

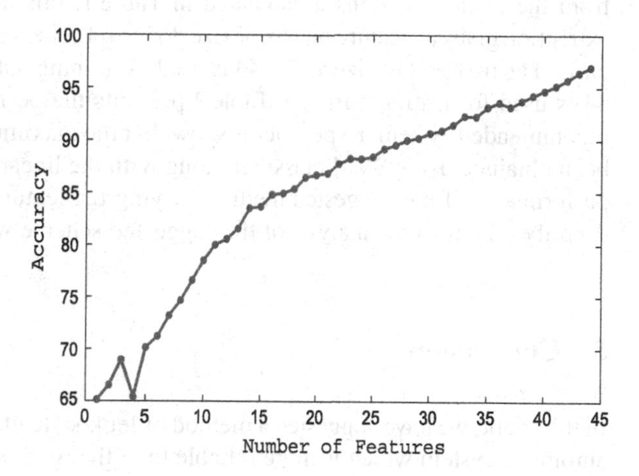

Fig. 5 Plot of accuracy with the increase number of features

paper takes into consideration of true positive rate (TPR), true negative rate (TNR), and accuracy as performance measures which are defined as follows,

$$TPR = \frac{TP}{TP + FN}, \quad TNR = \frac{TN}{TN + FP}, \quad Accuracy = \frac{TP + TN}{TP + FP + TN + FN}$$

where TP = true positive, FP = false positive, TN = true negative, FN = false negative.

4.1 Results and Discussion

ALL is diagnosed by the appearance or lack of unhealthy leukocytes. Therefore, leukocytes must be distinguished as “abnormal” or “healthy” cells in blood samples for the detection of ALL. So, a total number of 865 individual nuclei were obtained by the sub-imaging process. After finding number of nucleus, the next step is to

Table 3 Comparison of accuracy (%) with the other existing scheme

Classifier	90 features [12]	Proposed method (44 features)
Naive Bayes	81.66	96.27
k-NN	83.46	83.27
BPNN	58.7	95.36
SVM-L	**89.76**	**96.97**

extract the texture features using GLRL matrix from the sample images. The total number features extracted along all the directions is 44. The texture features extracted from the nucleus regions is tabulated in Table 1. Finally, the result of the feature extraction gives us feature matrix of size 865×44. The feature matrix is split into two parts. The first part of size 435×44 is used as training data and the rest of size 430×44 is used for testing purpose. Table 2 presents the performance evaluation of the recommended system. Experiments show that the maximum accuracy of 96.97% has been obtained using SVM classifier along with the linear kernel. Figure 5 shows the performance of the suggested method varying the features number. Finally, Table 3 describes the relative analysis of the suggested scheme with the existing scheme.

5 Conclusion

In this work, we have suggested a method of leukocyte identification by designing an automatic system which is more reliable than the work done by operators manually and is computationally less expensive. The proposed method can efficiently detect the leukocytes present in a blood smear and can classify lymphoblast with great precision, as indicated by our results. The proposed method incorporates the GLRL features to classify the lymphoblast cells and shows an accuracy of 96.97% with SVM classifier.

References

1. Mishra, S., Majhi, B., Sa, P.K.: A survey on automated diagnosis on the detection of Leukemia: a hematological disorder. In 3rd IEEE International Conference on Recent Advances in Information Technology (RAIT), pp. 460–466 (2016)
2. Madhloom, H.T., Kareem, S.A., Ariffin, H.: An image processing application for the localization and segmentation of lymphoblast cell using peripheral blood images. J. Med. Syst. **36**(4), 2149–2158 (2012)
3. Mohapatra, S., Patra, D., Satpathy, S.: An ensemble classifier system for early diagnosis of acute lymphoblastic leukemia in blood microscopic images. Neural Comput. Appl. **24**(7–8), 1887–1904 (2014)

4. Scotti, F.: Robust segmentation and measurements techniques of white cells in blood microscope images. In Proceedings of the IEEE Conference on Instrumentation and Measurement Technology (IMTC), pp. 43–48 (2006)
5. Halim, N.H.A., Mashor, M.Y., Hassan, R.: Automatic blasts counting for acute leukemia based on blood samples. Int. J. Res. Rev. Comput. Sci. **2**(4) (2011)
6. Galloway, M.M.: Texture analysis using gray level run lengths. Comput. Graph. Image Process. **4**(2), 172–179 (1975)
7. Mishra, S., Majhi, B., Sa, P.K., Sharma, L.: Gray level co-occurrence matrix and random forest based acute lymphoblastic leukemia detection. Biomed. Signal Process. Control **33**, 272–280 (2017)
8. Otsu, N.: A threshold selection method from gray-level histograms. Automatica **11**(285–296), 23–27 (1975)
9. Tang, X.: Texture information in run-length matrices. IEEE Trans. Image Process. **7**(11), 1602–1609 (1998)
10. Bishop, C.M.: Pattern recognition. Mach. Learn. **128**, 1–58 (2006)
11. ALL-IDB Dataset for ALL Classification. http://crema.di.unimi.it/~fscotti/all/
12. Mishra, S., Sharma, L., Majhi, B. and Sa, P.K.: Microscopic image classification using DCT for the detection of Acute Lymphoblastic Leukemia (ALL). In Proceedings of International Conference on Computer Vision and Image Processing (CVIP), pp. 171–180 (2017)

Personalization of Test Sheet Based on Bloom's Taxonomy in E-Learning System Using Genetic Algorithm

Mukta Goyal and K. Rajalakshmi

Abstract In E-learning systems during tutoring, evaluating the learning status of each learner is essential, and tests are a usual method for such evaluation. However, the quality of these test items depends upon the degree of difficulty, discrimination, and estimated time. While constructing the test sheet, the selection of appropriate test items is also important. This paper aims to provide a method to generate the test sheets based on the different learning levels of Bloom's taxonomy using genetic algorithm. The questions are initially categorized into six different learning levels based on the keywords given by Bloom's taxonomy. These six different learning levels are assigned difficulty degrees from 0.1 (lowest) to 0.6 (highest). Then, the different number of questions is generated for all these six different levels of learning using genetic algorithm. The numbers of questions are generated in such a manner that the total difficulty degree of all the questions is equal to the target difficulty degree given by the instructor. Based on the test sheet generated, learner performances are being analyzed, and by doing so, even weaker learners are able to attend low-level questions and perform well in the examination. This would also help to analyze at which level of learning learners are facing problem so that effort can be made to improve their learning status.

Keywords Bloom's taxonomy · E-test · E-learning environment
Genetic approach

M. Goyal (✉) · K. Rajalakshmi
Jaypee Institute of Information Technology, Noida, India
e-mail: mukta.goyal@jiit.ac.in

K. Rajalakshmi
e-mail: k.rajalakshmi@jiit.ac.in

P. K. Sa et al. (eds.), *Recent Findings in Intelligent Computing Techniques*,
Advances in Intelligent Systems and Computing 708,
https://doi.org/10.1007/978-981-10-8636-6_42

1 Introduction

In today's assessment system, test sheet construction is important and challenging issue to conduct a good test for various assessment requirements. Moreover, from teacher's perspective, a good test is required for teachers to verify the true knowledge of the learner. It also recognizes the learners' learning bottlenecks [1]. Generally, the test sheet can be constructed manually or randomly to select test items from the item bank which are unable to meet multiple assessment requirements simultaneously. However, in the present scenario, the major issues during the designing of an E-test are the lack of suitable classification of test assignments according to knowledge level of learners, limited types of test assignments incorporating different E-learning environments, and the lack of realization of different assessment systems [2].

In the literature, some of these issues are addressed as follows. Blooming Biology Tool (BBT) [3] is used for the assessment, which reflects to enhance the study skills and metacognition. To optimize the difficulty degree of the generated test sheet in automatic test sheet generator (ATG), an effective coding method and new heuristic genetic modification were used [4]. To improve the efficiency of framing optimal test sheet from item repositories to address multiple assessment criteria, an immune algorithm (IA) is applied in [5]. A particle swarm optimization (PSO) is used to generate the dynamic questions which satisfy the multiple assessment requirements for each learner to build the select tailored questions from item bank [6]. In addition, particle swarm optimization-based approach is used to generate the serial test sheets from voluminous item repository to meet multiple assessment criteria [7]. Dynamic programming and clustering method are also used to generate a multiple criteria test sheet [8].

However, it is to be noted that here the major issues need to be focused to generate the E-test [9–11] are

(i) To set a high-quality test, such that the learning status of a learner in an E-learning environment can be determined.
(ii) To improve the learner's performance in an E-learning environment through dynamic E-test.
(iii) Instructor should identify the learning level that learners are facing problems so that instructor should provide that content through which learners can make effort to improve their learning status.

However, the preliminarily solution is by selecting an appropriate test item when designing a test sheet. Specifically, the conventional test methodology may include questions with varying difficulty levels, to get the overall aspect about the learner's ability. This conventional method is less efficient to address dynamic learning ability of learners. Therefore, the enhancement and complexity to set high-quality test lies in, such that, the test items selected, should be from different level of learning ability of learner, instead of conventional method based on difficulty level of questions. The advantage of such enhanced design is that it allows learners to select a suitable ability value, which includes questions with difficulty level.

This paper aims to provide a method to generate the test sheets based on the different learning levels of Bloom's taxonomy using genetic algorithm. Section 2 provides background information about Bloom's taxonomy and genetic algorithm. Section 3 elaborates the methodology of integrating Bloom's taxonomy, genetic algorithm for setting efficient E-test paper. Section 4 discusses the experiments and results obtained. Section 5 concludes with discussion of the proposed solution.

2 Background Material

This section explains Bloom's taxonomy to determine the difficulty level of test sheet and genetic algorithm for automatic generation of test sheet.

2.1 *Cognitive Level of Bloom's Taxonomy*

A multitier model of thinking is classified into six cognitive levels of complexity which are defined by Bloom's taxonomy [12]. These are known as knowledge, comprehensions, and applications, considered as the lowest level, whereas analysis, synthesis, and evaluation are to be known the highest level. Each level is a part of higher level. For example, if the learner is working at the application level, then it is assumed that he also learned the material at the knowledge and comprehension levels. Figure 1 shows the cognitive domain of Bloom's taxonomy.

Table 1 shows the sample questions of data structure course with related to the cognitive domain of Bloom's taxonomy.

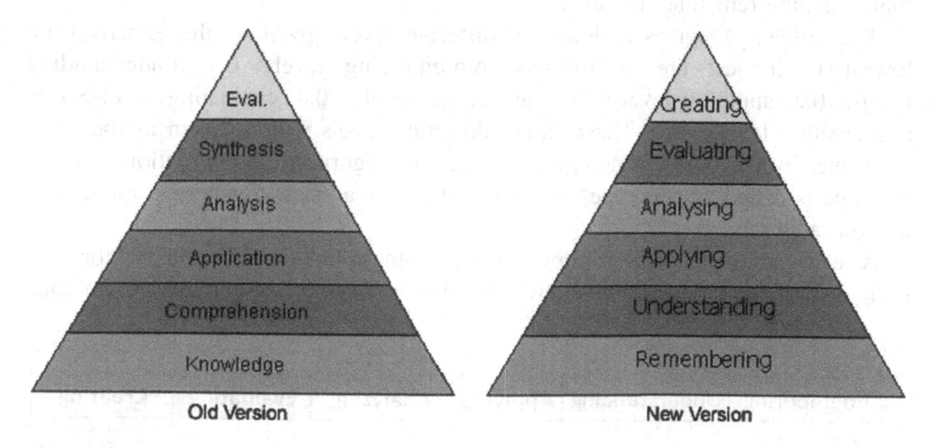

Fig. 1 Bloom's taxonomy—the cognitive domain [14]

Table 1 Mapping of the questions of data structure with Bloom's taxonomy domain

Remembering	Describe the process of creating heap with respect to heap sort technique with the help of suitable example
Understanding	Convert it into prefix and postfix expressions ((a + b) + c * (d + e) + f) * (g + h)
Applying	Show that if a black node has just one child, then that must be red
Analyzing	The execution of recursion is slower than iterative solution. Illustrate with example
Evaluating	Compare bubble sort and insertion sort with respect to working principle and time complexity
Creating	Design a sorting algorithm for red, white, or blue elements, stored in an array. Sorting should be done in such a way that all the reds come before all the whites and all the whites should come before all the blues

2.2 *Genetic Algorithm*

A genetic algorithm (GA) [13] is a search heuristic that is inspired by the process of natural evolution. It works in nondeterministic manner to generate the optimized solution to the problem using techniques: mutation, selection, and crossover. The chromosomes are defined as a population of strings also known as gene type of genome. The operator's selection, crossover, and mutation are applied to this population of strings to provide the optimized solution.

3 Methodology

Figure 2 shows the representation of chromosomes. Each gene is represented as questions of one of the learning levels of Bloom's taxonomy. Here first gene represents the remembering level.

The difficulty degrees assigned to different levels given by the experts from lowest to highest are as follows: remembering level-> 0.1, understanding level-> 0.2, applying level-> 0.3, analyzing level-> 0.4, evaluating level-> 0.5, and creating level-> 0.6. These target difficulty levels will be taken as the input parameter by the fitness function of the genetic algorithm. A population of chromosome is called as genotype, and it is created as a class to generate initial population randomly.

After generating the initial population, the fitness function is evaluated for each individual in that population. Here, the fitness function is defined as the total

Remembering	Understanding	Applying	Analyzing	Evaluating	Creating

Fig. 2 Chromosome of six genes

difficulty level of the test sheet represented by the potential solution and total no. of questions represented by the solution. The total difficulty level (dd) of the test sheet represented by the potential solution is

$$\begin{aligned} dd = & \text{numRemember} * 0.1 + \text{numUnderstand} * 0.2 + \text{numApply} * 0.3 \\ & + \text{numAnalyze} * 0.4 + \text{numEvaluate} * 0.5 + \text{numCreate} * 0.6 \end{aligned} \tag{1}$$

Then, it takes the difference between the total difficulty level (dd) and the target difficulty level which is given by the expert. If the difference is zero, then the fitness will be equal to 1; otherwise,

$$\text{Fitness} = (1/1 + \text{difference}) \tag{2}$$

The value of fitness lies in the range from 0 to 1. The best bit chromosome is used for reproduction. New individuals are created through crossover and mutation operations to give birth to offspring; here, 80% crossover probability and 0.001 mutations are used. The algorithm is run up to 20 evolutions. After reaching the maximum number of the evolutions, the required number of questions from each learning level is generated.

4 Results

The different test sheets are generated for the different target levels given by the instructor for checking the redundancy in the test sheet. Here, five test sheets for each of the different target difficulty levels (1.0, 1.5, 2.0, 2.5, 3.0, 3.5, 4.0, 4.5, and 5.0) are generated. Table 2 shows the evolution time by varying population size and maximum number of the evolutions.

The test sheet of the target difficulty degree is generated based on the different learning level questions. The experiment is done by generating different test sheets for the same as well as different target difficulty level for checking the redundancy in the questions. We have generated 45 test sheets 5 each for different difficulty

Table 2 Evolution time with respect to population size and number of evolution

Population size	Evolution time (in ms)			
	No. of evolution = 1	No. of evolution = 5	No. of evolution = 10	No. of evolution = 20
10	16	78	109	234
20	46	156	296	624
30	62	296	421	1217
40	78	390	717	1638
50	109	484	936	1903

levels from 1.0 to 5.0. As we increase the difficulty degree, the number of questions increases because the questions from lower levels will be more and less from higher levels.

5 Conclusion

A high-quality test is the imperative criterion for determining the learning status of learners and for improving their performance. Their performance can be increased by composing the test sheet that includes questions from different learning levels. By doing this, even weaker learners are able to attend at least low-level questions and perform well in the examination. This would also help us to analyze at which level the learners are facing problems so that we can make efforts to improve their learning status.

References

1. Hwang, G.J., Yin, P.Y., Yeh, S.H.: A tabu search approach to generating test sheets for multiple assessment criteria. IEEE Trans. Educ. **49**(1), 88–97 (2006)
2. Sokolova, M., Totkov, G.: About test classification in e-Learning environment. CompSys **5**, 16–17 (2005)
3. Crowe, A., Dirks, C., Wenderoth, M.P.: Biology in bloom: implementing Bloom's taxonomy to enhance student learning in biology. CBE-Life Sci. Educ. **7**(4), 368–381 (2008)
4. Al Sadoon, M.E., Abdul Wahhab, R.S., Manama, B.: A new computer-based test system: an innovative approach in E-learning. In: The Proceedings of Applied Mathematics and Informatics (2010). ISBN: 978–960-474-260-8
5. Lee, C.L., Huang, C.H., Lin, C.J.: Test-sheet composition using immune algorithm for e-learning application. In: International Conference on Industrial, Engineering and Other Applications of Applied Intelligent Systems. Springer, Berlin, Heidelberg, pp. 823–833 (2007)
6. Cheng, S.C., Lin, Y.T., Huang, Y.M.: Dynamic question generation system for web-based testing using particle swarm optimization. Expert Syst. Appl. **36**(1), 616–624 (2009)
7. Yin, P.Y., Chang, K.C., Hwang, G.J., Hwang, G.H., Chan, Y.: A particle swarm optimization approach to composing serial test sheets for multiple assessment criteria. Educ. Technol. Soc. **9**(3), 3–15 (2006)
8. Hwang, G.J.: A test-sheet-generating algorithm for multiple assessment requirements. IEEE Trans. Educ. **46**(3), 329–337 (2003)
9. Hwang, G.J., Lin, B.M., Tseng, H.H., Lin, T.L.: On the development of a computer-assisted testing system with genetic test sheet-generating approach. IEEE Trans. Syst. Man Cybern. Part C (Appl. Rev.) **35**(4), 590–594 (2005)
10. Huitt, W.: Bloom et al.'s taxonomy of the cognitive domain. In: Educational Psychology Interactive. Valdosta, GA, Valdosta State University. http://chiron.valdosta.edu/whuitt/col/cogsys/bloom.html (2008). Last Accessed May 15, 2008
11. http://it.coe.uga.edu/wwild/pptgams/resources/bloom_questions.pdf
12. http://www.en.wikipedia.org/wiki/Bloom's_Taxonomy
13. http://en.wikipedia.org/wiki/Genetic_algorithm
14. http://trackit.arn.regis.edu

Energy-Efficient Mechanisms for Next-Generation Green Networks

Vinodini Gupta and Padma Bonde

Abstract Explosion of new technologies and increasing digital dependency of users has led to revolutionary changes in wireless networking. With this, mobility, seamless communication, security, and better service quality have become the prime characteristics of new-generation networks (NGNs). However, achieving desired performance metrics and maintaining cost efficiency in power-constrained networking scenario is quite challenging which further degrade the end-user satisfaction. Apart from socioeconomic concern, ever-increasing power consumption is equally alarming from environmental perspective as well. This has called upon for the need of energy harvesting techniques to make NGNs greener. With the vision, the paper focused on power-constrained networking circumstances, limiting factors, and possible developments to enhance the energy usage in the forthcoming communication technologies.

Keywords Green networking · Wireless networks · Energy harvesting
Network performance · Throughput · Quality of service (QoS)
Quality of experience (QoE)

1 Introduction

The increasing use of Internet, need of faster data transmissions, and uninterrupted services has greatly revolutionized the preexisting information and communication technology (ICT) standards. Network performance, better service quality, and higher end-user satisfaction are the most important factors for profitable deployment of networks. Also, being ubiquitous, it demands to be highly secured.

V. Gupta (✉) · P. Bonde
Computer Science and Engineering Department, Shri Shankracharya Technical Campus, Bhilai, India
e-mail: gupta.vinu9@gmail.com

P. Bonde
e-mail: bondepadma@gmail.com

P. K. Sa et al. (eds.), *Recent Findings in Intelligent Computing Techniques*,
Advances in Intelligent Systems and Computing 708,
https://doi.org/10.1007/978-981-10-8636-6_43

However, attaining all these features simultaneously under limited power is highly challenging which calls up for the need of efficient energy harvesting in NGNs.

In the recent years, energy harvesting (EH) is gaining its roots as the most prominent and compelling research issue. The recent advancements in ICT have attracted many users toward wireless networking from almost every sphere, leading to an application-oriented networking causing additional power dissipation in NGNs. However, long-run environmental degradation due to these technologies calls up for an environmental-friendly and sustainable green technology (GT). It will further improve the network performance and end-user satisfaction and reduce the electricity cost.

Accounting for the energy characteristics of NGNs, this paper presented a brief review of existing energy-efficient mechanisms and the prevailing issues and analyzed various optimization techniques for resolving the power constraint issues. Section 2 discussed the various challenges of energy harvesting. Section 3 provided an overview of the power dissipation characteristics of ubiquitous networks. An outlook of energy harvesting model for NGNs is presented in Sect. 4. All the factors that incur power consumption are discussed in Sect. 5. Section 6 presented a brief overview of optimization techniques to reduce the overall energy consumption. Further, Sect. 7 described the currently existing energy-efficient techniques for NGNs. Section 8 provided experimental analysis of the existing energy-conserving mechanisms. Finally, Sect. 9 concluded the paper.

2 Power Consumption at Base Station and Mobile Host in Ubiquitous Networks

During communication, power is consumed at various levels like at base stations, wireless host, or intermediate components. However, maximum fraction of power is consumed at base station. The internal architectural design of base stations depicting the power consumption phenomenon is discussed below [1] (Fig. 1).

Power level at base stations is affected by the transceiver chain. Mathematically, it can be formulated as follows

$$P_{out} = (NTRX^{*}((P_{max}/\eta_{PA} \cdot (1-\sigma_{feed})) + P_{RF} + _{PBB}))/(1-\sigma_{DC})(1-\sigma_{MS})(1-\sigma_{cool}). \quad (1)$$

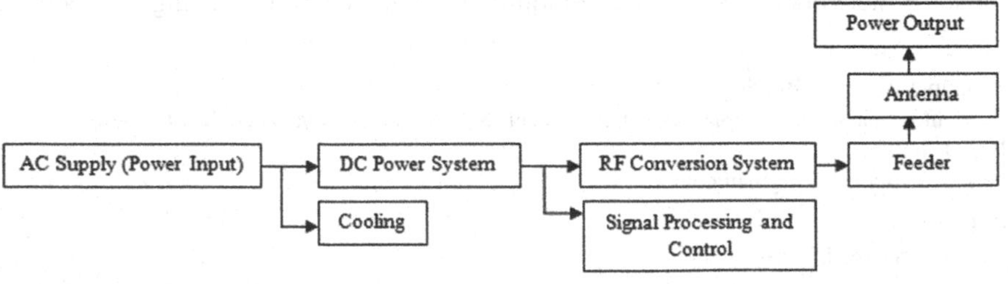

Fig. 1 Power consumption mechanism

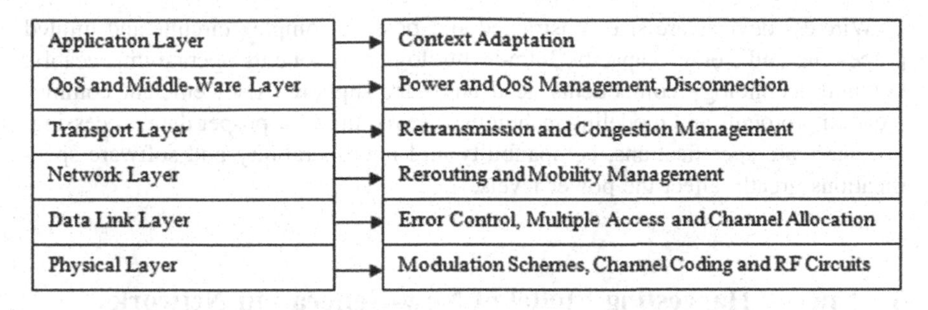

Fig. 2 Protocol stack in power management

where P_{out} is the output power, NTRX is the number of transceivers, P_{max} is the maximum power, η_{PA} is the active-mode efficiency, and σ_{feed}, σ_{DC}, σ_{MS}, and σ_{cool} are power coefficients of feeder, DC converter, mobile station, and cooling apparatus, respectively. P_{RF} and P_{BB} are the power of RF converter and power of backbone network, respectively.

Wireless devices can be either active or idle. So, mathematically, power consumption can be governed as follows

$$P_{active} = N_{tx}[P_{tx}(T_{on-tx} + T_{st}) + P_{out}T_{on-tx}] + N_{rx}[P_{rx}(T_{on-rx} + T_{st})]). \quad (2)$$

$$P_{radio} = P_{active} + T_{idle}P_{idle} + T_{sleep}P_{sleep}). \quad (3)$$

where P_{active} is the active-mode output power, P_{out} is output transmitted power, and P_{radio} is the overall power. N_{tx} and N_{rx} are the average rates of transmitter usage and receiver usage, respectively. P_{tx} and P_{rx} are the average power consumed by the transmitter and the receiver, respectively. $T_{on\text{-}tx}$ and $T_{on\text{-}rx}$ are transit on time and receive on time, respectively. T_{st} is transceiver start-up time. T_{idle} and T_{sleep} are the idle time and sleep time, respectively. P_{idle} and P_{sleep} are the power consumed during idle mode and sleep mode, respectively.

For reducing the power consumption, a new layer called the middleware layer [2, 3] is being introduced at the protocol stack as shown in Fig. 2.

3 Factors Affecting Power Dissipation in New-Generation Networks

The functionality, designing and topology of access networks and core networks differ and change dynamically. In real time, network conditions like frequent data rates, transmission policy, scalability are highly unpredictable. Thus, network architecture and conditions affect the energy levels to great extent.

Wireless devices are size-constrained and possess complex circuits and limited processing and storage capacity. Intra-technology components operate impeccably, but inter-technology components need auxiliary support. Also, efficient communication protocols and modeling techniques are required for proper data processing. So, hardware specifications, compatibility and interoperability and software specifications greatly affect the power levels.

4 Energy Harvesting Model of New-Generation Networks

Energy harvesting (EH) is becoming an essential technology for realizing perpetual and uninterrupted networking. The basic EH model [4, 5] is diagrammatically depicted in Fig. 3.

It comprises energy harvester to process ambient energy. It is stored in the form of charges at storage unit and gets converted into usable form through AC–DC converter. Microprocessor distributes the power among network components and stores it temporary at data buffers during idle period. Backbone networks provide backup support to base stations through power grid.

5 Issues in Energy Harvesting in New-Generation Networks

Throughput and QoS standards are important parameters in deciding the vitality of network. Also, different end-user applications consume different amount of power. This degrades the overall performance due to battery constraints. So, throughput maximization and QoS and QoE maintenance are intricate issues.

Mobility causes security threats, interference, signal fluctuations, and performance issues. Also, the mobile host often loses connection due to battery drainage. Moreover, incorporating authentication mechanisms increases communication latency causing extra power usage. So, managing mobility, security, and seamless connections are tough.

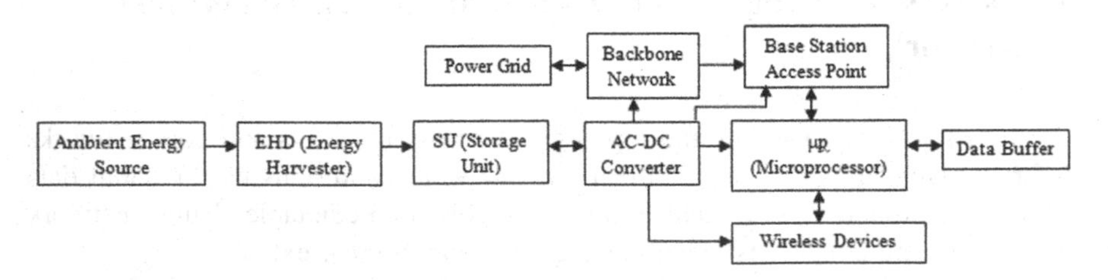

Fig. 3 Energy harvesting model

High mobility causes interference and fast battery drainage and affects the received signal strength (RSS). Also, frequency reusing raises interference and idle period causing extra power usage. This demands channel reusability for proper bandwidth utilization. So, bandwidth utilization, signal strength, and interference management are exigent tasks.

NGNs are highly dynamic and demand continuous resource usage. They must be scalable enough to support both centralized and distributed communications. Also, for better services, higher data rates are desirable. These factors cause enormous power consumption. Also, lower transmission power degrades performance, whereas higher values increase energy usage. So, managing scalability, high data rates, transmission power optimization, and infrastructure usage are intricate issues.

Ambient energy is sparse and sporadic, and its maintenance incurs infrastructure and cost overheads. Also, EH algorithms suffer various imperfections which make them unfit for real-time scenario. For better QoS and QoE standards, network must be fault tolerant [6, 7] and everlasting. Hence, ambient energy management, EH imperfection reduction, and robustness are exigent issues.

Heavy load causes congestion, data loss, and high latency, switches devices to idle mode, and degrades the performance. As complexity increases, data routing becomes tough and unpredictable causing communication overheads. Thus, proper traffic scheduling [7] and overhead reduction is tough in power-constrained networking.

6 Power Optimizing Techniques in New-Generation Networks

To optimize energy, selective switch off technique [8, 9] is most effective. It demands software redesigning, hardware reconfiguration, and routing protocol adaption to sustain standby mode. Different components have different power requisite as per their design. So, power ration of all the components [10] must be well optimized. Since large hop count raises communication latency and power usage, least hop count is admissible.

Efficient network designs for core networks [9], access networks, and metro networks [10] can be achieved by re-engineering the protocol stack. Several redesigning strategies and load adaptive and traffic engineering techniques have been introduced in [11] to achieve the goal of power-efficient backbone networks. At base stations, utmost power dissipates in transmission phase. So, base station transceiver (BST) optimization strategies like standby radio units, passive cooling, efficient rectifiers have been introduced.

Routing protocols must be energy efficient during activity and inactivity of nodes. Gouda et al. [12] and Kanakaris et al. [13] analyzed various energy-efficient routing protocols in ad hoc networks. Link failure causes channel congestion and increases load causing heavy power consumption. Thus, reducing link failures is highly desirable.

For power optimization, [11, 14] introduced certain techniques like wavelength grooming, waveband grooming, cell area revision to handle both static traffic and dynamic traffic properly. Load sharing can be achieved by varying link configurations or link rates. Addis et al. [6] and Dharmaweera et al.[11] presented configuration techniques like single link and bundled link and link rate techniques like single link rate (SLR), mixed link rate (MLR), and adaptive link rate (ALR).

Isotropic antenna [8] increases power dissipation, especially in unicast communications. So, the use of directional antenna reduces interference and data retransmission and enhances energy efficiency of networks. Cognitive radio [8] can reduce the interference and are capable of bonding multiple channels together to attain higher transmission speed. So they can be used to reduce the overall transmission power. Session management technique called beacon [8] reduced the duty cycle for optimizing the energy rations of the network enormously. While ambient energy [4] sources are intermittent and sparse, they can still meet the extra energy needs of NGNs. Hence, these resources must be properly harvested.

7 Overview of Existing Energy Conservation Techniques

To achieve the goal of green networking, [15] proposed an energy-efficient load-balancing technique. Similarly, [16] proposed a threshold-based transmission policy for correlated energy sources. For improvising the energy efficiency of small-scale wireless networks, a piezoelectric cane is used in [17]. A Markov decision process (MDP) was introduced in [18] to improve the energy necessities of self-organizing networks. Wang et al. [19] proposed a random online control policy based on the Lyapunov optimization technique to maximize the energy efficiency of wireless networks. A contention-based energy-efficient protocol was introduced in [20] to enhance the performance during intra-cluster communication. Orumwense et al. [21] analyzed the energy requirements in cognitive radio networks and investigated the various energy harvesting techniques for such networks. In [22], both energy spectrum and energy harvesting techniques are collectively configured to reduce the energy rations.

Wireless devices have limited size, battery, and operational capabilities. Hence, various routing protocols like minimum battery cost routing (MBCR), min–max battery cost routing (MMBCR) [23], and conditional min–max battery cost routing (CMMBCR) have been introduced to enhance battery lifetime. Also, various approaches have been proposed in [24] to enhance the energy efficiency in MANETs. Power-aware routing optimization (PARO) model [25] optimized packet forwarding energy in ad hoc networks.

Accounting for energy provisions of optical networks, [26] discussed various reduction techniques to make optical fibers more energy efficient. Rouzic et al. [27] discussed the solutions and innovative ideas toward power-efficient optical networks. The joint transmission scheme [28] optimally allocated time and power among the base stations. Heddeghem et al. [29] utilized analytical power model to analyze the power-saving techniques in backbone networks. Similarly, [30] proposed an energy-saving algorithm based on sleep mode methodology.

8 Discussions and Analysis

Zhou et al. [15] proposed a load-balancing algorithm for internetworking between LTE and low-powered Wi-fi networks. Through signal quality adjustments, the algorithm achieved 13.11% energy efficiency and also enhanced resource utilization and user experience. The PARO model proposed in [25] followed on-demand approach to reduce the overall point-to-point power consumption. The AODVEA protocol [31] performed local forwarding decisions over energy threshold values. However, the modified version was found to deliver optimized result as compared to original algorithm. Sinai et al. [32] proposed a delay-aware approach for efficient cross-layer resource allocation to enhance the energy productivity.

The survey in [29] revealed that to achieve power efficiency, either device power rating or traffic needs to be reduced. The energy-saving algorithm proposed in [30] evaluated decision index for switching off the nodes based on the stress centrality values. In this approach, neither traffic pattern nor parameter was required for decision making. It proved 50% more efficiency but faced implementation issues regarding testbed emulation and number of iterations. The study conducted at [21] revealed that cognitive radio is highly energy demanding because of spectrum sensing, reconfigure ability, higher QoS delivery, and network discovery activities. Simulation results in [22] proved the suitability of FreeNet in alleviating network congestion, upgrading network capability, and broadband provisioning in urban areas. Wang et al. [19] maximized the data rate and energy efficiency any prior knowledge about the channel by reducing battery imperfections of storage units.

The solutions in [27] improved the energy efficiency of optical networks and supported sustainable growth by reducing the impact of power constraints. The study at [26] revealed that bandwidth in optical networks can be enhanced up to 66% through thin client paradigm. The energy-efficient MAC protocol in [20] considered partial clustering and improved energy usage without any additional delays. The study at [33] revealed that threshold-based switching reduced 50% of the power consumption and interference.

An integrated holistic approach discussed in [34] enhanced the overall service quality, application standards, throughput, network capacity, and global productivity in a concurrent fashion. The power consumption model in [32] enhances the performance at data link layer by considerably improvising the effective capacity (EC) and effective energy efficiency (EEE) during transmission. The scheme in [18] conserves 53% power over average consumption, 73% over low traffic, and 23% over high traffic patterns.

9 Conclusions

The basic objective of the paper is to explore the hidden potentials for energy harvesting in NGNs for attaining energy optimization. This paper discussed the peculiar issues and challenges of wireless energy management. It focused over the affairs and factors contributing power dissipation in NGNs. It also analyzed currently existing techniques to provide future directions to the new researchers. The research will serve as a framework for coping up with mobility security and connectivity requirements. In light of this, novel energy-efficient mechanisms can be formulated to enhance the service quality and user satisfaction simultaneously.

References

1. Zheng, R., Hou, J.C, Li, N.: Power management and power control in wireless networks. At NSF under Grant No. CNS 0221357 (2005)
2. Cengiz, K., Dag, T.: A review on the recent energy-efficient approaches for the Internet protocol stack. EURASIP J. Wirel. Commun. Netw. (2015)
3. Havinal, R., Attimarad, G.V., Giriprasad, M.N.: A review of energy consumption issues in various layers of protocol stack in MANET's. Int. J. Innov. Emerg. Res. Eng. (IJIERE) **2**(2) (2015)
4. Faustine, A.A., Lydia, A., Geetha, K.S.: A survey on energy harvesting in wireless communication. Int. J. Innov. Res. Comput. Commun. Eng. (IJIRCCE) **3**(9) (2015)
5. Muhammad Mazhar Abbas, M.M, Muhammad, Z., Khalid Saleem, K., Nazar Abbas Saqib, N.A., Mahmood, H.: Energy harvesting and management in wireless networks for perpetual operations. J. Circuits Syst. Comput. **24**(3) (2015)
6. Addis, B., Capone, A., Carelloa, G., Gianolia, L.G, Sans, B.: Energy management in communication networks: a journey through modelling and optimization glasses. Comput. Commun. (2015)
7. Addis, B., Capone, A., Carello, G., Gianoli, L.G., Sansò, B.: Energy management in communication networks: a journey through modelling and optimization glasses. Research Gate Online Publications, July 2015
8. Guo, J., Jain, R.: Energy efficiency in wireless networking protocols. In: current issues in Wireless and Mobile Networking. http://www.cse.wustl.edu/~jain/cse574-14/index.html
9. Vereecken, W., Heddeghem, W.V., Deruyck, M., Puype, B., Lannoo, B., Joseph, W., Colle, D., Martens, L., Demeester, P.: Power consumption in telecommunication networks: overview and reduction strategies. IEEE Commun. Mag. (2011)
10. Zhang, Y., Chowdhury, P., Tornatore, M., Mukherjee, B.: Energy efficiency in telecom optical networks. IEEE Commun. Surv. Tutor. **12**(4), Fourth Quarter (2010)
11. Dharmaweera, M.N., Parthiban, R., Sekercio˘glu, A.Y.: Toward a power-efficient backbone network: the state of research. IEEE Commun. Surv. Tutor. **17**(1), First Quarter (2015)
12. Gouda, B.S., Dass, A.K., Narayana, L.K.: A comprehensive performance analysis of energy efficient routing protocols in different traffic based mobile ad-hoc networks. IEEE (2013)
13. Kanakaris, V., Ndzi, D., Azzi, D.: Ad-hoc networks energy consumption: a review of the ad-hoc routing protocols. J. Eng. Sci. Technol. Rev. (JESTR) (2010)
14. Ratheesh, R., Vetrivelan, P.: Power optimization techniques for next generation wireless networks. Int. J. Eng. Technol. (IJET) (2016)
15. Zhou, F., Feng, L., Yu, P., Li, W.: Energy-efficiency driven load balancing strategy in LTE-Wi-Fi interworking heterogeneous networks. In: IEEE Wireless Communications and

Networking Conference (WCNC)—Workshop—Self Organizing Heterogeneous Networks (2015)
16. Abad, M.S.H., Gunduz, D., Ercetin, O.: Energy harvesting wireless networks with correlated energy sources. EC H2020-MSCA-RISE- programme, grant number 690893, Tubitak, grant number 114E955 and British Council Institutional Links Program, grant number 173605884
17. Hiroki Sato, H., Yachi, T.,: Harvesting electric power with a cane for radio communications. In: 4th International Conference on Renewable Energy Research and Applications, Nov 2015
18. Kim, J., Kong, P.Y., Song, N.O., Rhee, J.K.K., Araji, S.A.: MDP based dynamic base station management for power conservation in self-organizing networks. In: IEEE WCNC Track 3 (Mobile and Wireless Networks). (2014)
19. Wang, X., Ma, T., Zhang, R., Zhou, X.: Stochastic online control for energy-harvesting wireless networks with battery imperfections. IEEE Trans. Wirel. Commun. (2016)
20. Amin Azari, A., Miao, G.: Energy efficient MAC for cellular-based M2M Communications. IEEE, Dec 2014
21. Orumwense, E.F., Afullo, T.J., Srivastava, V.M.: Achieving a better energy-efficient cognitive radio network. Int. J. Comput. Inf. Syst. Ind. Manag. Appl. (IJCISIMA) **8**, 205–213 (2016)
22. Ansari, N., Han, T.: FreeNet: Spectrum and Energy Harvesting Wireless Networks. In IEEE Network, January/February (2016)
23. Mahajan, S., Kaur, B., Pandey, V.: A study on power efficient techniques in various layers of protocol stack in MANETs. Int. J. Comput. Sci. Inf. Technol. (IJCSIT) **5**(3), 3748–3753 (2014)
24. Aggarwal, K.: Review of energy conservation approaches in MANET. Int. J. Adv. Res. Comput. Sci. Softw. Eng. **5**(4) (2015)
25. Gomez, J., Campbell, A.T., Naghshineh, M., Bisdikian, C.: Conserving transmission power in wireless ad hoc networks. At National Science Foundation (NSF) under the Wireless Technology Award ANI-9979439 and IBM Research
26. Vereecken, W., Heddeghem, W.V., Puype, B., Colle, D., Pickavet, M., Demeester, P.: Optical networks: how much power do they consume and how can we optimize this? IEEE Xplore (2010)
27. Rouzic, E.L., Idzikowski, F., Tahiri, I.: TREND towards more energy-efficient optical networks. Research Gate Online Publications, Apr 2013
28. Velkov, Z.H., Nikoloska, I., Karagiannidis, G.K., Duong, T.Q.: Wireless networks with energy harvesting and power transfer: joint power and time allocation. IEEE Signal Process. Lett. **23**(1) (2016)
29. Heddeghem, W.V., Parker, M.C., Lambert, S., Vereecken, W.: Using an analytical power model to survey power saving approaches in backbone networks. Research Gate Online Publication (2012)
30. Federico Patota, F., Chiaraviglio, L., Bella, F., Deriu, V., Fortunato, S., Cuomo, F.: DAFNES: a distributed algorithm for network energy saving based on stress centrality. Comput. Netw. (2016)
31. Ket, N., Hippargi, S.: Modified AODV energy aware routing for optimized performance in mobile ad-hoc networks. In: IEEE WISPNET (2016)
32. Sinai, M., Zappone, A., Jorswieck, E.A., Azmi, P.: A novel power consumption model for effective energy efficiency in wireless networks. IEEE Wirel. Commun. Lett. (2015)
33. Alam, A.S., Dooley, L.S., Poulton, A.S., Ji, Y.: Energy savings using an adaptive base station to relay station switching paradigm. IEEE (2012)
34. Slavisa Aleksic, S.: Energy-Efficient Communication Networks for Improved Global Energy Productivity. Springer, In Telecommunication System (2012)

Takagi Sugeno Fuzzy for Motion Control of Ball on Plate System

Vinodh Kumar Elumalai, V. Keerthi Prasath, B. Khizer Ahamed, Rajeev Gupta and Sameer Kumar Mohapatra

Abstract This paper presents the Takagi Sugeno (TS) fuzzy scheme for stabilization and motion control of ball on plate system. Plant nonlinearity and inter-axis coupling are the two major challenges which enhance the complexity of controller design for ball on plate system. Hence, in this paper, we utilize the TS fuzzy-based input–output mapping to deal with the nonlinearities and model variation associated with the ball on plate system, which emulates the concept of visual servo control (VSC). The motivation for using TS fuzzy model is that it can capture the dynamics of a nonlinear plant model with fewer fuzzy rules and yield better accuracy compared to Mamdani fuzzy. Moreover, as TS fuzzy is a multimodal approach, it offers a systematic way to derive the fuzzy rules from the given input–output data. The performance of the control scheme is validated through simulation, and the tracking results prove that the TS fuzzy scheme can offer precise tracking control of ball on plate system.

Keywords TS fuzzy · Ball on plate system · Visual servo control
Under-actuated system

V. K. Elumalai (✉) · V. Keerthi Prasath · B. Khizer Ahamed · R. Gupta · S. K. Mohapatra
School of Electrical Engineering, VIT University, Vellore 632014, India
e-mail: vinodhkumar.e@vit.ac.in

V. Keerthi Prasath
e-mail: keerthiprasath.v2015@vit.ac.in

B. Khizer Ahamed
e-mail: bkhizer.ahamed2015@vit.ac.in

R. Gupta
e-mail: rajeev.gupta2014@vit.ac.in

S. K. Mohapatra
e-mail: sameerkumar.m2014@vit.ac.in

P. K. Sa et al. (eds.), *Recent Findings in Intelligent Computing Techniques*,
Advances in Intelligent Systems and Computing 708,
https://doi.org/10.1007/978-981-10-8636-6_44

1 Introduction

Visual servo control, a multidisciplinary research area, has attracted considerable attention in both academia and industry for its vast applications ranging from autonomous vehicle navigation, robot control to factory automation [1]. Non-contact measurement, versatility, and accuracy are the key features that attract visual feedback for closed-loop control. The ball on plate system, a typical benchmark plant for VSC, is the extension of classical ball on plate system which is widely used in the control engineering laboratories to assess the efficacy of various control algorithms. As the ball on plate system is a highly nonlinear, multivariable, and under-actuated system, designing a control algorithm for stabilization and tracking is always challenging. In the literature, several control algorithms have been reported for position control of ball on plate system. For instance, using Euler estimator to determine the position of the ball, Park and Lee [2] put forward a sliding mode control algorithm to deal with the variation in the surface characteristics of plate and mass of the ball. To minimize the aftereffects of friction between the plate and ball, Wang et al. [3] proposed a disturbance observer-based friction compensation scheme and proved that design avoids limit cycle.

Even though several results have been reported on classical control design for ball on plate system, the use of fuzzy logic for controlling the dynamics of ball on plate has not been much explored. Hence, this paper aims to investigate the efficacy of TS fuzzy scheme for tracking application of ball on plate system. The key reason for using the TS fuzzy is that it does not require the accurate plant model for control implementation and is more suitable for nonlinear plant dynamics.

2 Ball on Plate System

The ball on plate system consists of a metal plate, a ball, an USB-based overhead camera, and two servo motors which control the X and Y directions of the plate. The plate is attached to the servo motors through 2 DoF gimbals such that it can swivel about any direction. The objective is to control the tilt angle of the plate through servo load gears so that the ball can be positioned at desired location. As the X and Y coordinates of ball on plate system have similar dynamics, the key equations that govern the dynamics of X-direction control are given here for brevity. Interested readers can refer [4] for detailed modeling of ball on plate system.

The translational and rotational forces acting on the ball due to gravity and moment of inertia are given by

$$F_{x,t} = m_b g \sin \alpha(t) \tag{1}$$

$$F_{x,r} = \frac{J_b \frac{d^2}{dt^2} x(t)}{r_b^2} \tag{2}$$

where m_b is the mass of the ball, g is gravity, α is the plate angle, J_b is the moment of inertia, and r_b is the radius of the ball. Applying Newton's law, we can write the following system dynamic equation.

$$m_b \frac{d^2}{dt^2} x(t) = F_{x,t} - F_{x,r} \tag{3}$$

Hence,

$$m_b \frac{d^2}{dt^2} x(t) = m_b g \sin \alpha(t) - \frac{J_b \frac{d^2}{dt^2} x(t)}{r_b^2} \tag{4}$$

From Fig. 1, the relation between the plate angle and servo angle can be described as

$$\sin \alpha(t) = \frac{2 r_{arm} \sin \theta_l(t)}{L_t} \tag{5}$$

where L_t is the length of the table, r_{arm} is the gear radius, and θ_l is the servo angle. Substituting (5) into (4), we can obtain the following nonlinear equation of motion of the ball.

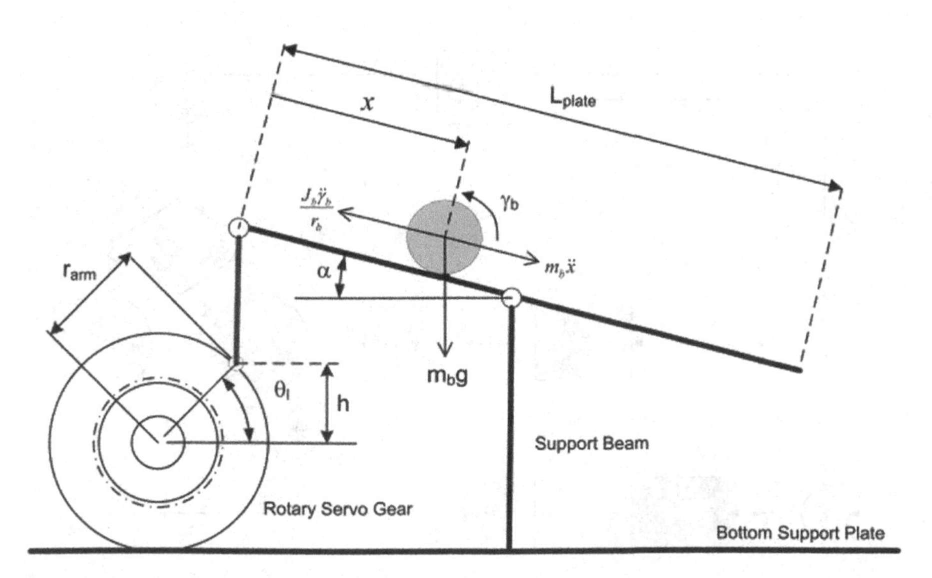

Fig. 1 Schematic of ball on plate system

$$\frac{d^2}{dt^2}x(t) = \frac{2m_b g\theta_l r_{arm} r_b^2}{L_t(m_b r_b^2 + J_b)} \tag{6}$$

We utilize TS fuzzy scheme, as illustrated in Fig. 2, to deal with the nonlinear dynamics of the plant. The plate image captured by the USB camera is used to determine the X and Y coordinates of the plate using background subtraction image processing technique. The control objective is to position the ball to the desired trajectory by controlling the servo angle through fuzzy controller. In the next section, the TS fuzzy scheme and the fuzzy rules are given in brief.

3 Takagi Sugeno Fuzzy Control

Takagi Sugeno fuzzy system, a special class of functional fuzzy system, is a multimodal approach that implements a nonlinear interpolation in the linear mapping. Compared to Mamdani fuzzy, TS fuzzy can model highly nonlinear system with fewer rules and better accuracy [5, 6]. Moreover, as the TS fuzzy has linear function of inputs as the consequents, it is computationally efficient and suitable for nonlinear plants. Consider a nonlinear system

$$\dot{x} = f(x) + g(x)u \tag{7}$$

where $x = [x_1, x_2, \ldots, x_n]^T$ is the state vector, $f(x)$ and $g(x)$ are the nonlinear functions of the system, and u is the control input. The TS fuzzy utilizes the input–output

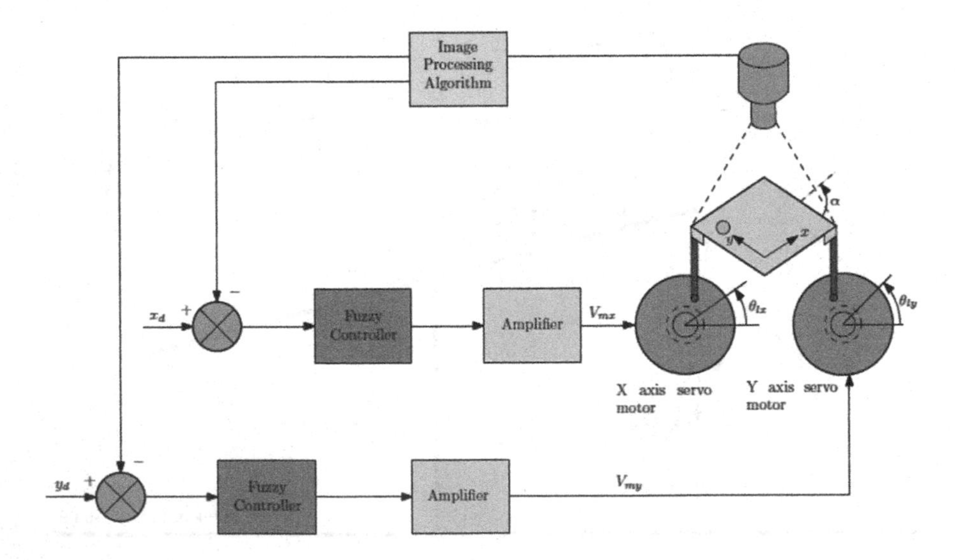

Fig. 2 Block diagram TS fuzzy control for ball on plate system

relationship to design a fuzzy rule-based model [7]. In general, the rule in the TS fuzzy is of the form:

$$\mathrm{R}^{\mathrm{i}}\text{: If } a_1 \text{is } Z_1^i, a_2 \text{ is } Z_2^i, \ldots . a_m \text{ is } Z_m^i$$

$$\text{then } y^i = a_0^i + a_1^i s_1 + a_2^i s_2 + \cdots + a_m^i s_m \tag{8}$$

where a_i represents the input models, y^i denotes the inferred output of the ith fuzzy submodel, Z_m^i is the fuzzy set, s_i is the consequent variable, and y^i is the output of the ith fuzzy implication. Sugeno-type fuzzy inference system (FIS) utilizes the following weighted average method to calculate the crisp output [8].

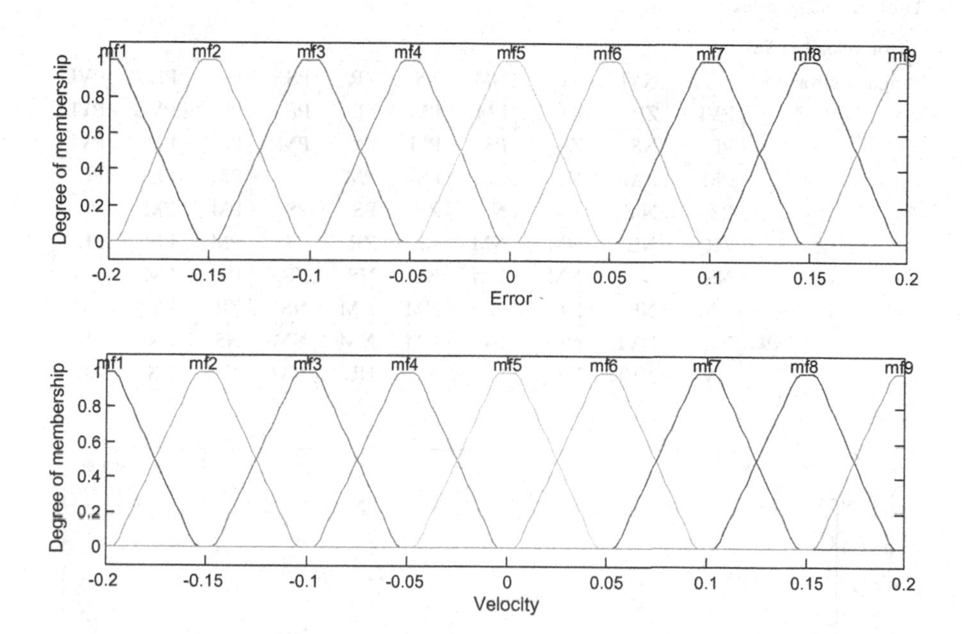

Fig. 3 Input membership functions of TS fuzzy

Table 1 Fuzzy membership and output parameters

Fuzzy	Error and velocity coordinates	Output
NVL: negative very large	[−0.245 −0.205 −0.195 −0.155]	−30
NL: negative large	[−0.195 −0.155 −0.145 −0.105]	−20
NM: negative medium	[−0.145 −0.105 −0.095 −0.055]	−10
NS: negative small	[−0.095 −0.055 −0.045 −0.005]	−5
ZR: zero	[−0.045 −0.005 0.005 0.045]	0
PS: positive small	[0.005 0.045 0.055 0.095]	5
PM: positive medium	[0.055 0.095 0.105 0.145]	10
PL: positive large	[0.105 0.145 0.155 0.195]	20
PVL: positive very large	[0.155 0.195 0.205 0.245]	30

$$y = \sum_{i=1}^{n} W^i y^i / \sum_{i=1}^{n} W^i \tag{9}$$

where n indicates the no of fuzzy rules, and W^i represents the degree of firing the *i*th rule, defined as

$$W^i = \prod_{j=1}^{m} \mu_{Z_j} a_j \tag{10}$$

Table 2 Fuzzy rules

Error velocity (Δe)										
Angular error (e)		NVL	NL	NM	NS	ZR	PS	PM	PL	PVL
	PVL	ZR	PS	PM	PM	PL	PL	PL	PVL	PVL
	PL	NS	ZR	PS	PM	PM	PM	PL	PL	PVL
	PM	NM	NS	ZR	PS	PM	PM	PM	PL	PL
	PS	NM	NM	NS	ZR	PS	PS	PM	PM	PL
	ZR	NL	NM	NM	NS	ZR	PS	PM	PM	PL
	NS	NL	NM	NM	NS	NS	ZR	PS	PM	PM
	NM	NL	NL	NM	NM	NM	NS	ZR	PS	PM
	NL	NVL	NL	NL	NM	NM	NM	NS	ZR	PS
	NVL	NVL	NVL	NL	NL	NL	NM	NM	NS	ZR

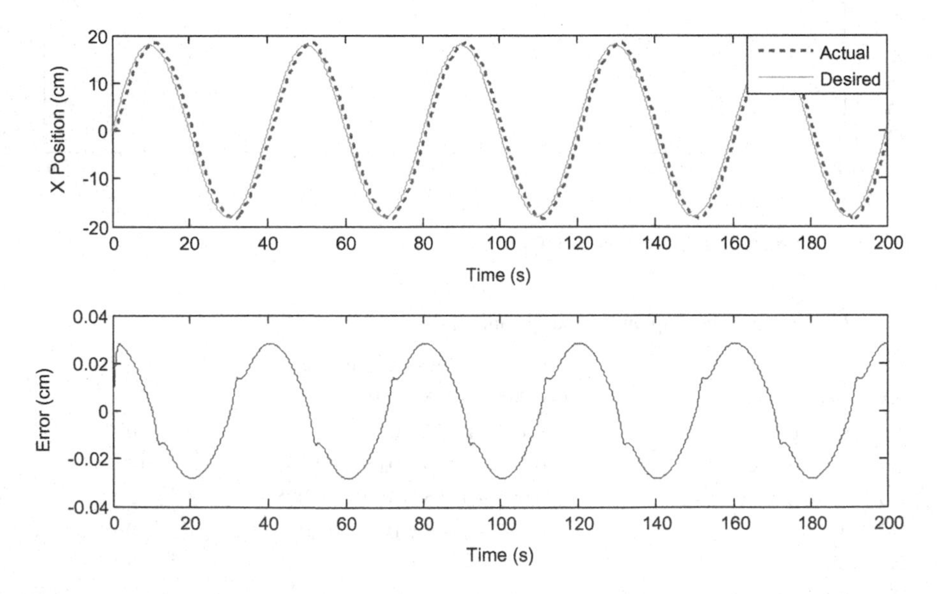

Fig. 4 X-coordinate tracking response and error

μ_{Z_j} is the membership function of the antecedent fuzzy set Z_i. The TS fuzzy controller input is

$$u = f(e, \dot{e}) \tag{11}$$

where $e = r - y$ with r as the reference input and y as the actual output.

4 Results and Discussion

The motion control performance of TS fuzzy is assessed using MATLAB/Simulink 2015B. For the ball on plate system, the error and velocity of error are the input membership functions for TS fuzzy and servo angle is the crisp output. Figure 3 illustrates the input membership functions, and Tables 1 and 2 give the fuzzy parameters and rules, respectively.

4.1 Command Tracking Response

To validate the reference tracking performance of the TS fuzzy-based ball on plate system, a sinusoidal trajectory with an input frequency of 0.02 Hz is given as a test signal for both X and Y axes. Figures 4 and 5, which illustrate the command tracking performance of X and Y coordinates along with the tracking errors, highlight that the fuzzy scheme closely tracks the input command and results in a tracking error of 0.05 cm peak to peak in both the axes. Figures 6 and 7 depict the

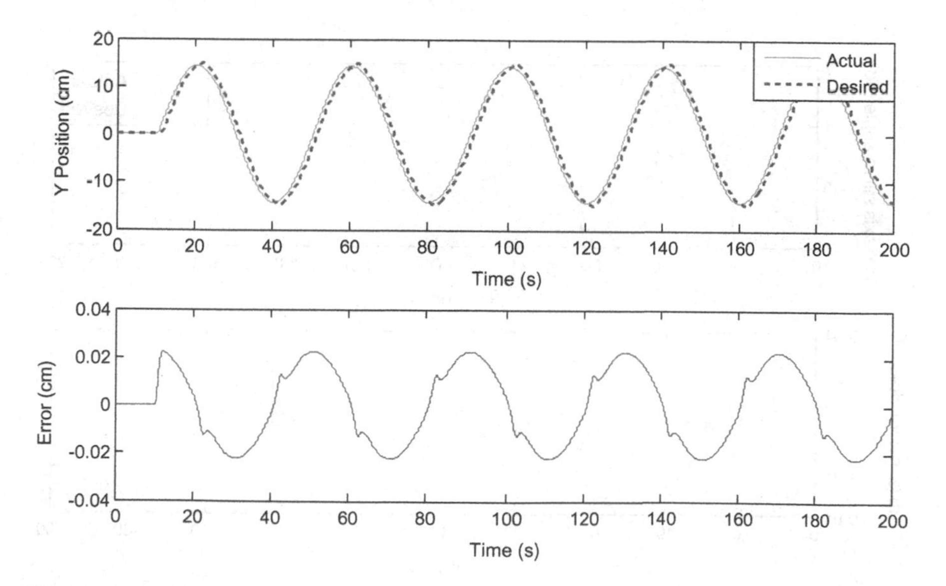

Fig. 5 Y-coordinate tracking response and error

respective servo angles of the two coordinates (θ_x, θ_y) and motor voltages (V_m, V_y) supplied to X and Y coordinates. It can be noted that the fuzzy scheme implements precise control over servo angle so as to capture the minimum change in ball position.

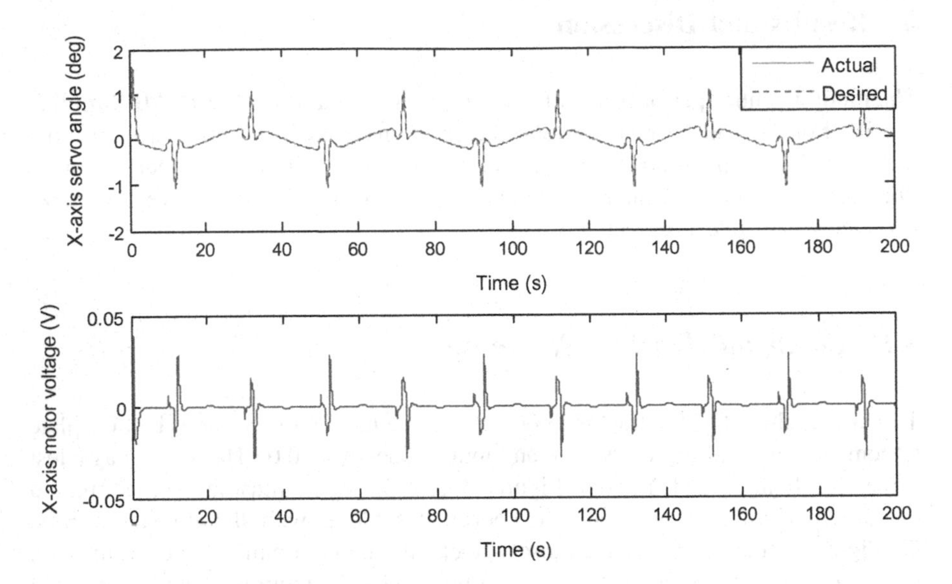

Fig. 6 X-coordinate servo angle and motor voltage

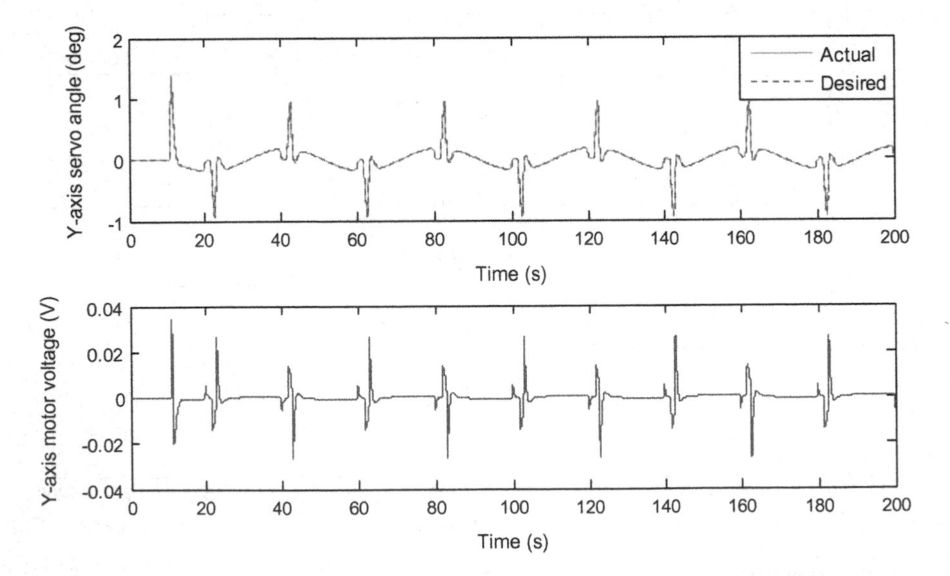

Fig. 7 Y-coordinate servo angle and motor voltage

5 Conclusions

This paper has presented a TS fuzzy scheme for stabilization and motion control of ball on plate system. Unlike the conventional control approaches which focus more on modeling the plant and using that model for designing a controller, fuzzy control acts as an artificial decision maker that does not require the accurate plant model for controller implementation. Moreover, fuzzy control can also accommodate non-linear dynamics of the plant and offer precise control based on the judicious selection of fuzzy rules. Motivated by these key aspects, we have synthesized a TS fuzzy scheme and assessed the performance of a ball on plate system for reference tracking application. Simulation results prove that the TS fuzzy can offer precise control even if the plant model has nonlinear dynamics.

References

1. Siradjuddin, I., Behera, L., McGinnity, T.M., Coleman, S.: Image-based visual servoing of a 7-DOF robot manipulator using an adaptive distributed fuzzy PD controller. IEEE/ASME Trans. Mechatron. **19**(2), 512–523 (2014)
2. Park, J.H., Lee, Y.J.: Robust visual servoing for motion control of the ball on a plate. Mechatronics **13**(7), 723–738 (2003)
3. Wang, Y., Sun, M., Wang, Z., Liu, Z., Chen, Z.: A novel disturbance observer based friction compensation scheme for ball and plate system. ISA Trans. **53**(2), 671–678 (2014)
4. Subramanian, R.G., Elumalai, V.K., Karuppusamy, S., Canchi, V.K.: Uniform ultimate bounded robust model reference adaptive PID control scheme for visual servoing. J. Franklin Inst. **354**(4), 1741–1758 (2017)
5. Cavallaro, F.: A Takagi-Sugeno fuzzy inference system for developing a sustainability index of biomass. Sustainability **7**, 12359–12371 (2015)
6. Nikdel, P., Hosseinpour, M., Badamchizadeh, M.A., Akbari, M.A.: Improved Takagi-Sugeno fuzzy model-based control of flexible joint robot via Hybrid-Taguchi genetic algorithm. Eng. Appl. Artif. Intell. **33**, 12–20 (2014)
7. Passino, K.M., Yurkovich, S.: Fuzzy Control. Addison-Wesley Longman, Inc, California (1998)
8. Husek, P., Narenathreyas, K.: Aircraft longitudinal motion control based on Takagi-Sugenofuzzy model. Appl. Soft Comput. **49**, 269–278 (2016)

Human Cancer Classification and Prediction Based on Gene Profiling

H. N. Megha and R. H. Goudar

Abstract Cancer is one of the larger families of diseases; it is a collection of 100s of diseases that involve abnormal cell to grow and spread to the other parts of the body which leads to worldwide death of the human being. There are different types of cancer, so for those we need to identify and classify all those different types. In the field of bioinformatics, the main important thing is cancer diagnosis, so we need to do it by selecting the subset of feature gene. In cancer diagnosis, the greatest significance is classifying the different types of tumors. By providing an accurate prediction for various types of tumors, we can provide a better treatment for cancer patients and also it reduces the toxicity on patients. The main important thing in this paper is to differentiate between cancer subtypes by creating different methodologies. This paper explains different methodologies, based on gene profiling for the classification and prediction for different types of human cancer. The proposed methodology in this paper is a combination of symmetrical uncertainty (SU) and genetic algorithm (GA).

Keywords Human cancer prediction and classification · Gene profiling
Symmetrical uncertainty (SU) · Genetic algorithm (GA) · Gene expression programming · Feature gene selection · Cancer diagnosis

1 Introduction

Cancer is a collection of many diseases which involve abnormal growth of cell/ abnormal cell division without control in present which spread to all other parts of the body. All tumors are not cancerous. This uncontrolled growth of the cell causes

H. N. Megha (✉) · R. H. Goudar
Department of Computer Netwok Engineering, Center for P.G. Studies, Visvesvaraya Technological University, Belagavi 590018, Karnataka, India
e-mail: meghahn6@gmail.com

R. H. Goudar
e-mail: rhgoudar.vtu@gmail.com

P. K. Sa et al. (eds.), *Recent Findings in Intelligent Computing Techniques*, Advances in Intelligent Systems and Computing 708,
https://doi.org/10.1007/978-981-10-8636-6_45

a lump called a tumor to form, and it spreads to the other parts of the body through blood and lymph system. There are more than 10 million patients in the world, and the cancer mortality rate is extremely high. Symptoms of cancer are not obvious, and the lesions are so small, so it is not so easy to find this cancer disease. The symptoms which indicate cancer are loss of weight, prolonged cough, abnormal bleeding, lump, etc.

There are different types of cancer, In that Each cancer has its feature of gene expression profiling, so the key to decide cancer samples classification and identification, we need to extract/select the DNA, from that we should select a collection of gene feature/markers in the form of tens of thousands which is measured from the selected/extracted DNA.

The main two aspects of gene expression profiling to identify and classify the tumor are as follows.

1. In the current era, even though there is availability of gene expression data, it is not totally appropriate for the practical application and hence the final decision lies in the hands of medical specialist. Using gene expression data from cancer patients by data screening, it reduces the time and payment and helps to extract the feature genes directly and use them in identification and diagnosis of cancer diseases.
2. DNA microarray data is of high dimension and is of small sample size, highly redundant and is imbalance in distribution. Thus, the way of extracting feature genes from high-dimensional data is of greater importance. Cancer gene profiling consists of a huge number of genes; however, most of the genes are not eligible to participate in the processes of identification and classification.

The correlated gene information increases computational overhead in classification, due to which the improvement lags behind. So, gene profiling analysis is required.

The above two aspects are used to identify and classify tumors in human using gene profiling.

The greatest significance in diagnosis of cancer tumors is classification of different types of tumor. By providing accurate prediction for various types of tumors, we can predict a better treatment for cancer patients and also it reduces the toxicity on patients. To predict the cancer disease, one of the important techniques is used which is called as "DNA microarray". For the analysis of gene expression in a large variety of experimental researches, the above technique is effectively used. There is a pre-processing step in DNA microarray data analysis; i.e., for the purpose of cancer classification, we need to select appropriate feature genes. From two samples cells taken from the blood of the suspected person, we can analyze gene expression levels of several thousands of genes by using a technique called DNA microarray. Depending on the test results, we can investigate the disease, like progress in diseases, extract diagnosis, drug reaction and improvement after treatment should be done.

This paper tells about the classification problem in human cancer disease by using gene profiling. It represents a new methodology to analyze the DNA microarray datasets, and efficiently, it classifies and predicts the different human cancer diseases.

(1) First methodology used here is symmetric uncertainty (SU) for feature selection in gene.
(2) Second methodology used is genetic algorithm (GA) for feature reduction in gene.
(3) Final methodology is gene expression programming (GEP); this is used to classify different types of human cancer diseases.

All the above 3 methodologies improves the accuracy in classification, For classifying the cancer disease by minimizing the number of feature gene and by avoiding the genetic algorithm being trapped in local optimum.

2 Proposed Methodology

The proposed methodology consists of classification and prediction of various types of human cancers and is shown in Fig. 1.

The details of Fig. 1 are explained below:

1. **Initial Sample**: The blood samples are collected from the suspects for further processing. These samples act as test data.
2. **Gene Profiling**: To understand the complete scenario of cellular function, we need to analyze thousands of genes activities at a time. To perform this activity, genes are extracted from initial sample to decide about the cancer-causing genes.
3. **Feature Selection**: From the extracted genes in the gene profiling, to select vital feature, the symmetrical uncertainty (SU) method is used.
4. **Feature Reduction**: For this process, genetic algorithm (GA) is performed on selected features of the genes obtained from the feature selection.
5. **Classification and prediction**: To classify and predict the type of cancer, gene expression programming (GEP) method is used.

1. **Feature Selection**
   There are two different kinds of feature selection, namely

   i. Filter method,
   ii. Wrapper method.

Filter Method is one of the most common methods used for feature selection based on symmetrical uncertainty (SU). The genetic algorithm population can be calculated by the addition and deletion of feature genes. It is used to detect the improvement or ranking of every feature separately. It is used for classification process. It is independent of their learning algorithm; it is very simple, fast, and scalable for computational process. This approach is used for feature selection and is carried out at a time. The intermediate result of this process is provided as input

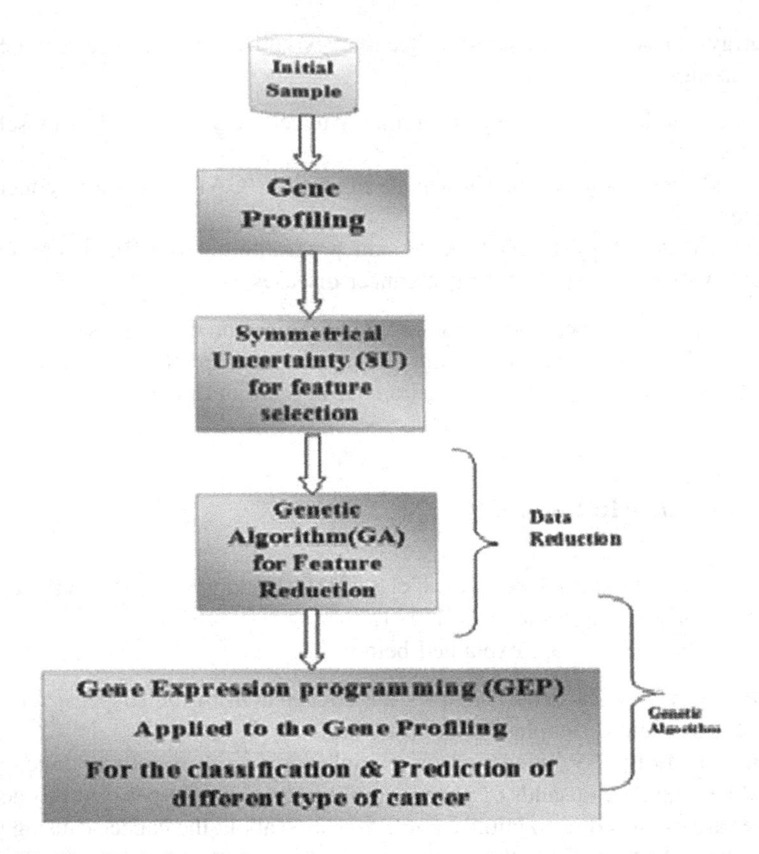

Fig. 1 Classification and prediction system

for different classifier. There is lots of feature selection/ranking technique that are introduced; some of them are correlation-based feature selection (CFS), principal component analysis (PCA), gain ratio (GR), information gain (IG), symmetrical uncertainty (SU), etc. Some of the technique does not perform feature selection instead they are used for feature selection ranking. Hence, in this paper, I am considering symmetrical uncertainty for feature selection.

Wrapper Method: It uses the classification method to calculate/measure the feature set importance, so the feature selection is completely dependent on the use of classifier model. This wrapper model is very costlier for the big dimensional database because in the form of computational complexity and time, the set of features of genes should be evaluated along with the classifier algorithm used here. It is very good in performance than filter model because the process of feature selection is optimized for the classification algorithm used in this technique.

Symmetrical uncertainty (SU): This technique is used for the selection of vital features from the input of gene profiling. Selected feature is used to choose a subset of related feature for powerful classification of data from the original feature set.

Table 1 List of different algorithms used in feature selection, feature reduction, classification, and prediction regarding the human cancer types based on gene profiling

Methodologies	Algorithms used
• Feature selection	• Symmetrical uncertainty
• Feature reduction	• Genetic algorithm
• Classification and prediction	• Gene expression profiling

2. **Feature Reduction**

Genetic algorithm (GA): This technique is used for feature reduction which is selected by symmetrical uncertainty. Genetic algorithm is an optimization algorithm which is used for the processes of biological evolution, and by modifying chromosomes population, it solves the problem of this algorithm. One value was assigned for each chromosome that is related to its achievement in solving the queries.

3. **Classification and Prediction**

Gene Expression Programming (GEP): This technique is used for the classification and prediction of different types of human cancer disease. It is the branch of genetic algorithm (GA). DNA microarray technology is used to predict the cancer disease in most of the experimental research centers. To classify that cancer disease, gene selection is one of the most important phases in biological field (Table 1).

3 Related Work

Salem et al. [1] proposed a classification of human cancer based on gene expression profile. New methodologies based on gene expression which combines both information gain (IG) and deep genetic algorithm (DGA) are used to classify the human cancer.

Xuekun et al. [2] have done a research based on gene expression profiles for cancer type identification. A feature from gene module was selected and then put to the test set which includes four types of cancer samples which is carried out by SVM and improved relief algorithm, which showed excellent performance to select and classify the cancer types.

Soto et al. [3] proposed analysis of gene expression for lung cancer based on approach of technology merging. A new approach was used which merges datasets from different microarray technologies from which a bigger dataset is obtained to analyze different genes using gene expression.

Chretien et al. [4] proposed method of selection of genes based on expression among various patients which help in finding new genetic markers for specific pathology. Relevant genes are clustered and computed using a LASSO estimator, and the most severe tumor state is finally provided.

Tarek et al. [5] proposed a cancer classification ensemble system based on gene expression profiling. The classification is carried out on molecular level investigation which is accurate for diagnosis of different cancer type. This system not only increases the performance, but also the confidence of the results because of the understanding of the functioning of the genes and the interaction between the gene in normal and abnormal conditions.

Sota et al. [6] proposed a method which compares several mathematical transformations across many datasets including nonparametric Z scaling (NPZ).

4 Conclusion

It is necessary to confirm the type of cancer and the stages of development before the treatment and therapies to cure cancer. A new method was proposed to detect and diagnose, predict and classify the different types of cancer. The symmetrical uncertainty method is used for feature selection. Feature reduction is carried out by genetic algorithm, and to classify and predict the different types of cancer, gene expression programming is used. The above three steps thus help in prediction and classification of different types of cancer in human being.

5 Declaration

Authors have obtained all ethical approvals from appropriate ethical committee and approval from the subjects involved in this study.

References

1. Salem, H., Attiya, G., El-Fishawy, N.: Gene expression profiles based human cancer diseases classification, pp. 181–187 (2015)
2. Xuekun, S., Han, Z., Yaoting, L., Yahui, H., Shaochong, X., Peijiang, Z.: Research about feature genes selection for cancer type identification based on gene expression profiles, pp 8563–8568 (2015)
3. Soto, J.M., Ortuno, F.M., Rojas, I.: Integrative gene expression analysis of lung cancer based on a technology-merging approach (2015)
4. Chretien, S., Guyeux, C., Boyer-Guittaut, M.: Investigating gene expression array with outlier and missing data in bladder cancer, pp. 994–998 (2015)
5. Tarek, S., Elwahab, R.A., Shoman, M.: Cancer classificatio ensemble system based ongene expression profiles (2016)
6. Sota, Y., Seno, S., Takenaka, Y., Noguchi, S., Matsuda, H.: Comparative analysis of transformation methods for gene expression profiles in breast cancer datasets, pp 328–333 (2016)

Fuzzy-Based Classification for Cervical Dysplasia Using Smear Images

Mithlesh Arya, Namita Mittal and Girdhari Singh

Abstract Cervical dysplasia is the second most cause of the death in the females. A Pap smear test is the most efficient and prominent screening method for the detection of dysplasia in cervical cells. Pap smear is time-consuming, and sometimes, it is an erroneous method. Automated and semi-automated systems can be used for cervical cancer diagnosis and treatment. In our proposed approach, we are segmenting image first; the RGB image transformed into L * a * b * format. Then, using K-means clustering technique image has segmented into background and cytoplasm. Thresholding and the morphological operations have used to segment nucleus only from the second cluster. The shape-based features of the nucleus have been extracted. In the classification phase, fuzzy C-mean (FCM) has been used for clustering. Principle component analysis (PCA) is used to find the most prominent features. The classification of Pap smear images is based on the Bethesda system. The approach has performed on a dataset obtained from pathologic laboratory containing 150 Pap smear images. Performance evaluation has been done using Rand index (RI). The RI of fuzzy C-mean is 0.933, and using PCA, it is 0.95.

Keywords Smear images · K-means clustering · The Bethesda system
Fuzzy C-mean · Principal component analysis

M. Arya (✉) · N. Mittal · G. Singh
Department of Computer Science and Engineering, Malaviya National Institute of Engineering, Jaipur 302017, Rajasthan, India
e-mail: mithlesharya@gmail.com

N. Mittal
e-mail: mittalnamita@gmail.com

G. Singh
e-mail: girdharisingh@rediffmail.com

P. K. Sa et al. (eds.), *Recent Findings in Intelligent Computing Techniques*,
Advances in Intelligent Systems and Computing 708,
https://doi.org/10.1007/978-981-10-8636-6_46

1 Introduction

According to a survey in the year 2015 on cervical dysplasia, a total number of detected cases were 1,22,500 and out of them, 67,400 lost their life [1]. In India, cervical cancer is the most common cause of female mortality. According to the latest census, female population aged 15 years and more is 432 million in India and this is the number which is at risk of acquiring cervical cancer. The peak incidence of detecting cancer is 54–59 years of age. There are many reasons for developing cervical cancer like lack of awareness, early marriage, prolonged use of contraceptive pills, multiple partners, poor hygiene, low immunity [2]. Infection of human papillomavirus (HPV) is strongly associated with cervical cancer. Vaccinations against many strains of HPV including HPV 16 and 18 are available in the market, but due to lack of awareness, these preventive measures are not in very much use. Although the government of India started many awareness programs, it is still in an early phase.

Pap's smear (Papanicolaou test) test is most commonly used screening test. Pap's smear test was first demonstrated by the scientist George Papanicolaou in 1940 [3]. Pap test helps in detecting precancerous changes in the cervical cells. In Pap smear cells are scrapped from the cervical cell lining, and then, cells are spread over the glass slide. Cells obtained are mostly from the superficial layer. The cell-laden slides are then stained with a dye called methylene blue and allowed to dry. A stained slide containing cervical cells and other cells is examined under the microscope. Normally, the nucleus to cytoplasm ratio is 1:4–1:6, but in precancerous cells, the ratio gets disturbed; that is, the nucleus size becomes many times of that normal nucleus size. Limitations of this procedure are that it is very time-consuming as well as a lot of experience is required to classify the cells according to their morphological findings. We are using a system of classification which is approved and recently been updated by the association of pathologists that is the Bethesda system (TBS) [4]. The Bethesda system (TBS) is used for reporting of cancerous and precancerous stages for Pap smear results. According to TBS, cervical dysplasia is categorized into three levels:

1. Normal
2. Low-grade squamous intraepithelial lessons (LSIL)
3. High-grade squamous intraepithelial lessons (HSIL).

2 Literature Work

There have been many studies on cervical dysplasia. For single cell and multiple cells, many automated and semi-automated systems have been proposed. These proposed methods can be categorized using these three factors: (1) type of segmentation algorithms used for identification of ROI, (2) features extracted, and (3) type of classification method used.

For the segmentation, many techniques have been proposed. The paper [5] has used Gaussian mixture model (GMM) and expectation–maximization (EM) algorithm for the segmentation of nucleus and cytoplasm both. In paper [6, 7], thresholding value and morphological closing methods have been used. In paper [8], GMM and EM algorithm with K-mean clustering have been used for segmentation. In paper [9], J 1.44 C has been used for segmentation and preprocessing. J image is an application for image processing. In paper [10], watershed algorithm has been used to identify the area of background, nucleus, and cytoplasm. In paper [11], multiple morphological operations and Gaussian function have been used to extract the ROI.

The literature has categorized into shape-based and textural-based features. In paper [6, 8], shape-based features like area, perimeter, major axis, minor axis, compactness, and N/C ratio have been extracted. In paper [12], comparative analysis of single and multiple features has done. In paper [10, 11], gray-level co-occurrence matrix (GLCM), Haralick, Gradient and Tamura-based features have been extracted.

The literature reflects the cell classification mainly focusing on a single cell and multiple cells into normal and abnormal classes. Smear level classification is comparatively difficult. In paper [13], single-cell classification into two classes using SVM has been compared with ANN and KNN. In paper [14], decision tree has been used to classify into four classes. In paper [15], minimum distance classifier and KNN have been used for single-cell classification into two classes.

3 Generated Dataset

In our study, we are capturing images using the high-resolution digital camera (Leica ICC50 HD) which is mounted on a microscope (Leica BX 51) in the Department of Zoology at the Rajasthan University, Jaipur. The images are stored in a digital format with tiff extension. Magnification of images can be done at various scales like 10×, 20×, 40×, and 100×. We are using 40× magnifications. Images obtained are displayed at the resolution of 2560 × 1920 with 24 bits color depth. In our study, 150 Pap smear images have been collected which contain at least 700 cells. For the validation of our work, DTU/HERLEV Pap smear benchmark dataset [16] which has been collected by the Department of Pathology at HERLEV University Hospital. The dataset consists of 917 images which are classified into seven classes. The first three classes correspond to normal cells, and the remaining four classes correspond to abnormal cells. Cell distribution is mentioned in Table 1 in the dataset.

Table 1 DTU/Herlev benchmark dataset description

Type	Number of cells
Superficial squamous epithelial	97
Intermediate squamous epithelial	70
Columnar epithelial	74
Mild dysplasia	182
Moderate dysplasia	146
Severe dysplasia	197
Carcinoma in situ	150

4 Proposed Method

4.1 Segmentation

The steps of our proposed work have been represented in Fig. 1. Preprocessing and segmentation tools have been used to extract the nucleus from the multiple cells' image. For obtaining efficient segmentation results, we have used following steps. The RGB color image is converted into L * a * b * format because lots of colors like pink, red, and blue are presented in the colored image. The L * a * b * color space enables to easily visually distinguish colors from each other. The median filter has been used to remove noise. K-means clustering function has been used to separate the objects by clustering and separate out by Euclidean distance. These two clusters segment the image into background and cytoplasm. From these two segments, we have selected only second segment and global threshold value has been

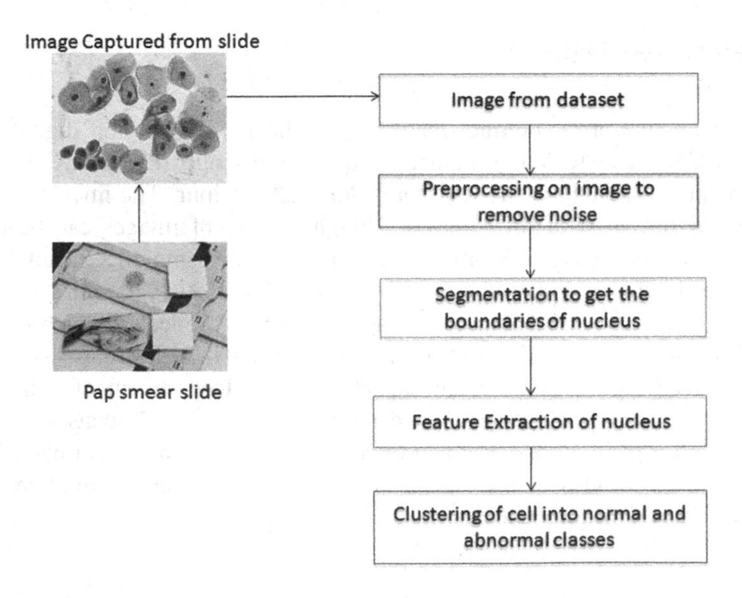

Fig. 1 Flowchart of the proposed method

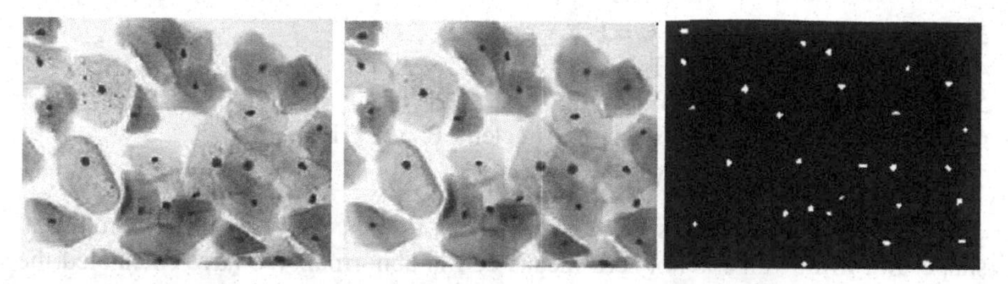

Fig. 2 Normal cells—original, preprocessed, and segmented

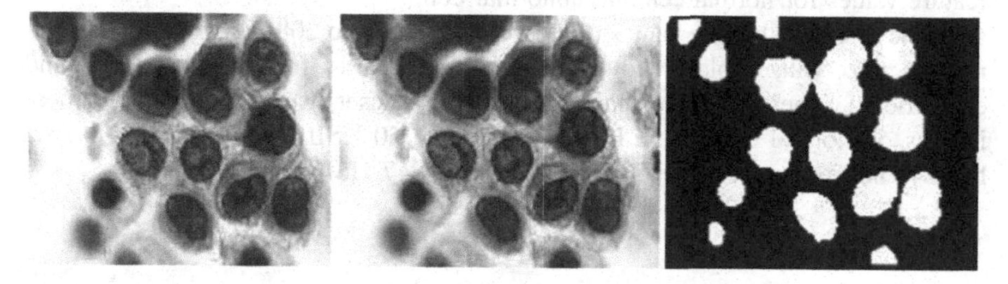

Fig. 3 Abnormal cells—original, preprocessed, and segmented

used to convert it into the gray image and after that separate out the nucleus from the cytoplasm. To find the exact boundary of nucleus, morphological operations like dilation and erosion with disk have been used. In the last step, the mask has been subtracted from the actual image to get the exact nucleus. Figures 2 and 3 show the preprocessed and segmented images of normal cell and abnormal (HSIL) cell.

As we can see in figure, the size and shape of the nucleus have been changed. Normal cells have small and round nucleus, but when the normal cell is converted into abnormal cell, its nucleus size is increased and shape becomes oval or elliptical.

4.2 Extracted Features

Feature extraction is a process for transferring most relevant information from the original dataset into a low-dimensional space. In our work, the feature extraction is applied for converting microscopic images into quantitative and parametric values.

The segmented image gives a number of the nuclei in the smear image. We have extracted these six features [17] of the nucleus for further classification.

1. Area of nucleus in terms of pixel (A)
2. Perimeter of nucleus in terms of pixel (P)

3. Compactness of nucleus $C = P^2/A$
4. Major axis of nucleus (H)
5. Minor axis of nucleus (W)
6. Ratio of minor and major axes of nucleus $R = W/H$
7. Eccentricity $E = \{(H^2 - W^2)/W^2\}^{1/2}$

Eccentricity [18] value is zero means cell is normal and its nucleus is round in shape. But if the value is not zero means cell is abnormal. We have calculated the above-mentioned features of all nuclei in the single smear image. After that, calculate the mean value of all features for the single image. Tables 2 and 3 show the feature values for normal cell and abnormal cell.

Table 2 shows that the values of area vary from 55 to 208. Eccentricity value is near to zero. But, Table 3 shows the area value varies from 1400 to 2700 and eccentricity values are more than zero. In our dataset, we have 150 Pap smear images, 50 normal images, 50 LSIL images, and 50 HSIL images. Herlev dataset has 241 normal images, 328 LSIL images, and 347 HSIL images.

4.3 Classification

Our proposed method has classified cervical dysplasia according to the Bethesda system into three classes: normal, LSIL, and HSIL using shape-based features of nuclei only. In classification phase, we have used fuzzy-based clustering techniques. By using the fuzzy C-mean method, we have made three clusters. PCA has applied on extracted features of nuclei to get the most prominent features.

Table 2 Normal cell area, perimeter, major axis, and eccentricity

Area	Perimeter	Manor axis	Minor axis	Eccentricity
55.5494	17.4786	11.0243	8.3602	0.8596
112.3112	31.2199	15.0895	11.0530	0.9294
208.0573	42.1356	15.4119	11.8190	0.8369
117.7223	26.7150	10.2624	7.4882	0.9371
65.9688	20.5998	11.5373	8.8187	0.8436

Table 3 Abnormal cell area, perimeter, major axis, minor axis, and eccentricity

Area	Perimeter	Manor axis	Minor axis	Eccentricity
2775.9630	277.6528	48.0584	22.6244	1.8741
1980.6102	195.8354	41.2767	18.8192	1.9521
1413.7889	145.7198	37.9165	19.3573	1.6843
2142.7917	247.2871	45.7400	20.0471	2.0508
1481.5333	215.5227	62.6146	27.6389	2.0328

Table 4 Confusion matrix of FCM

Normal	LSIL	HSIL	
48	2	0	Normal
2	47	1	LSIL
0	5	45	HSIL

5 Result and Evaluation

Fuzzy C-mean clustering technique has been used for classification of data. Three clusters have been made for normal, LSIL, and HSIL cells. We have used 100 iterations for the better performance of FCM. This external criterion has been used for quality of clustering. Rand index (RI) measures the percentage of decision that is correct.

$$RI = (TP + TN/TP + FP + TN + FN) \quad (1)$$

where TP is true positive values, FP is false positive values, FN is false negative values, and TN is true negative values.

Confusion matrix of FCM has been shown in Table 4. PCA values of seven features are 74.7188, 17.8757, 5.3276, 1.8830, 0.1490, 0.0302, and 0.0156. Out of these seven, only first four are important. Table 5 shows the confusion matrix using PCA. Rand index value of fuzzy C-mean is 0.933. Using PCA, Rand index of fuzzy C-mean has been increased and it is 0.953. For Herlev, dataset RI value is 0.925, and using PCA, it is 0.946. Figure 4 shows the classes using two clusters of normal and abnormal cells, and Fig. 5 shows the three clusters of normal, LSIL, and HSIL.

Table 5 Confusion matrix of FCM using PCA

Normal	LSIL	HSIL	
48	2	0	Normal
2	47	1	LSIL
0	2	48	HSIL

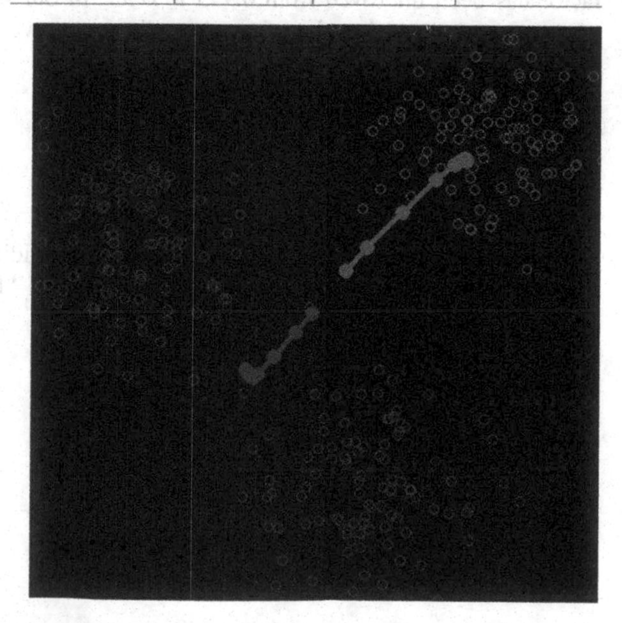

Fig. 4 Normal and abnormal classes

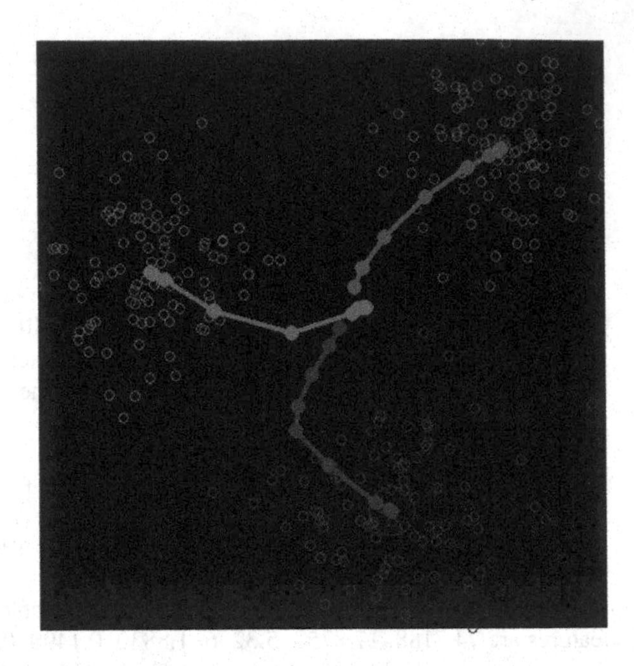

Fig. 5 Normal, LSIL, and HSIL classes

6 Conclusion

The proposed method in this study for the detection of cervical cancer cells has given good results. In segmentation, the median filter has worked better for removal of noise and debris. Fuzzy-based clustering technique FCM has been used for the clustering of cells into three classes normal, LSIL, and HSIL. Rand index obtained for FCM is 0.933. One more factor PCA has been used for extracting the most prominent features of the nucleus. The performance has been increased, and the RI factor has been improved from 0.933 to 0.95. In our future work, we will try to extract cytoplasm and nucleus from the cells so that we will get more features, and hence, PCA factor will give better results.

7 Declaration

We would like to thank Dr. Archana Parikh for providing us the Pap smear slides for our real dataset from her pathology laboratory "Parikhs Pathology Center, Jaipur."

References

1. Sreedevi, A., Javed, R., Dinesh, A.: Epidemiology of cervical cancer with special focus on India. Int. J. Women's Health **7**, 405–414 (2015)
2. Nordqvist, C.: Cervical Cancer: Causes, Symptoms and Treatments. http://www.medicalnewstoday.com/articles/159821.php. Last Accessed 12 June 2016
3. Tan, S.Y., Tatsumura, Y.: George Papanicolaou (1883–1962): discoverer of the Pap smear. Singap. Med. J. **56**(10), 586 (2015)
4. Bethesda system article. https://en.wikipedia.org/wiki/Bethesda_system
5. Ragothaman, S., Narasimhan, S., Basavaraj, M.G., Dewar, R.: Unsupervised segmentation of cervical cell images using Gaussian mixture model. In: Proceedings of the IEEE Conference on Computer Vision and Pattern Recognition Workshops, pp. 70–75 (2016)
6. Athinarayanan, S., Srinath, M.V., Kavitha, R.: Computer aided diagnosis for detection and stage identification of cervical cancer by using Pap smear screening test images. ICTACT J. Image Video Process. **6**(4), 1244–1251 (2016)
7. Arya, M., Mittal, N., Singh, G.: Cervical cancer detection using segmentation on Pap smear images. In: Proceedings of the International Conference on Informatics and Analytics. ACM (2016)
8. Lakshmi, G.K., Krishnaveni, K.: Multiple feature extraction from cervical cytology images by Gaussian mixture model. In: World Congress on Computing and Communication Technologies (WCCCT), pp. 309–311 (2014)
9. Divya Rani, M.N., Narasimha, A., Harendra Kumar, M.L., Sheela, S.R.: Evaluation of pre-malignant and malignant lesions in cervico vaginal (PAP) smears by nuclear morphometry. J. Clin. Diagn. Res. **8**(11), FC16 (2014)
10. Pradeep, S., Kenny, K., Allwin, S.: Detecting prominent textural features for cervical cancer lesions using outlier detection and removal techniques. CIS J. **5** (2014). ISSN 2079-8407
11. Kenny, S.P.K., Allwin, S.: Determining optimal textural features for cervical cancer lesions using the Gaussian function. jiP **1**, 1 (2014)
12. Kumar Kenny, S.P., Victor, S.P.: A comparative analysis of single and combination feature extraction techniques for detecting cervical cancer lesions. ICTACT J. Image Video Process. **6**(3) (2016)
13. Athinarayanan, S., Srinath, M.V.: Classification of cervical cancer cells in PAP smear screening test. ICTACT J. Image Video Process. **6**(4) (2016)
14. Paul, P.R., Bhowmik, M.K., Bhattacharjee, D.: Automated cervical cancer detection using Pap smear images. In: Proceedings of Fourth International Conference on Soft Computing for Problem Solving, pp. 267–278 (2015)
15. Otsu, N.: A threshold selection method from gray-level histograms. Automatica **11**(285–296), 23–27 (1975)
16. Martin: Papa-Smear (DTU/HERLEV) Databases & Related Studies (2003)
17. Chankong, T., Theera-Umpon, N., Auephanwiriyakul, S.: Automatic cervical cell segmentation and classification in Pap smears. Comput. Methods Prog. Biomed. **113**(2), 539–556 (2014)
18. Mahanta, L.B., Nath, D.C., Nath, C.K.: Cervix cancer diagnosis from pap smear images using structure based segmentation and shape analysis. J. Emerg. Trends Comput. Inf. Sci. **3**(2), 245–249 (2012)

References

Human Hand Controlled by Other Human Brain

Vemuri Richard Ranjan Samson, U. Bharath Sai, S. Pradeep Kumar, B. Praveen Kitti, D. Suresh Babu, G. M. G. Madhuri, P. L. S. D. Malleswararao and K. Kedar Eswar

Abstract Presently a days, world confronting a noteworthy reason for the brain illnesses, in each five youngsters one kid has enduring with cerebrum issues today, i.e., twenty rate of world population is experiencing diverse brain maladies. Here we present another innovation, i.e., brain to human peripheral interface (BHPI), which is not the quite same as brain–computer interface (BCI) and brain to brain interface (BBI). The brain-dead patient cannot do day-by-day work with their own muscles, and they require outer assistance from others. In this paper, we have utilized the ECG sensor to get the ECG and EMG signals from the peripherals of the subject and by this, ECG and EMG signals are used to control the brain-dead patient's hand movement through the electrical muscle simulator. Therefore, this technique can be utilized to lessen the endeavors of the brain-dead patient.

Keywords Brain to human peripheral interface (BHPI) · Brain–computer interface (BCI) · Brain to brain interface (BBI) · ECG · EMG · Electrical muscle simulator

1 Introduction

In the previous years, the analysis of brain wave signals is used to move the mechanical parts. Here, this paper explains how to control the human peripherals using other human brain. Weakened people can work home apparatus or electronic gadgets

V. R. R. Samson (✉) · U. Bharath Sai · S. Pradeep Kumar · B. Praveen Kitti · D. Suresh Babu
G. M. G. Madhuri · P. L. S. D. Malleswararao · K. Kedar Eswar
Department of ECE, Potti Sriramulu Chalavadi Mallikarjunarao College of Engineering and Technology, Vijayawada, Andhra Pradesh, India
e-mail: richardssamsonvemuri@gmail.com

U. Bharath Sai
e-mail: ultimatebharath@gmail.com

S. Pradeep Kumar
e-mail: satpradeep@gmail.com

P. K. Sa et al. (eds.), *Recent Findings in Intelligent Computing Techniques*,
Advances in Intelligent Systems and Computing 708,
https://doi.org/10.1007/978-981-10-8636-6_47

utilizing instruments like remote, mouse, voice, and reassure, and these information systems can be efficiently used by favored people. But here the condition is quite different, the patient has no control on his muscles so we have to provide the necessary brain signals to control their muscles.

The brain is an amazing and complex organ. 80 billion neurons inside your brain send logical messages and chemical messages to peripherals of your body. But it is shame that one out of five, i.e., 20% of entire world has neurological disorders and zero cure for these disorders. The major brain diseases that are faced by the world population are Parkinson's disease, brain tumor, mental disorder, autism, meningitis, and brain dead.

The brain to human peripheral interface is a good innovation in the sense of brain-dead patients and in the fields of extracting the brain wave signals. Here the chemical messages and the logical messages send from the brain to the hand of the subject. The ECG and EMG signals are caught at the hand of the subject by using the ECG sensor AD8232. The ECG and EMG [1] signals are used to control the patient hand by utilizing the electrical muscle simulator.

2 Related Work

EEG method has been utilized by a considerable lot of analysts for controlling of various devices like wheelchair, robotic arm, game controllers, animation movies. Kazno Tanaka prepared an algorithm to identify the patterns from EEG signal to control left and right position of wheelchairs [2]. Junichi Miyata proposed an algorithm in light of coordinates for straight and corner movement of wheelchair [3]. Brice Rebasamen built up a wheelchair using P300 BCI for predefined locations [4]. Naisan Yazdani developed a wheelchair which can move left, right, backward, and forward by using eight electrodes which are kept in the predefined location on the cerebrum thereby he implemented 3D virtual environment for capturing EEG signals [5].

3 Proposed Methodology

3.1 System Overview

The proposed methodology uses the subject raw human brain, ECG sensor, and the electrical muscle simulator.

The ECG sensor has three electrodes that are placed on the hand of the subject. The EEG signals from the cortex are passed through nervous system, and they make to shift the hands. In this paper, the motor nerves are responsible to make closing and opening of the fingers. When the subject closes the fingers, then the ECG and EMG waves are detected at that time, and by this ECG and EMG waves, we have to

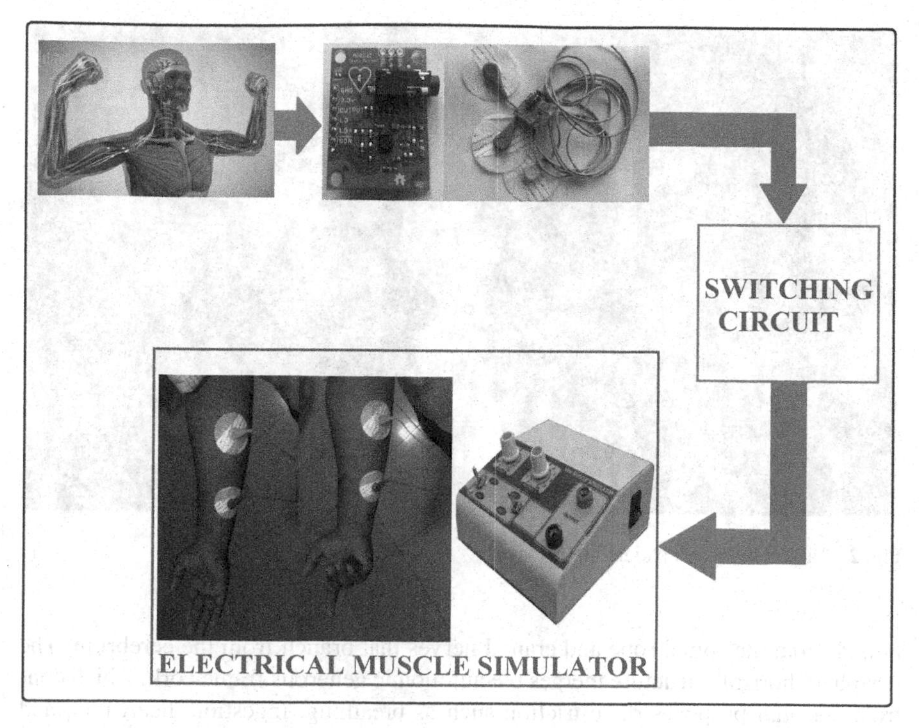

Fig. 1 Proposed design

make to switch the electrical muscle simulator. The Fig. 1 illustrates the proposed design.

3.1.1 Brain Anotomy

Limitlessly, nothing on the planet can be isolated and the human cerebrum. The 1.2 kg and 2% of our body weight organ control all body limits including getting and decoding information from the outside world and passing on the focal point of the cerebrum and soul. Information, imaginative vitality, feeling, and memories are a couple the distinctive things facilitated by the cerebrum. The cerebrum gets information through different sensors, for instance, find, see, touch, taste, and hearing. The cerebrum builds up the got data from the various sensors and shapes a basic message. The cerebrum controls our body progress of the arms and legs, encounters, memory, and talk. It moreover picks how a human response to different conditions, for instance, stress by arranging our heart and breathing rate [6, 7] (Fig. 2).

As it is known, the material structure is another major system in the human body. The material structure disengaged into central and periphery systems. The central material framework is made out of two standard parts which are the cerebrum and spinal line. The periphery material framework is made out of spinal nerves that

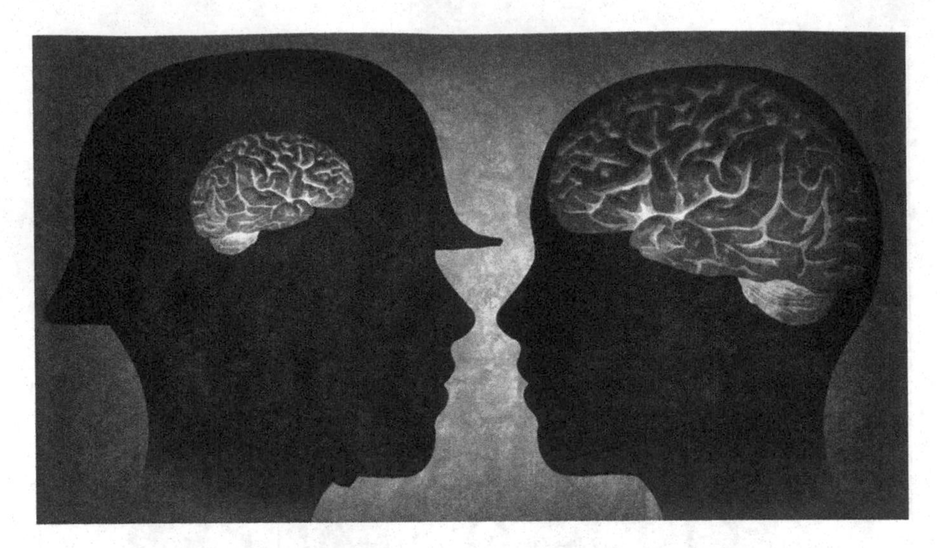

Fig. 2 Subject brain and patient brain

branch from the spinal rope and cranial nerves that branch from the cerebrum. The periphery horrible structure merges the autonomic generous framework, which controls essential purposes of restriction such as breathing, ingestion, heart rate, and landing of hormones [8].

3.1.2 ECG Sensor

The AD8232 Heart Rate Monitor module is used to look at the electrocardiogram and furthermore electromyogram. ECG sensor has three leads that can be set at various positions on the human body like on trunk, on hands.

The electrical course of action of the heart controls the time and spread of electrical banners through the heart muscle at once in a while separations and loosens up to pump blood in the midst of each heartbeat. The sound heart will have a methodical development depolarization. Depolarization is a sudden electrical change within the cell gets the opportunity to be particularly positive for a brief time frame. Depolarization starts with pacemaker cells in the sinoatrial nerve spreads out through the atrioventricular center into the store of this fiber and into the purkinje fibers spreading down and to the other side through the ventricles not at all like each and every other nerve cells that oblige lift to fire [9]. The sinoatrial nerve can be seen as self-faltering or self-ending it, again and again, gives depolarization discharge and after that repolarizes and fire again it can be appeared differently in relation to working of loosening up oscillator in electronics. To be sure, maintained fabricated pacemakers the loosening up oscillator is used, which goes about as the sinoatrial nerve.

Fig. 3 ECG sensor

The layers of living cells acts like charge capacitors one of the refinements is that among capacitors and film is that the capacitors can be charged and discharged fundamentally speedier than living cells. This is credited to the direct mechanical nature of ionic nature in living cells. The electrodes set at patient body recognize the little changes of electrode potential on the skin that rises up out of the heart muscle depolarizing in the midst of accomplish heartbeat not in the slightest degree like in a customary tripled ECG in which tend electrodes put on patients extremities and trunk.

Here for straightforwardness, we use simply fire terminals put on extremities or just on trunk. The degree extent of the voltage is only 100 V to around 5 mV. For this wander, we used AD8232 single lead heartbeat screen unit related with the Arduino PC. This pack is useful in light of the way that it has all the crucial things i.e. m the body, the terminal and wires.

The Arduino board is used just to give 3.3 V power supply to the ECD board and trade information to PC. It is possible to use the power supply on the board from 2 AA batteries and to relate on oscilloscope on which a comparable picture can be seen. A convenient workstation without relationship with the AC control connector is used for security reasons.

The AD8232 (Fig. 3) is an organized circuit made by straightforward contraptions a significant drawn-out period of time earlier. It contains everything that is imperative to separate open up and channel the little biopotential banner in no basic conditions. The heart screen unit is in like manner used for EMG. EMG recording of the electrical activity conveyed by skeleton muscles is used as a demonstrative framework.

3.1.3 Electrical Muscle Simulator

The brain sends electrical impulses via the nervous system to the respective muscles. So it is natural for our body to trigger muscular activity using electrical impulses.

Muscle simulator (Fig. 4) has copied this natural body principle of electrical muscular activation. The key to this is a high-quality current modulation medium

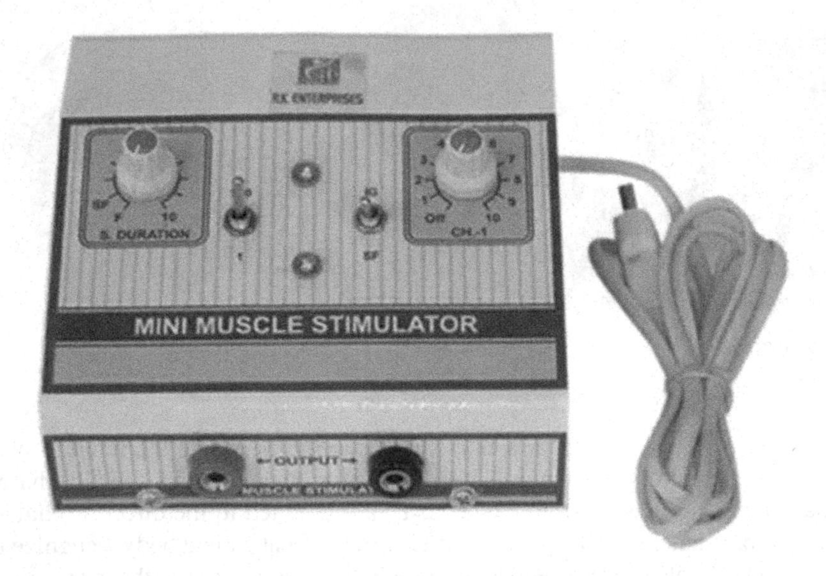

Fig. 4 Electrical muscle simulator

frequency (MMF). With this new form of current, we should trained and made possible in three different ways. First one of these the so called Neural modulation can be setup. Here the nerve pathways serve as carriers of the electrical signals. The motor nerves in the depths of the body are also reached. Accordingly, more muscle groups can be addressed especially those we can only workout difficulty when using conventional methods [9].

The second possible way to strengthen our muscle with a platonic involves a real revolution in EMS training. Modulated medium frequency can activate the muscles directly without irritating the nerves. The so called bio-modulation of the current is comparatively gentle but especially effective physiological method of muscle training, eventually it allows the digital technology of the amplitude imbibe gyro and mile to modulation. In addition this combination increases the effects of workout and makes possible the accurate adjustment of the training. So as to meet the training objectives. We are the muscle impact through the use of amplitude is concerned science speaks of quasi physiological activation of muscle. That means the targeted muscle contractions equal the contractions produced by the body itself.

4 Implementation

As shown in Fig. 5, here the EMG waves of the subject are outlined with particular activity. At the point, when the subject will close the fingers, then the EMG waves have colossal increment in the amplitude and frequency due to the muscle movement.

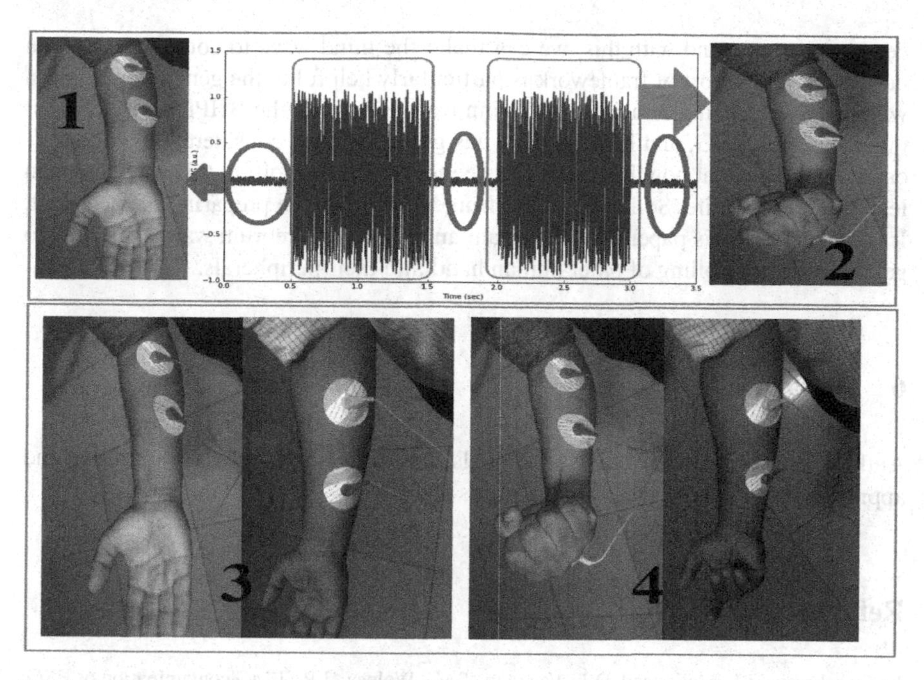

Fig. 5 1 EMG waves when subject opened the fingers. 2 EMG waves when subject closed the fingers. 3 Patient opens the fingers when there is no response from subject. 4 Patient automatically closes the fingers when subject closes the fingers

The EMG waves have low sufficiency and recurrence when the subject in the rest state. With this adequacy and recurrence levels, we will control the patient hand, i.e., when the subject anxiety the fingers to close, then the electrical muscle simulator quick to ON and it goes OFF when the subject is in rest position. Here the muscle movement is finished by utilizing cerebrum waves and in this manner, the subject mind controls understanding hand. The subject and the patient hands are delineated in the figure with their particular action. So due to these wafting action in muscle simulator, there happens action on contraction and expansion making the movement to the hand of the patient.

5 Conclusion

By and by the mind wave signs are dissected to control the mechanical parts like mechanical arm, wheel seat, and home mechanization gadgets. In any case, here this framework controls the human hand by utilizing alternate others human mind. Human hand controlled by other human cerebrum decreases the endeavors of the mind-dead patient and gives another lovely life to them. This framework is additionally helps for the future work to know the phrasing of comprehension the human cerebrum waves, and with this, we can make the mind wave to control the human peripherals. This sort of framework is particularly helpful to the general population who are experiencing mind dead. The innovative work of the BHPI with the assistance of EEG, ECG, and EMG signals has gotten a lot of consideration since they can help the general population who has neurological turmoil. In this way, they can lead their typical life. Solidness of the framework relies on preparing of the client. It is normal that this paper illuminates a man about the cerebrum waves and how to get them for controlling of other human hand and their peripherals.

6 Declaration

Authors have obtained all ethical approvals from appropriate ethical committee and approval from the subjects involved in this study.

References

1. Goncharova, I.I., McFarland, D.J., Vaughan, T.M., Wolpaw, J.R.: EMG contamination of EEG: spectral and topographical characteristics. Clin. Neurophysiol. **114**, 1580–1593 (2003)
2. Tanaka, K., Matsunaga, K., Wang, H.O.: Electroencephalogram-based control of an electric wheelchair. IEEE Trans. Robot. **21**(4) (2005)
3. Miyata, J., Kaida, Y., Murakami, T.: Coordinate-based power-assist control of electric wheelchair for a caregiver. IEEE Trans. Ind. Electron. **55**(6) (2008)
4. Rebsamen, B., Guan, C., Zhang, H., Wang, C., Teo, C., Ang Jr., M.H., Burdet, E.: A brain controlled wheelchair to navigate in familiar environments. IEEE Trans. Neural Syst. Rehabil. Eng. **18**(6), 590–598 (2010); technologies state-of-the-art and challenges. Frontiers Neurosci. **4**, 115 (2010)
5. Yazdani, Naisan: Fatemah Khazab. Sean Fitzgibbon, Martin Luerssen Towards a brain-controlled Wheelchair Prototype (2010)
6. Gilbert, S., Dumontheil, I., Simons, J., Frith, C., Burgess, P.: Wandering minds: the default network and stimulus-independent thought. Sci. Mag. **315**(5810), 393–395 (2007)
7. Buckner, R., Andrews-Hanna, J., Schacter, D.: The brains default network: anatomy, function, and relevance to disease. Ann. N. Y. Acad. Sci. **1124**, 138 (2008)
8. Looney, D., Kidmose, P., Mandic, P.: Ear-EEG: user-centered and wearable BCI. Brain-Comput. Interface Res. Biosyst. Biorobot. **6**, 41–50 (2014)
9. de Oliveira Melo, M., Arago, F.A., Vaz, M.A.: Neuromuscular electrical stimulation for muscle strengthening in elderly withknee osteoarthritis—a systematic review. Complement Ther Clin Pract. **19**(1), 27–31 (2013)

A Novel Noise Adjustment Technique for MIMO-OFDM System Based on Fuzzy-Based Adaptive Method

Darshankumar C. Dalwadi and Himanshu B. Soni

Abstract In this paper, we have presented the novel noise adjustment technique for MIMO-OFDM system on fast time-varying multipath fading channel using fuzzy-based method. The investigated parameters are bit error rate, size of the antenna, and the types of modulation method. We have proposed the method with respect to Rayleigh fading channel. We have also compared the proposed fuzzy-based adaptive method with the conventional SKF, DKF, and Nonlinear Kalman filter method. In the proposed method, the lowest value of bit error rate is achieved compared to conventional method.

Keywords Multiple input multiple output (MIMO) · Orthogonal frequency division multiplexing (OFDM) · Bit error rate (BER) · Single Kalman filter (SKF) · Double Kalman filter (DKF) · Fuzzy Adaptive Method (FAM)

1 Introduction

MIMO-OFDM is the latest wireless physical layer technologies which are used in the current 4G wireless mobile standards 3GPP-LTE, WiMAX, and high-speed WLAN standards. Such 4G mobile standards provide the higher data rates (>100) Mbps through MIMO-OFDM system and used in applications such as HDTV on demand, high-speed Internet, and broadcast video. MIMO-OFDM is used in IEEE 802.16, 802.11n, 802.11ac and plays a major role in 802.11ax and also in fifth-generation (5G) mobile phone systems.

D. C. Dalwadi (✉)
Electronics and Communication Department, Gujarat Technological University, Ahmedabad, India
e-mail: darshan242@yahoo.co.in

H. B. Soni
G. H. Patel College of Engineering & Technology, Vallabh Vidyanagar, Gujarat, India
e-mail: sony_himanshu@yahoo.com

P. K. Sa et al. (eds.), *Recent Findings in Intelligent Computing Techniques*, Advances in Intelligent Systems and Computing 708,
https://doi.org/10.1007/978-981-10-8636-6_48

In recent wireless communication system, there is a demand for high bandwidth and efficient signal transmission [1]. For that MIMO technique is adopted. The high bandwidth is achieved by simulcasting the transmit signal through number of antennas. A diversity technique used in the MIMO system is space-time block coding (STBC) [2]. To transmit the signal uniquely for that orthogonal STBC technique is adopted. Orthogonal STBC is providing significant capacity gains with respect to spatial diversity and channel coding technique. Conventional studies assumed that the channels are flat in nature and constant with respect to time. But if the channels are fast fading, i.e., if the velocity of the mobile is very high (more than 80 km/h), then the MIMO system requires an efficient equalizer [3] to eliminate inter-symbol interference (ISI). For that frequency-selective channels are adopted. To overcome this scenario, the combination of MIMO-OFDM system is used with the advantages are high bandwidth, removing ISI effect, and spatial diversity [4].

2 System Model

In this section, we have discussed the block diagram of MIMO-OFDM transmitter and receiver. We have discussed the various parts of the MIMO-OFDM system.

2.1 MIMO-OFDM Transmitter Block Diagram

Figure 1 shows the basic block diagram of MIMO-OFDM transmitter. According to the properties of the orthogonal space time block coder, let transmitted signal is given by,

$$G_t(n) = [G_{t,1}(n), G_{t,2}(n), G_{t,3}(n), \ldots, G_{t,m}(n)]^T \tag{1}$$

where m is the number of symbols.

The orthogonal space time block coder matrix is given by,

$$D(G_t(n)) = \sum_{m=1}^{L} (A_m Re[G_{t,m}(n)] + jB_m Im[G_{t,m}(n)]) \tag{2}$$

where (A_m, B_m) are mth fixed time slot × no. of tx antenna matrices.

2.2 MIMO-OFDM Receiver Block Diagram

Figure 1 shows the basic block diagram of MIMO-OFDM receiver. The received signal can be expressed as,

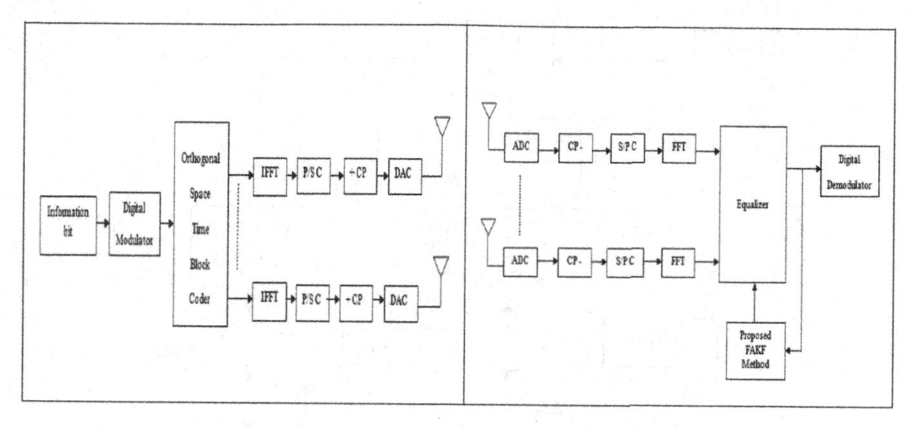

Fig. 1 Block diagram of MIMO-OFDM transmitter and MIMO-OFDM receiver

$$R_t(n) = D(G_t(n))H_t(n) + P_t(n) \tag{3}$$

For n $= 0, 1, 2, \ldots, N_c - 1$, where $R_t(n)$ is the receiver matrices with $T_s \times N_r$ and $H_t(n)$ is the frequency response of channel (no. of tx ant. × no. of rx ant. matrix) of the tth OFDM block symbol on the nth subcarrier and $P_t(n)$ is the noise with $T_s \times N_r$ matrix.

3 Proposed Fuzzy-Based Adaptive Noise Adjustment Technique

In this section, we have discussed the channel estimation and mathematical analysis of proposed fuzzy-based noise adjustment technique.

3.1 Mathematical Model of Proposed Noise Adjustment Technique

In the conventional Kalman filter, [5] it assumes that there is a prior information of the [6] S_p and S_q. Practically in most applications S_p and S_q covariance are not known. The role of covariance S_p and S_q in the Kalman filter algorithm is to adjust the Kalman gain in such a way that it controls the bandwidth of the filer as the process and the measurement errors vary. In fuzzy-based adaptive filter method, the covariance S_p and S_q are adjusted in such a way that to generate minimum error (Fig. 2).

Figure 2 shows the flowchart of the proposed fuzzy model. Measurement noise covariance matrix are adjusted S_q in two steps: First, having available the innovation sequence i_k, its theoretical covariance is given by,

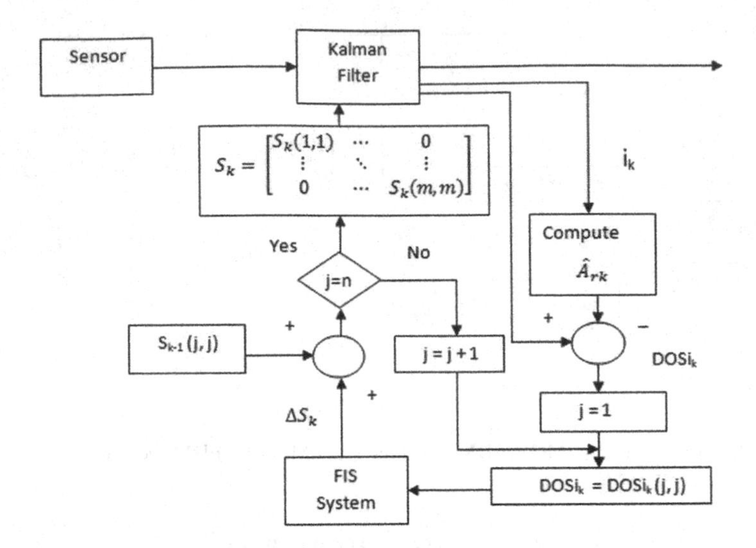

Fig. 2 Flowchart of noise adjustment of proposed fuzzy method

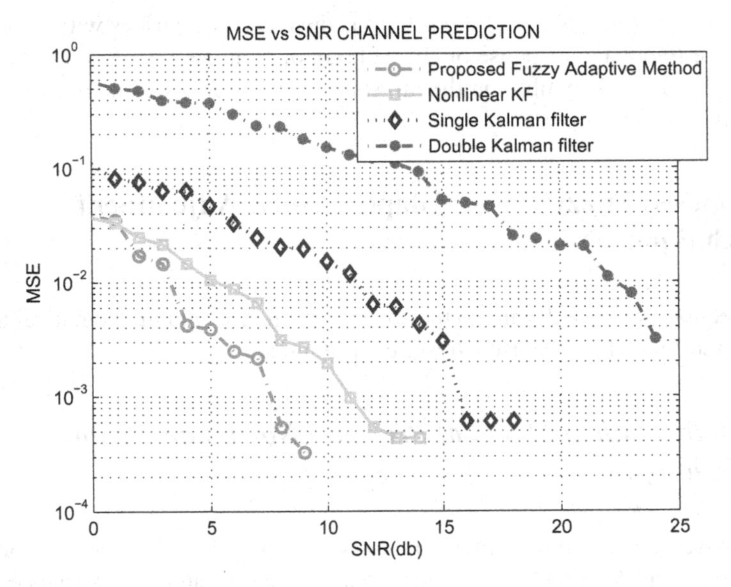

Fig. 3 MSE versus SNR for MIMO-OFDM system with FAM, SKF, DKF, and Nonlinear KF (Prediction)

$$S_k = H_k P_k H_k^T + Sq_k \tag{4}$$

where,

$$P_k = E[(g_k - \hat{g}_k)(g_k - \hat{g}_k)^T] \tag{5}$$

Second, if it is found that the original covariance i_k has any mismatching with its ideal value, then a Fuzzy system is used to derive adjustments for S_q. The objective of these adjustments is to correct this mismatching.

The following discussion is related to the adjustment of S_q: Given the availability of the residual sequence i_k, its original covariance $\hat{A}_r$ is approximated by its expected value or mean value inside a moving estimated block of size N,

$$\hat{A}_{rk} = \frac{1}{N}\sum_{j=j_0}^{k} i_j i_j^T \tag{6}$$

where, j_0 is the staring sample inside the estimated block. Now variable called the Degree of Similarity (DoSi) is defined to detect the size of the mismatching between S and $\hat{A}_r$, this is:

$$DoSi_k = S_k - \hat{A}_{rk} \tag{7}$$

It can be noted from equation of S_k that an increment in S_q will increment S and vice versa.

The following discussion is related to Degree of Similarity (DoSi) analysis: From DoSi equation, three adaptive rules are defined: If $DoSi \cong 0$, (this means S matches perfectly to original value) then maintain S_q unchanged. If $DoSi > 0$, (this means S is greater than its original value) then decrease S_q. IF $DoSi < 0$, (this means S is smaller than its original value) then increase S_q. The correction in S_q is made in this way:

$$\mathrm{Sq_k(j,j)} = \mathrm{Sq_{k-1}(j,j)} + \Delta Sq_\mathrm{k} \tag{8}$$

where ΔSq_k is the tuning factor that is either additive or subtractive from the element (j, j) of Sq. In this way, we can also adjust the process noise covariance.

4 Simulation Results and Discussion

In this section, we have presented the results of proposed noise adjustment technique for MIMO-OFDM system. In this section, we have discussed the simulation results of MSE versus SNR and BER versus SNR of MIMO-OFDM system with proposed FAM, SKF, DKF, and nonlinear Kalman filter method (Fig. 3).

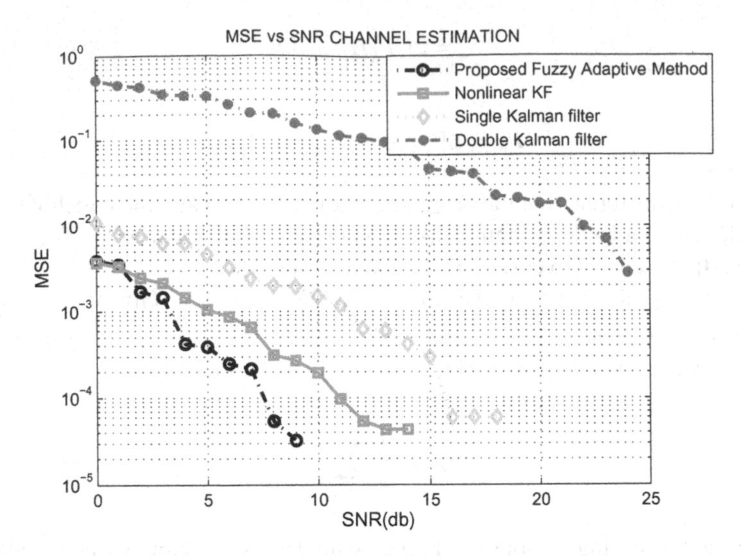

Fig. 4 MSE versus SNR for MIMO-OFDM system with FAM, SKF, DKF, and Nonlinear KF (Estimation)

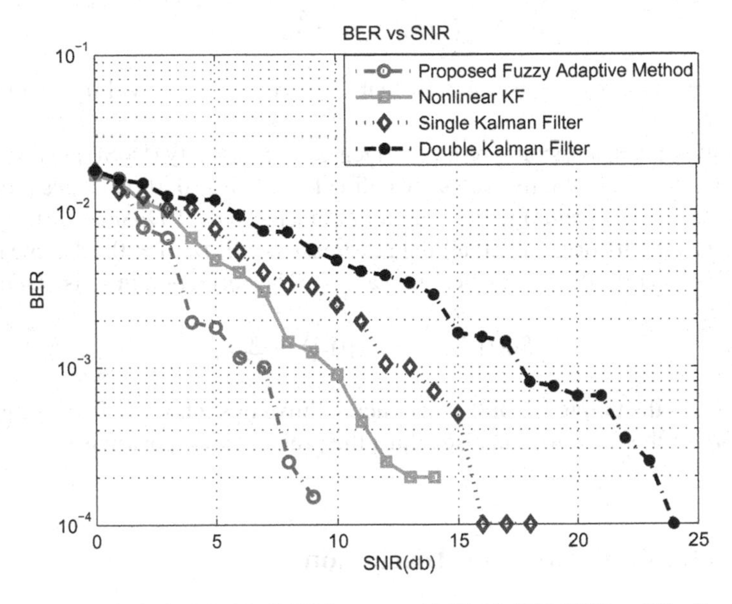

Fig. 5 BER versus SNR for MIMO-OFDM system with FAM, SKF, DKF, and Nonlinear KF

5 Conclusion

Simulation results show that lowest BER value is achieved in proposed FAM method compared to SKF and DKF and nonlinear Kalman filter method. As the order of the Kalman filter is increased, the BER and MSE values are also increased. The proposed FAM works better for nonlinear system compare to the other Kalman filter (Figs. 4 and 5).

References

1. Dalwadi, D.C., Soni, H.B.: A novel channel estimation technique of MIMO-OFDM system based on modified Kalman filter. Indian J. Sci. Technol. **9**(36) (2016)
2. Dalwadi, D.C., Soni, H.B.: A novel discrete spreading scheme with RC filter for PAPR reduction in OFDM system using multiple interleaver. Int. J. Appl. Eng. Res. **10**(17), 37806–37812 (2015)
3. Karami E, Shiva M.: Blind MIMO channel tracking using decision-directed maximum-likelihood estimation. IEEE Trans. Veh. Technol. **56**(3) (2007)
4. Hanzo, L., Akhtman, Y., Wang, L., Jiang, M.: MIMO-OFDM for LTE, Wi-Fi and WiMAX. Wiley
5. Kalman, R.E.: A new approach to linear filtering and prediction problems. Trans. ASME-J. Basic Eng. **82**(Series D), 3545 (1960)
6. Komninakis, C., Fragouli, C., Sayed, A., Wesel, R.: Multi-input multi-output fading channel tracking and equalization using Kalman estimation. IEEE Trans. Signal Process. **50**(5), 1065–1076 (2002)

5 Conclusion

[illegible]

References

[illegible]

A Novel Hierarchical Clustering Algorithm for Online Resources

Amit Agarwal and Rajendra Kumar Roul

Abstract The importance of hierarchical clustering in data analytics is escalating because of the exponential growth of digital content. Often, these digital contents are unorganized, and there is limited preliminary field knowledge available. One of the challenges in organizing these huge digital contents is the computational complexity involved. Aiming in this direction, we have proposed an efficient approach whose aim is to improve the efficiency of traditional agglomerative hierarchical clustering method that is used to organize the data. This is done by making use of disjoint-set data structure and a variation of Kruskal's algorithm for minimum spanning trees. The disjoint sets represent the clusters, and the elements inside the sets are the records. This representation makes it easy to efficiently merge two clusters and to easily locate the records in any cluster. For evaluating this approach, the algorithm is tested on a sample input of 50,000 records of unorganized e-books. The experimental results of the proposed approach show that e-resources can be efficiently clustered without compromising the clustering performance.

Keywords Agglomerative · Clustering · Hierarchical · Kruskal's algorithm
Minimum spanning tree · Prim's algorithm · Single linkage

1 Introduction

Clustering means partitioning a given set of elements into homogeneous groups based on given features such that elements in the same group are more closely related to each other than the elements in other groups [1, 2]. It is one of the most significant unsupervised learning problems, as it is used to classify a dataset without any previous knowledge. It deals with the problem of finding pattern in a collection of

A. Agarwal (✉) · R. K. Roul
BITS-Pilani, K.K. Birla Goa Campus, Pilani, India
e-mail: amitagarwal.17081996@gmail.com

R. K. Roul
e-mail: rkroul@goa.bits-pilani.ac.in

P. K. Sa et al. (eds.), *Recent Findings in Intelligent Computing Techniques*,
Advances in Intelligent Systems and Computing 708,
https://doi.org/10.1007/978-981-10-8636-6_49

unclassified data. There is no specific algorithm for clustering. This is because the notion of what constitutes a cluster can vary significantly, and hence, the output will vary. Many research works have been done in this domain [3–7].

Hierarchical clustering [8] is one of the most popular methods of clustering, which aims to build a hierarchy of clusters. At the bottom-most level, all objects are present in different clusters, while at the topmost level, all objects are merged into one cluster. Hierarchical clustering can be done in two ways: agglomerative and divisive. In agglomerative clustering, initially all objects are placed in different clusters and the clusters are joined as we move up the hierarchy. So it is a bottom-up approach. In divisive clusters, initially all objects are placed in a single cluster and the clusters are divided as we go down the hierarchy. So it is a top-down approach. Single-linkage clustering is a technique for performing hierarchical clustering in which the similarity among two groups of objects is determined by two objects (one in each group) that are most similar to each other. One of the major challenges in hierarchical agglomerative single-linkage clustering is the computational complexity involved. In 1973, Sibson [9] proposed SLINK, a single-linkage clustering algorithm which had a time complexity of $\mathbf{O(n^2)}$ and space complexity of $\mathbf{O(n)}$. In this paper, an alternative approach for single-linkage clustering is proposed based on minimum spanning trees [10] having same space and time complexity as the SLINK algorithm.

The paper can be organized on the following lines: Sect. 2 discusses the proposed approach to cluster the e-books. The experimental work has been carried out in Sect. 3, and finally, in Sect. 4, we have concluded the work.

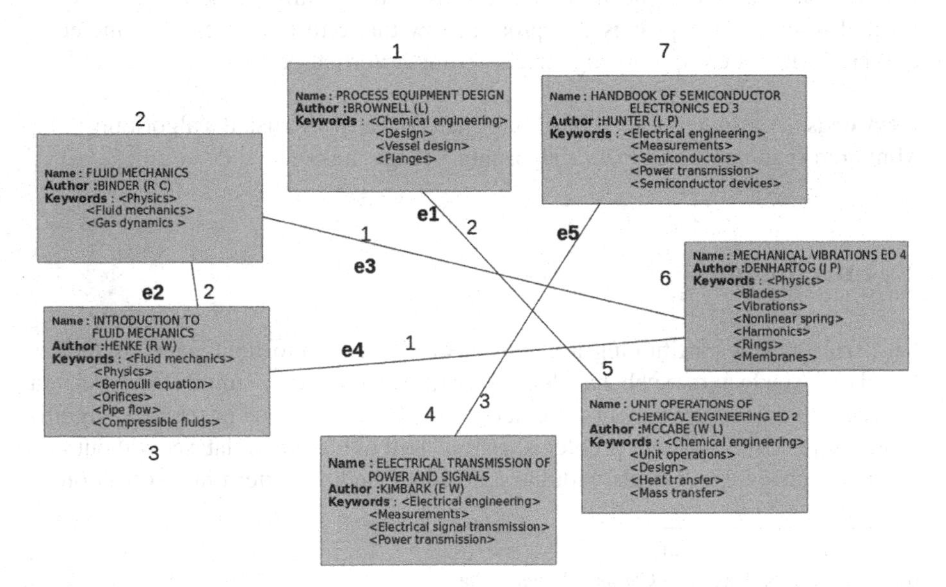

Fig. 1 Graph on books

2 The Proposed Approach

2.1 *Problem Statement*

Consider a corpus of *N* documents. All the documents are preprocessed by extracting the keywords using Natural Language toolkit.[1] Now, our entire data is a set *X* of N records (documents). Mathemetically, $\mathbf{X} = \{\mathbf{x_1}, \mathbf{x_2}, \ldots, \mathbf{x_N}\}$ where each $\mathbf{x_i} = \{\mathbf{x_{i1}}, \mathbf{x_{i2}}, \ldots, \mathbf{x_{ik_i}}\}$ is a set of k_i keywords (features); the aim of clustering is to divide the set *X* into clusters $C_1, C_2, C_3, \ldots$, where minimum similarity between any two records in a cluster is not less than a particular threshold set by the user. The dataset is represented in form of a graph where nodes represent the records and edges represent the similarity between the records. So edge weight is directly proportional to the similarity between the records directly connected by that edge. We denote **G(X) = (V, E)** as the undirected complete graph of X. Now, the edge weights can be derived by finding the number of features that are common to two records. The weight function is represented as $\mathbf{w(x_i, x_j)}$.

Mathematically:

$\mathbf{V = X}$
$\mathbf{E = \{(x_i, x_j), i \neq j, x_i \in X, x_j \in X\}}$
w: E→N (set of natural numbers)

Figure 1 demonstrates the graphical representation of a sample data of e-book records. The nodes represent the books, and the features are represented by the keywords. The weight of the edge connecting two nodes (books) is equal to number of keywords that are common to the two nodes (books).

Book Name	Author	Keywords
PROCESS EQUIPMENT DESIGN	BROWNELL (L)	<Chemical engineering><Design><Vessel design><Flanges>
FLUID MECHANICS	BINDER (R C)	<Physics><Fluid mechanics><Gas dynamics >
ENGINEERING MANUAL	PERRY J H	<Mechanical engineering><Mechanics><Machine design data><Thermodynamics><Fluid mechanics>
ELECTRICAL TRANSMISSION OF POWER AND SIGNALS	KIMBARK (E W)	<Electrical engineering><Electrical power transmission><Electrical signal transmission><Power transmission>
HEAT TRANSFER VOL 1	JAKOB (M)	<Mechanical engineering><Heat transfer><Conduction><Convection><Thermal radiation>
SOLID FUEL REACTORS	DIETRICH (J R)	<Nuclear engineering><Solid fuel reactors><Fast breeder reactors><Reactor physics><Sodium technology>
FLUID FUEL REACTORS	LANE (J A) ET AL ED	<Nuclear engineering><Fluid fueled reactors><Homogeneous reactors><Aqueous suspensions>
MECHANICAL VIBRATIONS ED 4	DENHARTOG (J P)	<Physics><Blades><Vibrations><Nonlinear spring><Harmonics><Rings><Membranes><Plates><Torsion>
HANDBOOK OF SEMICONDUCTOR ELECTRONICS ED 3	HUNTER (L P)	<Electronic engineering><Measurements><Semiconductors><Power transmission><Semiconductor devices>

Fig. 2 Unorganized input data

[1] http://www.nltk.org/.

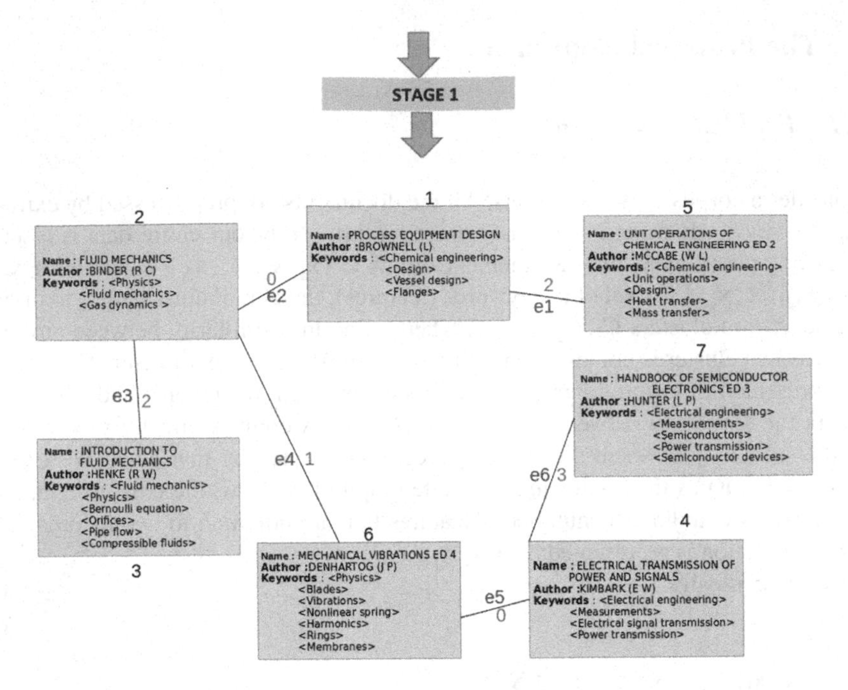

Fig. 3 Maximal Spanning Tree

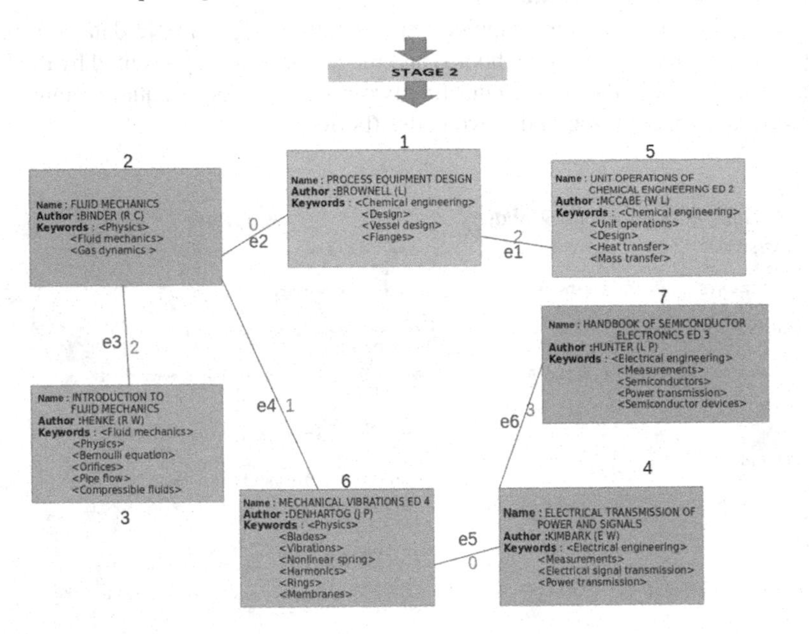

Fig. 4 Clusters after applying Kruskal's algorithm

2.2 Algorithm Overview

The algorithm comprises of two major stages: In the first stage, we construct a maximum spanning tree of the given dataset (Fig. 2) using Prim's algorithm. In the second stage, the records are merged and clusters are formed by applying Kruskal's algorithm to the maximum spanning tree obtained in the first stage.

2.3 Constructing the Maximum Spanning Tree (MST)

A maximum spanning tree is a tree which connects all the vertices of a connected, undirected graph such that the sum of edge weights of the tree is maximum.

The maximum spanning tree of the graph (Fig. 3) can be formed using Prim's algorithm for minimum spanning trees [11]. The algorithm is described as follows:

1. Assign each vertex-v of the graph these 3 quantities :
   - C[v] : Highest cost of a connection to vertex v
   - Parent[v] : the node with which v has the highest cost connection
   - Visited[v] : Indicates whether v has been included in MST.
2. Make 2 sets :
   - S : set of vertices(v) sorted in descending order with C[v] as the key.
   - E : set of edges of MST
3. Initialization :
   - C[v] $\leftarrow -\infty, \forall v \in V$
   - Visited[v] $\leftarrow$ *false*, $\forall v \in V$
   - Parent[v] $\leftarrow$ *null*, $\forall v \in V$
   - S = all vertices
   - E = ϕ
4. Repeat these steps until $S = \phi$:
   (a) Find and then remove the first vertex v from S.
   (b) Set Visited[v] $\leftarrow$ true.
   (c) If Parent[v] $\neq$ null , add (Parent[v],v) to the set E.
   (d) Loop over all vertices u , s.t u $\neq$ v and Visited[u]= false :
      i. If w(u,v) > C[u] :
         A. Set C[u] $\leftarrow$ w(u,v)
         B. Set Parent[u] $\leftarrow$ v

When the set S becomes empty, set E will contain the edges of Maximal Spanning Tree of the graph.

2.4 Merging of Records into Clusters

In this stage, the edges of maximum spanning tree are used to form clusters using a slight modification of Kruskal's algorithm [12]: The algorithm is stopped when the edge weight becomes less than a particular lower threshold for similarity value: $w_{crtical}$ which is specified by the user. The algorithm makes use of disjoint-set data structure which maintains a collection of disjoint sets having some finite number of elements within them. The disjoint sets represent the clusters, and the elements inside the sets are the records. Every set has a representative element which is used to uniquely identify that set. This representation makes it easy to efficiently merge two clusters and to easily locate the records in any cluster because of two useful operations supported by this data structure:
Find: It takes an element as input and determines which set the particular element belongs to.
Find(v):

1. If Parent[v] $\neq$ v then Parent[v] $\leftarrow$ Find(Parent[v])
2. Return Parent[v]

Union: It takes two disjoint sets as input and joins them into a single set.
Two optimization techniques: Union by rank and path compression have been employed which reduce the time complexity of Union and Find operations to a small constant.
Union(u, v):

1. $set_u \leftarrow$ Find(u)
2. $set_v \leftarrow$ Find(v)
3. If $Size[set_u] < Size[set_v]$ *then* $Parent[set_u] \leftarrow set_v$
4. If $Size[set_u] > Size[set_v]$ *then* $Parent[set_v] \leftarrow set_u$
5. If $Size[set_u] = Size[set_v]$ then
   (a) $Parent[set_u] \leftarrow set_v$
   (b) $Size[set_v] \leftarrow Size[set_v] + 1$

The modified Kruskal's algorithm is described as follows:

1. Associate each vertex v with 2 quantities :
   – Parent[v] : the parent of v in the set.
   – Size[v] : the size of disjoint set which has v as its representative element.
2. Make N(number of records) sets $S_1, S_2, S_3 \ldots S_N$, each set representing a cluster.
3. Initialization :
   – $Parent[v] \leftarrow v, \forall v \in \mathbf{V}$
   – $Size[v] \leftarrow 1, \forall v \in \mathbf{V}$
   – $S_i = \phi, \forall i \in \{1, 2, \ldots, N\}$
4. Sort the edges in set E (set containing edges of Maximal Spanning tree) in decreasing order of edge weight.

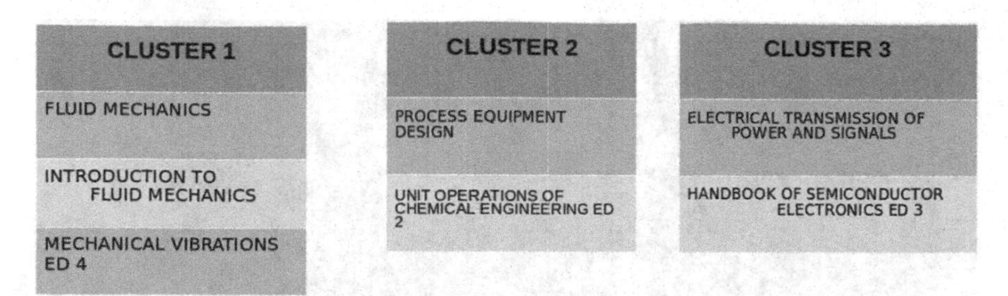

Fig. 5 Generated clusters

5. Iterate over edges e(u, v) in set E till edge weight of $(e - w(u, v)) > w_{crtical}$:
   (a) *Union(u, v)*
6. For all vertices v :
   (a) $i \leftarrow Find(v)$
   (b) Add vertex v to set S_i

At the end, each non-empty set S_i will represent a cluster containing records of similar type. Figure 4 shows the graph structure after the clustering process. Boxes having the same color are part of the same cluster.

3 Experimental Results

The proposed algorithm was tested on a sample input of 50,000 records of unorganized books.[2] The records were taken as input in the following format:
Book Name, Author Name, $< keyword_1, keyword_2, \ldots >$ where keywords are a set of words which describe the book. These keywords are used as features for each record (book). A total of 541 clusters were obtained with each cluster containing books of similar subject category. Figure 5 shows a sample of the clusters obtained after running the algorithm on the dataset.

The time taken and space utilized to cluster different sizes of input data are shown in Figs. 6 and 7. The analysis shows that the time complexity of the proposed algorithm is $\mathbf{O(n^2)}$ and the space complexity is $\mathbf{O(n)}$. Table 1 shows the size of dataset and corresponding hypothetical computational cost required with respect to a typical desktop PC with a CPU speed of **2.4 GHz** which can perform approximately $\mathbf{2.4 * 10^9}$ operations per second.

[2]http://www.gutenberg.org/dirs/.

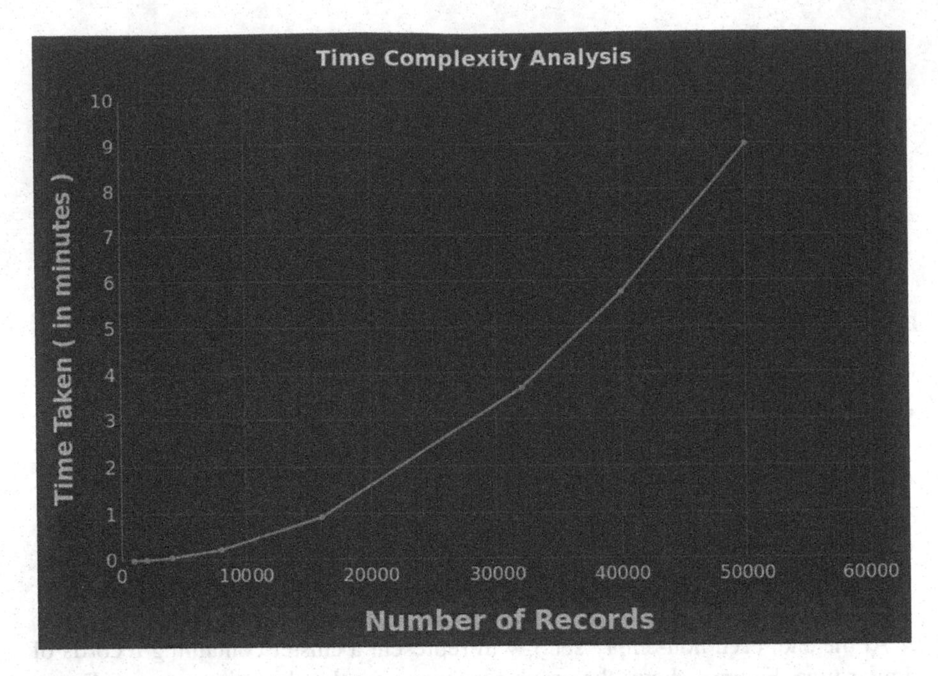

Fig. 6 Time complexity

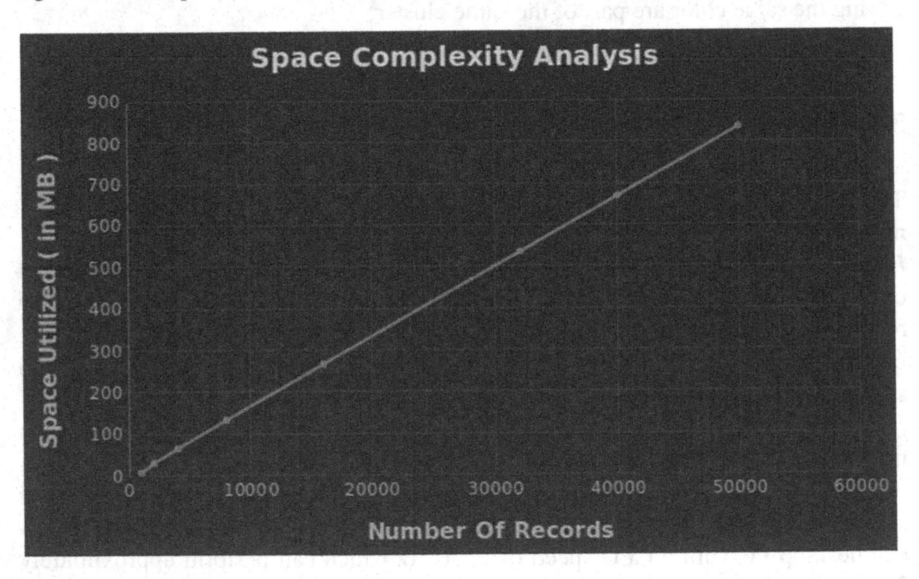

Fig. 7 Space complexity

Table 1 Computational complexity

Size of the dataset	Time required
10^4	0.04 s
10^5	4 s
10^6	7 min
10^7	11.5 h
10^8	48 days
10^9	13 years

4 Conclusion

The proposed algorithm employs MST algorithm in two stages: In the first part of the algorithm, a maximal spanning tree is constructed out of the given dataset using Prim's algorithm. In the second part, this maximal spanning tree is fed as input to a variation of Kruskal's algorithm which then generates the clusters. Since the space complexity of the algorithm is $\mathbf{O(n)}$, the algorithm is efficient in terms of space requirement to compute clusters even for large datasets. However, the time complexity of constructing a maximum spanning tree is $\mathbf{O(n^2)}$, which hinders the application of MST in case of large datasets.

The future work is to find an approach which captures the similar information like MST in a better time and space complexity.

References

1. Everitt, B., Landau, S., Leese, M.: Cluster Analysis. ser. Hodder Arnold Publication. Wiley (2001). https://books.google.co.in/books?id=htZzDGlCnQYC
2. Gan, G., Ma, C., Wu, J.: Data clustering: theory, algorithms, and applications. SIAM (2007)
3. Calandriello, D., Niu, G., Sugiyama, M.: Semi-supervised information-maximization clustering. Neural Netw. **57**, 103–111 (2014)
4. Baghshah, M.S., Afsari, F., Shouraki, S.B., Eslami, E.: Scalable semi-supervised clustering by spectral kernel learning. Pattern Recogn. Lett. **45**, 161–171 (2014)
5. Roul, R.K., Nanda, A., Patel, V., Sahay, S.K.: Extreme learning machines in the field of text classification. In: 2015 16th IEEE/ACIS International Conference on Software Engineering, Artificial Intelligence, Networking and Parallel/Distributed Computing (SNPD), pp. 1–7. IEEE (2015)
6. Jain, A.K., Murty, M.N., Flynn, P.J.: Data clustering: a review. ACM Comput. Surv. (CSUR) **31**(3), 264–323 (1999)
7. Roul, R.K., Varshneya, S., Kalra, A., Sahay, S.K.: A novel modified apriori approach for web document clustering. In: Computational Intelligence in Data Mining-Volume 3, pp. 159–171. Springer (2015)
8. Murtagh, F., Contreras, P.: Algorithms for hierarchical clustering: an overview. Wiley Interdiscip. Rev.: Data Min. Knowl. Discov. **2**(1), 86–97 (2012)
9. Sibson, R.: Slink: an optimally efficient algorithm for the single-link cluster method. Comput. J. **16**(1), 30–34 (1973)

10. Gower, J.C., Ross, G.: Minimum spanning trees and single linkage cluster analysis. Appl. Stat. 54–64 (1969)
11. Prim, R.C.: Shortest connection networks and some generalizations. Bell Labs Tech. J. **36**(6), 1389–1401 (1957)
12. Kruskal, J.B.: On the shortest spanning subtree of a graph and the traveling salesman problem. Proc. Am. Math. Soc. **7**(1), 48–50 (1956)

Quality Control System for Pathogen Detection

Sonali Kulkarni, D. M. Verma and R. K. Gupta

Abstract This paper elaborates idea of quality control system for pathogen detection. Contamination of food with different bacteria is a serious threat happened nowadays. It is specially regulating cost-effective quality control systems for processed foods. Quality control system is used for the detection of pathogen, bacteria from processed food which is packed and checking by over long time. The actual role of quality control system in food industry is important for the detection of bacteria like salmonella or e-coli from the preserved food. Here, we are considering chocolate as product from which we can try to find out bacteria salmonella. Which will protect our health from food infection and food contamination? We can make different sensors for the detection of the pathogen from food or from water.

Keywords Surface plasmon resonance · Biosensors · Foodborne pathogen Quality control system · Pathogen · Salmonella · Food safety

1 Introduction

In modern trends and technology, quality control system plays an important role. It provides and assures market that product is free from pathogen like salmonella or e-coli. Surface plasmon technology will give support to make advance trends to test food product like chocolate, bakery products, water. Which will help us to protect our health from food contamination and diseases? The technique used to test

S. Kulkarni (✉) · R. K. Gupta
Department of E & Tc, T E C, Nerul, Navi Mumbai, India
e-mail: sonalik15@gmail.com

R. K. Gupta
e-mail: rajivmind@gmail.com

D. M. Verma
SAMEER, IIT Campus, Powai, Mumbai, India
e-mail: dmverma@yahoo.com

P. K. Sa et al. (eds.), *Recent Findings in Intelligent Computing Techniques*,
Advances in Intelligent Systems and Computing 708,
https://doi.org/10.1007/978-981-10-8636-6_50

bacterial contamination is surface plasmon technology and laboratory-based technique is plate count as microbiological testing.

Method. Quality control department which gives food testing methods. Q.C. Dept. includes the Process Lab., Microbiology Lab., and Packaging Lab [1]. The Q.C. Dept. involves routine analytical tests on raw materials, in-process materials, and finished goods which can be termed as measurable quantitative physical standard by which its value is judged and Objectives:

(1) Q.C. ensures that the manufacturing practice is ethical, and all the relevant tests are carried out.
(2) Thus beneficial for both the consumers and the industry.
(3) Also, developing procedures that aid in meeting consumer demands and quality consciousness both within and outside the organization.
(4) To evaluate quality status of raw materials, product used in manufacturing and the finished goods.

1.1 Surface Plasmon Resonance (SPR) Concept Used for the Detection of Bacteria (Pathogen)

Surface plasmon resonance technique is widely used in industry as advanced detection phenomenon. This technique is used for speed of analysis and also for high accuracy for getting testing results. Surface plasmon resonance is transducer which is usually designed by using prism coupling of incident light onto an optical substrate that is coated with a semi-transparent noble metal under conditions of angular position. When the flow of liquid is passed from flow channel, at same time angular position of incident angle which is to be a critical angle measures because of different indices of prism. At middle of flow channel, more binding occurred because of sensor chip with gold-plated surface having ligand and antibody which are present in buffer solution in which chocolate is dissolved which we want to test [1, 2] (Fig. 1).

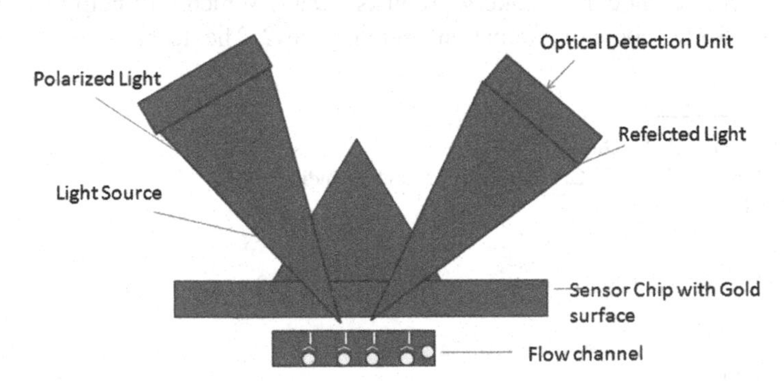

Fig. 1 Surface plasmon resonance technology concept

2 Method 1: Microorganism Technical Laboratorical Pathogen Test

2.1 Microbiological Analysis

Each complete microbial examination takes 3 to 4 consecutive working days (48- to 72-h incubation periods) to complete. After identification of microorganisms and their number of count, this information is recorded in the record book. The whole batch of finished product is held in store till book. The whole batch of finished product is held in store till the microbial result comes [2].

2.2 Microbial Analysis

To collect samples from each day's production and The microbiological analysis of all edible materials i.e. raisin, apricot, cashew, almond etc and also raw material like milk powder, cocoa powder, untreated cocoa mass, treated cocoa mass etc. and all finished products has to be carried out. Chemical analysis of chocolate has to be carried out.

2.3 Study of Some Equipment Used for Microbiological Analysis of Autoclave

There are two types of autoclave. Autoclave for routine laboratory sterilization is kept for 15 lbs that gives a temperature of 1210 °C for 15–20 min.

(a) Pressure cooler type,
(b) Gravity displacement autoclave.

2.4 Sterilization of Media/Glassware

All media and glassware should be sterilized in an autoclave at 121 °C for 15 min. The autoclave should be hot before sterilization, as this helps to hasten the microbiology work. The autoclave should be loaded using asbestos hand gloves. Once the autoclave is charged, the lid of the autoclave should be closed and bolts are tightened. Opposite bolt should be tightened simultaneously to prevent steam leakage. Wait for about 15 min. till the steam starts coming out. The valve should then be closed, and the steam pressure should be allowed to reach 15 lbs/sq. inch. The autoclave should be kept at this pressure for 15–20 min. for media and glassware.

2.5 Preparation of Culture Media

Media preparation: The appropriate amount of distilled water should be taken into the flasks using a measuring cylinder. The ingredients should be weighed in appropriate amounts and added to the flask.

2.6 Isolation of Microorganism (Culturing Method and Maintenance of Culture)

Isolation of procedure in which a given species of organisms present in sample or specimen is obtained in pure culture in order to characterize and study a particular organism.

2.7 Spread Plate Technique

Principle: In the spread plate method, small volume of (0.1–0.25 ml) suspension of microorganisms is placed on the center of an agar plate and spread over the surface of agar using a sterile glass rod. The glass rod is normally sterilized by dipping in alcohol and flaming to burn off the alcohol. By spreading the suspension over the plate as even layer of cells is established so that individual microorganisms are separated from other organism in the suspension and deposited at the discrete location [3].

3 Method 2: Daily Microbial Examination

The purpose of routine microbiological examination is to ensure that raw material and finished products are safe for human consumption with respect to bacteria, mold, and yeast. Being a multinational chocolate giant, it should perform its microbiological test and analysis within the plant itself rather than relying on others. Sample from all three shifts of the previous day as well cocoa powder is checked for their microbial count like ***Salmonella*** bacteria, etc., indicator microorganisms. Prior to any analysis, one should ensure the sterilization or disinfection of their hands, the laboratory, and all apparatus. Each complete analysis or microbial examination takes 3 to 4 consecutive working days (48- to 72-h incubation periods) to complete, and full examination consists of an isolation of salmonella [4].

Aim: Isolation of salmonella.

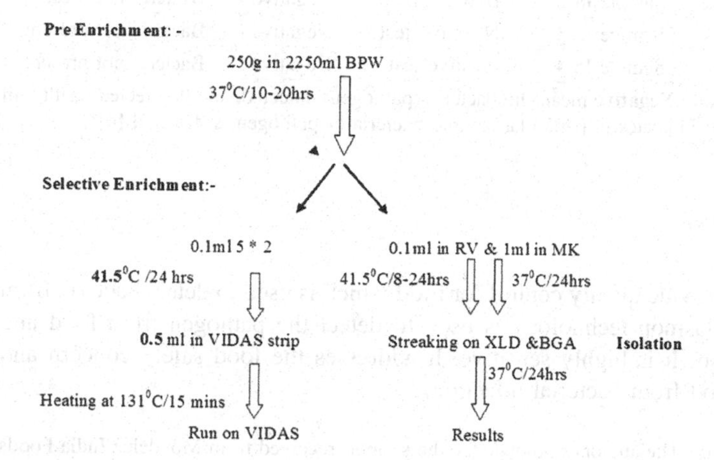

4 Results

The isolation of salmonella has been done with VIDAS system, below-mentioned procedure followed for testing of particular bacteria like mentioned as salmonella which spread foodborne disease and which infects body which is harmful to human health. The quality control system for particular pathogen detection is used to prevent food and food product contamination due to temperature and contents used for making product (Tables 1 and 2).

Table 1 Test is for (SLM) Section A from VIDAS

Position	Sample name	Result	Test remark
Position slot 1	Sample ID-1	Negative	Bacteria not present
Position slot 2	Sample ID-2	Negative	Bacteria not present
Position slot 3	Sample ID-3	Negative	Bacteria not present
Position slot 4	Sample ID-4	Negative	Bacteria not present
Position slot 5	Sample ID-5	Negative	Bacteria not present
Position slot 6	Sample ID-6	Negative	Bacteria not present

Note In Table 1 result, **Negative** means no bacteria (pathogen) detected, and if detected result will mention **Positive**, it indicates in particular sample bacteria or pathogen is detected [4]

Table 2 Test is for salmonella (SLM)—Section B

Position section B	Sample name	Test name	Result	Test remark
Position slot 1	Sample Id 1	Sterility test	Negative	Bacteria not present
Position slot 2	Sample Id 2	Blank test	Negative	Bacteria not present
Position slot 3	Sample Id 3	Negative test	Negative	Bacteria not present
Position slot 4	Sample Id 4	Positive test	Negative	Bacteria not present

Note In Table 2 result, **Negative** means no bacteria (pathogen) detected, and if detected result will mention **Positive,** it indicates in particular sample bacteria or pathogen is detected [4]

5 Conclusion

Biosensor can provide quality control for food which is used to detect bacteria from food. Surface plasmon technology is used to detect the pathogen from food and clinical base also. It is highly sensitive. It addresses the food safety concern and takes care of food from bacterial infection.

Acknowledgements The authors acknowledge the support received from Mondelez India Foods Pvt. Ltd (Formally Cadbury India Ltd), Thane, for providing necessary measurement facilities.

References

1. Sharma, H., Agarwal, M., Goswami, M., Sharma, A., Roy, S.K., Rai, R., Murugan, M.S., Rich, R.L., Myszka, D.G.: Advances in surface plasmon resonance biosensor analysis. Curr. Opin. Biotechnol. **11**, 54–61 (2000)
2. Malmqvist, M.: BIACORE: an affinity biosensor system for characterization of biomolecular interactions. Biochem. Soc. Trans. **27**, 335–340 (1999)
3. Lackmann, M.: Isolation and characterization of "orphan-RTK" ligands using an integrated biosensor approach. Methods Mol. Biol. **124**, 335–359
4. Millopore, Muller: Biological Raw Material. Merck & Millipore Publication (2005)

Dolphin Echolocation and Fractional Order PID-Based STATCOM for Transient Stability Enhancement

Shiba R. Paital, Prakash K. Ray and Asit Mohanty

Abstract This paper investigates the application of fractional order PID (FOPID) controller to study and analyze transient stability in a single machine infinite bus (SMIB) system. For enhancement of transient stability and damping of low-frequency oscillations in a SMIB system, a shunt FACTS controller called static synchronous compensator (STATCOM) has been incorporated. In this paper, emphasis was given to the function of FOPID-based STATCOM controller for stability analysis in a SMIB system. A new heuristic optimization technique called dolphin echolocation (DE) was proposed for selecting optimal stabilizer parameters. A comparison based on nonlinear simulations was done between conventional PID controller and FOPID controller with small change in load and system parameters. Simulation results clearly signify the superiority of the proposed DE-based optimization technique.

Keywords FOPID controller · PID controller · Stability · STATCOM DE

1 Introduction

Modern power system demands for better quality power at comparatively lower cost with subsequently lesser environmental impacts. Power system oscillations, mainly low-frequency oscillations, are one of the leading concerns in every power

S. R. Paital (✉) · P. K. Ray
Department of Electrical and Electronics Engineering,
IIIT Bhubaneswar, Bhubaneswar, India
e-mail: shiba.paital@gmail.com

P. K. Ray
e-mail: pkrayiiit@gmail.com

A. Mohanty
Department of Electrical Engineering, CET Bhubaneswar, Bhubaneswar, India
e-mail: asithimansu@gmail.com

P. K. Sa et al. (eds.), *Recent Findings in Intelligent Computing Techniques*,
Advances in Intelligent Systems and Computing 708,
https://doi.org/10.1007/978-981-10-8636-6_51

industry [1]. For overcoming issues related to small signal stability, a power system device called power system stabilizer (PSS) is used. PSS applies modulating signals to excitation systems for adding rotor oscillations damping instantly followed by a disturbance. But PSS has its own limitations related to selection of control parameters, load uncertainties, especially when damping interarea oscillations. Therefore, nowadays power electronics-based devices called flexible AC transmission system (FACTS) controllers are evolved as a solution to these problems [2–4]. FACTS devices have the ability of controlling power flow in the transmission line, provide necessary active and reactive power compensation, and provide excellent voltage support, damping interarea oscillations, etc. FACTS controllers can be connected in series and shunt as well depending upon the requirement. From various FACTS devices, static synchronous compensator (STATCOM) is a shunt-connected FACTS controller which compensates reactive power, regulates bus voltage, and also improves the dynamic stability of the system [5–8]. STATCOM supplies the lagging as well as leading reactive power required by the system. Previously, soft computing-based techniques like GA, BFO, PSO, FF, and SA are used for selecting optimal parameters for designing STATCOM-based damping controllers [9–11]. In this paper, a novel soft computing-based approach called dolphin echolocation (DE) [12] is proposed for selecting optimal parameters of damping controllers for SMIB with STATCOM.

2 Power System Configuration and Its Modelling

2.1 Single Machine Infinite Bus System with STATCOM

A SMIB system with STATCOM considered in this study is shown in Fig. 1a. In this figure, a synchronous generator feeds continuous power to the load side through a transformer. The power system is also incorporated with STATCOM through a coupling transformer [13–15]. In Fig. 1, v_t and v_b represent generator and infinite bus voltage, respectively, x_s represents the reactance of the coupling transformer, C_{DC} corresponds to DC link capacitor value, m is the modulation index, and ϕ is the phase angle in PWM. A linearized Heffron-Phillip model of SMIB with STATCOM was also given in this study as shown in Fig. 2. Exchange of reactive power can be controlled through controlling the amplitudes of three-phase output voltage (V_s) and the utility bus voltage (V_m). The nonlinear control equations of the system are as follows.

$$V_m = kmV_{DC}(\cos\alpha_S + j\sin\alpha_S) = kmV_{DC}\angle\alpha_S \tag{1}$$

$$\delta = \omega_0(\omega - 1) \tag{2}$$

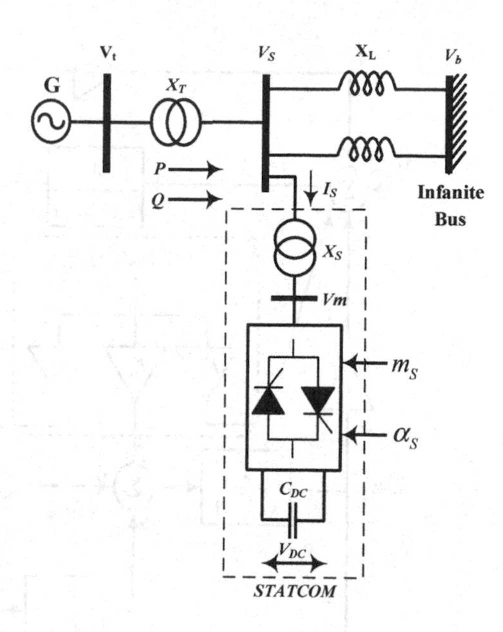

Fig. 1 SMIB system with STATCOM

$$\dot{\omega} = \frac{1}{M}(P_m - P_e - D(\omega - 1)) \tag{3}$$

$$\dot{E}_q^{'} = \frac{1}{T_{d0}^{'}}\left(E_{fd} - E_q^{'} - \left(x_d - x_d^{'}\right)i_d\right) \tag{4}$$

$$\dot{E}_{fd} = \frac{1}{T_A}\left(K_A\left(v_{ref} - v_t + u\right) - E_{fd}\right) \tag{5}$$

$$\dot{V}_{dc} = \frac{m}{C_{dc}}\left(I_{sd}\cos\psi + I_{sq}\sin\psi\right) \tag{6}$$

where ω and ω_0 are the angular speed and base speed (in p.u), respectively; δ is the load angle; P_m and P_e are the mechanical and electrical power, respectively; M is moment of inertia; D is the damping ratio coefficient; $E_q^{'}, E_{fd}$ represents internal voltage and field voltage of the machine; x_d and $x_d^{'}$ are the d-axis reactance and transient reactances, respectively; i_d is the d-axis current, $T_{d0}^{'}$ is the open-circuit transient time constant; K_A and T_A are the field circuit gain and time constants, respectively; v_{ref} and v_t represent the reference voltage and terminal voltage, respectively; u is the controller parameter; m is the modulation index; I_{sd} and I_{sq} are the d-axis and q-axis STATCOM current, respectively; ψ is the phase difference [15, 16]. Deviation in speed $(\Delta\omega)$ is chosen as input to the controller. Here, output of the stabilizer is compared with the reference value, and the resulting output is fed to the proposed controller. The modulation index (m_S) and phase angle (α_S) are the control inputs to STATCOM.

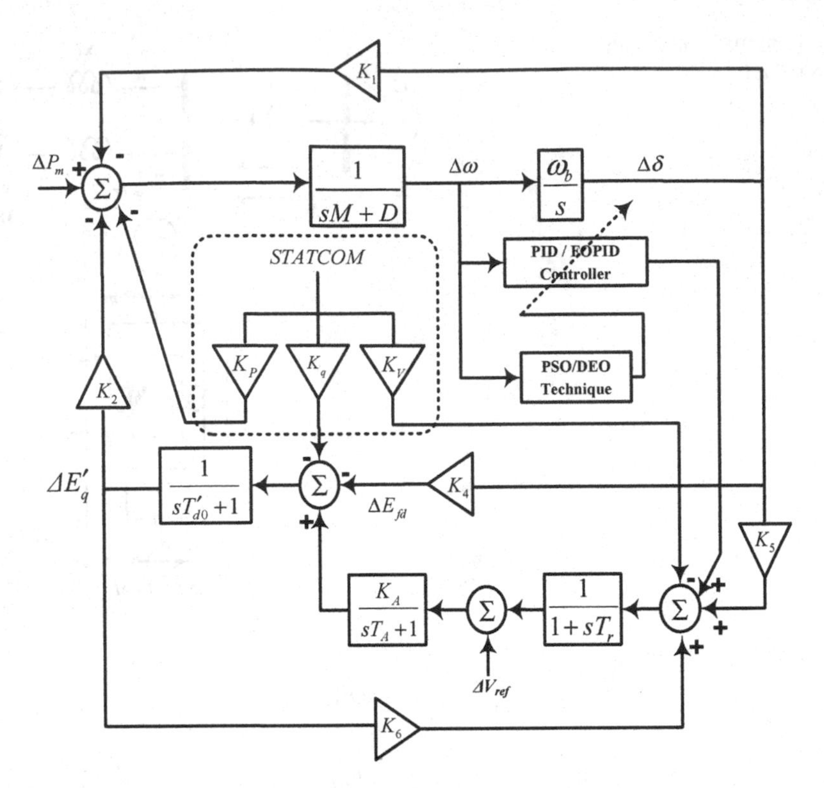

Fig. 2 Linearized model of SMIB system with STATCOM

3 Proposed Controller and Objective Function

3.1 Fractional Order PID (FOPID) Controller

The fractional order PID controller ($PI^{\lambda}D^{\mu}$) is introduced by Podlubny in 1999 [17]. It is a generalized version of the conventional PID controller. In addition to the conventional K_P, K_I, K_D gains, two additional fractional operators integrating order (λ) and derivative order (μ) in the range of [0, 1]. This makes this controller more flexible, versatile, and robust for improving the dynamic performance of the controller.

The transfer function of the of the proposed $\left(PI^{\lambda}D^{\mu}\right)$ controller is given as:

$$TF_{FOPID} = K_P + \frac{K_I}{s^{\lambda}} + K_D s^{\mu}, \qquad (\lambda, \mu > 0) \tag{7}$$

The equation of the $\left(PI^{\lambda}D^{\mu}\right)$ controller in time domain is given by,

$$u(t) = K_P e(t) + K_I D^{-\lambda} e(t) + K_D D^{\mu} e(t) \tag{8}$$

$PI^{\lambda}D^{\mu}$ controller finds the optimal solution in a five-dimensional search space $(K_P, K_I, K_D, \lambda, \mu)$. This ability makes the proposed $PI^{\lambda}D^{\mu}$ controller more robust, efficient than the conventional PID controller which makes this controller suitable for many real-time applications and many real-time process control systems [18]. The objective function of FOPID controller can be derived using integral squared error (ISE) as given below:

$$J(K_P, K_I, K_D, \lambda, \mu) = \int_0^t [\Delta\omega]^2 dt \tag{9}$$

Minimize J considering the following constraints.

$$\begin{aligned} &K_P^{\min} \le K_P \le K_P^{\max}; K_I^{\min} \le K_I \le K_I^{\max}; K_D^{\min} \le K_D \le K_D^{\max} \\ &\lambda^{\min} \le \lambda \le \lambda^{\max}; \mu^{\min} \le \mu \le \mu^{\max} \end{aligned} \tag{10}$$

The gains of FOPID can be optimized by soft computing techniques like PSO and DE which are explained as follows.

4 Optimization Techniques

4.1 Dolphin Echolocation Optimization (DE)

Particle swarm optimization is a population-based evolutionary algorithm that relies on the social behavior of the swarm as schooling of fish, flocking of birds, etc. in search of the best location in a multi-dimensional search space. In PSO, every particle in the swarm updates its position and velocity according to its individual best position (p_{best}) and global best value (g_{best}) [3]. Similarly, dolphin echolocation optimization (DE) is a metaheuristic optimization technique [12] that imitates dolphin's hunting process, i.e., locating, discriminating, and handling of their preys through their biosonar systems. The term echolocation illustrates the flying ability of bats for discriminating between obstacles and preys through generating sound waves in the form of clicks and listening to the echoes returned with high-frequency clicks. The process of echolocation is best studied in case of bottlenose dolphins. Dolphins initially search for their preys randomly. Once the prey was detected, the dolphin increases its clicks for concentrating others to the discovered location. This method of echolocation by dolphins is explored through following steps:

1. Randomly initialize the location of dolphins.
2. Determine predefined probability (PP) of the loop using the following equation.

$$PP(Loop_i) = PP_1 + (1 - PP_1)\frac{Loop_i^{Power} - 1}{(Loops\ Number)^{Power} - 1} \quad (11)$$

3. Determine fitness of each location such that location with best fitness value will get higher values.
4. Assign fitness of each location to its neighbors according to a symmetric triangular distribution or any symmetric distribution.
5. Add all devoted fitness values to form accumulative fitness.
6. Add a small value of ε to AF matrix. The value of ε is defined as per the fitness value.

$$AF = AF + \varepsilon \quad (12)$$

7. Find the best location achieved and set its AF to zero.
8. Calculate the probability by normalizing AF as:

$$P_{ij} = \frac{AF_{ij}}{\sum_{i=1}^{MaxAj} AF_{ij}} \quad (13)$$

where P_{ij} is the probability of the ith alternative to appear in the jth dimension; AF_{ij} is the accumulative fitness of the ith alternative to be in the jth dimension; Max Aj is the maximum number of alternatives available for the jth dimension.

9. Select PP (Loopi) percent of next step locations from best location dimensions. Distribute other values according to P_{ij}.
10. Repeat steps 2–8 for as many times as the loops number.

5 Simulation Results and Discussions

The simulation results for deviation in voltage and angular frequency are presented in this section for studying the stability of SMIB with use of STATCOM as FACTS controller in combination with conventional FOPID controller. As there is chance of more deviations in the above parameters, the gains of the FOPID controller are optimized by PSO and DE techniques for improving transient stability performance under different operating conditions like step and random variation in load demand. The SMIB system considered for the study is developed in MATLAB/Simulink environments. The parameters of the considered SMIB system and optimal gains of

FOPID controller using PSO and DE techniques are presented in Appendix section in Tables 2 and 3.

5.1 Analysis of Stability for Variations in Load

This subsection presents the voltage and angular frequency deviations in the considered SMIB and presented in Fig. 3a, b. It is observed from the results that a sudden increase in load leads to mismatch in reactive power generation and load demand, and as a result, the voltage and angular frequency deviate from the rated value as reflected from the simulated result of figure. But, due to the presence of the FACTS device STATCOM, the reactive requirement is supplied into the system in order to improve the voltage profile of the system. Again, as the FOPID gains are

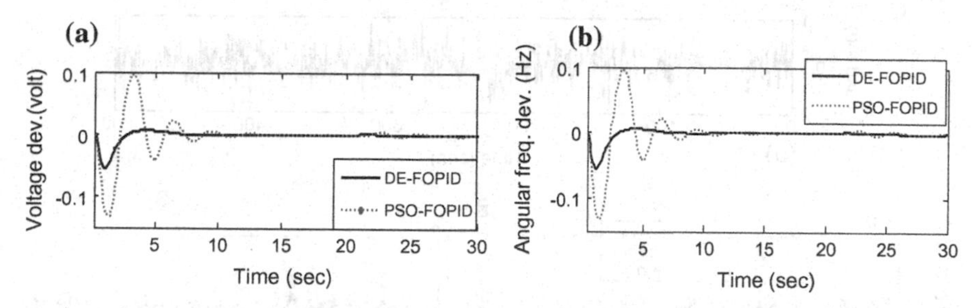

Fig. 3 Stability analysis in SMIB for step increase in loads **a** Voltage deviation **b** Angular frequency

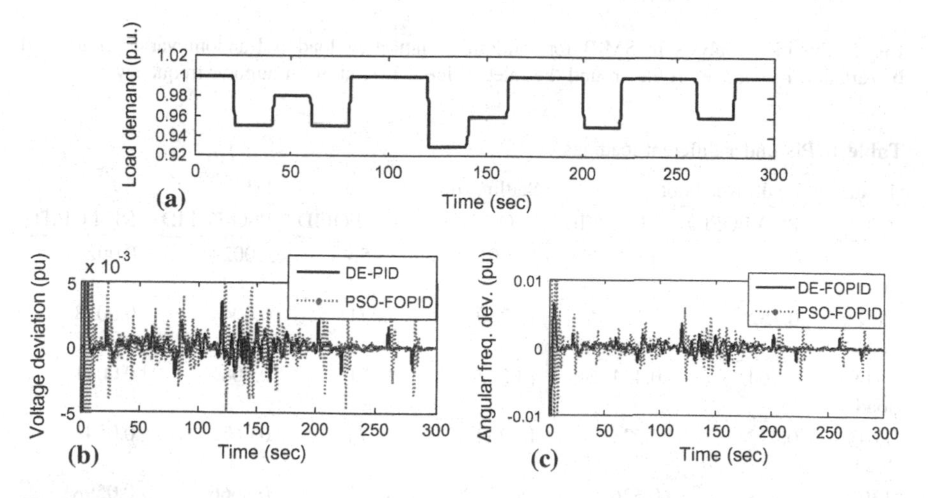

Fig. 4 Stability analysis in SMIB for continuous step variation in load **a** Load variation with step increase and decrease **b** Voltage deviation **c** Deviation in angular frequency

being optimized by the PSO and DE algorithms, the performance of STATCOM with PSO and DE-optimized FOPID controller shows better damping of the oscillation as compared to the conventional FOPID controller. Further, a continuous step increase and decrease in load is considered as disturbance into the SMIB system, and the effect of the STATCOM and the proposed controllers is analyzed from Fig. 4a–c. Again, the variation in voltage and angular frequency for random change in load is analyzed from Fig. 5. From the simulated results, it has been noticed that DE-FOPID, PSO-FOPID controllers show better performance in terms of the indices peak overshoots and settling time as compared to the conventional FOPID controller.

The performances of the conventional FOPID, PSO-optimized FOPID, and DE-optimized FOPID are compared in terms of the performance indices such as peak overshoot, settling time, and integral square error under step and random load

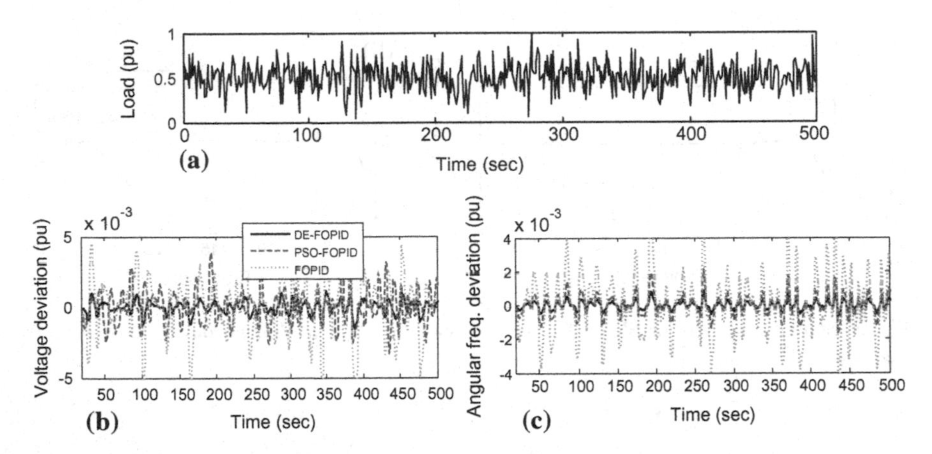

Fig. 5 Stability analysis in SMIB for random variation in load **a** Random variation in load **b** Variation in voltage profile around the rated value **c** Deviation in angular frequency

Table 1 PIs under different loadings

Load/ PIs	Peak overshoot		Settling time		ISE	
	PSO-FOPID	DE-FOPID	PSO-FOPID	DE-FOPID	PSO-FOPID	DE-FOPID
80% load	0.148	0.322	13.001	5.456	0.0029	0.0027
70% load	0.152	0.422	13.200	5.471	0.0044	0.0028
110% load	0.164	0.434	13.267	5.871	0.0086	0.0041
120% load	0.173	0.528	14.234	6.174	0.0672	0.0541
130% load	0.178	0.536	14.355	6.182	0.0666	0.0598

changing conditions. It is observed that DE-FOPID shows comparatively better performance as compared to the conventional FOPID and PSO-optimized FOPID controllers, and the comparative analysis is presented in Table 1.

6 Conclusion

This paper has studied the stability in SMIB system under different operating conditions such as step and random variation in load demand. The reactive power injection into the system is done by a STATCOM controller. The FOPID controller parameters are being optimized by PSO as well as DE. The effectiveness of the above controllers was simulated in MATLAB, and their performance in terms of controlling the voltage and angular speed of the SMIB system was evaluated. Both graphical and quantitative analyses are performed to test the efficiency of the proposed controllers under considered operating conditions. Hence, based on the study, it is concluded that DE-optimized FOPID performs better in comparison with FOPID and PSO-optimized FOPID controllers under small as well as large or random variation in load demand.

Appendix

Alternator: MVA = 500; V = 11 kV; f. = 50 Hz; R_a = 0.025 pu; X_a = 0.2 pu.
See Tables 2 and 3.

Table 2 Parameters of Philip-Heffron model of SMIB

Parameters	Values	Parameters	Values	Parameters	Values
M	5 s	D	4.0	K_3	0.3600
x'_d	0.32	K_A	10	K_4	−0.0626
T'_{d0}	5.55	T_A	0.01	K_5	0.4674
C_{DC}	1.0	V_{DC}	1.0	K_6	0.9565
K_1	0.9439	ω_b	314.15	x_q	1.6
K_2	1.2243	X_d	1.9	X_L	0.3

Table 3 Optimal values of gains using PSO and DE

Techniques	K_p	K_i	K_d	μ	λ
DE-FOPID	1.1482	0.7222	0.7560	0.3325	0.3857
PSO_FOPID	0.7825	0.2952	0.8964	0.7092	0.7689

References

1. Padiyar, K.R.: Power System Dynamics: Stability and Control, 2nd edn. B.S. Publications, Hydrabad. (2002)
2. Kundur, P.: Power System Stability and Control. McGraw-Hill (1994)
3. Saadat, H.: Power System Analysis. McGraw-Hill (1999)
4. Paital, S.R., Patra, S., Singh, A.K., Mohanty, A., Ray, P.K.: Reactive power compensation using PSO controlled UPFC in a microgrid with a DFIG based WECS. In: IEEE INDICON, pp. 1–5 (2015)
5. Panda, S.: Differential evolution algorithm for SSSC-based damping controller design considering time delay. J. Franklin Inst. **348**(8), 1903–1926 (2011)
6. Abido, M.A.: Design of PSS and STATCOM-based damping stabilizers using genetic algorthims. In: Proceedings of the IEEE General Meeting on Power Engineering Society, Canada, Montreal, pp. 1–8 (2006)
7. Padiyar, K.R., Prakash, V.S.: Tuning and performance evaluation of damping controller for a STATCOM. Int. J. Electr. Power Energy Syst. **25**(2), 155–166 (2003)
8. Magaji, N., Mustafa, M.W.: Optimal location and selection of FACTS devices for damping oscillations. IEEE International Conference on Power and Energy, Malaysia, pp. 1339–1344 (2008)
9. Jumaat, S.A., Musirin, I., Othman, M.M., Mokhlis, H.: Evolutionary particle swarm optimization (EPSO) based technique for multiple SVCs optimization. In: IEEE International Conference on Power and Energy, (PECon) 2012, Malaysia
10. Wang, Y.P., Hur, D.R., Chung, H.H., Watson, N.R., Arrillaga, J., Matair, S.S.: A genetic algorithms approach to design an optimal PI controller for static var compensator. In; International Conference on Power System Technology (PowerCon 2000) 3(2000)n, pp. 1557–1562
11. Yang, X.S.: Firefly algorithm, Lévy flights and global optimization. In: Bramer M et al. (eds.) Research and Development in Intelligent Systems, vol. XXVI, pp. 209–18. Springer, London (2010)
12. Kaveh, A., Farhoudi, N.: A new optimization method: Dolphin echolocation. Adv. Eng. Softw. **59**, 53–70 (2013)
13. Yu, Q., Li, P., Liu, W., Xie, X.: Overview of STATCOM technologies. In: IEEE international conference on electric utility deregulation, restructuring and power technologies, Apr 2004
14. Kothari, M.L., Patra, J.C.: Design of STATCOM controllers with energy storage system using GEA. In: Proceedings of the 37th Annual North American Power Symposium, pp. 260–266 Oct 2005
15. Bamasak, S.M., Abido, M.A.: Damping improvement of power system oscillation using STATCOM. In: 15th PSCC. Session 32. Paper 5, Aug 2005
16. Syed, F.F., Rahim, A.H. and Ba-Khashwain, J.M.: Robust STATCOM controller design using PSO Based automatic loop-shaping procedure. In: Proceedings of IEEE Conference on Control Applications. pp. 440–445, Aug 2005
17. Podlubny, I.: Fractional-order systems and PID-controllers. IEEE Trans. Autom. Control **44**, 208–214 (1999)
18. Mohanty, A., Viswavandya, M., Mohanty, S.: An optimised FOPID controller for dynamic voltage stability and reactive power management in a stand-alone micro grid. Electr. Power Energy Syst **78**, 524–536 (2016)

Part V
Emerging Techniques in Computing

Design of Decision-Making Techniques Using Improved AHP and VIKOR for Selection of Underground Mining Method

B. Bhanu Chander, Amit Kumar Gorai and S. Jayantu

Abstract Decision making is a process of selecting the best alternative from the pool of alternatives. This selection process depends on many influencing parameters; these parameters may be beneficiary or non-beneficiary. The proposed decision-making techniques were implemented in the appropriate underground mining method selection process. The selection of underground mining method depends on various geo-mining parameters such as technical, physical, mechanical, and economic parameters. In the proposed work, the selection of best mining method for bauxite deposit was implemented using Improved AHP and VIKOR. Improved AHP technique was considered to determine the weights of the influencing parameters. The proposed techniques can consider the association among the influencing parameters and alternatives. Results obtained by the proposed techniques were compared with the results obtained by other researchers for the bauxite deposit. The results showed that the suitable mining method for the specified criteria of the bauxite mine was conventional cut and fill.

Keywords Decision making · Improved AHP · VIKOR · Underground mining methods · Influencing parameters

1 Introduction

The process of selecting a choice from the available options is known as decision making. For effective decision making, a person must predict the suitable alternative or option, based on influencing factors of the alternatives. This type of problem is known as multi-attribute decision making (MADM). To solve the specific multi-attribute decision-making problem, there are many MADM techniques available. The most popular MADM techniques are WSM, AHP, WPM, VIKOR, TOPSIS, ELECTRE, etc. To deal with imprecise data, these classical MADM

B. B. Chander (✉) · A. K. Gorai · S. Jayantu
Department of Mining Engineering, NIT Rourkela, Rourkela 769008, Odisha, India
e-mail: bhanuchanderbalusa@gmail.com

P. K. Sa et al. (eds.), *Recent Findings in Intelligent Computing Techniques*,
Advances in Intelligent Systems and Computing 708,
https://doi.org/10.1007/978-981-10-8636-6_52

techniques are not desired. Fuzzy MADM techniques can deal the problems of those imprecise data. Such fuzzy MADM techniques are Improved AHP and VIKOR.

This paper deals with the selection of underground mining method for bauxite deposit. Underground mining is known as extraction of any minerals below the surface of the earth, whereas opencast mining refers to the mineral extraction above the earth surface. The extraction of minerals in underground depends on many parameters, which includes geological, technical, physical, mechanical, and economic parameters. The selection of suitable mining method is a type of multi-attribute decision-making problem. In this paper, Improved AHP (analytic hierarchy process) and VIKOR were proposed to solve the underground mining method selection problem. In the proposed work, influencing parameter weights were calculated by AHP. These calculated weights are used in both Improved AHP and VIKOR. The proposed techniques give the accurate results while comparing to other MADM techniques.

Naghadehi et al. (2009) proposed fuzzy-AHP technique for selecting suitable underground excavation method for bauxite deposit [1]. Jamshidi et al. (2009) proposed AHP technique for selecting the mining method for bauxite ore deposit [2]. Mikaeil et al. (2009) proposed FAHP and TOPSIS for the selection of underground mining method for bauxite mineral [3] deposit. Ataei et al. (2008) proposed TOPSIS for selecting a mining method for bauxite deposit [4]. Ataei et al. (2013) proposed Monte Carlo AHP to an underground mining method for bauxite deposit [5].

2 Proposed Techniques

2.1 AHP

Analytic hierarchy process (AHP) is the most popular multi-attribute decision-making technique, developed by Satty in 1980. In AHP, the problem is decomposed into a number of levels: The first level defines the goal of the problem, the second level defines criteria, next levels (if exist) define the sub-criteria, and the last level defines the alternatives. The AHP technique requires the pair-wise comparison between all the criteria of a problem and between all the alternatives of each criterion. AHP can deal with qualitative and quantitative factors [6]. The steps in the procedure of AHP are defined as follows:

Step 1: Define the goal, criteria, and alternatives for the decision-making problem.

Step 2: Define the relative importance matrices for each level or set of variables of the given problem. The relative importance matrices are used to determine the weights of the criteria parameters. The following sub-steps can be used for determining the weights of the criteria parameters.

- Define the relative pair-wise comparison matrices for variables at different levels based on the Satty's scale (1–9).
- Determination of the normalized weight (w_j) of all the attributes using Eqs. (1) and (2) [6].

$$GM_j = \left[\prod_{j=1}^{M} b_{ij}\right]^{1/M} \quad (1)$$

And,

$$wj = GM_j / \sum_{j=1}^{M} GM_j \quad (2)$$

- Determination of the maximum eigenvalue (λ_{max}) of the pair-wise comparison matrices.
- Determination of the consistency ratio (CR), where CR = consistency index (CI)/random index (RI). CI can be determined as $\lambda_{max} - M/(M-1)$, and RI can be determined based on the number of criteria in the given problem. The general condition of CR should be less than or equal to 0.1.

Step 3: Construction of the pair-wise matrices between the alternatives of every criterion.

Step 4: Determination of the overall score for each alternative as:

- Multiplying each criterion weight with corresponding alternative value (normalized) in the decision table.
- The overall score is determined by taking the summation of all the scores. The result is the score of every alternative, and the highest score alternative will be the best one for the given problem.

2.2 *Improved AHP*

AHP can be improved by eliminating the pair-wise comparison matrices between the alternatives of every criterion. Usually, AHP considers the pair-wise comparison matrices between the alternatives of every criterion. The number of comparison matrices in Improved AHP has been reduced so that the time taken to evaluate the best alternative has been reduced. Improved AHP normalizes the criteria values in a proper way [7]. The Improved AHP follows the following procedure for selecting the best alternative.

Step 1: Define the decision variables at each level of the hierarchy. It consists of the evaluation alternatives, criteria, and sub-criteria parameters.

Step 2: Normalize the data in the decision table, which follows the below two rules.

- If the criteria are beneficiary criteria, place '1' where the maximum value exists and divide the remaining criteria by the maximum value.
- If the criteria are non-beneficiary criteria, place '1' where the small value exists and divide the small value with the remaining criteria values.

Step 3: Determine the weights of the criteria. This step is similar to Step 2 of AHP.

Step 4: Find out the scores of the selected alternatives by multiplying the weights of the criteria with a corresponding normalized value of the criteria obtained in Step 2 and sum all criteria values over every alternative as shown in Eq. (3). Here, w_j is the related weight of each criterion and m_{ij} is the related normalized value of the alternative.

$$\text{Score} = \sum_{j=1}^{M} w_j \left(m_{ij}\right) \tag{3}$$

Step 5: Rank the alternatives based on their scores. The higher the score, the alternative is most suitable; the lower the score, the alternative is the worst.

2.3 VIKOR

VIKOR is known as compromise ranking method, and it is developed by Yu and Zeleny. The compromise ranking of the VIKOR is the feasible solution. This solution is closest to the ideal solution. VIKOR gives suitable alternative based on the influencing parameters; due to this, it is one of the multi-attribute decision-making techniques [6]. The following steps can be used for VIKOR to determine the score of the alternatives.

Step 1: Define a decision table, which consists of the selected alternatives, selected criteria, and their corresponding values.

Step 2: Determine the ranking measures E_i and F_i as shown in Eqs. (4), (5), (6), and (7), respectively [6]. Where M is the number of criteria.
For beneficiary criteria

$$E_i = \sum_{j=1}^{M} w_j \left[\left(m_{ij}\right)_{max} - m_{ij}\right] / \left[\left(m_{ij}\right)_{max} - \left(m_{ij}\right)_{min}\right] \tag{4}$$

For non-beneficiary criteria

$$E_i = \sum_{j=1}^{M} w_j\left[(m_{ij}) - (m_{ij})_{min}\right] / \left[(m_{ij})_{max} - (m_{ij})_{min}\right] \quad (5)$$

For beneficiary criteria

$$F_i = \max \text{ of } w_j\left[(m_{ij})_{max} - m_{ij}\right] / \left[(m_{ij})_{max} - (m_{ij})_{min}\right] \quad (6)$$

For non-beneficiary criteria

$$F_i = \max \text{ of } w_j\left[(m_{ij}) - (m_{ij})_{min}\right] / \left[(m_{ij})_{max} - (m_{ij})_{min}\right] \quad (7)$$

Step 3: Determine the measures of P_i using Eq. (8). In the equation, the E_{imax} and E_{imin} are calculated from the E_i values. Similarly, the F_{imax} and F_{imin} are determined from F_i values. The value of v is 0.5

$$Pi = v\left[\frac{Ei - Ei, \min}{Ei, \max - Ei, min}\right] + (1 - v)\left[\frac{Fi - Fi, \min}{Fi, \max - Fi, min}\right] \quad (8)$$

Step 4: The best alternative is decided by the least P_i value. Rank the alternatives as per their scores from least to highest. The least P_i is selected, if the following conditions are satisfied.

- $P(A_k) - P(A_1) \geq 1/(N-1)$, here A_k and Al are first and second alternatives, respectively. Where N defines the number of alternatives.
- A_k should be best ranked in E and/or F.

If any of the above conditions are not satisfied, then compromised solutions can be proposed. They are as follows:

- A_k and A_1, when the second condition is not satisfied.
- A_k, A_1, A_2, …, A_p, if the first condition is not satisfied. A_p is determined by $P(A_p) - P(A_1)$ approximately equal to $1/(N-1)$.

3 Application of Proposed Techniques

As mentioned, underground mining method selection is a multi-attribute decision-making problem. The most suitable mining method selection depends on many parameters. To solve this problem, Improved AHP and VIKOR are proposed. The selected mining methods are conventional cut and fill (CCF), shrinkage stoping (SHS), mechanized cut and fill (MCF), bench mining (BM), sub-level stoping (SLS), and stull stoping (SS). The influencing parameters for the selected mining methods are identified as a dip (P1), shape (P2), thickness (P3), hanging wall RMR (P4), ore grade (P5), footwall RMR (P6), technology (P7), depth (P8), ore uniformity (P9), ore zone RMR (P10), dilution (P11), production (P12), and recovery (P13).

Decision table for the selected mining methods (Alternatives) and influencing parameters (Criteria) is defined in Table 1 [3]. Here, the original data is defined in 5-point scale.

3.1 Weights Calculation of Parameters Using AHP

Before applying Improved AHP and VIKOR techniques, the parameter weights have to be calculated using AHP. As mentioned in Sect. 2, the weights are calculated using Satty's 9-point scale. The pair-wise comparison matrix for the influencing parameters was shown in Table 2 [3].

The pair-wise comparison matrix shown in Table 2 was used to determine normalized weights for the influencing parameters, and these weights are shown in Table 3.

λ_{max} value for the pair-wise comparison matrix of influencing parameters was determined as 13.719. CI value was determined as 0.059. RI for 13 parameters is 1.56 as in [7]. Hence, CR = CI/RI, and which is found as 0.038. As mentioned in Sect. 2, here, CR found less than 0.1 and which is acceptable. So the weights of the parameters can consider for selecting the best underground mining method.

3.2 Application of Improved AHP

As mentioned in the procedure of Improved AHP, first step is to normalize the decision table data. The normalized data of the decision table (Table 1) is shown in Table 4.

After calculating the normalized values with corresponding weights of the criteria and sum over the alternatives, the resultant data is shown in Table 5.

Table 1 Decision table [3]

	P1	P2	P3	P4	P5	P6	P7	P8	P9	P10	P11	P12	P13
CCF	5	3	2	3	1	4	2	5	4	2	4	4	4
MCF	2	3	2	3	2	4	2	3	4	2	4	4	4
SHS	2	1	2	3	2	2	2	2	1	2	2	1	1
SLS	2	1	2	1	5	1	2	2	1	2	2	2	1
BM	1	1	1	1	1	1	2	1	2	2	1	1	1
SS	1	1	2	2	1	2	2	1	1	2	1	1	1

Table 2 Pair-wise comparison matrix [3]

	P1	P2	P3	P4	P5	P6	P7	P8	P9	P10	P11	P12	P13
P1	1	3	1	1	7	3	3	5	3	1	7	7	7
P2	1/3	1	1/3	1/3	5	1	1	3	1	1/3	5	5	5
P3	1	3	1	1	7	3	3	5	3	1	7	7	7
P4	1	3	1	1	7	3	3	5	3	1	7	7	7
P5	1/7	1/5	1/7	1/7	1	1/5	1/5	1/3	1/5	1/7	1	1	1
P6	1/3	1	1/3	1/3	5	1	1	3	1	1/3	5	5	5
P7	1/3	1	1/3	1/3	5	1	1	3	1	1/3	5	5	5
P8	1/5	1/3	1/5	1/5	3	1/3	1/3	1	1/3	1/5	3	3	3
P9	1/3	1	1/3	1/3	5	1	1	3	1	1/3	5	5	5
P10	1	3	1	1	7	3	3	5	3	1	7	7	7
P11	1/7	1/5	1/7	1/7	1	1/5	1/5	1/3	1/5	1/7	1	1	1
P12	1/7	1/5	1/7	1/7	1	1/5	1/5	1/3	1/5	1/7	1	1	1
P13	1/7	1/5	1/7	1/7	1	1/5	1/5	1/3	1/5	1/7	1	1	1

As observed from the Table 5, the score for CCF (conventional cut and fill) is 0.986, followed by MCF (mechanized cut and fill), which has the score of 0.882. Hence, the Improved AHP says, conventional cut and fill is the suitable mining method for the bauxite deposit.

3.3 *Application of VIKOR*

As mentioned in the procedure of the VIKOR, based on the decision table (Table 1), E, F, and P values for each alternative were determined, and same are shown in Table 6.

After arranging the E, F, and P values in ascending order, respectively, the result data is shown in Table 7.

As observed from the values of Pi, 0 is the least value among the alternatives, and this is the score of conventional cut and fill. This result also satisfies the compromise condition of VIKOR as specified in the procedure. The second followed the best alternative is found as mechanized cut and fill method with the score of 0.4482.

4 Results Comparison

After observing the results of Improved AHP and VIKOR techniques, the most suitable underground mining method is the conventional cut and fill followed by mechanized cut and fill. The obtained results are also compared with the results of

Table 3 Weights of the criteria

Influencing parameters	P1	P2	P3	P4	P5	P6	P7	P8	P9	P10	P11	P12	P13
Weights	0.156	0.068	0.156	0.156	0.016	0.068	0.068	0.032	0.068	0.156	0.016	0.016	0.016

Table 4 Normalized decision table

	P1	P2	P3	P4	P5	P6	P7	P8	P9	P10	P11	P12	P13
CCF	1	1	1	1	0.2	1	1	1	1	1	1	1	1
MCF	0.4	1	1	1	0.4	1	1	0.6	1	1	1	1	1
SHS	0.4	0.33	1	1	0.4	0.5	1	0.4	0.25	1	0.5	0.25	0.25
SLS	0.4	0.33	1	0.33	1	0.25	1	0.4	0.25	1	0.5	0.5	0.25
BM	0.2	0.33	0.5	0.33	0.2	0.25	1	0.2	0.5	1	0.25	0.25	0.25
SS	0.2	0.33	1	0.66	0.2	0.5		0.2	0.25	1	0.25	0.25	0.25

Table 5 Ranks of the mining methods by Improved AHP

Alternative	Score	Rank
CCF	0.986	1
MCF	0.882	2
SHS	0.710	3
SLS	0.603	5
BM	0.483	6
SS	0.613	4

Table 6 Scores determined by VIKOR

Alternatives	E	F	P
CCF	0.0166	0.0166	0
MCF	0.1460	0.1171	0.4482
SHS	0.3820	0.1171	0.6088
SLS	0.5431	0.1562	0.8583
BM	0.7513	0.1562	1.0005
SS	0.5170	0.1562	0.8405

Table 7 Scores arranging in ascending order for E, F, and P

E	F	P
0.0166	0.0166	0
0.1460	0.1171	0.4482
0.3820	0.1171	0.6088
0.5170	0.1562	0.8405
0.5431	0.1562	0.8583
0.7513	0.1562	1.0005

other techniques used by earlier researchers. The comparison with other techniques is shown in Table 8. Hence, Improved AHP and VIKOR proved the efficient MADM techniques for best alternative selection.

Table 8 Comparison of results with MADM techniques

Author	MADM technique	Result
Naghadehi	Fuzzy-AHP	CCF
Jamshidi	AHP	CCF
Ataei	TOPSIS	CCF
Mikaeil	Fuzzy-AHP and TOPSIS	CCF
Ataei	Monte Carlo AHP	CCF
Present study	Improved AHP and VIKOR	CCF

5 Conclusion

The proposed multi-attribute decision-making techniques indicate that the model is efficient for the selection of best alternative when multiple criteria are involved in the selection process. In this study, the proposed techniques (Improved AHP and VIKOR) are efficiently implemented for selection of the most suitable underground mining method for excavation of bauxite deposit. The proposed techniques showed that the conventional cut and fill method is the suitable mining method for excavation of minerals of bauxite ore deposit.

References

1. Naghadehi, M.Z., Mikaeil, R., Ataei, M.: The application of fuzzy analytic hierarchy process (FAHP) approach to selection of optimum underground mining method for Jajarm Bauxite Mine. Iran. Expert Syst. Appl. **36**(4), 8218–8226 (2009)
2. Jamshidi, M., et al.: The application of AHP approach to selection of optimum underground mining method, case study: Jajarm Bauxite Mine (Iran). Arch. Min. Sci. **54**(1), 103–117 (2009)
3. Mikaeil, R., et al.: A decision support system using fuzzy analytical hierarchy process (FAHP) and TOPSIS approaches for selection of the optimum underground mining method. Arch. Min. Sci. **54**(2), 349–368 (2009)
4. Ataei, M., et al.: Suitable mining method for Golbini No. 8 deposit in Jajarm (Iran) using TOPSIS method. Min. Technol. **117**(1), 1–5 (2008)
5. Ataei, M., Shahsavany, H., Mikaeil, R.: Monte Carlo analytic hierarchy process (MAHP) approach to selection of optimum mining method. Int. J. Min. Sci. Technol. **23**(4), 573–578 (2013)
6. Rao, R.V.: Decision making in the manufacturing environment: using graph theory and fuzzy multiple attribute decision making methods. Springer Science & Business Media (2007)
7. Rao, R.V.: Multiple attribute decision making in the manufacturing environment. In: Decision Making in Manufacturing Environment Using Graph Theory and Fuzzy Multiple Attribute Decision Making Methods. Springer, London, pp. 1–5 (2013)

Computing Model to Analyze Voting Rate Failure in Election System Using Correlation Techniques

K. S. Gururaj and K. Thippeswamy

Abstract Conversion from non-engineering domain to engineering is one the major challenges for the engineering community. Utilization of the collected data in science and engineering may certainly resolve some of the minor flaws in the society. In this connection, this paper tries to design a model to analyze voting rate failure in election system by applying Pearson correlation coefficient on the data collected during elections. Further, it concludes with the need of technology-based solution to minimize the voting rate failure in the election systems.

Keywords Analysis on election system · Voting rate failure
Pearson correlation coefficient

1 Introduction

Usually in democratic countries, the electing of right leader in the elections is one of the major responsibilities of each and every citizen and it is crucial part of the system; in turn, it affects the development of the country. Due to some reasons, part of the community will reject in the involvement of the election process. Rejection due to several reasons may be because of literacy rate, development rate of the city, population count, rainfall rate, and so on. This rejection may lead to electing of person who may not be fit for the position. Certainly that causes many problems and may lead to spoil/corrupt the system. In this connection, there are many statistical approaches available to analyze the issues. This paper tries an engineering perspective model to analyze the voting rate failure in the election process for electing the

K. S. Gururaj (✉)
Computer Science & Engineering, Visveswaraya Technological University, Belagavi, India
e-mail: gururaj.k.s79@gmail.com

K. Thippeswamy
Department of Computer Science & Engineering, PG Center, Visvesvaraya Technological University, Mysuru, India
e-mail: thippeswamy@vtu.ac.in

P. K. Sa et al. (eds.), *Recent Findings in Intelligent Computing Techniques*,
Advances in Intelligent Systems and Computing 708,
https://doi.org/10.1007/978-981-10-8636-6_53

right leader and proposes a technology-based revolutionary solution for the success of the election, thereby the development of the country.

2 Review of Literature

2.1 Data Collection

Table 1 shows the election data of various Lok Sabha elections held in India and corresponding census data starting from 1951 till date [1, 2] (Since the emphasis is given to the model rather than the data, certain data are simulated). Few census columns may be considered for the analysis of voting rates, namely literacy, population, rainfall rate, and gross development product. Hence, for every two elections, the features may be common. Literacy is one of the major features of the society and plays an important role in elections. This information is found during the census conducted on every starting of the new decade, i.e., 1951, 1961, 1971, 2011 [3]. As per the literature, a literate is one who is aged 7 and above and can read and write with understanding of the language. A person who can only read but cannot write is not literate. Before 1991, children below 5 years of age were necessarily treated as illiterates [4]. Population is another major factor which may impact on voting rates [5]. Mentioned population is in the form of its growth, i.e., percentage of increase or decrease, compared to previous measure.

2.2 Mobile Usage Statistics

Figure 1 shows the analysis of the smartphone users penetration in India by www.statista.com [6] also indicating the eventual growth in the smartphone usage by year-wise then predicting that around 39% of people may use smartphone by the year 2019.

One of the most popular daily news magazines [7] has published "India has become the second-biggest smartphone market in terms of active unique smartphone users, crossing 220 million users, surpassing the US market, according to a report by Counterpoint Research." in one of its article on February 3, 2016.

Migration of the people from one place to another normally occurs due to the need of food, marriage, community, and job these days at national (between districts and states) [8, 9] and international level (between countries) [10]. Certainly, this migration causes many problems in the system. Due to migration, one may not concentrate properly on the rights during the election choosing right person as leader of the society. This may be due to lack of time, financial condition, transportation, or distance. Studies have shown that migration is rapidly increasing during these recent years [10, 11]. The net migration is the overall estimation of migration obtained

Table 1 Voting rates of various Lok Sabha elections and corresponding census data (literacy and population rates are simulated for the designing of the model)

Election	Year	Voting rate	Literacy rate	Population rate
1st	1952 [1]	61.2	0.058	1.32
2nd	1957 [2]	62.2	0.056	0.07
3rd	1962	55.42	0.032	0.132
4th	1967	61.33	0.049	0.02
5th	1971	55.29	0.165	0.12
6th	1977	60.49	0.128	0.123
7th	1980	56.92	0.127	0.06
8th	1984	63.56	0.239	0.09
9th	1989	61.95	0.951	0.15
10th	1991	56.93	0.324	0.18
11th	1996	57.94	0.15	0.16
12th	1998	61.97	0.345	0.13
13th	1999	59.99	0.764	0.12
14th	2004	48.74	0.6348	15.75
15th	2009	58.19	0.6348	15.75
16th	2014	66.4	0.7279	13.63

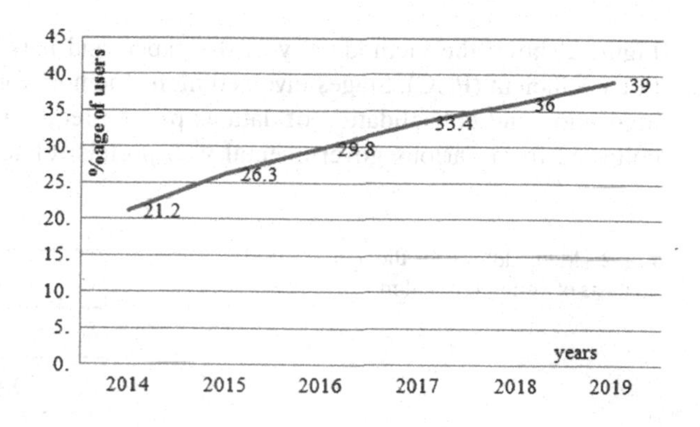

Fig. 1 Share of mobile phone users that use a smartphone in India from 2014 to 2019*. *Courtesy* www.statista.com [6]

by adding immigration and subtracting emigration quantities [11] for every 1000 persons.

2.3 Pearson Correlation Coefficient [12]

Pearson correlation coefficient is originally designed by Karl Pearson during 1878 but introduced by Francis Galton during 1880. It is the widely usable method in the field of science, engineering, and research to check linear dependencies between two variables.

The degree of the dependencies ranges from −1 to +1. Here, −1 and +1 indicate that two variables are negatively and positively related, whereas 0 indicates that variables are not related to each other. Following is the equation used to find Pearson correlation coefficient (r), where x and y are two different variables with set of values.

$$r = \frac{N\sum xy - \sum x \sum y}{\sqrt{(N\sum x^2 - (\sum x)^2)(N\sum y^2 - (\sum y)^2)}} \quad (1)$$

where, N = number of pairs of scores
$\sum xy$ = sum of the products of paired scores
$\sum x$ = sum of x scores
$\sum y$ = sum of y scores
$\sum x^2$ = sum of squared x scores
$\sum y^2$ = sum of squared y scores

3 Methodology

Figure 2 shows the methodology of the paper, and it is based on Pearson correlation coefficient (PCC). Stages involved in the methodology are data collection, normalization and consolidation of data as per the requirement of PCC [12]. Data is collected from various governmental web portals related to elections and census.

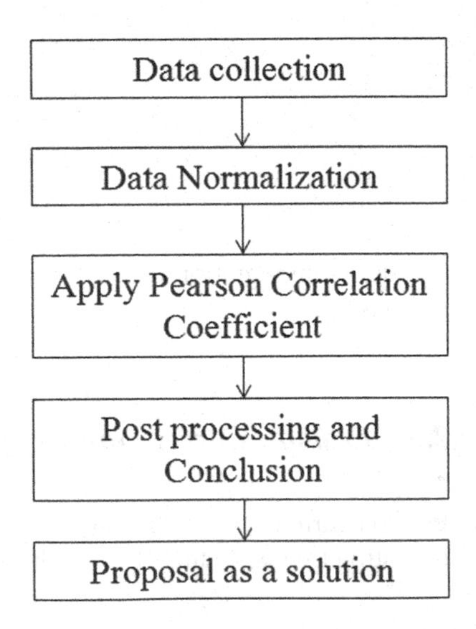

Fig. 2 Methodology for the analysis of voting rate failure

Table 2 Correlation coefficients

	Literacy rate	Population rate
Voting rate	0.138195095	−0.224588

Normalization of data is required to convert data into a particular standard as represented in Table 2.

4 Results and Discussion

Table 2 and Fig. 3 represent the correlation coefficients between the percentage of various elections and considered features namely literacy, population, and rainfall rates. Observations reflect that the coefficient between elections and literacy is higher as compared to others. The next highest coefficient is between elections and rainfall. This depicts that when people are more educated or literate, certainly the voting rates will be higher. Hence, our system should emphasize more on education by opening schools and colleges at interior places of the country. System may introduce certain new projects like free education for everyone till they complete basic education (may be from 1st standard to SSLC). This may resolve observed problem, and literacy may reduce the voting rate failure. Population is moving toward inverse correlation with respect to elections. This may be due to the manner in which elections are organized. But earlier when population was under control, the voting rate in the election was good and now it is completely inverse, i.e., less voting rate with more population. Hence, this observation indicates that when the population rate is less, obviously the rate of voting will be good.

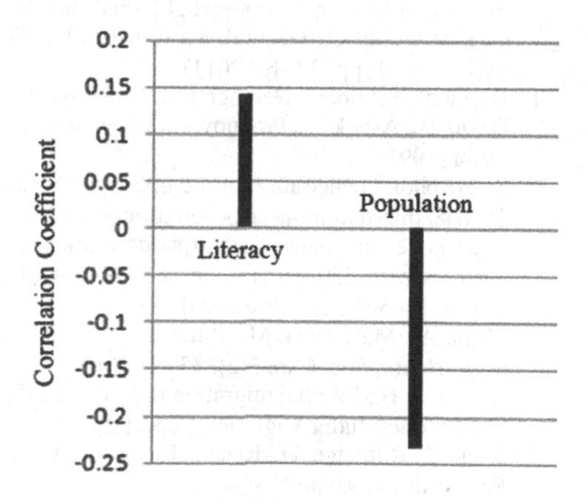

Fig. 3 Correlation comparative graph between various parameters of election

5 Conclusion

This paper has tried to design a simple statistical model to analyze the reason behind voting rate failure in election system. This paper has considered only two available random variables, i.e., literacy and population rates. Conclusion of the paper as per statistics is "literacy rate should be more than the population rate to decrease the degree of voting rate failure." Literacy may enhance voting rate, but sometimes it may be the reason for voting failure, due to abundant analysis on the requirement of voting at constituency where there is no development. As per the literature and statistics, migration may also lead to same. Hence, this paper proposes a revolutionary- and technology-based model, which may enhance voting rate in election system by adopting secured cloud access through mobile applications. Analysis on the need of technology-based model for elections in the existing system is certainly important discussion and a challenging too. As per the literature, smartphone usage is rapidly and exponentially increasing day by day. Hence, this paper is a building model for the further research and development activities toward the need of technology in the election systems.

Acknowledgements We would like to thank Election Commission of India, Census of India, and Government of Uttarakhand for providing data and relevant information through Web media. We would like to express our gratitude and thanks to VTU-PG Center Mysuru and GSSS Institute of Engineering and Technology for Women, Mysuru, for their support.

References

1. Voting rates in loksabha elections from the official website of government of Uttarakhand. http://ceo.uk.gov.in/pages/view/165/169-voting-percentage-in-various-lok-sabha-elections. Accessed 23 Oct 2016
2. Voting rates in loksabha elections by Election Commission of India. http://eci.nic.in/eci_main1/SVEEP/VoterTurnoutHighlightsLokSabha2014.pdf. Accessed 05 Nov 2016
3. Chandramouli, C., General, R.: Census of India. In: Provisional Population Totals, Paper I, Series, vol. 1, pp. 39–40 (2011)
4. Dasgupta, S.: Social Demography. Pearson Education India (2011)
5. David, B., Ashok, L., Prannoy, R.: India Decides: Elections 1952–1995. Books & Things, New Delhi (1995)
6. Smart phone penetration in the upcoming years in India. https://www.statista.com/statistics/257048/smartphone-user-penetration-in-india/. Accessed 09 Dec 2016
7. India is second biggest smartphone market. http://www.thehindu.com/news/cities/mumbai/business/with-220mn-users-india-is-now-worlds-secondbiggest-smartphone-market/article8186543.ece. Accessed 09 Dec 2016
8. Mitra, A., Murayama, M.: Rural to urban migration: a district-level analysis for india. Int. J. Migr. Health Soc. Care **5**(2), 35–52 (2009)
9. Bhagat, R.B.: Internal migration in India: are the underclass more mobile. Migration, Identity and Conflict: India Migration Report (2011)
10. Stone, J., Rutledge, D., Rizeva, P., Smith, A., Hou, X.: The Wiley Blackwell Encyclopedia of Race, Ethnicity, and Nationalism (2016)
11. Net migration rate. http://www.indexmundi.com/g/g.aspx?c=inŹv=27. Accessed 09 Dec 2016
12. Stigler, S.M.: Francis Galton's account of the invention of correlation. Stat. Sci. 73–79 (1989)

A Robust Watermarking Scheme for 3D Polygon Mesh Models Using Mean Curvature

Babulal Hansda, Vikas Maheshkar and Sushila Maheshkar

Abstract With growing advancement in digital technology, multimedia technology advanced so rapidly. Now images and videos are used in 3D form for architecture, computer games, medical diagnosis and maps. In our proposed algorithm, 3D mesh models are watermarked on the basis of vertex curvature for showing robustness and imperceptibility on mesh models. On the basis of curvature value of each vertex, the object mesh model is partitioned into two bins. One part consists all the vertices which has convex curvature, and other part contains the rest of the vertices. Vertices are modified in such a way that the watermarked mesh model cannot be distinguished from the original mesh model. With human visualization the watermark mesh model is imperceptible. Watermarked mesh models show better results on resistivity over attacks as like smoothing attack, subdivision attack, noise attack and cropping attack. *abstract* environment.

Keywords 3D watermarking · Mean curvature · Watermark attack
Root mean square (rms) error

B. Hansda · S. Maheshkar
Department of Computer Science and Engineering, Indian Institute of Technology (Indian School of Mines), Dhanbad, India
e-mail: babulal.hansda@gmail.com
URL: http://www.iitism.ac.in

V. Maheshkar (✉)
Division of Information Technology, Netaji Subhas Institute of Technology, New Delhi, Delhi, India
e-mail: vikas.maheshkar@gmail.com
URL: http://www.nsit.ac.in

S. Maheshkar
e-mail: sushila_maheshkar@yahoo.com

P. K. Sa et al. (eds.), *Recent Findings in Intelligent Computing Techniques*, Advances in Intelligent Systems and Computing 708,
https://doi.org/10.1007/978-981-10-8636-6_54

1 Introduction

As the world digitalized so rapidly, images, videos and audio qualities improve day by day. 2D images and videos are being replaced with 3D images and videos. 3D models are used for architecture as in computer-aided design (CAD), 3D computer games, 3D images in medical diagnosis for better treatment and 3D maps to view places from any angle. 3D watermark is necessary on these models for authentication and privacy prevention to ensure copyright protection. Imperceptibility and robustness are important requirements for 3D watermark [4]. 3D models contain basically face values and vertices values. Face values indicate how vertices interconnected with each other, i.e. topology of models and vertices values tells about position of vertices in space. Vertices are the most probable values for embedding watermark on 3D models.

In this paper, we propose a non-blind algorithm based on mean curvature of vertices. Geometric property of vertices can be used for watermarking on 3D models [6–8]. Curvature can be used for embedding watermark. Some of the areas are there on model where surfaces are curve or flat. By using this property, watermark can be done on curve surfaces so that minimum distortion can be achieved.

2 Proposed Work

Proposed algorithm embeds watermark on 3D mesh object. According to the curvature-based watermarking, there are three methods to calculate curvature, i.e. mean curvature, Gaussian curvature and root mean curvature [5]. Our proposed algorithm is based on mean curvature. For both watermark embedding and watermark extraction, the mean curvature is required.

2.1 *Watermark Embedding*

Vertices are values in which watermark can be embedded. Curvature-based watermarking is used in the proposed algorithm. Mean curvature C_m is calculated by the following formula:

$$C_m = \frac{\sum_{i=1}^{n} \theta(v_{i,j})}{\frac{1}{3}\sum_{k=1}^{n2} A_k} \tag{1}$$

$\theta\ (v_{i,j})$ denotes the angles between the normal of the two adjacent triangles, and $A_k (k = 1, 2, \ldots, n)$ denotes the triangle area corresponding to the one-ring neighbourhood of vertex. According to mean C_m of each vertex, vertices are divided into two bins. One bin contains all vertices which have values greater than zero, and the other bin contains rest of the vertices [1]. Watermark is embedded on vertices which have curvature values greater than zero.

Algorithm 1 Watermark Embedding:

Require: 3D object model
Ensure: 3D watermarked model
1: Read object model and get the values of vertices and faces.
2: Calculate mean curvature of each vertex.
3: On the basis of values of mean curvature, vertices can be divided into bins
i) B1 bin contains the vertices of mean curvature value as greater than zero.
ii) B2 bin contains the rest of the vertices.
4: Watermark image (w) is taken as binary image. And take values in single column matrix form.
5: Now B1 bin vertices can be watermarked by taking any one axis as from the formula
$B_1' = B_1 + \eta \times w$
Where B_1' is the new vertices value B_1 is original vertices values W is the watermark η is constant value
6: Arrange the vertices with the new values as per the original sequence of vertices.
7: Convert the vertices with their face values transform in model object file.

2.2 Watermark Extraction

As the proposed algorithm is non-blind watermark, we need original 3D models for extraction. The mean curvature of original mesh models is calculated. Watermarked mesh model vertex is divided on the basis of original mesh model mean curvature of vertex. The curvature with values greater than zero is used for extraction of watermark. By the help of original mesh model and watermarked mesh model, the watermark is extracted.

Algorithm 2 Watermark Extraction:

Require: 3D watermarked model
Ensure: Watermark image
1: First three steps are same for extraction process.
2: Watermarked model and original model both are converted into bins $B1'$, $B2$ and B1, B2 as from the embedding process.
3: Extraction of watermark can be gained by formula as
$$w' = (B_1' - B_1)/\eta$$
Where B_1', B_1 are the watermarked and original vertices, w' is recovered watermark, η is constant.
4: Arrange the vertices in matrix two dimensions as the gain watermark is in column form.
5: Recover watermark binary image is compared with the original binary image.

3 Experimental Results and Discussion

We have performed experiments for our proposed algorithm on different 3D model objects using MATLAB with minimum three iterations on each model. Model objects contain vertices and faces. These models are Bunny, Dragon, Hand, Venus. Watermark embedding is applied on models having high number of vertices. So that more number of vertex can be obtained for watermark. Watermark image is taken of size 64×64 (4096 bits). The threshold value for this experiment is $\eta = 0.01$. This threshold can be minimized to get more secure and imperceptible watermarking. Root mean square (RMS) error, Hausdorff distance, PSNR and correlation values are used as performance evaluation metrics.

3.1 Imperceptibility

Higher PSNR value, Hausdorff distance indicated high imperceptibility. With higher values after different attack, PSNR value in Table 1, RMS value in Table 2, correlation value in Table 4 and Hausdorff distance in Table 3 shows better result.

3.2 Robustness

Robustness of the proposed scheme is evaluated under different attacks like smoothing, subdivision, cropping and noise attack. Figure 1 shows watermarked mesh under

Table 1 PSNR values after different attacks

Object models	Bunny	Dragon	Armadillo	Hand	Venus
Without attack	72.604	75.486	121.41	74.227	78.628
Subdivision attack	67.278	68.342	66.251	63.726	72.463
Smoothing attack	60.605	60.038	68.302	59.055	66.341
Cropping attack	32.93	40.612	46.543	31.799	30.744
Noise attack	5.3423	5.8141	6.2279	5.8224	3.6495

Table 2 RMS values (10^{-3}) under different attacks

Object models	Bunny	Dragon	Armadillo	Hand	Venus
Without attack	0.103	0.055	0.083	0.071	0.040
Subdivision attack	0.335	0.357	0.139	0.201	0.152
Smoothing attack	0.537	0.672	0.262	0.587	0.250
Cropping attack	5.438	2.229	3.563	6.730	12.421
Noise attack	1.006	1.660	1.436	0.849	0.716

Table 3 Hausdorff distance values after different attacks

Models	Bunny	Dragon	Armadillo	Hand	Venus
Without attack	0.001	0.001	0.001	0.001	0.001
Subdivision attack	0.0658	0.0383	0.3276	0.0059	0.0155
Smoothing attack	0.0104	0.0103	0.4286	0.0097	0.0074
Cropping attack	0.1195	0.1454	10.204	0.1528	0.4318
Noise attack	0.0010	0.0081	2.6238	0.0010	0.0155

Table 4 Correlation values after different attacks

Object models	Bunny	Dragon	Armadillo	Hand	Venus
Without attack	1	1	1	1	1
Subdivision attack	0.9577	0.8935	0.7810	0.8834	0.8912
Smoothing attack	0.8768	0.9996	0.7810	0.8832	0.9997
Cropping attack	0.9997	0.9997	0.7804	0.8807	0.9957
Noise attack	0.00023	0.13	0.13	−0.0242	0.0062

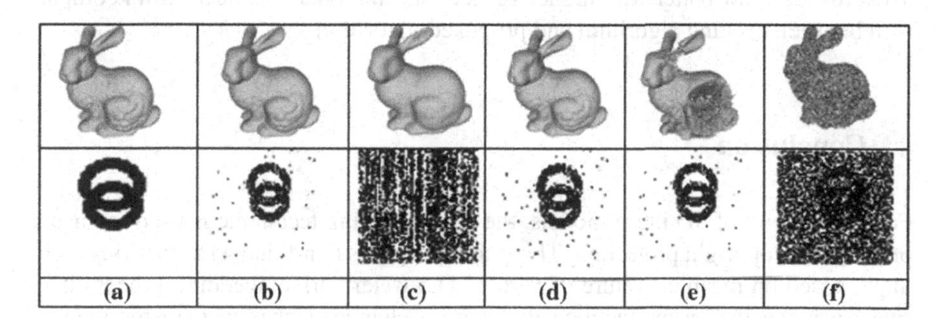

Fig. 1 Watermarked bunny and extracted watermark under different attacks. **a** Watermarked model, **b** smoothing attack, **c** simplification attack, **d** subdivision attack, **e** cropping attack, **f** noise attack

different attacks and the corresponding extracted watermarks. Except simplification attack, watermark is distinctly visible in all cases, thus the proposed watermarking scheme is robust against these attacks. Tables with different parameter show the attack is sustainable for watermarked models. We can observe that correlation value from Table 4 is tending towards 1 which shows the results are better.

3.3 Comparative Analysis

Proposed watermarking algorithm is compared with Garg et al. [1]. The result shows better with respect to correlation and Hausdorff distance values in some attacks.

Table 5 Comparison for correlation, RMS and Hausdorff distance

	Garg et al. [1]			Proposed algorithm		
	Bunny	Dragon	Armadillo	Bunny	Dragon	Armadillo
No. of vertices	34,834	35,000	30,995	34,835	50,000	106,289
No. of faces	69,451	70,216	61,986	69,653	100,000	212,574
Correlation without attack (%)	100	100	100	100	100	100
Hausdorff distance without attack	0.4138	0.0011	0.3289	0.001	0.001	0.001
RMS without attack (10^{-3})	0.024	0.028	0.030	0.098	0.055	0.083
Correlation smoothing attack (%)	62	45	41	95	89	78
Hausdorff distance smoothing attack	0.5381	0.0013	0.3418	0.0134	0.0103	0.4286
RMS smoothing attack (10^{-3})	0.053	0.013	0.034	0.577	0.672	0.262
Correlation subdivision attack (%)	89	91	89	87	99	78
Hausdorff distance subdivision attack	0.5714	0.0016	0.3862	0.0690	0.0386	0.3276
RMS subdivision attack (10^{-3})	0.061	0.023	0.038	0.112	0.357	0.139

Most results show better with higher vertices mesh models. Table 5 shows comparison between existing algorithm and proposed algorithm.

4 Conclusion

For the security of 3D mesh models, the watermarking technique is used to ensure privacy and copyright protection. This paper proposes non-blind watermarking technique based on mean curvature of vertex. The watermark embedding process is in such a way that the vertex fraction values cannot change much more to distort the vertices from its original coordinates. The experimental results show that the proposed algorithm is imperceptible in respect to visual masking on 3D watermarked models. Watermarked mesh models can resist smoothing, simplification, subdivision, noise and cropping attacks. For better result with respect to visibility and robustness of 3D watermarked object, the vertices must be in higher density.

References

1. Garg, H., Arora, G., Bhatia, K.: Watermarking for 3-D polygon mesh using mean curvature feature. In: 2014 International Conference on Computational Intelligence and Communication Networks (CICN), pp. 903–908. IEEE (2014)
2. Zhan, Y.Z., Li, Y.T., Wang, X.Y., Qian, Y.: A blind watermarking algorithm for 3D mesh models based on vertex curvature. J. Zhejiang Univ. Sci. C **15**(5), 351–362 (2014)
3. Medimegh, N., Belaid, S., Werghi, N.: A survey of the 3D triangular mesh watermarking techniques. Int. J. Multimed. **1**(1) (2015)

4. Dong, L., Fang, Y., Lin, W., Seah, H.S.: Perceptual quality assessment for 3D triangle mesh based on curvature. IEEE Trans. Multimed. **17**(12), 2174–2184 (2015)
5. Mokhtarian, F., Khalili, N., Yuen, P.: Curvature computation on free-form 3-D meshes at multiple scales. Comput. Vis. Image Underst. **83**(2), 118–139 (2001)
6. Ohbuchi, R., Masuda, H., Aono, M.: Watermaking three-dimensional polygonal models. In Proceedings of the Fifth ACM International Conference on Multimedia, pp. 261–272. ACM (1997)
7. Ohbuchi, R., Masuda, H., Aono, M.: Watermarking three-dimensional polygonal models through geometric and topological modifications. IEEE J. Sel. Areas Commun. **16**(4), 551–560 (1998)
8. Ohbuchi, R., Mukaiyama, A., Takahashi, S.: Watermarking a 3D shape model defined as a point set. In: 2004 International Conference on Cyberworlds, pp. 392–399. IEEE (2004)
9. Jabra, S.B., Zagrouba, E.: A new approach of 3D watermarking based on image segmentation. In: IEEE Symposium on Computers and Communications, ISCC 2008, pp. 994–999. IEEE (2008)
10. Benedens, O.: Geometry-based watermarking of 3D models. IEEE Comput. Graph. Appl. **1**, 46–55 (1999)
11. Yu, Z., Ip, H.H.S., Kwok, L.F.: A robust watermarking scheme for 3D triangular mesh models. Pattern Recogn. **36**(11), 2603–2614 (2003)

Design Space Exploration for Architectural Synthesis—A Survey

R. Shathanaa and N. Ramasubramanian

Abstract Design Space Exploration (DSE) is the process of exploring design alternatives and it is iterative in nature. It includes a vast set of design choices and relies largely on the decision of the architect. The number of design constructs for a particular design is huge and it exponentially increases the problem complexity. To speed up the exploration process, many algorithms that find the optimal designs automatically have been proposed. This work presents a survey on the different techniques proposed to solve the Design Space Exploration problem by reducing the design space-time in order to provide good insight to researchers for further exploration.

Keywords Design Space Exploration · Pareto-optimal · High-level design Design metrics

1 Introduction

System designers develop circuit models using high-level languages such as C or C++ since they offer higher levels of abstraction. This makes verification of the model and its functionality easy. This also makes reusability of the code possible. For developing the corresponding hardware, hardware designers must analyze the high-level code and select a suitable hardware for the given code. An important challenge in converting this high-level design to equivalent hardware design is the many design possibilities that need to be considered. The solution spaces for the design problem involve multiple design metrics like area, execution time, resource usage, power, cost, and multiple design parameters like the number and type of processing cores, memories, interconnection network, scheduling, and arbitration policies [1]. The relation between the design metrics of interest is very difficult to establish, due to various aspects such as application parallelism, dynamic application behavior, and

R. Shathanaa (✉) · N. Ramasubramanian
National Institute of Technology, Tiruchirapalli 620015, India
e-mail: shathanaaraj@gmail.com

P. K. Sa et al. (eds.), *Recent Findings in Intelligent Computing Techniques*, Advances in Intelligent Systems and Computing 708,
https://doi.org/10.1007/978-981-10-8636-6_55

resource sharing. A single modeling approach or analysis tool that can cope with all the challenges of modern hardware design is yet to be designed. The architecture designed should be optimized to achieve best trade-offs in the selected metrics of interest.

Design Space Exploration involves optimizing the design by selecting components that minimize or maximize the metrics of interest as needed. This optimization problem has multiple, often conflicting objectives that need to be achieved during designing. For example, the optimization problem may need to minimize power consumption under an execution time constraint or vice versa. The optimal designs are called Pareto-optimal designs. Pareto front consists of all the design solutions that exist at the efficient frontier. Exploring the Pareto set is easier compared to exploring the entire design space which may take large amount of time. This exploration problem is exponential in nature. A simple method of exploration is the brute force approach. Brute force approach will be able to find all the Pareto-optimal solutions but it will take long running time even for small designs.

The remainder of this paper is organized as follows: Sect. 2 describes the various techniques that have been proposed for Design Space Exploration. Section 3 describes the common evaluation metrics used for Design Space Exploration. The comparison of different methods for different parameters is presented in Sects. 4, and 5 concludes the paper.

2 Design Space Exploration Techniques

Different techniques have been proposed for solving the Design Space Exploration problem. The recent advancements in solving this problem are discussed below.

Learning-based methods Machine learning offers a solution to identify the optimal designs by sampling only a portion of interest from the entire design space. Learning-based methods guide the exploration process by predicting the quality of the solution before the actual synthesis. Meta-heuristics is a technique that tries to reduce the design space. These heuristics search among alternate design solutions and permits locally non-optimal solutions so that ultimately global or better sub-optimal solutions can be found. Mahapatra and Schafer [2] used the Simulated Annealing meta-heuristic based on decision tree learning. By taking into consideration only the synthesis directives given by the system designers that result in achieving optimal value of one of the parameters given in the design constraint, design space reduction is achieved. The drawback of such meta-heuristics is that they take long time for execution. The quality of results is directly proportional to the number of design alternatives considered for exploration, and hence, it is very important to consider correct parameters. This consideration was taken in to account and the algorithm learned the combinations of attributes which minimize the total cost and thus reduced the search space.

Rival Penalized Competitive Learning (RPCL) was proposed to address the problem of clustering analysis. Liu and Schafer [3] used RPCL to classify which designs need to be synthesized to find the true Pareto-optimal designs. A score sheet which predicts whether a given design can lead to smaller designs than the smallest design found is maintained. Solutions that do not lead to smaller designs are pruned. Since pruning techniques work by restricting the design space, it may lead to entire exploration regions being missed. Most of the existing heuristics from machine learning frameworks try to model how the design objective functions respond to the tunable parameters in the designs. But the focus of such learning model is obtaining good designs. To address this issue [4] focused on losing good designs due to learning inaccuracy. This is done by actively eliminating non-Pareto designs. Semi-Supervised Learning (SSL) is a methodology for exploiting unlabeled data to improve the prediction accuracy. Chen et al. [5] proposed Co-Training Model Tree (COMT) which is a semi-supervised learning model, where two models label unlabeled data for each other. To achieve good accuracy, this method requires large pool of training labels. Also, the accuracy decreases with number of iterations since promising unlabeled examples which are valid for enhancing prediction accuracy eventually exhaust.

The above learning models had focused on predicting the performance of any given design configuration, whereas [6] considered two design alternatives and predicted which among the two would perform best. Here, Design Space Exploration is considered as a problem requiring a partial or total order of items, according to an item quality metric and a ranking technique was proposed for ordering the designs. But this model is limited to only pairwise predictions. Liu and Carloni [7] introduced Transductive Experimental Design which consists of training and refinement stages. This iterative-refinement framework relies on the accuracy of the approximate learning model created out of initial directives given in high-level design. However, this method could find only an approximate Pareto set. Also, it does not consider power and data pipelining. Sinaei and Fatemi [8] used high-level simple analytical models for application mapping and evaluation to prune the design space followed by Simulated Annealing to perform Design Space Exploration of the pruned design space.

Though the above works use learning approaches and present good results, using machine learning design heuristics is risky. This is because some actual optimal Pareto designs may be incorrectly excluded from the predicted output. As said earlier, the goal of existing machine learning frameworks is to model how the design objective functions respond to the tunable parameters in the designs. It does not move toward the eventual goal of seeking the good designs.

Exploration types Exploring the design choices can be done sequentially [9], iteratively [10], or hierarchically [11]. Exploring every design choice one-by-one sequentially will take huge amount of time. To handle this, [9] classified the choices into subgroups and then explored each group sequentially. This way the design space could be reduced considerably. The reduction in design space is done by employing probabilistic methods. An iterative, recursive, and multi-platform method is proposed [10], to discover the proper set of high-level synthesis directives for efficient

Design Space Exploration. This work focused on area reduction as the design constraint. Similar to [9], separate exploration of design choices was done by [11] but by using a hierarchical method. Additionally, it could detect whether any changes had been made in the behavioral description between two consecutive runs of the exploration and only explored the functions that have changed since the previous run. Incremental exploration is supported since the results of the previous exploration are reused. Cyclic Redundancy Check (CRC) had been used to look for changes in the source code for incremental exploration and exhaustive search had been used for exploring every function. But, in the case of larger functions this can lead to long execution time. The main advantage of this method is that when the behavioral description is modified in a subsequent run, only the functions that have changed since the last run need to be explored, thus reducing the running time.

Evolutionary algorithm Optimization problems can be solved using an evolutionary algorithm which is a generic population-based meta-heuristic optimization algorithm. An evolutionary algorithm mimics biological evolution, such as reproduction, gene mutation, recombination, and selection. Possible solutions are considered as individuals in a population, and a fitness function is defined that determines the quality of the solution. Evolution of the population then takes place after the repeated application of the above operators [20]. A genetic algorithm which is a class of evolutionary algorithm was used by [12] and [13] for DSE. Genetic algorithms use crossover and mutation to search the space of possible solutions. Sarkar et al. [12] addressed the issue of exploring an optimal scheduling for control data flow graphs driven through guided selection of loop unrolling factor and datapath resource configuration during high-level synthesis. Integration of multi-objective optimization models was done using Non-dominated Sorting Genetic Algorithm in [13] to explore the design space using a set of rules. For this purpose, a number of good designs are grouped into candidate populations according to the optimization criteria. But a main challenge to be addressed while using genetic operations such as mutation to create new candidate solutions is checking the feasibility of the new solutions.

Population-based stochastic optimization method Swarm intelligence (SI) is the collective behavior of decentralized, self-organized systems, natural, or artificial [21]. Design Space Exploration had been formulated as an optimization problem and explored using swarm intelligence algorithms like particle swarm optimization (PSO) [14, 16, 17, 19], bacterial foraging optimization (BFO) [15, 18], and ant colony optimization [9]. Sengupta and Mishra [14] used integrated particle swarm optimization for power-execution time trade-off during architectural synthesis of data-intensive applications. While using swarm-based algorithms, particle velocity should be maintained to avoid over drift. Also, particle positions should be within boundary. These conditions are met using clamping algorithm to manage excessive velocity outburst during searching and another algorithm to restrict boundary constraints violation in [14, 16, 17]. Adaptive mechanisms such as resource fixing and step size fixing to make sure that the design solutions meet with the boundary conditions are used in [15]. A temperature-dependent bacterial foraging optimization algorithm-based exploration process with power and execution time as design metrics during high-level synthesis which models the behavior of bacteria was proposed in [15]. In [17], PSO had been used for fully automated parallel exploration of optimal datapath and unrolling factor. For this, area and performance were considered as conflicting objectives. It tried to find the execution delay of control data flow graph (CDFG) after unrolling the loops, based on resource constraint, without unrolling the entire graph. Sengupta and Bhadauria [18] proposed exploration algorithm driven by bacterial foraging optimization algorithm. Here change in the direction of exploration is made easy using tumble or swim actions of the bacteria if the current exploration has been found ineffective. PSO was used for the generation of an optimal fault secured datapath structure by considering power and execution time as conflicting objectives for high-level synthesis in [19]. Mishra and Sengupta [16] used PSO for DSE in a similar manner but for application-specific processor design. Carrion Schafer [9] also used PSO for exploration of pruned design space. Sengupta and Bhadauria [18] and Sengupta and Bhadauria [19] does not provide any methods to control velocity and boundary conditions.

3 Performance Metrics

Some common parameters used to measure the performance of the Design Space Exploration algorithm are given below:

ADRS (Average Distance from Reference Set) Given a reference Pareto set $\Pi = \{\pi_1, \pi_2, \ldots \mid \pi_i = (a, t), a \in A, t \in T\}$ and an approximate Pareto set $\Lambda = \{\lambda_1, \lambda_2, \ldots \mid \lambda_i = (a, t), a \in A, t \in T\}$

$$ADRS(\Pi, \Lambda) = \frac{1}{(|\Pi|)} \sum_{\pi \in \Pi} \min_{\lambda \in \Lambda} \delta(\pi, \lambda)$$

where

$$\delta\left(\pi = (a_\pi, t_\pi), (a_\lambda, t_\lambda)\right) = \max\left\{0, \frac{a_\lambda - a_\pi}{a_\pi}, \frac{t_\lambda - t_\pi}{t_\pi}\right\}$$

Note that the lower the value of ADRS (Π, Λ), the closer the approximate set Λ to the reference set Π.

Pareto Dominance Pareto dominance is given by the ratio between the total number of points in the Pareto set considered for evaluation, also present in the reference Pareto set. Higher value of dominance indicates better Pareto Set.

Cardinality The cardinality indicates the number of dominating designs found by each method. A high cardinality denotes a larger choice of solutions. A high value should be considered to be positive, although it needs to be interpreted carefully with the rest of the results.

IPC IPC denotes the average number of instructions executed per clock cycle.

Table 1 Comparison of algorithms using learning-based algorithms

S.no	Method	Details	Performance metrics used	Improvement achieved
1	Learning-based method	Randomized selection	ADRS	ADRS improved by 21.3%
2		Fast simulated annealer	Running time, Pareto dominance, cardinality	Running time decreased by 36%, dominance improved 12.5%, 4.7% improvement in cardinality
3		Pruning with adaptive windowing	Running time, Pareto dominance, cardinality	170% improvement in ADRS, 38% improvement in dominance, cardinality 19.5% improved
4		Non-Pareto elimination	ADRS	2.24X synthesis speedup, 129% improvement in ADRS
5		DSE using unlabeled design configurations	Performance (IPC), power (Watt)	36% increase in performance, 79% reduction in power consumption
6		Ranking approach	Reliability	31.77% better configuration, 38.13% fewer, incorrect, pairwise predictions
7		Heuristic mapping	Accuracy, uniformity, extent	Improved accuracy. Less uniformity and more extent

Table 2 Comparison of different exploration types

S.no.	Method	Details	Performance metrics used	Improvement achieved
1	Exploration method	Sequential: probabilistic multiknob acceleration	ADRS, Pareto dominance, cardinality	12.2 acceleration, average ADRS of 1.7%, dominance decrease by 10%, cardinality 3% decrease
2		Iterative: area-oriented iterative exploration	HW efficiency	4X lower flip-flops, reduced lookup tables by 32%, overhead of 38% in compute time, HW 50% more efficient
3		Hierarchical: incremental exploration	ADRS, dominance, number of runs, running time	0.43% improvement in ADRS, dominance 22.8% improved, 585% decrease in number of runs, running time decreased by 61%

Table 3 Comparison of algorithms using evolutionary algorithms

S.no	Method	Details	Performance metrics used	Improvement achieved
1	Evolutionary alg.	Exploration of loop unrolling factor and datapath	Exploration runtime, quality of result (cost)	Less runtime, 66% improvement in cost
2		Non-dominated Sorting Genetic Algorithms (NSGA)	Two-tailed wilcoxon tests	Above 90% fulfillment of constraints, reduced cost and increase in resource usage

Table 4 Comparison of algorithms using population-based methods

S.no	Method	Details	Performance metrics used	Improvement achieved
1	Population-based optimization	Integrated PSO	Quality of Results (QoR), Exploration Time (ET)	QoR and ET improved 21% & 80% resp
2		BFOA	QOR, ET	QoR improved by 27% & ET reduced by 44%
3		Multi-objective PSO	QOR, ET	QoR improved by 9% & ET reduced by 90%
4		Adaptive, exploration	QOR, ET	QoR improved by 30% & ET reduced by 92%
5		BFOA	QOR, ET	QoR improved by 9% & ET reduced by 26%
6		PSO	Cost	Final solution proposed with lower cost

Accuracy (D-metric) The distance of the resulting non-dominated set to the Pareto-optimal front. This should be minimal.

Uniformity (*Δ*-metric) The solutions should be well distributed (in most cases uniform).

Extent (∇-metric) The non-dominated solutions should cover a wide range for each objective function value.

Quality of Result Quality of Result is given by,

$$QOR = \frac{1}{2}\left(\frac{P_T}{P_{max}} + \frac{T_E}{T_{max}}\right)$$

where

P_T is power consumed by a design combination,
T_E is execution time taken by a design combination,
P_{max} is the maximum allowed power,
T_{max} is the execution time limit.

Hardware Efficiency Hardware efficiency is measured as,

$$HW_{eff} = \frac{Speed_{norm}}{Area_{norm}}$$

where $Speed_{norm}$ and $Area_{norm}$ are normalized execution speed and area occupied, respectively.

4 Comparison of Various Techniques

Tables 1, 2, 3, and 4 compare the various techniques used for Design Space Exploration. The techniques are compared based on the evaluation metrics used and the performance improvement achieved. From this comparison, it can be seen that learning-based methods achieve better performance in terms of ADRS values, whereas swarm intelligence-based methods outperform others in terms of execution speed.

5 Conclusion

This paper surveyed the different techniques used for Design Space Exploration. It was seen that none of the existing techniques achieve performance in par with complete exploration of the design space. It was found that an exploration algorithm that can achieve performance equivalent to complete exploration of design space with a practical execution time needs to be developed.

References

1. Kritikakou, A., Catthoor, F., Goutis, C.: Scalable and Near-Optimal Design Space Exploration for Embedded Systems. Springer International Publishing (2014)
2. Mahapatra, A., Schafer, B.C.: Machine-learning based simulated annealer method for high level synthesis design space exploration. In: Proceedings of the Electronic System Level Synthesis Conference (ESLsyn). IEEE (2014)

3. Liu, D., Schafer, B.C.: Efficient and reliable high-level synthesis design space explorer for FPGAs. In: 26th International Conference on Field Programmable Logic and Applications (FPL), pp. 1–8 (2016)
4. Meng, P., Althoff, A., Gautier, Q., Kastner, R.: Adaptive threshold non-pareto elimination: rethinking machine learning for system level design space exploration on FPGAs. In: Proceedings of Design, Automation & Test in Europe Conference & Exhibition (DATE), pp. 918–923 (2016)
5. Chen, T., Chen, Y., Guo, Q., Zhou, Z.H., Li, L., Xu, Z.: Effective and efficient microprocessor design space exploration using unlabeled design configurations. ACM Trans. Intell. Syst. Technol. (TIST) **5**(1), 20 (2013)
6. Chen, T., Guo, Q., Tang, K., Temam, O., Xu, Z., Zhou, Z.H., Chen, Y.: Archranker: a ranking approach to design space exploration. In: ACM/IEEE 41st International Symposium on Computer Architecture (ISCA), pp. 85–96, June 2014
7. Liu, H.Y., Carloni, L.P.: On learning-based methods for design-space exploration with high-level synthesis. In: Proceedings of the 50th Annual Design Automation Conference, pp. 1–7, May 2013
8. Sinaei, S., Fatemi, O.: Novel heuristic mapping algorithms for design space exploration of multiprocessor embedded architectures. In: 24th Euromicro International Conference on Parallel, Distributed, and Network-Based Processing (PDP), pp. 801–804 (2016)
9. Carrion Schafer, B.: Probabilistic multiknob high-level synthesis design space exploration acceleration. IEEE Trans. Comput. Aided Des. Integr. Circuits Syst. **35**(3), 394–406 (2016)
10. da Silva, J.S., Bampi, S.: Area-oriented iterative method for design space exploration with high-level synthesis. In: 2015 IEEE 6th Latin American Symposium on Circuits & Systems (LASCAS), pp. 1–4 (2015)
11. Schafer, B.C.: Hierarchical high-level synthesis design space exploration with incremental exploration support. IEEE Embedded Syst. Lett. **7**(2), 51–54 (2015)
12. Sarkar, P., Sengupta, A., Naskar, M.K.: GA driven integrated exploration of loop unrolling factor and datapath for optimal scheduling of CDFGs during high level synthesis. In: 2015 IEEE 28th Canadian Conference on Electrical and Computer Engineering (CCECE), pp. 75–80 (2015)
13. Abdeen, H., Varró, D., Sahraoui, H., Nagy, A.S., Debreceni, C., Hegedüs, Á., Horvth, Á.: Multi-objective optimization in rule-based design space exploration. In: Proceedings of the 29th ACM/IEEE International Conference on Automated Software Engineering, pp. 289–300 (2014)
14. Sengupta, A., Mishra, V.K.: Integrated particle swarm optimization (i-PSO): an adaptive design space exploration framework for power-performance tradeoff in architectural synthesis. In: Fifteenth International Symposium on Quality Electronic Design, pp. 60–67 (2014)
15. Sengupta, A., Bhadauria, S.: Automated exploration of datapath in high level synthesis using temperature dependent bacterial foraging optimization algorithm. In: IEEE 27th Canadian Conference on Electrical and Computer Engineering (CCECE), pp. 1–5 (2014)
16. Mishra, V.K., Sengupta, A.: MO-PSE: adaptive multi-objective particle swarm optimization based design space exploration in architectural synthesis for application specific processor design. Adv. Eng. Softw. **67**, 111–124 (2014)
17. Sengupta, A., Mishra, V.K.: Swarm intelligence driven simultaneous adaptive exploration of datapath and loop unrolling factor during area-performance tradeoff. In: IEEE Computer Society Annual Symposium on VLSI, pp. 106–111 (2014)
18. Sengupta, A., Bhadauria, S.: Exploration of multi-objective tradeoff during high level synthesis using bacterial chemotaxis and dispersal. Elsevier J. Procedia Comput. Sci. **35**, 63–72
19. Sengupta, A., Bhadauria, S.: User power-delay budget driven PSO based design space exploration of optimal k-cycle transient fault secured datapath during high level synthesis. In: Sixteenth International Symposium on Quality Electronic Design, pp. 289–292 (2015)
20. Banzhaf, W., Nordin, P., Keller, R., Francone, F.: Genetic Programming—An Introduction. Morgan Kaufmann, San Francisco (1998)
21. Beni, G., Wang, J.: Swarm intelligence in cellular robotic systems, proceed. In: NATO Advanced Workshop on Robots and Biological Systems, Tuscany, Italy, June 2630 (1989)

A Novel Distributed Algorithm with Bitcoin Incentives for Phylogeny Analysis

B. J. Bipin Nair

Abstract Phylogenetic tree reconstruction plays a key role in the field of bioinformatics for the understanding of evolutionary history of species and species traits. However, computation of such trees, especially for enormous data sets, is a challenging task. Maximum Parsimony method generates accurate results, although it is highly resource intensive. In this paper, we present a novel algorithm that improves over the original Maximum Parsimony method. Following a bottom-up approach of the divide-and-conquer methodology, the algorithm divides the input data set into sub-data sets. Sub-trees corresponding to the sub-data sets are generated and analysed in parallel via multiple threads of execution running over multiple cores of multiple processors. Finally, they are merged to generate the optimal evolution tree. Cluster cell formation is encouraged with a novel idea of utilizing bitcoins to reward participating users. We show that our method has reduces time complexity and analyse the impact of bitcoins on user participation willingness. The proposed algorithm is programming language independent.

Keywords Phylogenetic tree · Bioinformatics · Maximum parsimony
Bitcoin · Parallel algorithm · Distributed algorithm

1 Introduction

Evolution of new organisms is driven by three major factors: diversity—different individuals carry different variants of the same basic blueprint, mutation—DNA sequence changes due to deletion/insertion of DNA segments and selection bias—survival of the fittest. A phylogenetic tree is used to depict hypothetical evolutionary relationships among different organism species, or among related nucleic acid or

B. J. Bipin Nair (✉)
Department of Computer Sciences, Amrita School of Arts and Sciences,
Amrita Vishwa Vidyapeetham, Mysuru, India
e-mail: bipin.bj.nair@gmail.com

P. K. Sa et al. (eds.), *Recent Findings in Intelligent Computing Techniques*,
Advances in Intelligent Systems and Computing 708,
https://doi.org/10.1007/978-981-10-8636-6_56

amino acid sequences. It is widely used in biological domains such as multiple sequence alignment (MSA), protein function prediction and pharmacology.

The reconstruction of phylogenetic trees (Hill [1]) is one of the most critical aspects of evolutionary study. Modern phylogenetic tree analysis uses molecular information like DNA and protein sequences. As multitudes of sequences are being obtained each passing day, inferring evolutionary history and construction of phylogenetic trees have become major problems in computational biology. The major challenge of phylogenetic tree reconstruction (Brocchieri [2]) is the simplification of the branching process. In a tree, branching indicates evolutionary relationships and the branch length indicates the degree of relatedness between species and sequences corresponding to nodes at the endpoints of branches. The internal nodes are the hypothetical ancestors of a particular group of Operational Taxonomic Units (OTUs). There are two broad approaches for computing phylogenetic trees based on molecular sequencing data: distance-based and character-based.

Distance-based methods rely entirely on the computation of a matrix of pairwise distances between aligned sequences to construct a tree. Neighbour-Joining (NJ) method (Saitou and Nei [3]), Fitch–Margoliash (FM) method (Fitch and Margoliash 1967), FastME (Desper and Gascuel 2002) and many other distance methods fall into this group. The NJ algorithm follows a greedy strategy in an attempt to construct an optimum phylogenetic tree by minimizing the sum of all branch lengths. Although distance-based approaches provide fast and reasonable approximations of phylogenetic trees, unsatisfactory performance on large data sets proves to be an Achilles' heel. Moreover, the loss of evolutionary information when a sequence alignment is converted to pairwise alignment (Steel 1988) and the incapability to use information about local high-similarity and high-variation regions that occur across multiple sub-trees are further drawbacks.

Character-based methods scrutinize each column of a sequence alignment individually and determine the best tree from a large number of possible trees to represent the given data. Methods based on this approach include Maximum Parsimony (MP) method [4] and Maximum Likelihood (ML) method (Felsentein 1981). MP method follows the Occam's razor principle in predicting the evolutionary tree that minimizes the number of iterations required to generate the observed mutations in nucleic acid or amino acid sequences from their common ancestral sequences. PTree (Gregor [5]) is a stochastic-based MP approach for phylogenetic tree reconstruction. ML method uses a probabilistic model of rates of evolutionary change in sequences to generate an evolutionary tree model and then adjusts the model until a best fit of the data set is obtained. Character-based algorithms take into account every minor detail and change in taxa sequences as a result of columnwise hypothesis testing. Hence, these algorithms are considered "information rich" and lead to comparatively accurate trees (Steel and Penny, 2000). However, the MP method is an NP-hard problem. As the number of input sequences in the data set increases, the total number of possible trees that have to be generated and analysed increases exponentially. Thus, MP method is computationally very expensive. ML has unknown complexity (Steel 1994), thus making it an impractical implementation.

This paper presents a novel parallel version of the MP algorithm for constructing phylogenetic trees on systems based on the × 86 hardware architecture. Optimal phylogenetic tree reconstruction using MP approach for large data sets of sequences is a challenging task, especially for those involving thousands of taxa. The objective of the parallel algorithm is to utilize the computational capability available in modern computers when clustered together in a distributed architecture. Cluster computing is widely recognized for the benefits it presents for performance applications, and is extremely useful in domains like bioinformatics. The wide-scale clustering proposed is made viable with bitcoin incentives for participating systems. Bitcoin is a digital currency that is produced and rewarded to people for lending their personal computer processing power for solving the algorithmic and mathematical problems.

The rest of this paper is organized as follows. Section 2 elucidates the problem statement. Section 3 describes the tools and method employed in the multi-threaded algorithm and its execution. Section 4 presents the results obtained on the input test data set in a descriptive fashion. Section 5 concludes the material presented in the paper and outlines future work that is feasible.

2 Problem Statement

MP method results in a quite accurate phylogenetic tree representation, however, which is unviable due to its heavy computational cost and lack of scalability. The parallel approach to MP implementation discussed in this paper aims at distributing the workload among a cluster of computers for much faster construction of an accurate phylogenetic tree, all the while retaining the advantages inherent in the MP method. The proposed method has the following features:

- A qualitative method for the division and a quantitative method for the merging of data sets,
- Calculation of sub-trees over multiple computational threads,
- Bitcoins are the incentive that drives the creation of large widespread computational clusters,
- Reduction in algorithmic computational time and order of growth.

3 Tools and Methodology

The application development framework used to implement the algorithm is the NET Framework 3.5 by Microsoft. The algorithm discussed here has been implemented in Visual Studio 2012 using the C# programming language.

However, the algorithm can be implemented in any high-level programming language of choice. The tool used to generate the visualization of the phylogenetic tree is Graphviz 2.38. C# is a good choice for selection of programming language as most computers in the world today run the Windows operating system. Furthermore, it provides strong multi-threading support. In Unix-/Linux-based systems, the Mono framework can be used to support the running of C# applications.

In a phylogenetic tree, leaves represent taxa (genes, individuals, strains and species) being compared. Internal nodes are hypothetical ancestral units that have no physical counterpart. In a rooted tree, the path from the root represents an evolutionary path (root represents the common ancestor). An unrooted tree specifies relationships among taxa, but not an evolutionary path. The proposed algorithm generates unrooted trees to depict relationships among taxa only and not an evolutionary path describing a chain of occurrences. The MP method evaluates all possible tree topologies for the minimal, hence optimal tree. However, the number of unrooted trees to be evaluated drastically increases with the number of OTUs as shown in Table 1 below.

In a phylogenetic tree, leaves represent taxa (genes, individuals, strains, species) being compared. Internal nodes are hypothetical ancestral units that have no physical counterpart. In a rooted tree, the path from the root represents an evolutionary path (root represents the common ancestor). An unrooted tree specifies relationships among taxa, but not an evolutionary path. The proposed algorithm generates unrooted trees to depict relationships among taxa only and not an evolutionary path describing a chain of occurrences. The MP method evaluates all possible tree topologies for the minimal, hence optimal tree. However, the number of unrooted trees to be evaluated drastically increases with the number of OTUs as shown in Table 1 below.

Table 1 Possible trees for MP method analysis

Number of OTUs	Number of unrooted trees
2	1
3	1
4	2
5	15
6	105
7	954
9	135,135
10	34,459,425
15	2.13E15

The number of unrooted trees (Nu) that can be constructed from n OTUs is given by

$$\mathbf{Nu} = (\mathbf{2n} - \mathbf{5})!/(\mathbf{2exp}(\mathbf{n} - \mathbf{3}))\,(\mathbf{n} - \mathbf{3})! \quad (1)$$

The proposed algorithm implements a heuristic divide-and-conquer technique to reduce the enormous number of computations that would otherwise be necessary. The input data set is split into sub-data sets, each containing of about 4-5 taxa that are closest together, as determined by a qualitative analysis measure beforehand. This bottom-up approach leads to the computation of optimal MP sub-trees based on the sub-data sets. The process is performed over multiple cores of a processor, and multiple processors over a cluster. Sub-tree computation is designed to run as individual threads of execution. Thus, the analysis of large data sets in a reasonable time limit is achievable. Finally, the global tree is reconstructed on the sub-data sets using recursive function calls to merge the sub-trees to obtain an optimal tree on the full data set. The recursive function calls are monitored by a control processor designated for the same.

The implementation of the proposed algorithm is given below in the form of pseudocode. The algorithm was implemented in C# to get the output.

1. Initialize algorithm -
   Set D= {D0, …, Dn-1} (dataset of molecular sequences), n= number of sequences, T= starting tree, nproc= number of available processors, ctrlproc= control process
2. Split dataset into sub-datasets of 4-5 taxa using qualitative measures such as distance or similarity measures.
3. Populate each sub-dataset into common shared memory among the clustered systems

```
 String s;
 Create new StreamReader While ( s=reader.ReadLine() )
 Create new string array of length 2 Initialise values array to s.Split
Initialise k to 0
 Foreach string item in values If item is not equal to ""
         Set k of finalvalues to item EndIf EndForeach
        dataGridView1.Rows.Add with finalvalues Set flag to 1
      EndWhile
```

4. Create threads corresponding to each sub-dataset and assign threads to available, idle cores and processors in the cluster (nproc).
   Using System.Threading

5. Each thread carries out the following subtree computation:

```
Initialise taxona, taxonb, taxonc, taxond For i=0 to taxona.length
If (position i in a = position i in b) & (position i in c = position i in d) & (position i in a != position i in c) Changeb++
Else if (position i in a = position i in c) & (position i in b = position i in d) & (position i in a != position i in b) Changec++
Else if(position i in a = position i in d) & (position i in c = position i in b) & (position i in a != position i in c)
        Changed++ EndIf EndFor
If (Math.Max(changed, Math.Max(changeb, changec)) == changeb)
    string text = "graph { " + taxon1 + " -- I1; " + taxon2 + " -- I1; I1 -- I2; "
    + taxon3 + " -- I2; " + taxon4 +
    " -- I2; I1 [shape=Msquare]; I2 [shape=Msquare]; }"
Else if (Math.Max(changed, Math.Max(changeb, changec)) == changec)
    string text = "graph { " + taxon1 + " -- I1; " + taxon3 + " -- I1; I1 -- I2; " +
    taxon2 + " -- I2; " + taxon4 +
    " -- I2; I1 [shape=Msquare]; I2 [shape=Msquare]; }"
Else if (Math.Max(changed, Math.Max(changeb, changec)) == changed)
        string text = "graph { " + taxon1 + " -- I1; " + taxon4 + " -- I1; I1 --
        I2; " + taxon3 + " -- I2; " + taxon2 +" -- I2; I1 [shape=Msquare]; I2
        [shape=Msquare]; }" EndIf
    Else if (Math.Max(changed, Math.Max(changeb, changec)) == changed)
        string text = "graph { " + taxon1 + " -- I1; " + taxon4 + " -- I1; I1 --
        I2; " + taxon3 + " -- I2; " + taxon2 +
        " -- I2; I1 [shape=Msquare]; I2 [shape=Msquare]; }EndIf
    File.WriteAllText("graph.gv", text) Process.Start("graph.gv") Return
    subtree to ctrlproc
```

6. An external bitcoin agent keeps track of threads computed by individual users in a cluster and awards them bitcoins for their respective contribution to the computation process.
7. Each thread returns the subtree to the control process ctrlproc which merges the subtrees by calling a recursive function

    Merge(subtree,T)
8. End

4 Results

The input data set used to verify the algorithm consists of complete and small sub-unit rDNA gene sequences from Trypanosoma cruzi (SansonAll data set by Sanson GF 2002), a strain of protozoa. The data set consists of 37 different variants of aligned sequences. The resultant tree was tested for topological accuracy with respect to a guide tree provided with the data set. The computations were done on a computer with 3.1GHz Intel i5 processor and 8 GB of main memory. The execution threads utilized the four physical cores available in the processor to obtain the results (Figs. 1 and 2).

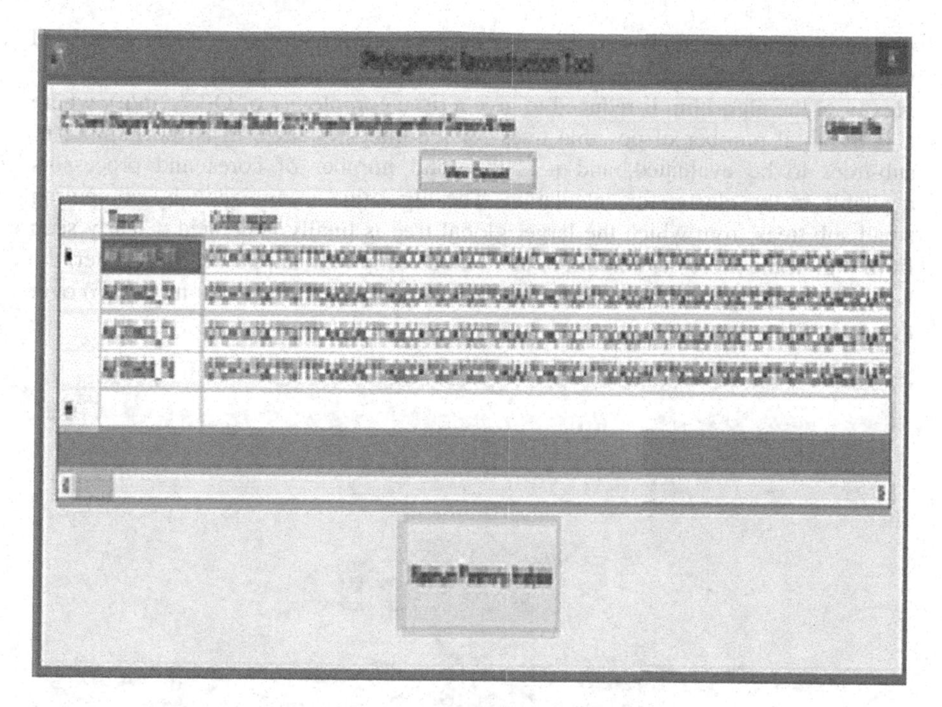

Fig. 1 Split data set running on single thread of execution

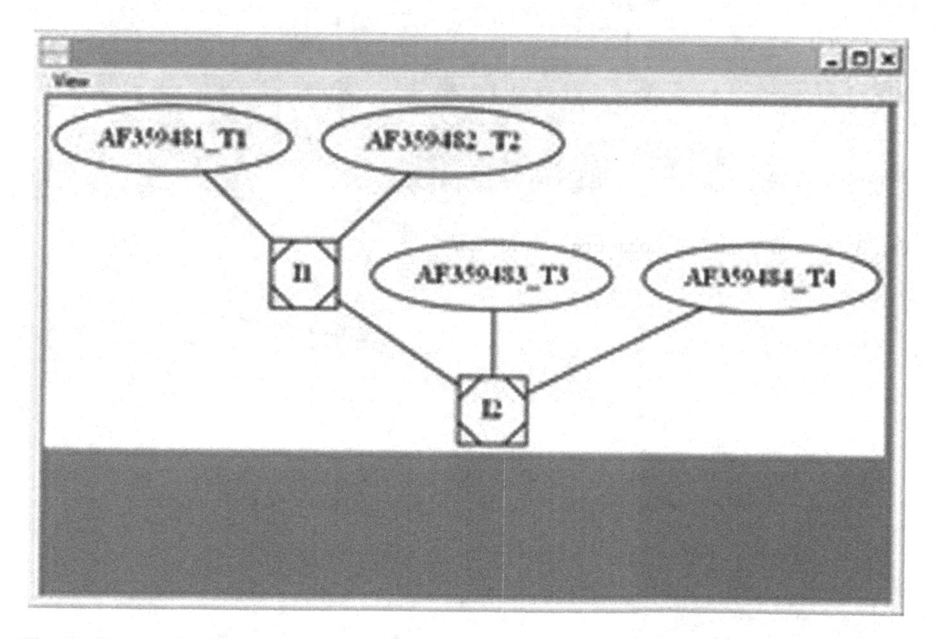

Fig. 2 Generated sub-tree based on MP calculation

The above obtained result in Fig. 3 is in correlation with the guide tree provided with the data set. Thus, the algorithm produces accurate results. The cost complexity of the algorithm is reduced to give a time complexity of O(k*t*m/n), where k is the total number of possible trees for t number of taxa, m is the number of sub-trees to be evaluated, and n is the total number of cores and processors available in the cluster for calculation. The algorithm minimizes t by computing small sub-trees from which the larger global tree is finally generated. Clearly seen from Fig. 4, the more the amount of combined computational power in a cluster, the better the running time of the algorithm. Space complexity is O(k*t*m + t*m) over the entire cluster unit.

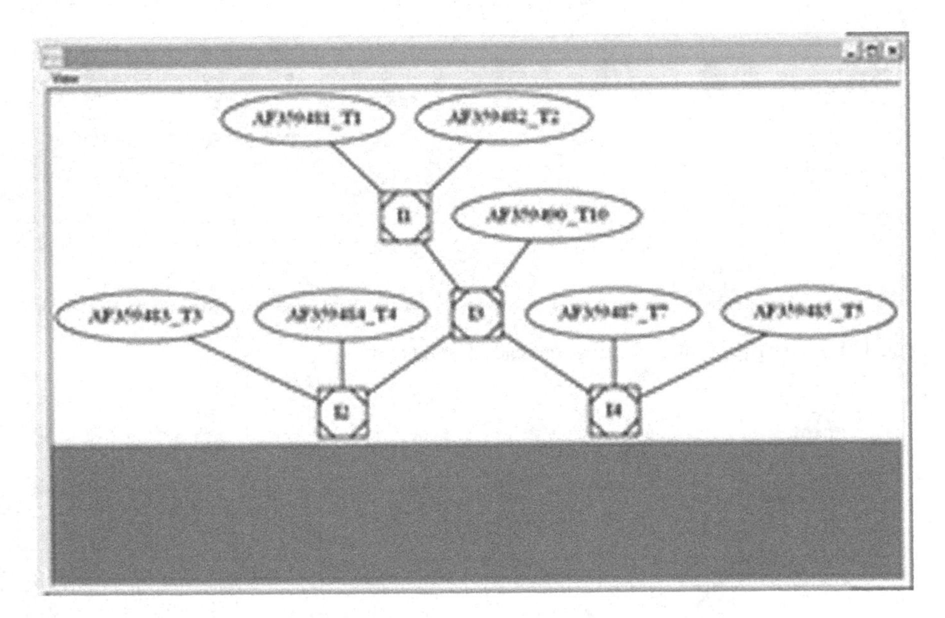

Fig. 3 Recursive optimal global tree construction

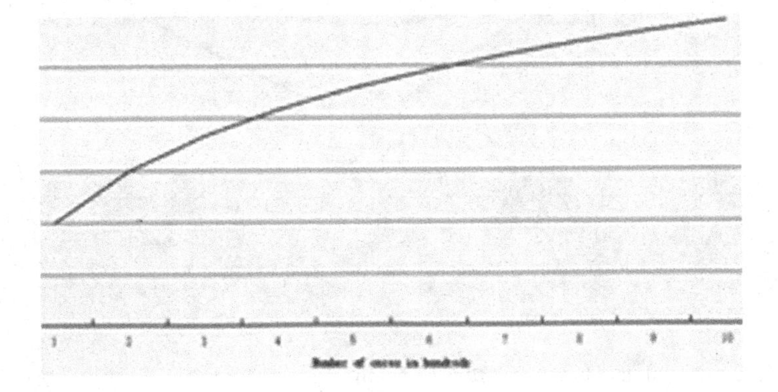

Fig. 4 Algorithm time complexity w.r.t number of cores

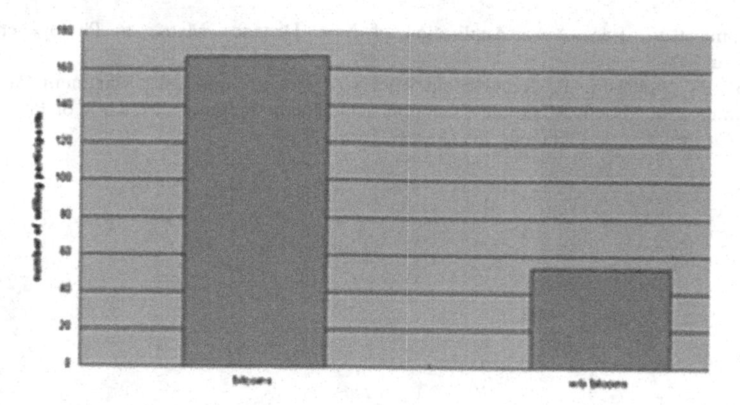

Fig. 5 Poll survey for cluster participation willingness

The viability of cluster computing over a large scale is made possible by an incentive mechanism of bitcoin rewards for joining the computational cluster. The increase seen in the number of willing cluster participants is 150%. Thus, bitcoins are an effective way of setting up highly distributed cluster systems (Fig. 5).

5 Conclusion

An efficient multi-threaded algorithmic approach for the construction of phylogenetic trees is presented here. Its supremacy over the original MP method lies in its ability to utilize the computational capability of cluster computing. Based on the results, the algorithm provides accurate results that a bioinformatician can use. Further work can focus on methods to reduce the number of sub-trees generated by using heuristic methods to remove sub-optimal trees from analysis.

References

1. Hill, T1., Lundgren, A., Fredriksson, R., Schiöth, H.B.: Genetic Algorithm for Large-Scale Maximum Parsimony Phylogenetic Analysis of Proteins
2. Brocchieri, L.: Phenotypic and Evolutionary Distances in Phylogenetic Tree Reconstruction (2013)
3. Saitou, N., Nei, M.: The Neighbor-Joining Method: a New Method for Reconstructing Phylogenetic Trees (1987)
4. Camin, J. H., Sokal, R. R.:A method for deducing branching sequences in phylogeny. Evolution, (1965)
5. Gregor, I., Steinbrück, L., McHardy, A.C.: PTree: Pattern-Based, Stochastic Search for Maximum Parsimony Phylogenies
6. Misawa, K., Tajima, F.: New Weighting Methods for Phylogenetic Tree Reconstruction using Multiple Loci (2012)

7. Lakshmi, P.V., Rao, A.A.: Application of New Distance Matrix to Phylogenetic Tree Construction
8. Saitou, N., Imanishi, T.: Relative efficiencies of Fitch-Margoliash, Maximum Parsimony, Maximum-Likelihood, Minimum-Evolution, and Neighbor-Joining Methods of Phylogenetic Tree Construction in Obtaining the Correct Tree (1989)

An Approach for Predicting Structure and Ligand Interaction as Well as Comparing Five Species of Homo Family Blood Protein Sequences Using MSA Technique

B. J. Bipin Nair, S. Saikrishna and Arun P. Prabhan

Abstract We introduce a model for protein structuring as well as multiple sequence alignment (MSA) technique. In the field of bioinformatics, we are dealing with protein 3^0 structure modelling to identify the sequence variation in five types of homo family blood protein. Our tool is to predict the 3^0 structure of blood protein in 3D structure representation. In our work, we select blood proteins which contain three components: albumin, globulin, fibrinogen and three sets of sequence of each component for comparative analysis using MSA technique. In comparison, we find the match, mismatch and gap from that result and calculate the efficiency. Here, we are using a sequence alignment algorithm to compare five different homo species blood proteins. In this work, we combine sequence alignment algorithm and 3D structure prediction which interacts with the ligand.

Keywords Blood protein · MSA · Homo family · 3D structure

1 Introduction

Homo is the genus that includes Homo sapiens, Homo habilis, Homo-neanderthal enosis, Homo-Australopithecus and Homo erectus. The Hominidae whose members are known as great apes or hominids are taxonomic family of primates that include seven extant species. Protein structure prediction is to predict the 3^0 structure of

B. J. Bipin Nair (✉) · S. Saikrishna · A. P. Prabhan
Department of Computer Science, Amrita School of Arts and Sciences,
Mysuru Campus, Amrita Vishwa Vidyapeetham, Mysuru, India
e-mail: bipin.bj.nair@gmail.com

S. Saikrishna
e-mail: saisureshkrishna@gmail.com

A. P. Prabhan
e-mail: arunperuvelil007@gmail.com

P. K. Sa et al. (eds.), *Recent Findings in Intelligent Computing Techniques*,
Advances in Intelligent Systems and Computing 708,
https://doi.org/10.1007/978-981-10-8636-6_57

blood protein. Homology modelling is a method to build 3D model for protein's multiple sequence alignment algorithms which is used in our work to compare the number of homo protein sequences. MSA is sequence alignment of the more number of organisms. With the help of our 3D predicting tool, we are predicting the 3D structure from the Fasta File format which is uploaded and the result is generated. We are selecting data set as blood protein, and it contains three components: albumin, globulin and fibrinogen. We are selecting three sets of blood sequence from the various homo family of each component for comparative analysis. In comparison, we find the match–mismatch and gap and calculate the optimal alignment. To find the interaction between the protein and a ligand, molecular docking studies are performed. A drug molecule that can interact, bind and control the function of biological receptors helps to cure blood-related disease. The ligand is an interacting chemical molecule which interacts with blood proteins and with ligand molecules. Five blood protein sequences of fibrinogen are selected from the homo family and compared these sequences to find out the optimality in various species of a homo family using MSA. The tertiary structure of the protein is a folded structure of amino acids. Protein tertiary structure changes to the quaternary structure when changes occur in the amino acid sequence because the folded structure will become unfolded. The existing works predict the 3D structure of the protein from its amino acid sequence, but there is no availability of sequence comparison. So we come to a conclusion that to develop an approach we needed to compare the protein sequence and predict the 3D structure. Using the system, we can predict the disease caused by the organism as well as the structural change in 3D structure and analyses the protein–ligand interaction.

2 Flow Diagram

See Fig. 1.

2.1 Algorithm

The algorithm we used for the MSA is below.

AlinA—the first input sequence. AlinB—second input sequence.

5 text boxes used in order to take five homo species protein sequences.

Algorithm for accepting the sequence and comparing multiple sequences.

for i = 0 **to len**(A) P(i, 0) ← k * i
for j = 0 **to len**(B) P(0, j) ← k * j
for i = 1 **to len**(A) **for** j = 1 **to len**(B) {Mat ← P(i − 1, j − 1) + S(A_i, B_j)
Del ← P(i − 1, j) + k Ins ← P(i, j − 1) + k P(i, j) ← **max**(Mat, Ins, Del)}

AlinA ← "" AlinB ← "" i ← **len**(A) j ← **len**(B)
while (i > 0 | j > 0){**if** (i > 0 **&** j > 0 **&** P(i, j) == P(i − 1, j − 1) + O(A_i, B_j))
{AlinA ← A_i + AlinA AlinB ← B_j + AlinB i ← i − 1 j ← j − 1}
else if (i > 0 **&** P(i, j) == P(i − 1, j) + k) {AlinA ← Ai + AlinA AlinB ← "−" + AlinB
i ← i − 1} **else** {AlinA ← "−" + AlintA AlintB ← Bj + AlinB j ← j − 1}}

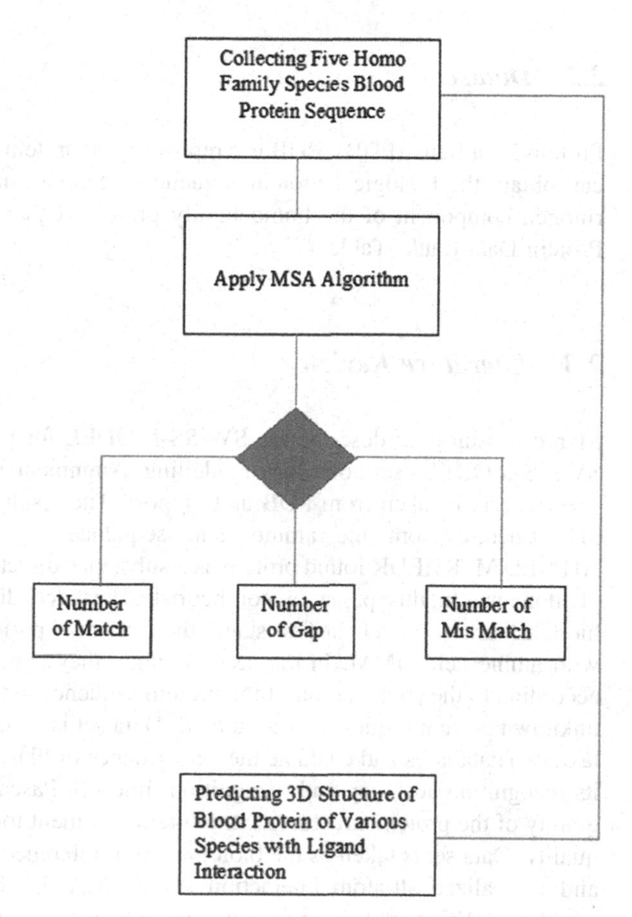

Fig. 1 Flow diagram of the system, which describes the system functioning and the modules of the system and its work flow

Table 1 Blood protein sequence data of various homo family species

Species	Sequence
Human being	ILTIPQSLDSWWMENITSGFLGP
Homo erectus	LVLQAGFFLLTRSLDSWLVLQA
Gorilla	MENITSGFLGPLIPQSLDSWLVL
Bonnet macaque	MENITCTFLGPILTIPQSLDSWL
Rhesus monkey	MENITSGFLGMRHIAHTQRCLS

2.2 Dataset

Protein Data Bank (PDB): PDB is a repository of protein sequences, from where we can obtain the biological protein sequences of organisms. The blood protein fibrinogen component of the homo family protein sequences are obtained from the Protein Data Bank (Table 1).

2.3 Literature Review

Marco-Biasini et al describe the SWISS-MODEL for protein structure modelling. SWISS-MODEL uses homology modelling. Alignment algorithm used is TMScore. The data set is taken from PDB and uniport. The result of the paper is the protein 3D structure from the amino acid sequence [1]. Whereas researchers by AHMED M. KHEDR found protein as a substance directly connected the biological phenomena. In this paper, a non-heuristic fast decoding algorithm is based on hidden Markov model. In first stage, the unknown protein sequences are matched with amino acid HMM. In the second stage, they concatenate the previous array according to the composition of the protein sequence in the database and the overall unknown protein sequence is computed. Data set is collected from PDB. They will take two databases and evaluate the performance of FDA and Viterbi algorithm, and its recognition accuracy and recognition time [2]. Pascal Benkert et al described a quality of the protein structures. Structure assessment tools are used to measure the quality. Data set is taken as the biological unit reference set. It compares traditional and normalized all-atom interaction [3]. D.W.A. Buchan et al. developed UCL which is a Web portal using various methods, which generate the structural quality of proteins from the primary structure. PSIPRED, a secondary structure prediction method, is used in UCL. Results are in the form of protein 3D visualization [4]. Georgina Mirceva et al. studied about a large number of protein structure molecules that are determined based on FSSP technology. FSSP, Viterbi algorithms, and classification algorithm are used in this project. The data set for the project is SCOP 1.73 database. Alex Bateman et al developed Pfame which contains a collection of protein families and domains. Tracking language modelling techniques from speech recognition and domain comparison tool is used. The result is in the form of structure image [5]. Robert c. Edgar helps to predict the multiple sequences and to predict the genetic analysis. [6]. Bino John's and Andrej Sali's research revealed that protein structure modelling minimizes the errors in the alignment of a modelled sequence. Iterative alignment algorithm is used [7]. Helen M Berman et al used PDB as a repository for biological macromolecules and protein sequences. JCE algorithm is used for the pairwise structure. Results are the crystal structure of different protein families [8]. The HMM structure is using protein modelling similar to that of a weight matrix. Viterbi algorithms are used for modelling. Data set is taken from FSSP. Results are high degree quality of target structure alignment. In the research work of Kevin Karplusy et al [9]. Bipin Nair B J in this research identified rare genetic mutation in human DNA sequence by using string matching algorithm [10]. Bipin Nair B J et al's research revealed that method for training the system which inputs relevant protein sequences to find out the locations of splice junctions in DNA sequences.

3 Methodology

To predict the 3^0 structure of blood protein: Protein structure homology modelling is used to generate 3D model for protein. To view it in the 3D structure, we used to gather the codon table information for protein. The quaternary structure or clustered structure of a protein of many individual protein chains into a final structure. Protein tertiary structure changes to the quaternary structure when the changes occur to the amino acid sequence and because of that folded structure will become unfolded. In this work, we propose the model for protein structure modelling using MSA technique. We are selecting blood protein, and it contains three components: albumin, globulin and fibrinogen. We are selecting these three set of sequence, and we do comparative analysis. In comparison, we find the match–mismatch from that result and calculate the most efficient match using the most matching percentage, and we are modelling the tertiary and quaternary structures as well as the protein–ligand interaction.

4 Experimental Result

The model accepts the blood protein sequence of human beings and predicts the tertiary and quaternary structures of blood protein. And comparing the normal protein sequence and disease caused a sequence of a person. We implemented modified version of Needleman–Wunsch algorithm to find out the sequence alignment. To view it in a 3D structure, we use the codon table information for

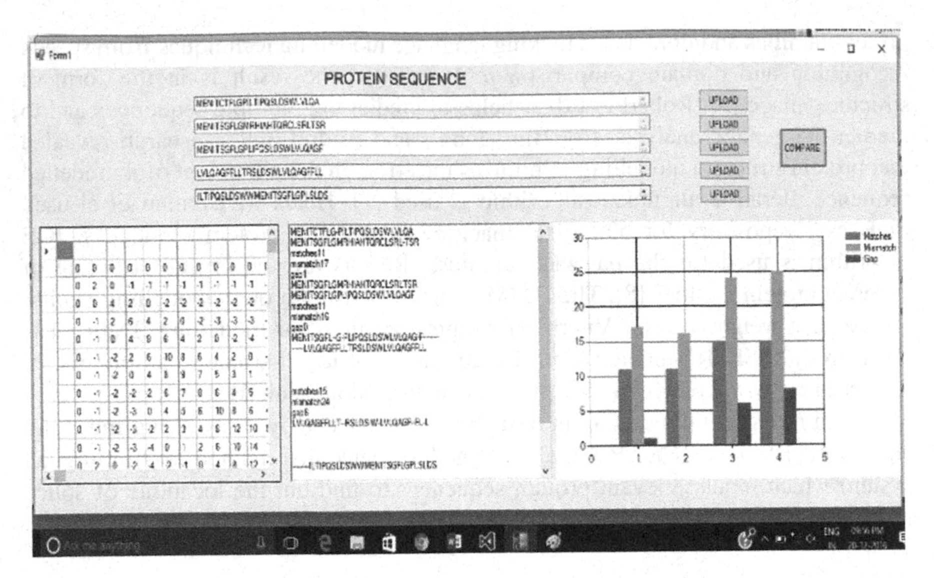

Fig. 2 Predicting the multiple sequence alignment of five different blood proteins of homo family

protein. In codon table, it consists of amino acids like alanine, valine, thymine, cytosine.

Using our proposed methodology, it will be helpful for a doctor without doing general experiments, and it will help the doctor to predict the structural change in the protein (Fig. 2).

Steps in getting the MSA alignment:

Here, we had selected five different blood protein fibrinogen components of Homo family, in which with the help of customized Needleman–Wunsch algorithm. In this work, the prediction and comparison of the match, mismatch, and gap of the different sequences and display the matrix representation. And the graphical representation is displayed.

Multiple sequence alignment algorithms are used in our paper. Comparing the multiple sequences of protein and from which predicting the protein 3D structure. Protein is a collection of amino acid sequences, and even it can be viewed in primary structure, secondary structure and tertiary structure. To view it in the 3D structure, we used to gather the codon table information of protein. In codon table, it consists of alanine, valine, thymine and cytosine. But the way of 3D representation is difficult. Even the existing works provide less accuracy to view the 3D structure with its each codon way of representation (Figure 3).

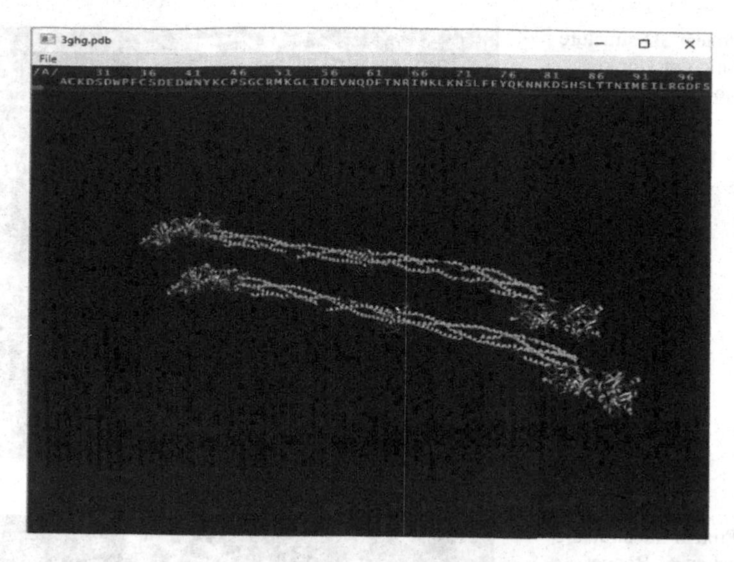

Fig. 3 Protein 3D structure representation of human being blood fibrinogen structure

With the help of our 3D predicting tool, we are predicting the 3D structure, the Fasta format is uploaded, and we generate the result (Figs. 4, 5 and 6).

All the structures are predicted with the help of our tool that we have developed, and the Fasta is selected in .pdb format.

We are selecting data set as blood protein, it contains three components: albumin, globulin, fibrinogen and selecting 3 set of sequence of each component for comparative analysis. In comparison, we are finding the match-mismatch and gap and calculate the optimal alignment. Using the optimal alignment, we are modelling the tertiary structure.

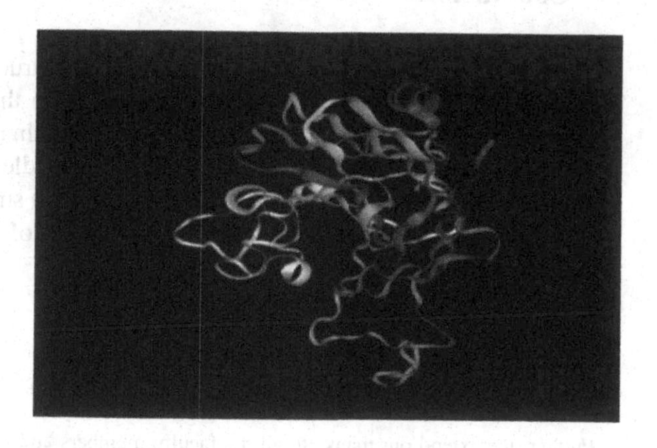

Fig. 4 Protein 3D structure representation of rhesus monkey blood protein

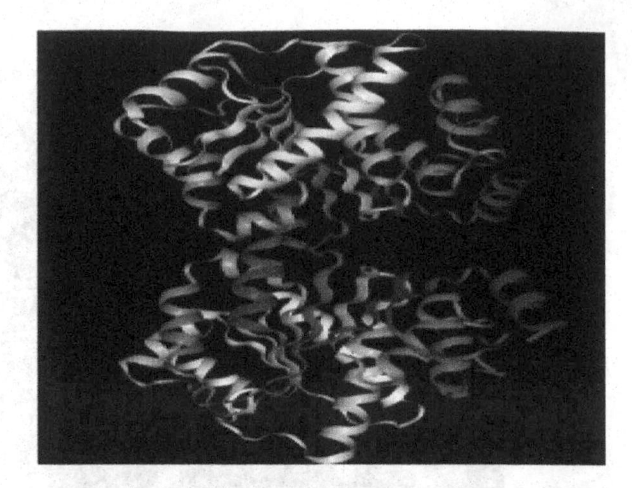

Fig. 5 Protein 3D structure representation of Homo erectus blood protein

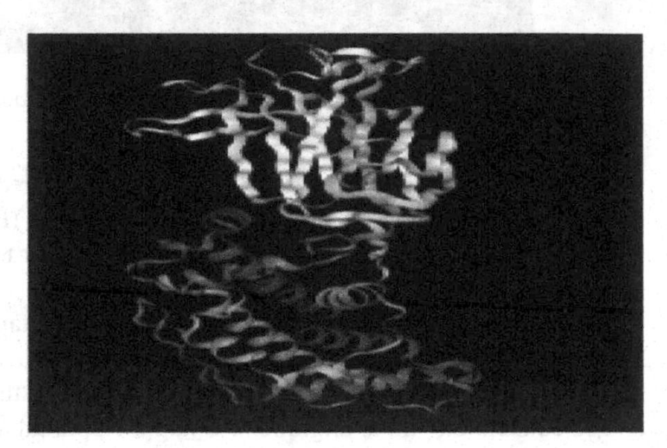

Fig. 6 Protein 3D structure representation of Bonnet macaque blood protein

5 Conclusion

In bioinformatics, we are dealing with protein 3^0 structure modelling to identify various types of diseases among human beings from the blood sample. Here, we summarize the way of predicting the structure of protein and various blood proteins. In Phase 1, we implemented modified version of Needleman–Wunsch algorithm to find out the sequence alignment. To view it in the 3D structure, we used the codon table information of protein. Codon table consists of amino acids like alanine, valine, thymine, cytosine.

Acknowledgements We would like to express our heartfelt gratitude to our guide Mr. Bipin Nair BJ, Faculty Associate, Department of Computer Science, Amrita Vishwa Vidyapeetham, Mysuru campus, for his valuable suggestions and excellent guidance rendered throughout this research.

Further, we extend our thanks to all the faculty members and technical staff of our department and ICTS for their suggestions, support and providing resources at needed times.

Finally, we express our sincere gratitude to our parents and friends who have been the embodiment of love and affection which helped us to carry out the research in a smooth and successful way.

Appendix: A Short Explanation of Research-Relevant Methods

Multiple sequence alignment (MSA) algorithms are used in our work to compare the number of homo protein sequences. MSA is sequence alignment of the more number of organisms to align the sequence.

Homo family-Homo is the genus that includes Homo sapiens, Homo habilis, Homo-neanderthal enosis, Homo-Australopithecus and Homo erectus. The Hominidae whose members are known as great apes or hominids are taxonomic family of primates that include seven extant species.

References

1. Biasini, M., Bienert, S., Waterhouse, A., Arnold, K., Studer, G., Schmidt, T., Schwede, T.: "SWISS-MODEL: modelling protein tertiary and quaternary structure using evolutionary information." Nucleic acids Res. gku340 (2014)
2. Khedr, A.M.: Improvement of recognition speed protein tertiary structure prediction using hidden Markov model. Kuwait J. Sci. Eng **38**(2A), 147–161 (2011)
3. Benkert, P., Biasini, M., Schwede, T.: Toward the estimation of the absolute quality of individual protein structure models. Bioinformatics **27**(3), 343350 (2011)
4. Buchan, D.W., Ward, S.M., Lobley, A.E., Nugent, T.C.O., Bryson, K., Jones, D.T.: Protein annotation and modelling servers at University College London. Nucleic. Acids Res. gkq427 (2010)
5. Edgar, R.C.: MUSCLE. Nucleic Acids Res. CA 9494 (2004)
6. Kim, D.E., Chivian, D., Baker, D.: Protein structure prediction and analysis using the Robetta server. Nucleic Acids Res. **32**(suppl 2), W526–W531 (2004)
7. John, B., Sali, A.: Comparative protein structure modeling by iterative alignment, model building and model assessment. Nucleic Acids Res. **31**(14), 3982–3992 (2003)
8. Berman, H.M., Battistuz, T., Bhat, T.N., Bluhm, W.F., Bourne, P.E., Burkhardt, K., Fagan, P.: The protein data bank. Acta Crystallogr. Sect. D: Biol. Crystallogr. **58**(6), 899–907 (2002)
9. Karplus, K., Sjölander, K., Barrett, C., Cline, M., Haussler, D., Hughey, R., Sander, C.: Predicting protein structure using hidden Markov models. Proteins Struct. Funct. Genet. **29**(s 1), 134–139 (1997)
10. Bipin Nair, B.J.: DNA sequence alignment using matching algorithm to identify the rare genetic mutation in various proteins. Int. J. Eng. Technol. (IJET). e-ISSN 0975-4024
11. Mirceva, G., Davcev, D.: HMM based approach for classifying protein structures. Int J Bio-Sci Bio-Technol **1**(1), 37–46 (2009)
12. Bipin Nair, B.J., Khamarudheen, K.S., Ranjitha, H.S.: An approach for identifying the presence of factor IX gene in DNA sequences using position vector ann. J. Theor. Appl. Inf. Technol. **87**(3), 396 (2016)

13. Athri, P., Wilson, W.D.: Molecular dynamics of water-mediated interactions of a linear benzimidazole– biphenyl diamidine with the DNA minor groove. J. Am. Chem. Soc. **131**(22), 7618–7625 (2009)
14. Shetty, A.C., Athri, P., Mondal, K., Horner, V.L., Steinberg, K.M., Patel, V., Zwick, M.E.: SeqAnt: a web service to rapidly identify and annotate DNA sequence variations. BMC Bioinform. **11**(1), 471 (2010)

Inferring Transcriptional Dynamics with Time-Dependent Reaction Rates Using Stochastic Simulation

Keerthi S. Shetty and B. Annappa

Abstract Gene transcription is a stochastic process. Single-cell experiments show that transcription occurs in bursty fashion. It has been shown that stochastic switching between promoter active (ON) and inactive (OFF) states results in bursts. However, increasing evidence on promoter switching between ON and OFF states suggests that promoter states' times can be non-exponential and exhibits multi-OFF mechanism. All these experimental facts motivate to present a multi-state promoter model of gene expression for characterizing promoter architecture. In this paper, we develop a model for promoter with arbitrary number of promoter OFF states and single ON state using Erlang-distributed ON/OFF times. In this paper, we use Monte Carlo approach of Expectation–Maximization (MCEM) that uses direct method of Gillespie's Stochastic Simulation Algorithm (SSA) to infer the unknown kinetic rates, number of promoter states and parameters representing estimation errors of inferred parameters. We use graphics processing units (GPUs) to accelerate the MCEM. Application of MCEM to time-series data of endogenous mouse glutaminase promoter shows that promoter switching between ON and OFF states effects bursting. The production of mRNAs becomes consistent as the number of OFF states increases and arrives at deterministic time intervals. The main objective is to analyse the impact of promoter switching on bursty production of mRNAs. Our analysis demonstrates that bursting kinetics depends on the promoter ON/OFF states.

Keywords Parameter inference · Mass action kinetics · Time-series data
MCEM · Multi-state promoter model

K. S. Shetty (✉) · B. Annappa
Department of Computer Science and Engineering,
National Institute of Technology Karnataka, Surathkal, Mangalore 575025, India
e-mail: keert.cs@gmail.com

B. Annappa
e-mail: annappa@ieee.org

P. K. Sa et al. (eds.), *Recent Findings in Intelligent Computing Techniques*,
Advances in Intelligent Systems and Computing 708,
https://doi.org/10.1007/978-981-10-8636-6_58

1 Introduction

Gene expression is a stochastic and dynamic process. It involves two main steps such as transcription and translation. It produces mRNAs and proteins through transcription and translation steps, respectively. Transcription plays a major role in all cellular functions. The irregularity of this process results in diseases such as cancer, diabetes and neurological disorders [1]. In particular, transcriptional bursting in gene expression is not well understood.

Recently, several works have provided proof for the synthesis of mRNAs [2–9] and proteins [10, 11] in bursts. Although the origins of the transcriptional burst remain poorly understood [12], it has been shown that stochastic switching between promoter active and inactive states leads to bursts [13–21]. The random telegraph model [22–24] is most commonly used model to analyse transcriptional bursting. This model has been used as the key model for several works to infer parameters from experimental data. In general, the assumption of random telegraph model is not valid because it involves multiple kinetic steps in promoter activation [25–27]. Recent single-cell experiments show that promoter exhibits multi-OFF mechanism and time spent in OFF state is non-exponential. All these experimental facts combined with above analysis motivate us to introduce a multi-state promoter model of gene expression.

In this paper, we model promoter with arbitrary numbers of OFF states and single ON state with Erlang-distributed ON/OFF times, while retaining synthesis and degradation as single-step reactions. The parameter inference of discrete-state stochastic systems is often performed by applying the principle of maximum likelihood. In this paper, we use Monte Carlo approach of EM (MCEM) method [28], which makes use of direct method of Gillespie's Stochastic Simulation Algorithm (SSA) [29] to generate exact system trajectories. For the efficient selection of trajectories and iterations, ascent-based MCEM [30] is used. We show that the mass action kinetics leads to simple procedure to infer unknown kinetic rates, states and parameters representing estimation errors. We use graphics processing unit (GPUs) to reduce computational time required by the algorithm. Applying this algorithm to endogenous mouse glutaminase promoter reveals that as the number of promoter OFF states increases, production rate of mRNA bursts is more consistent.

The remaining paper is organized as follows. Section 2 describes model formulation. Section 3 explains the algorithm to infer unknown kinetic rates and promoter states. Section 4 describes inferring parameters from experimentally observed time-series data. Finally, the conclusion has been drawn in Sect. 5.

2 Multi-state Promoter Model Formulation

2.1 Random Telegraph Model

The random telegraph model is represented by using biochemical reactions which are as follows:

$$\begin{aligned} DNA_{off} &\underset{k_{off}}{\overset{k_{on}}{\rightleftharpoons}} DNA_{on} \\ DNA_{on} &\xrightarrow{c_1} DNA_{on} + mRNA \\ mRNA &\xrightarrow{\gamma_m} \phi\,. \end{aligned} \tag{1}$$

In this model (1), promoter switches from OFF to ON and ON to OFF states with rates k_{on} and k_{off}, respectively. The mRNA production happens from the ON state with rate c_1. mRNAs live for an exponentially distributed time interval with mean lifetime $1/\gamma_m$, where γ_m is the mRNA degradation rate.

Model (2) represents transcriptional bursting of mRNAs:

$$\begin{aligned} DNA_{on} &\xrightarrow{c_1} DNA_{on} + B_m \times mRNA \\ mRNA &\xrightarrow{\gamma_m} \phi \\ B_m &\sim Geometric(c_3)\,. \end{aligned} \tag{2}$$

Basically, transcriptional bursting is represented by two parameters: c_1 and B_m denote the burst frequency and burst size, respectively. In this model formulation of (2) [31], mRNA bursts arrive at exponentially distributed time intervals with rate c_1. Each burst produces a geometrically distributed number of transcripts B_m with mean value $(1 - c_3)/c_3$ [32].

A more realistic representation of multiple, sequential OFF states in (1) is as follows:

$$\begin{aligned} DNA_{off_1} &\xrightarrow{k_1} DNA_{off_2} \\ &\vdots \\ DNA_{off_{N-1}} &\xrightarrow{k_N} DNA_{off_N} \\ DNA_{off_N} &\xrightarrow{k_{on}} DNA_{on} \\ DNA_{on} &\xrightarrow{k_{off}} DNA_{off_1}\,. \end{aligned} \tag{3}$$

Model (3) differs from model (1) in the distribution of time spent in OFF states. In contrast to (1), it is now non-exponential. It follows hypoexponential distribution (sum of exponential random variables) which approaches an Erlang distribution when switching rates are equal.

The representation of transcriptional bursting model (2) in terms of (3) requires generation of inter-burst arrival times. This is done making burst frequency time-dependent. In this time-dependent burst frequency model, we select different values for inter-burst arrival times. This is denoted as c_2 in the model.

3 Methods

3.1 *MCEM for Multi-state Promoter Model*

In this paper, we use reactions that follow mass action kinetics—i.e. where $a_j(X) = c_j h_j(X)$ with c_j a positive real kinetic constant and $h_j(X)$ a function that quantifies the number of possible ways reaction R_j can occur given system state X. If the reaction rate c_j is a constant, such a propensity is called time-homogeneous and is written as $a_j(X(t))$. The time-homogeneous propensity $a_j(X(t))$ changes only when the state changes. In case the reaction rate c_j is a time-dependent function, the propensity of a reaction R_j is denoted explicitly as $a_j(X(t), t)$ and it is called time-inhomogeneous because it depends on both state and time. In this paper, time-inhomogeneous propensity is used.

Single-cell time-series data are incomplete; it provides the number of molecules for a species at discrete time instances. The Expectation–Maximization (EM) [33] is an algorithm to calculate maximum likelihood given an incomplete data. In this paper, we use Monte Carlo approach of EM (MCEM) that uses direct method of SSA to simulate trajectories [34, 35]. We use the same technique as in [35] excluding cross-entropy method. MCEM produces better results for time-series data. To achieve convergence of MCEM algorithm, we increase the number of iterations.

3.2 *Optimizing Model Complexity*

In this paper, Akaike information criterion (AIC) [36] is used to compare complexity of different models. It gives lower value for models which best fits observed experimental data. It is given by

$$AIC = 2m - 2log(\hat{L}) , \tag{4}$$

where m denotes the number of unknown parameters from the model.

4 Results

4.1 Parameter Inference Using Time-Series Data

In this paper, we use glutaminase promoter actual time-lapse microscopy data from a reporter gene driven by a mammalian promoter (Suter et al. [20]). In performing glutaminase data inference, we used $K = 3000$ (number of trajectories) and $n = 100$ (number of iterations).

$$
\begin{aligned}
DNA_{off} &\underset{k_{off}}{\overset{c_1}{\rightleftharpoons}} DNA_{on} \\
DNA_{on} &\xrightarrow{k_m} DNA_{on} + mRNA \\
mRNA &\xrightarrow{0.924} \phi \\
mRNA &\xrightarrow{12.6} mRNA + Protein \\
Protein &\xrightarrow{1.98} \phi \,.
\end{aligned} \tag{5}
$$

$$
\begin{aligned}
DNA_{off_1} \longrightarrow \ldots \longrightarrow DNA_{off_{16}} &\underset{k_{off}}{\overset{c_1}{\rightleftharpoons}} DNA_{on} \\
DNA_{on} &\xrightarrow{k_m} DNA_{on} + mRNA \\
mRNA &\xrightarrow{0.924} \phi \\
mRNA &\xrightarrow{12.6} mRNA + Protein \\
Protein &\xrightarrow{1.98} \phi \,.
\end{aligned} \tag{6}
$$

$$
\begin{aligned}
DNA_{off_1} \longrightarrow \ldots \longrightarrow DNA_{off_{26}} &\underset{k_{off}}{\overset{c_1}{\rightleftharpoons}} DNA_{on} \\
DNA_{on} &\xrightarrow{k_m} DNA_{on} + mRNA \\
mRNA &\xrightarrow{0.924} \phi \\
mRNA &\xrightarrow{12.6} mRNA + Protein \\
Protein &\xrightarrow{1.98} \phi \,.
\end{aligned} \tag{7}
$$

The unknown kinetic parameters of the models (5), (6), (7) were c_1, c_2, k_{off}, k_m. These models include the mRNA degradation, protein translation and degradation reactions with 0.924, 12.6 and 1.98 [20], respectively. Model (5) represents one OFF

Table 1 Bursty parameter inference using glutaminase data

No. of states	Parameter inference			
	c_2	c_1	k_{off}	k_m
1	0	2.12	5.28	73.90
16	3.40	19.58	3.55	70.48
26	4.75	40.75	3.19	66.13

Table 2 Optimizing model complexity

No. of states	Model complexity	
	Loglikelihood	AIC
1	−1737.991	3483.98
16	−1728.018	3464.03
26	−1727.88	3463.76

Table 3 Parameters representing estimation errors

No. of states	Estimation error		
	c_1	k_{off}	k_m
1	0.20	0.44	0.21
16	0.25	0.24	0.13
26	0.26	0.23	0.11

state model. Models (6) and (7) represent multi-state OFF to ON transitions. These models include bursting with the correct parameterization. It assumes random burst production. We initialize the unknown parameters of the model to 1. But c_3 was initialized to 0.5. We initialize unobserved initial promoter state and the number of mRNAs to DNA_{off} and 20, respectively. Models (5), (6), (7) are time-dependent; we selected values of time instances, and it is denoted as c_2. Table 1 represents unknown kinetic rates inferred for models (5), (6), (7). Table 2 represents calculating AIC from the likelihood. Table 3 shows parameters representing estimation errors of the inferred parameters.

4.2 *Comparison with One Off State and Multi-off State Model*

26 states promoter model shows best fit to the experimentally observed time-series data giving lower AIC value. It shows very less varying OFF state dwell times compared to models (5) and (6). This 26 OFF state promoter models also show more consistent rate of mRNA production. Models (5) and (7) produce burst size of ≈14

and ≈20 mRNA molecules from the ON state, respectively. We compare one OFF state model with 26 OFF state models. These results show that 26 state models produce more consistent rate of mRNA production compared to one OFF state model. It demonstrates that multi-OFF state model effects bursting. It produces consistent rate of mRNAs, as the number of OFF states increases.

5 Conclusion

In this work, we model multi-state promoter with arbitrary number of promoter OFF states and single ON state using Erlang-distributed ON/OFF times. Based on time-series data of endogenous mouse glutaminase promoter, we have inferred unknown kinetic rates, states and parameters representing estimation errors from the model using MCEM. Our results show that 26 OFF state models best fit the experimental time-series data. It shows very less varying OFF state dwell times. It produces consistent rate of mRNAs, as the number of OFF states increases. These results demonstrate that bursting kinetics depends on the promoter switching between ON and OFF times.

References

1. Lee, T.I., Young, R.A.: Transcriptional regulation and its misregulation in disease. Cell **152**(6), 1237–1251 (2013)
2. Golding, I., Paulsson, J., Zawilski, S.M., Cox, E.C.: Real-time kinetics of gene activity in individual bacteria. Cell **123**(6), 1025–1036 (2005)
3. Chubb, J.R., Trcek, T., Shenoy, S.M., Singer, R.H.: Transcriptional pulsing of a developmental gene. Curr. Biol. **16**(10), 1018–1025 (2006)
4. Raj, A., Peskin, C.S., Tranchina, D., Vargas, D.Y., Tyagi, S.: Stochastic mRNA synthesis in mammalian cells. PLoS Biolo. **4**(10) (2006)
5. So, L.H., Ghosh, A., Zong, C., Seplveda, L.A., Segev, R., Golding, I.: General properties of transcriptional time series in Escherichia coli. Nat. Genet. **43**(6), 554–560 (2011)
6. Taniguchi, Y., Choi, P.J., Li, G.W., Chen, H., Babu, M., Hearn, J., et al.: Quantifying E. coli proteome and transcriptome with single-molecule sensitivity in single cells. Science **329**(5991), 533–538 (2010)
7. Zong, C., So, L.H., Seplveda, L.A., Skinner, S.O., Golding, I.: Lysogen stability is determined by the frequency of activity bursts from the fate-determining gene. Mol. Syst. Biol. **6**(1) (2010)
8. Ochiai, H., Sugawara, T., Sakuma, T., Yamamoto, T.: Stochastic promoter activation affects Nanog expression variability in mouse embryonic stem cells. Sci. Rep. **4** (2014)
9. Senecal, A., Munsky, B., Proux, F., Ly, N., Braye, F.E., Zimmer, C., et al.: Transcription factors modulate c-Fos transcriptional bursts. Cell Rep. **8**(1), 75–83 (2014)
10. Cai, L., Friedman, N., Xie, X.S.: Stochastic protein expression in individual cells at the single molecule level. Nature **440**(7082), 358–362 (2006)
11. Yu, J., Xiao, J., Ren, X., Lao, K., Xie, X.S.: Probing gene expression in live cells, one protein molecule at a time. Science **311**(5767), 1600–1603 (2006)
12. Chubb, J.R., Liverpool, T.B.: Bursts and pulses: insights from single cell studies into transcriptional mechanisms. Curr. Opin. Genet. Dev. **20**, 478–484 (2010)

13. Blake, W.J., KAErn, M., Collins, J.J.: Noise in eukaryotic gene expression. Nature **422**, 633–637 (2003)
14. Raser, J.M., OShea, E.K.: Control of stochasticity in eukaryotic gene expression. Science **304**, 1811–1814 (2004)
15. Boeger, H., Griesenbeck, J., Kornberg, R.D.: Nucleosome retention and the stochastic nature of promoter chromatin remodeling for transcription. Cell **133**, 716–726 (2008)
16. Larson, D.R.: What do expression dynamics tell us about the mechanism of transcription? Curr. Opin. Genet. Dev. **21**, 591–599 (2011)
17. Mao, C., Brown, C.R., Boeger, H.: Quantitative analysis of the transcription control mechanism. Mol. Syst. Biol. **6**, 431 (2010)
18. Mariani, L., Schulz, E.G., Hofer, T.: Short-term memory in gene induction reveals the regulatory principle behind stochastic IL-4 expression. Mol. Syst. Biol. **6**, 359 (2010)
19. Miller-Jensen, K., Dey, S.S., Arkin, A.P.: Varying virulence:epigenetic control of expression noise and disease processes. Trends Biotechnol. **29**, 517–525 (2011)
20. Suter, D.M., Molina, N., Gatfield, D., Schneider, K., Schibler, U., Naef, F.: Mammalian genes are transcribed with widely different bursting kinetics. Science **332**(6028), 472–474 (2011)
21. Harper, C.V., Finkenstadt, B., White, M.R.: Dynamic analysis of stochastic transcription cycles. PLoS Biol. **9** (2011)
22. Peccoud, J., Ycart, B.: Markovian modeling of gene-product synthesis. Theor. Popul. Biol. **48**(2), 222–234 (1995)
23. Shahrezaei, V., Swain, P.S.: Analytical distributions for stochastic gene expression. Proc. National Acad. Sci. **105**(45), 17256–17261 (2008)
24. Dobrzyski, M., Bruggeman, F.J.: Elongation dynamics shape bursty transcription and translation. Proc. Nat. Acad. Sci. **106**(8), 2583–2588 (2009)
25. Pedraza, J.M., Paulsson, J.: Effects of molecular memory and bursting on fluctuations in gene expression. Science **319**(5861), 339–343 (2008)
26. Jia, T., Kulkarni, R.V.: Intrinsic noise in stochastic models of gene expression with molecular memory and bursting. Phys. Rev. Lett. **106**, 058102 (2011)
27. Xu, X., Kumar, N., Krishnan, A., Kulkarni, R.V.: Stochastic modeling of dwell-time distributions during transcriptional pausing and initiation. In: IEEE 52nd Annual Conference on Decision and Control (CDC), pp. 4068–4073. IEEE (2013)
28. Wei, G., Tanner, M.: AMonte-Carlo implementation of the EM Algorithm and the poor mans data Augmentation algorithms. J. Am. Stat. Assoc. **85**(411), 699–704 (1990)
29. Gillespie, D.T.: Exact stochastic simulation of coupled chemical reactions. J. Phys. Chem. **81**(25), 2340–2361 (1977)
30. Caffo, B.S., Jank, W., Jones, G.L.: Ascent-based monte carlo expectation-maximization. J. R. Stat. Soc. Ser. B **67**(2), 235–251 (2005)
31. Gillespie, D.T.: Stochastic simulation of chemical kinetics. Annu. Rev. Phys. Chem. **58**, 35–55 (2007)
32. Evans, M., Hastings, N., Peacock, B.: Statistical distributions. 3rd edn. Wiley (2000)
33. Dempster, A.P., Laird, N.M., Rubin, D.B.: Maximum likelihood from incomplete data via the EM Algorithm. J. R. Stat. Soc. Ser. B (Methodological) **39**, 1–38 (1977)
34. Wilkinson, D.J.: Stochastic modelling for systems biology. Boca Raton: Taylor and Francis: Chapman and Hall/CRC Mathematical and Computational Biology Series (2006)
35. Daigle Jr., B.J., Roh, M.K., Petzold, L.R., Niemi, J.: Accelerated maximum likelihood parameter estimation for stochastic biochemical systems. BMC Bioinform. **13**, 68 (2012)
36. Singer, Z.S., Yong, J., Tischler, J., Hackett, J.A., Altinok, A., Surani, M.A., Cai, L., Elowitz, M.B.: Dynamic heterogeneity and DNA methylation in embryonic stem cells. Mol. Cell **55**, 319–331 (2014)

Taxonomy of Leaf Disease Detection and Classification

Manisha Goswami, Saurabh Maheshwari, Amarjeet Poonia and Dalpat Songara

Abstract Gaining foods from plant is the most difficult task when plant suffers from diseases like fungal, bacterial, and virus. Leaf is the most affected part of plant diseases. Although farmers are proficient to detect leaf disease, if crops are spread in dense area, then this task is more difficult. Image processing provides an easy technique to detect leaf disease using symptoms. In this paper, we present various method such as background removal, enhancing, segmentation, feature extraction, and classification. Some leaf disease detection, grading, and classification are also discussed.

Keywords Color model · Image segmentation · GLCM · SVM and BPNN

1 Introduction

Agriculture is called as backbone of our country. Approximately 70% population live in villages and depend on agriculture field. When disease affects the plant, there is a dilemma that how can we detect. Plant disease affects both production quality and quantity of crop, and then, growth rate will be decreased. Image processing provides an easy, fast, and optimistic solution to detect disease in plant by extracting feature of affected area of the leaf energy, homogeneity, entropy, color, and shape, etc. Grading can also apply using disease affected area calculation, so farmer can efficiently take its solution.

M. Goswami (✉) · S. Maheshwari · A. Poonia · D. Songara
Govt Women Engineering College, Ajmer, India
e-mail: manishagswmi@gmail.com

S. Maheshwari
e-mail: dr.masurabh@gmail.com

A. Poonia
e-mail: amar.gweca@gmail.com

D. Songara
e-mail: dalpatsongara@gmail.com

P. K. Sa et al. (eds.), *Recent Findings in Intelligent Computing Techniques*,
Advances in Intelligent Systems and Computing 708,
https://doi.org/10.1007/978-981-10-8636-6_59

Leaf disease detection using image processing is an increasing research area because it reduces our working time which farmer spent to detect disease. To detect leaf disease, first, we take a leaf image with some background and then apply some preprocessing steps to extract leaf and after then segment the leaf spot and extract its feature and use some classification technique to detect disease and use grading so farmer as early as apply disease solution. Steps of leaf disease detection are as follows:

- **Image Acquisition and Processing**: The test images are collected in database which are used for training and testing; then, apply image enhancement, leaf extraction, segmentation, and feature extraction (color, texture, shape, and morphological feature).
- **Grading**: We can grade leaf by using % of affected area calculation. It can divide into classes A, B, C, D, E, and risk also [1].
- **Classification**: To classify disease we can use SVM, artificial neural network, probabilistic neural network, back propagation neural network, fuzzy and k-nearest neighbor classifiers.

2 Literature Review

Plant disease symptoms can be visualized in different parts of plant which are stem, fruit, and leaves. These are reviews of methods to detect leaf disease using image processing (Fig. 1).

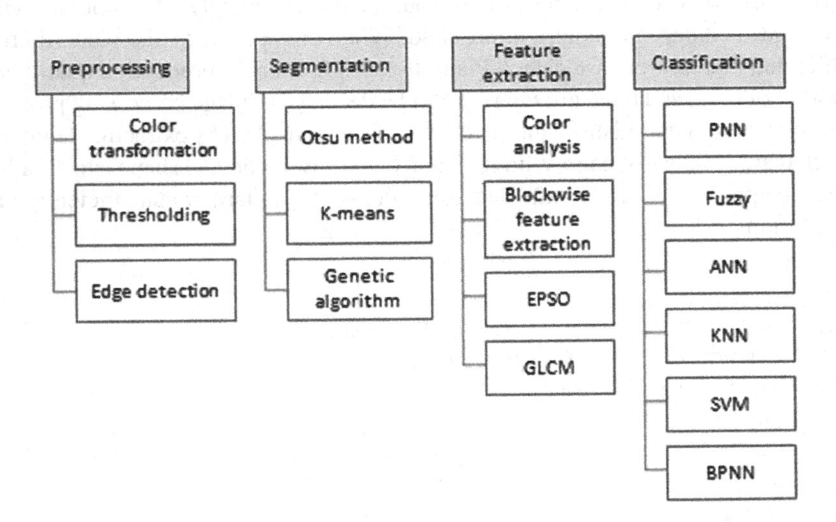

Fig. 1 Taxonomy of leaf disease detection steps

Table 1 Preprocessing techniques

S.No.	Method	Strength	Weakness
1.	Lab color transformation	It provides leaf details without vein	It is device independent not represent original color
2.	Threshold	Provide significant information and hide unwanted information	Only two classes are generated and are not used for multichannel images that cause sensitive to noise
3.	Edge detection	Provide sharp edges and low amount of noise	May not provide closed connected curve so edge linking is required

2.1 Preprocessing

- **Color Transformation**: The work done by Chaudhary et al. [2] and Athanikar and Badar [3] aims to improve color quality and remove vein information. Convert RGB to LAB color model.
- **Thresholding**: This method [4–7] has been implemented to enhance image and mask green pixel from image. Convert RGB image to grayscale image and apply gray level thresholding method then use median filter and morphological operation for noise removal and better image quality.
- **Edge detection**: The method proposed by Asfarian et al. [8] tries to obtain noise-free diseased paddy leaf legion. The cropped lesion convert into HSV color model then extract saturation component then use histogram equalization and Laplacian filtering. The method proposed by Revathi and Hemalatha [9] aims to resize and edge detection of leaf of cotton crop. First resize collected image in 150 * 150 then use canny with Sobel filter to detect edge (Table 1).

2.2 Segmentation

- **Otsu method**: This method [2, 10, 11] has been implemented which aims to detect diseased leaf area. Otsu is a threshold selection method which separates pixel into two classes.
- **K-means**: This method [1, 3, 6, 7, 12–14] has been implemented to detect diseased leaf spot. K-means clustering divides an image into k number of cluster; then, we can easily detect disease affected cluster and that part used for segmentation.
- **Genetic Algorithm**: The method proposed by Singh and Misra [5] and Revathi and Hemalatha [9] tries to examine diseased spot using a genetic algorithm. The

Table 2 Segmentation techniques

S.No.	Method	Strength	Weakness
1.	Otsu	Select optimal threshold. Best for small object size	More computation time. Contain only two classes foreground and background
2.	k-means	Fast, 3-D thresholding method	Difficult to identify cluster center
3.	Genetic algorithm	Large no. of data can be processed at the same time. Give a no. of optimum solution, not a single solution	High computation cost and takes much time

algorithm starts with random population of n chromosome and then calculates a fitness function for each chromosome. If new offspring is enhanced then old then replace it and find a segmented image (Table 2).

2.3 Feature Extraction

- **Color analysis**: The work done by Zambre and Patil [4] aims to extract feature of diseased leaf. Shape feature is extracted using blob analysis, and color feature is extracted by calculating Euclidean distance between two color points and after then take statistical analysis to extract best feature and remove redundancy of features.
- **Block-wise feature extraction**: The method proposed by Suman [15] and Padol and Yadav [6] tries to extract feature of diseased leaf spot using quantification. HSV color model is divided into 3 * 3 blocks and then calculate color features for each of H/S/V plane of nine blocks; then, texture features and color features are extracted for classification.
- **EPSO**: The method proposed by Revathi and Hemalatha [9] tries to extract features using enhanced particle swarm optimization method. Edge variance calculates using canny edge detector, color and texture variance also extracted. These features are sent to SVM, BPN, and fuzzy classifiers.

Table 3 Feature extraction techniques

S.No.	Method	Strength	Weakness
1.	Color analysis	Provide color pixel information	Take much time for computation
2.	Block-wise feature extraction	Boost computation efficiency and classification accuracy	Chance of detecting noisy data
3.	Enhanced particle swarm optimization	Randomly find disease matching pixel finally obtain best matching pixel	High computational complexity
4.	Gray level co-occurrence matrix	Simplicity, smaller length of feature vector, estimate second-order statistics feature	High computation cost and not enough to describe all texture feature

- **GLCM**: This method [1, 3, 11–13] has been implemented to extract texture feature from GLCM including contrast, homogeneity, correlation, and energy which have been calculated to detect type of disease and sent to classifier to automatically detect disease (Table 3).

2.4 Classification

- **PNN**: Asfarian et al. [8] proposed a method in which they try to detect and classify paddy leaf, brown spot, bacterial leaf blight, and tungro using probabilistic neural network with 83% accuracy.
- **Fuzzy**: Revathi and Hemalatha [9] proposed a method to classify leaf spot. They present a comparison of SVM, BPN, and fuzzy classification, and fuzzy classifier presents highest accuracy as 94%.
- **ANN**: The method proposed by Rastogi et al. [1] aims to detect and classify early scorch, cottony mold, tiny whiteness, and normal leaf with 94.67% accuracy.
- **KNN**: The method proposed by Eaganathan et al. [7] aims to detect and classify sugarcane leaf scorch disease using KNN classification method with 95% accuracy, where k is an integer value and finds k-nearest neighbor.
- **SVM**: The work done by Singh and Misra [5] and Padol and Yadav [6] classify disease of grape, rose, lemon, banana, and beans with 95.71 and 88.89% accuracy.
- **BPNN**: Sannakki et al. [14] proposed a method to diagnose and classify downy and powdery grape diseases with 100% accuracy (Table 4).

Table 4 Classification techniques

S.No.	Method	Strength	Weakness
1.	PNN	Tolerant of noisy input and instances classified by more than one output	Limited for small dataset if we increase dataset, it provides adverse impact on computational complexity of network
2.	Fuzzy	Simple to use. Individual treatment for each object. Easy for decision maker to express their ideas in natural language	Analysis is difficult and requires more feature
3.	ANN	It can handle large amount of dataset. Nonlinear model which provides high accuracy and noise tolerant	Require large time because learning process is slow
4.	KNN	Simple classifier works on basic pattern classification problem. Useful when small data available which is not trained	Slow classifier because it compute distance and sort training data for each predicted value if large no. of training data. Produce noisy data and if we change k value it changes predicted class label
5.	SVM	High accuracy and work well even if data is linearly separable	Speed and size more in training and testing. High complexity extensive more memory require in many cases
6.	BPNN	Applicable for large amount of data. Easy to implement and able to classify nonlinear data	Learning process is slow. Difficult to know how many neuron and layers are required

3 Conclusion and Future Work

Plant leaf diseases detection is an agricultural application. To solve a problem with better way it is understand the solution steps. This paper presents methods of leaf disease detection and classification with strength and weakness of these methods. To improve leaf disease detection accuracy, we can extract more features and then perform statistical analysis and remove repeated feature and use BPNN network.

References

1. Rastogi, A., Arora, R., Sharma, S.: Leaf disease detection and grading using computer vision technology & fuzzy logic. In: 2nd International Conference on Signal Processing and Integrated Networks (SPIN) (2015)
2. Chaudhary, P., Chaudhari, A.K., Cheeran, A.N., Godara, S.: Color transform based approach for disease spot detection on plant leaf. Int. J. Comput. Sci. Telecommun., 65–70 (2012)

3. Athanikar, G., Badar, P.: Potato leaf diseases detection and classification system. Int. J. Comput. Sci. Mob. Comput., 76–88 (2016)
4. Zambre, R.S., Patil, S.P.: Classification of cotton leaf spot disease using support vector machine. Int. J. Eng. Res. Appl., 92–97 (2014)
5. Singh, V., Misra, A.K.: Detection of plant leaf diseases using image segmentation and soft computing techniques. Inf. Process. Agric. (2016)
6. Padol, P.B., Yadav, A.A.: SVM classifier based grape leaf disease detection. In: Conference on Advances in Signal Processing (CASP), pp. 175–179 (2016)
7. Eaganathan, U., Sophia, J., Luckose, V., Benjamin, F.J.: Identification of sugarcane leaf scorch diseases using K-means clustering segmentation and K-NN based classification. Int. J. Adv. Comput. Sci. Technol. (IJACST), 11–16 (2014)
8. Asfarian, A., Herdiyeni, Y., Rauf, A., Mutaqin, K.H.: Paddy diseases identification with texture analysis using fractal descriptors based on Fourier spectrum. In: International Conference on Computer Control Informatics and Its Application, pp. 77–81 (2013)
9. Revathi, P., Hemalatha, M.: Cotton leaf spot diseases detection utilizing feature selection with skew divergence method. Int. J. Sci. Eng. Technol., 22–30 (2014)
10. Deya, A.K., Sharma, M., Meshram, M.R.: Image processing based leaf rot disease, detection of betel vine (Piper BetleL.). In: International Conference on Computational Modeling and Security, pp. 748–754 (2016)
11. Kakade, N.R., Ahire, D.D.: An implementation of grape leaf plant disease detection. IJARIIE, 527–535 (2015)
12. Rathod, A.N., Tanawala, B.A., Shah, V.H.: Leaf disease detection using image processing and neural network. Int. J. Adv. Eng. Res. Dev. (IJAERD) (2014)
13. Al-Hiary, H., Bani-Ahmad, S., Reyalat, M., Braik, M., ALRahamneh, Z.: Fast and accurate detection and classification of plant diseases. IJCA, 31–38 (2011)
14. Sannakki, S.S., Rajpurohit, V.S., Nargund, V.B., Kulkarni, P.: Diagnosis and classification of grape leaf diseases using neural networks. In: 4th ICCCNT (2013)
15. Suman, D.: Classification of paddy leaf diseases using shape and color features. IJEEE, 239–250 (2015)

Assessment of Objective Functions Under Mobility in RPL

Shridhar Sanshi and C. D. Jaidhar

Abstract Due to the technological advancement in Low-power and Lossy Network (LLN), the sensor node mobility has become a basic requirement. Routing protocol designed for LLN must ensure certain requirements in a mobile environment such as reliability, flexibility, scalability to name a few. To meet the needs of LLN, Internet Engineering Task Force (IETF) released the standard IPv6 Routing Protocol for LLNs (RPL). RPL depends on Objective Function (OF) to select optimized routes from source to destination. However, the standard did not specify which OF to use. In this study, performance analysis of different OFs such as Objective Function zero (OF0), Energy-based Objective Function (OFE), Delay-Efficient Objective Function (OFDE), and Minimum Rank with Hysteresis Objective Function (MRHOF) is carried out under different mobility models, which makes this study unique. The metrics used to measure the performance are latency, packet delivery ratio (PDR), and power consumption. Simulation results demonstrate that under different mobility models, MRHOF achieved better results in terms of PDR and power consumption, while OFDE shows better results in terms of latency compared to other OFs.

Keywords Internet of Things · Mobility models · MRHOF · Objective function OF0 · Power consumption · Routing protocol · RPL

1 Introduction

Internet of Things (IoT) technology is developed with the aim of making any object to interact with any other object in the world [1]. In this technology, Wireless Sensor Network (WSN) is an important paradigm which consists of several sensor

S. Sanshi (✉) · C. D. Jaidhar
Department of Information Technology, National Institute of Technology
Karnataka, Surathkal, Mangalore, India
e-mail: it15f03.sanshi@nitk.edu.in

C. D. Jaidhar
e-mail: jaidharcd@nitk.edu.in

P. K. Sa et al. (eds.), *Recent Findings in Intelligent Computing Techniques*,
Advances in Intelligent Systems and Computing 708,
https://doi.org/10.1007/978-981-10-8636-6_60

nodes, these sensor nodes gather surrounding environment information or contextual information based on the application, and the gathered information is sent over the Internet on behalf of the user [2]. In order to achieve pervasive or ubiquitous computing environment, WSN must be associated with an Internet Protocol (IP)-based networks [3]. However, WSN nodes are characterized by low computing capability, power constrained, reduced memory, and reduced radio coverage. Adding an IP protocol to these constrained nodes is a challenging task.

Initially, it was considered that the IP protocol is heavy to be processed by resource-constrained nodes [4]. Nevertheless, the advance in technology made one to rethink many misunderstandings about using IP protocol over resource-constrained nodes. IETF created the working group IPv6 over Low-power Wireless Personal Area Network (6LoWPAN), which later came up with a 6LoWPAN adaption layer that allows sensor nodes to have IPv6 address above the IEEE 802.15.4 medium access control layer to associate with the Internet [5]. However, 6LoWPAN required a routing protocol to deliver packets and most of the LLN configurations are more than one hop. To meet this requirement, IETF standardized RPL protocol [6] was developed by Routing Over Low-power and Lossy networks (ROLL) working group, which is capable of fulfilling the precise requirements of LLNs.

RPL is a fairly simple distance vector routing protocol. It organizes the network into a tree like structure know as DAG, the tree is rooted towards one node which is usually, DAG root using the OF. RPL is designed to choose the optimized path to the destination based on the defined OF, the OF uses routing metrics to construct DAG. The OF can use any of the routing metrics like expected transmissions (ETX), hop count, remaining energy, end-to-end delay, RSSI, local traffic, or any combination of these routing metrics. The OF influences the PDR, delay, power consumption, and other parameters. The choice of OF plays a crucial role to obtain better performance for the network scenario. Most of the researchers worked on RPL by considering a static scenario. However, due to the advancement in technology, many applications require mobility support for the sensor nodes [7], like devices carried by people (health care monitoring), mounted on moving vehicles (warehouse), or integrated with machines (industrial automation). These devices collect information and then send collected information in real time. Therefore, RPL protocol should provide guaranteed reliability to transmit data during mobility of the node.

In this work, we have considered four different OFs and are implemented in Contiki RPL to study the behavior under different mobility models. The primary goal of our simulation work is to ascertain which OF performs better in a mobile environment. The absence of studies focusing on different OFs with respect to mobile environment motivated this study. The OFs considered are MRHOF [8], OF0 [9], OFE [10], and OFDE [11]. For the mobility models, we have considered Gauss–Markov (GMM), Random Walk (RWK), and Random Waypoint (RWP) mobility models.

The rest of the paper is structured as follows: Sect. 2: describes pertinent contributions related to RPL under mobility. An overview of RPL and OFs is described in Sect. 3; Sect. 4: provides performance analysis of RPL and finally the study concludes in Sect. 5.

2 Related Work

In [12], the authors conducted a simulation work to analyze the performance of RPL routing protocol. The evaluation revealed that the RPL performs better in the network setup and bounded communication delays compared to other routing protocols. In [13], the authors investigated the mobility of the sink and developed a distributed and weighted moving strategy for the RPL protocol. To enhance leaf nodes network lifetime, the sink nodes are moved toward leaf nodes. The main idea is to select sink node and then moving the sink node toward routers which is having highest remaining energy, hop count, and neighbors count. The experiment results showed that the lifetime of the network has increased significantly. In [14], the authors analyzed and evaluated RPL under different scenarios in order to estimate the impact on attributes like energy, communication overhead, storage overhead, along with maximum hop count. Simulation results showed that when DAG root is in the middle, the hop count is reduced and likewise the convergence time. It is also noticed that the network performance improved when the number of DAGs are increased.

In [15], the authors evaluated the performance of the RPL protocol by considering the sink node as static and mobile. When the sink node is mobile, it follows different paths like, node passes near field, diagonally across the sensor field, and circulates around sensor field. The results are compared with different time frames with respect to PDR, power consumption, and latency. From the results, it is observed that the fixed sink outperforms the mobile scenarios in all the metrics and also it is revealed that some node had excessively high average power consumption while others were isolated. In [16], the authors investigated the performance of RPL in terms of two OFs, namely MRHOF and OF0 in the static scenario under the random and grid topologies. To study the impact of OF on PDR and power consumption, the packet reception ratio is varied under different topologies. The results reveal similar behavior for OFs. However, in the case of a low-density network, the MRHOF shows better results in terms of power consumption compared to the OF0. The authors have not considered the behavior of RPL in the mobile environment.

In [17], the authors analyzed the behavior of RPL by considering the default OF under different mobility models. The authors classified the mobility models into two entities and evaluated the RPL separately. The impact of different mobility models is investigated in terms of throughput, PDR, and lost ratio. From the results, it is showed that the group mobility models show better results compared to entity models and mobility models that have the straight impact on data transmission. The lack of studies focusing on OFs under mobility motivated us to do this work.

3 Overview of RPL

RPL organizes network into one or more tree-like structure known as DAG, which determines default routes between the nodes in the LLNs. Each of the DAG is rooted at a single node usually at the sink node, so the DAGs are known as

Destination-Oriented DAG. In DODAG, a node can associate with more than one parent node, in contrast to other tree-like structures. During topology construction, the router maintains a set of possible parents to become a next hop toward the DODAG root. The choice of preferred parent is done on the basis of OF, which depends on the routing metrics (e.g., delay, connectivity, energy, ETX) or constraints among other candidate nodes. ETX and hop count are the default routing metric provided by the standard, and their usage in RPL is published. The RPL is specifically designed to support the communication among LLNs, which provide communications for point-to-point, multipoint-to-point, and finally point-to-multipoint. In a multipoint-to-point, all the data from the distributed nodes are passed toward a sink node, which is the most commonly used scenario.

3.1 Objective Functions

We have considered four OFs for analyzing the behavior of RPL under the mobile scenario. Here, the OFs are briefly discussed:

OF0 is the default OF for the interoperability in the RPL. This does not use any routing metric to select the route toward sink node instead; it uses the rank of the node to select the optimized route to the sink node. The hop count is used to measure rank of a node. The hop count is defined as the distance from sensor node (sender) to sink node in terms of intermediate sensor nodes (hops). The rank of the sink node is zero and then increases with scalar value (step_of_rank) down the link toward the leaf nodes.

MRHOF is proposed by ROLL working group and is the default OF used for formation of DAG in the RPL. It is based on the additive metric along the route, by default, it uses ETX as link metric which is the expected transmissions required by the node to accomplish the task of delivering the packet to its destination. It distinguishes path that is more reliable, i.e., which require less number of packet transmissions. The value of ETX varies from one to infinity with one as 100% throughput. It is found that the throughput decrease as ETX value increases. As per [8], the value of the ETX is calculated by measuring the probability that a packet reaches the neighbor (D_f) successfully and the probability that an acknowledgment packet is successfully received (D_r).

$$ETX = \frac{1}{D_f \times D_r} \tag{1}$$

The node calculates the path metric to reach the destination through each of its neighbor as per [8]:

$$Pathcost = ETX(m) + MinPathcost(m) \tag{2}$$

where $ETX(m)$ is the ETX value for the neighbor m and $MinPathcost(m)$ is the advertised ETX value of neighbor m. The node selects the neighbor node with minimum *Pathcost* as its preferred parent. MRHOF also uses the minimum hysteresis to decrease churn in response to small variations in the routing metric.

OFE Since energy is considered as an important parameter in WSN, the authors in [10] proposed an OF considering nodes remaining energy as the routing metric in RPL. Here, PW_i represents path cost of ith node to the sink node. Path cost of the ith node is the minimum path cost between the preferred parent and its own energy. The sink node sets the value as MAX_{energy}. A node selects the neighbor that advertises the highest path cost value as a parent. As per [10], path cost is calculated as

$$PW_i = min[\max_{j\epsilon N_i}(PW_j), E_i] \tag{3}$$

where N_i is the set of neighbors toward the sink and E_i represents the energy of node i.

OFDE The authors in [11] proposed a routing metric to minimize the delay to reach sink node. The routing metric is defined as the cumulative sum of delay at every hop along the route toward sink node. The node chooses its preferred parent from its neighbors with the minimum sum of average delay advertised by a neighbor node along with delay to reach that neighbor node. As per [11], it can be expressed as

$$Average\_Delay = Average\_Delay_i + D_f \tag{4}$$

where $Average\_Delay_i$ is delay announced by ith neighbor, D_f indicates the forwarding latency between node and its ith neighbor.

Figure 1 shows the Modified RPL in Contiki Operating System. The program logic uses the objective function defined by the application to choose preferred parent among the neighbors.

The Algorithm 1 shows the generalized pseudocode to select the preferred parent among its neighbors.

Generalized algorithm for selection of Preferred Parent

```
Algorithm1 Preferred_Parent ()
  Input: NeighborTable of C
  Output: PreferredParent of C
  initialize: MAX_METRIC
  PreferredParrent = NULL
  begin
    for c in NeighborTable
      if (c.metric < MAX_METRIC) then
        MAX_METRIC = c.metric
        PreferredParent = c
      end if
    end for
    if (PreferredParent = NULL)
```

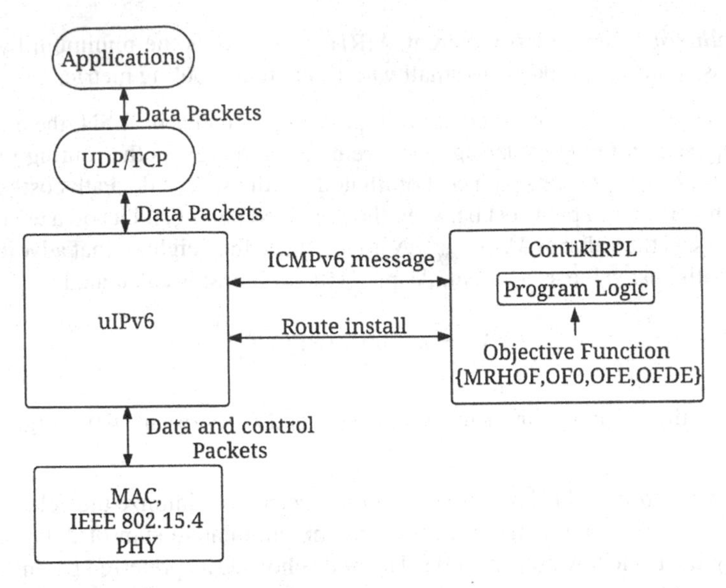

Fig. 1 Modified RPL in Contiki Operating System

```
      send(DODAG Information Solicitation message)
    else
      send(PreferredParent)
    end if
  end
```

4 Performance Evaluation

Simulation environment, parameters for the simulation, performance metrics, and results are described in this section.

4.1 Simulation Environment

We have used COOJA simulator provided by the Instant Contiki 2.6 [18] a well-known simulator for Internet of Things. COOJA simulator is also called as a cross-level emulator which operates on Contiki Operating System (OS). Contiki is the OS which is a lightweight open-source OS running on the sensor nodes. It supports standards like IPv6, 6LoWPAN, RPL to provide low-power communications.

Table 1 Simulation parameters

Simulation parameters	Value
Radio model	UDGM
Node density	30
Network size	100 m × 100 m
Transmission range	50 m
Mote type	Tmote sky
Mobility model	RWK, RWP, GMM
Simulation time	600 s

Table 2 BonnMotion software parameters

Settings	Simulation area	Minimum speed	Maximum speed	Mobile nodes
Value	100 m × 100 m	4 Km/h	6 Km/h	3, 6, 9, 12, 15

4.2 Simulation Parameters

To simulate the lossy behavior of the network, we have considered Unit Disk Graph Model (UDGM) [19] which is available in the simulator. The node transmission range is set to 50 m. RWK, RWP, and GMM mobility models are considered to analyze the performance of RPL with different OFs. The percentage of mobile nodes were varied (10, 20, 30, 40, and 50) in the simulation to investigate the behavior of RPL in terms of power consumption, latency, and PDR. A well-known BonnMotion software [20] is used in this work to generate mobility patterns. Table 1 shows the parameters used for simulation, and Table 2 presents the parameters utilized for generating mobility patterns in BonnMotion software.

4.3 Performance Metrics

The performance metrics analyzed for different OFs with respect to different mobility models are as follows:

Power consumption: To calculate power consumption, we have used power trace tool available in the COOJA simulator [21]. Power trace uses energy capsule structures to attribute energy consumption to events such as packet receptions and transmissions. It uses power state monitoring to calculate system power consumption.

PDR: It is calculated as the total number of packets that successfully reached the sink node to the sum of packets sent by each node in the network.

Latency: It is the time taken by the packets from the point of transmission to its reception at the sink node. The average latency is computed as the ratio of the sum of latency of each packet to the total number of packets that reaches successfully at the sink node.

4.4 Results and Discussion

The results are compared with varying percentage of mobile nodes in the network under different mobility models and are analyzed.

Power consumption: Fig. 2 shows the average power consumption of nodes during simulation under RWK, RWP, and GMM mobility models by varying mobile node count in the network respectively. Initially, all the sensor nodes are static, and the average energy consumption of nodes with the OFE performs better compared to the other OFs. But, an increment of mobile nodes in the network increases the

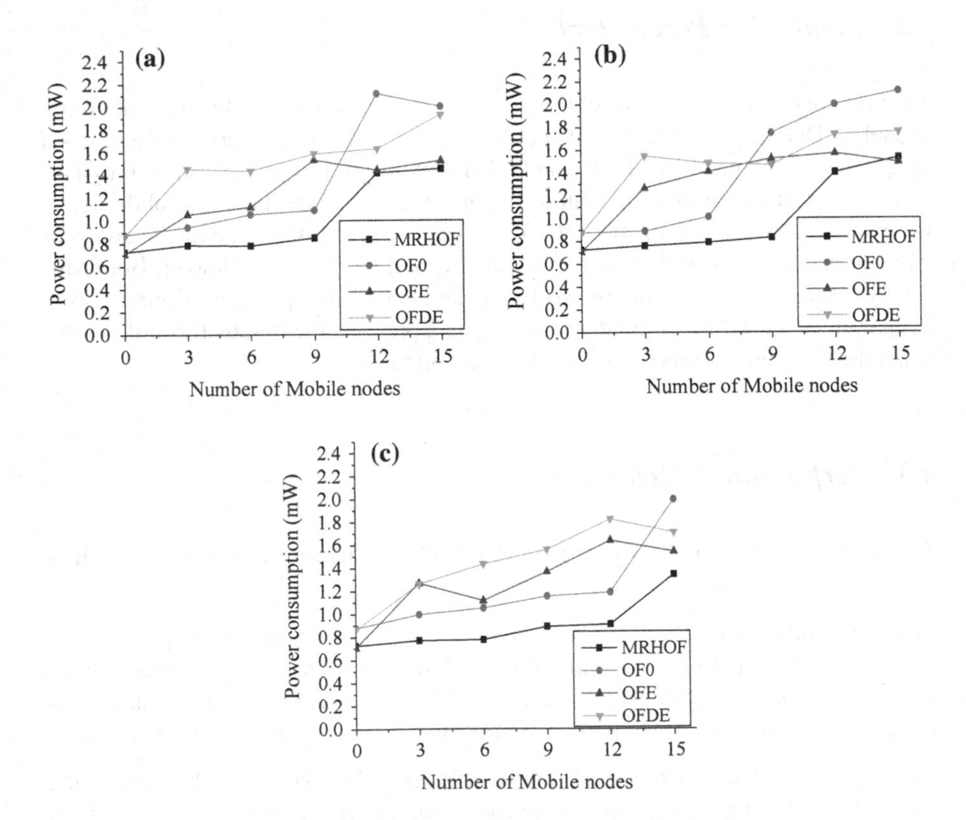

Fig. 2 **a** Power consumption under RWK **b** Power consumption under RWP **c** Power consumption under GMM

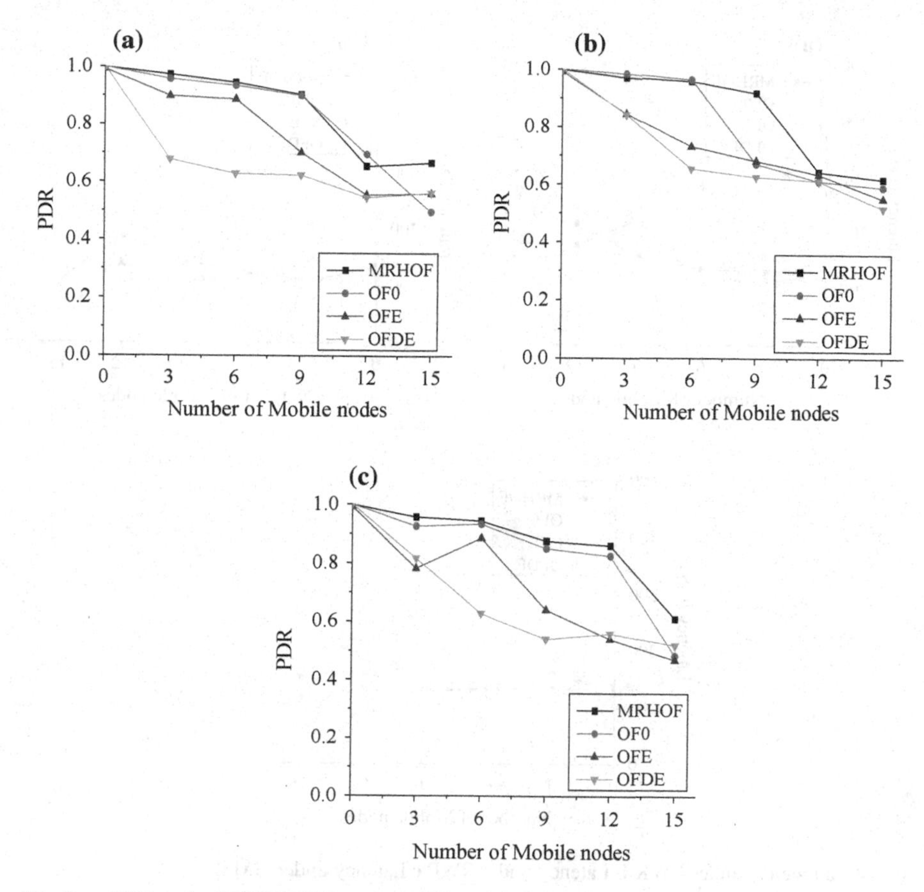

Fig. 3 **a** PDR under RWK **b** PDR under RWP **c** PDR under GMM

power consumption. The reason for power consumption is that nodes may require several retransmissions to successfully deliver packets to the sink node. For all the mobility models, MRHOF performs better in the case of mobile nodes as it requires less number of retransmission of packets compared to all other OFs.

PDR: Fig. 3 shows the PDR during the simulation under RWK, RWP, and GMM mobility models by varying mobile nodes count in the network respectively. Initially, all the sensor nodes are static, and all the OFs show similar behavior. But, an increment of mobile nodes decreases the PDR of all the OFs in all the mobility models. Again, MRHOF performs better in all the mobility models compared to the other OFs. The reason is that the other OFs do not consider the quality of links while choosing the preferred parent which leads to dropping off some packets.

Latency: Fig. 4 shows the behavior of latency during the simulation under RWK, RWP, and GMM mobility models by varying mobile nodes count in the network respectively. Initially, the OF0 performs better compared to other OFs as it is based

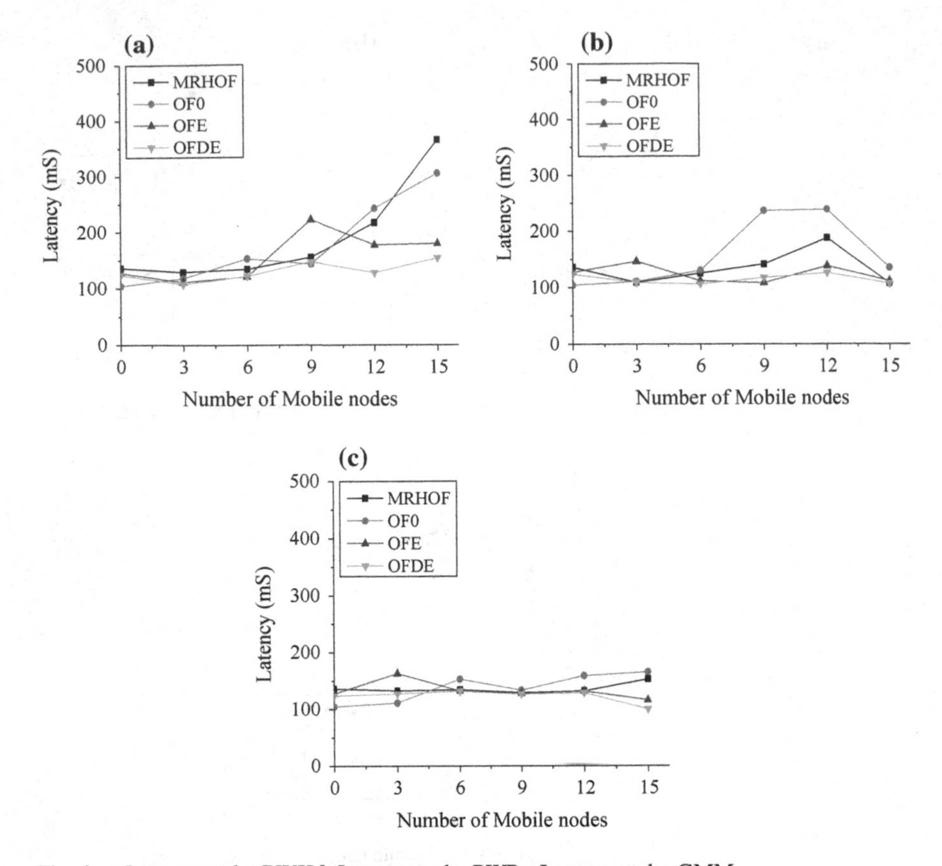

Fig. 4 **a** Latency under RWK **b** Latency under RWP **c** Latency under GMM

on the shortest path. But, as the mobile node count increases the latency also increases. This demonstrates that having shortest path does not guarantee the lower latency because the intermediate node may be congested due to the mobility of nodes. The OFDE performs better in all the mobility models as the next hop is selected based on lower latency to reach the sink node.

5 Conclusion

In this study, we have investigated the performance of RPL by considering different OFs, namely MRHOF, OF0, OFE, and OFDE under RWK, RWP, and GMM mobility models. The results demonstrated that under mobility, the OFs have a direct effect on PDR, power consumption, and latency. Furthermore, the MRHOF performs better in terms of PDR and power consumption. Whereas, OFDE achieved better results in terms of latency compared to other OFs in all the mobility models used in this work. Future study would aim to consider some more OFs with different mobility models and analyze the behavior of RPL.

References

1. Gubbi, J., Buyya, R., Marusic, S., Palaniswami, M.: Internet of Things (IoT): a vision, architectural elements, and future directions. Future Gener. Comput. Syst. **29**(7), 1645–1660 (2003)
2. Alcaraz, C., Najera, P., Lopez, J., Roman, R.: Wireless sensor networks and the Internet of Things: do we need a complete integration? In: Proceedings of the International Workshop on the Security of The Internet of Things (SecIoT) (2010)
3. Rodrigues, J.J., Neves, P.A.: A survey on IP-based wireless sensor network solutions. Int. J. Commun. Syst. **23**(8), 963–981 (2010)
4. Durvy, M., et al.: Making sensor networks IPv6 ready. In: Proceedings of the 6th ACM Conference on Embedded Network Sensor Systems, pp. 421–422. New York, ACM (2008)
5. Kushalnagar, N., Montenegro, G., Schumacher, C.: IPv6 over low-power wireless personal area networks (6LoWPANs): overview, assumptions, problem statement, and goals (No. RFC 4919) (2007)
6. Winter, T., Thubert, P.: RPL Author Team: RPL: IPv6 Routing Protocol for Low power and Lossy Networks, internet Draft draft-ietf-roll-rpl-11 (work in progress)
7. Ko, J., Chang, M.: Momoro: providing mobility support for low-power wireless applications. IEEE Syst. J. **9**(2), 585–594 (2015)
8. Gnawali, O., Levis, P.: The minimum rank with hysteresis objective function, RFC6719 (2012)
9. Thubert, P.: Objective function zero for the routing protocol for low-power and lossy networks (RPL) (2012)
10. Kamgueu, P.O., Nataf, E., Djotio, T.: Energy-based routing metric for RPL. Research Report Inria, RR-8208 (2013)
11. Gonizzi, P., Monica, R., Ferrari, G.: Design and evaluation of a delay-efficient RPL routing metric. In: 2013 9th International Wireless Communications and Mobile Computing Conference (IWCMC), pp. 1573–1577. IEEE (2013)
12. Accettura, N., Grieco, L.A., Boggia, G., Camarda, P.: Performance analysis of the RPL routing protocol. In: 2011 IEEE International Conference on Mechatronics (ICM), pp. 767-772. IEEE (2011)
13. Saad, L.B., Tourancheau, B.: Sinks mobility strategy in IPv6-based WSNs for network lifetime improvement. In: 2011 4th IFIP International Conference on New Technologies, Mobility and Security (NTMS), pp. 1–5. IEEE (2011)
14. Gaddour, O., Koubaa, A., Chaudhry, S., Tezeghdanti, M., Chaari, R., Abid, M.: Simulation and performance evaluation of DAG construction with RPL. In: Communications and Networking, Third International Conference, pp. 1–8. IEEE (2012)
15. Wadhaj, I., Kristof, I., Romdhani, I., Al-Dubai, A.: Performance evaluation of the RPL protocol in fixed and mobile sink low-power and lossy-networks. In: 2015 IEEE International Conference on Computer and Information Technology; Ubiquitous Computing and Communications; Dependable, Autonomic and Secure Computing; Pervasive Intelligence and Computing, pp. 1600–1605. IEEE (2015)
16. Qasem, M., Altawssi, H., Yassien, M.B., Al-Dubai, A.: Performance evaluation of RPL objective functions. In: IEEE International Conference on Computer and Information Technology; Ubiquitous Computing and Communications; Dependable, Autonomic and Secure Computing; Pervasive Intelligence and Computing (CIT/IUCC/DASC/PICOM), pp. 1606–1613. IEEE (2015)

17. Lamaazi, H., Benamar, N., Imaduddin, M.I., Habbal, A., Jara, A.J.: Mobility support for the routing protocol in low power and lossy networks. In: 30th International Conference on Advanced Information Networking and Applications Workshops (WAINA), pp. 809–814. IEEE (2016)
18. Contiki Operating System. http://www.contiki-os.org
19. Kuhn, F., Wattenhofer, R., Zollinger, A.: Ad-hoc networks beyond unit disk graphs. In: Proceedings of the 2003 Joint Workshop on Foundations of Mobile Computing, pp. 69–78. ACM (2003)
20. Aschenbruck, N., Ernst, R., Gerhards-Padilla, E., Schwamborn, M.: BonnMotiona mobility scenario generation and analysis tool. In: Proceedings of the 3rd International ICST Conference on Simulation Tools and Techniques (2010)
21. Dunkels, A., Eriksson, J., Finne, N., Tsiftes, N.: Powertrace: network-level power profiling for low-power wireless networks. SICS Technical Report T2011:03, ISSN: 10013154 (2012)

Author Index

P. K. Sa et al. (eds.), *Recent Findings in Intelligent Computing Techniques*,
Advances in Intelligent Systems and Computing 708,
https://doi.org/10.1007/978-981-10-8636-6

Printed in the United States
By Bookmasters